IET ENERGY ENGINEERING SERIES 182

Surge Protection for Low Voltage Systems

Other volumes in this series:

Surge Protection for Low Voltage Systems

Edited by
Alain Rousseau

The Institution of Engineering and Technology

Published by The Institution of Engineering and Technology, London, United Kingdom

The Institution of Engineering and Technology is registered as a Charity in England & Wales (no. 211014) and Scotland (no. SC038698).

© The Institution of Engineering and Technology 2022

First published 2021

The Institution of Engineering and Technology
Michael Faraday House
Six Hills Way, Stevenage
Herts, SG1 2AY, United Kingdom

www.theiet.org

British Library Cataloguing in Publication Data
A catalogue record for this product is available from the British Library

ISBN 978-1-83953-265-8 (hardback)
ISBN 978-1-83953-266-5 (PDF)

Typeset in India by MPS Limited
Printed in the UK by CPI Group (UK) Ltd, Croydon

Contents

8 Specific application rules   **371**
Ralph Brocke, Nicholas Kokkinos, Alain Rousseau and Antony Surtees

About the editor

Alain Rousseau is the president of SEFTIM, a company based in France but specialized in lightning and surge protection worldwide. His previous assignments include EDF, the French national utility. He is the Chairman of French AFNOR lightning protection and surge protection standard committees, the Chairman of the CENELEC lightning protection and surge protection standard committees as well as the Chairman of the IEC surge protection standard committees. He is a member of many standard committees for lightning and surge protection and especially in charge of risk assessment, thunderstorm warning system and surge-protective devices. He is running many seminars and trainings worldwide to share his 38 years of experience in that field.

Chapter 1

Introduction

Alain Rousseau[1]

1.1 Introductory words from Mitchell Guthrie (USA)

The link between protection against lightning and surge protection dates back to the use of electricity in the United States. Shortly after the installation of overhead lines as a part of the US telegraph system infrastructure, damages of atmospheric origin began to appear. A paper written in 1847 by Joseph Henry (for whom the SI unit of inductance is named), then a professor at what is now Princeton University, provided a description of a simple air-gap device for use in reducing damages to telegraph lines from the results of such atmospheric discharges. Henry's proposed design recommended the installation of a wire connected to the earth at the bottom of a telegraph pole that would be attached to the pole up to a point within 0.5 inches (13 mm). of the telegraph line. Within the next few decades, the development of the technology became more sophisticated. With the advent of the light bulb, overhead power lines provided additional challenges. Non-linear resistors, oxide film arrestors, the introduction of silicon carbide devices with nonactive and then active gaps to zinc-oxide devices ensued. With the development of sophisticated electronic systems and today's smart homes and offices, surge-protective devices (SPDs) for low-voltage applications require even greater sophistication and even smart SPDs as discussed in the following chapters of this book.

This book provides an excellent comprehensive discussion on the concept, the design and the implementation of surge protection for a spectrum of applications. Readers will find it to be a useful guide in the application of surge protection as well as the assessment of the risk of overvoltages and impulse currents due to exposure to direct and indirect lightning. It provides discussions of industrial and residential applications incorporating the selection of SPDs for power, telecom and signalling systems, including DC applications. It addresses power generation applications associated with nuclear power plants, wind turbines and photovoltaic (PV) installations. Other applications range from street lighting to data centres, electric vehicle and boat charging stations and cathodic protected oil and gas pipelines.

[1]SEFTIM, Vincennes, France

Of significance, the book provides an insightful description of the relationship between the various IEC standards that specify or recommend surge protection, including those developed by the IEC Technical Committees on Surge arrestors/low-voltage SPDs (TC37/SC37A), electrical installations and protection against electric shock (TC64) and lightning protection (TC81). The chapter on risk assessment logically lays out the progression of the most detailed risk assessment relating to surge protection, the lightning risk assessment published as IEC 62305-2. Chapter 4 provides a good example of how the detailed assessment of the Lightning Protection Committee (81) can be simplified with the SPD subcommittee (37A) providing a basis to specifically address the need for SPDs. From this guidance, the Electrical Installations Committee (64) developed their own simplified assessment for use in electrical installations where a lightning protection system is not installed. The relationships among the SPD subcommittee (37A), the Electrical Installations Committee (64) and PV installations committee (TC82) and among the SPD subcommittee (37A), the Electrical Installations Committee (64) and the Wind Turbines committee (TC88) are also explained.

Those looking for details specific to SPDs and their associated components will find this book particularly useful. It provides a comprehensive discussion on surge protection devices as well as the components that make up the devices and the associated nomenclature. It provides details on the technologies used and characteristics associated with specific topologies. Main parameters for SPDs are identified and discussed along with test techniques and waveforms to verify compliance with SPD standards.

I believe that engineers, specifiers and others interested in understanding the principles of surge protection, types of components available, surge protection standards and their application will find this book useful and will be a valuable addition to their library. The editor is an authority in the subject matter presented and a primary figure in the development of the standards discussed in the book as well as in their applications.

Mitchell Guthrie is a lightning and grounding specialist who serves as the Technical Advisor of the US IEC TC81 TAG, a US member of various IEC TC81 Working Groups, Maintenance Teams and Convenor of WG18 and ahG19. He served as the Chair of IEC TC81 from 2007 to 2017 and NFPA 780 from 1995 to 2005. He is also a member of IEC TC64/MT4.

1.2   Note from the editor

In 1949, C. W. Ceram, a German writer, published a book titled 'Gods, Graves, and Scholars' that became a best seller opening the history of archaeology to a general public. In the foreword of that book, the author declares that the matter may be difficult and instead of reading the chapters in a row and perhaps getting lost quickly, the readers could jump first to chapters that interest them most and then, when more confident on the topic, they can read the remaining chapters.

Such a warning remains valid for many technical books and we can consider that such a recommendation nicely applies to the present book.

Start where you want, get what you are looking for and then get in-depth knowledge with the remaining chapters and, if necessary, the bibliography provided for each chapter.

Of course, reading from introduction to last chapter remains possible. Chapters are organized in a logical progressive order:

- Lightning and surges (Manu Haddad): explain the surge environment that will apply to installations and SPDs.
- Risk assessment (Alain Rousseau): defines the statistical risk method that will determine the needed level of protection for structure, installation or equipment.
- Standard environment (Alain Rousseau): many standards address SPDs and this chapter is a guideline for finding the appropriate ones.
- Surge-protective components (Vincent Crevenat): these components are the active parts of the SPDs and presenting their characteristics helps understanding the SPD behaviour.
- SPDs (Hubert Bachl-Hesse with Ralf Hausmann for telecom and signalling SPDs): this part presents the SPDs, their main parameters and characteristics and the associated tests to demonstrate the performance and the safety aspects.
- Application rules (Alain Rousseau with Ralf Hausmann for telecom and signalling SPDs): this part presents how to select and use the SPDs, including installation rules. The basic rules are presented first and then a few of them are expanded for more demanding applications.
- Specific application rules (Nikolas Kokkinos with Ralph Brocke for DC applications, Alain Rousseau for nuclear applications and data centres, Tony Surtees for country specificities): when the previous chapter was addressing general applications, this chapter is devoted to specific applications especially covering the growing needs such as PV, DC and also a different approach in a few countries.
- New trends (Alain Rousseau with Qibin Zhou for smart SPDs and Ralph Brocke for specific SPD disconnector): this is a detailed glance on two new features associated with SPDs that may become a standard soon and additionally a state-of-the-art discussion on the benefit of multiple pulse testing.
- On-going issues and possible solutions (Alain Rousseau): this part tries to define the remaining shadow areas of SPD applications and what could happen in that field in future.

Note: Sometimes, a few chapters address the same topic from various perspectives. This has been accepted and even encouraged because the point of view may be slightly different from one author to another one and this different view angle helps the reader to familiarize with this topic.

1.3 Acknowledgements

The editor would like to thank warmly and deeply all the co-authors for their invaluable help in writing that book. Each of them is a specialist in his field and very often a specialist in many parts of that book. But it was necessary to make choices and allocate chapters or sub-chapters to someone. The sensibility of each of them was important in the author selection in order to find a good balance among users, manufacturers, laboratories and scientists. It was also important to provide a focus on the various points of view among various regions (Europe, USA and Asia). Outside the small group of authors, there are many other people who may have contributed. They very often contribute to the IEC standard committee in charge of SPD and are, of course, also very knowledgeable in that field. However, it has been decided to limit the number of writers, often a single writer for a full chapter, to keep a common style. This was an additional burden for them and for chapter coordinators, which they gladly accepted. Many thanks to Manu Haddad, Vincent Crevenat and Hubert Bachl-Hesse for having accepted the task to write Chapters 2, 5 and 6 respectively. Many thanks also to Ralf Hausmann for his contribution to Chapter 7 (Section 7.8), to Ralph Brocke for his contribution to Chapter 8 (Sections 8.1 and 8.2) and to Chapter 9 (Section 9.2), to Nikolas Kokkinos for coordinating Chapter 8 and writing Section 8.3 to 8.7, to Tony Surtees for his contribution to Chapter 8 (Section 8.10) and Qibin Zhou for his contribution to Chapter 9 (Section 9.1).

Many thanks to all our colleagues from standard committees involved in surge protection and especially IEC SC 37A, CENELEC TC 37A and AFNOR UF 37AB. Their permanent input and technical debates not only greatly influenced the standard's content but also this book.

Special thanks to Mitch Guthrie who accepted to write introductory words and reviewed most of my chapters and improved the text greatly, Gianfranco D'ippolito who bring relevant comments for Chapter 7 and Fernanda Cruz who made useful suggestions for many chapters.

Warm thanks for fruitful discussions regarding surge protection, all over the last 20 years, to Giovanni Battista Lo Piparo and Rick Gumley, two respected and inspiring specialists.

Many thanks to my colleagues at SEFTIM for challenging discussions and support over 20 years.

The author thanks the International Electrotechnical Commission (IEC) for permission to reproduce Information from its International Standards. All such extracts are copyright of IEC, Geneva, Switzerland. All rights reserved. Further information on the IEC is available from www.iec.ch. IEC has no responsibility for the placement and context in which the extracts and contents are reproduced by the author, nor is IEC in any way responsible for the other content or accuracy therein.

Alain Rousseau

1.4 Why a complete book dedicated to Surge Protective Devices?

The first idea behind the book was to share almost 40 years of experience in the low-voltage surge protection field with many applications over the world. When you design, define and apply SPDs of various types in various conditions, you collect a valuable know-how that needs to be shared: you meet simple cases, classical cases, complex cases, extreme cases and sometimes non-standardized cases. The surge protection is the aim and the SPD is just a mean to protect. If the needed SPD does not exist yet, you need to adapt to the situation. For that, you need to understand the problem, what could be the solutions and how to demonstrate they are efficient.

To transfer a bit of this experience, it was necessary to first give information to the reader on what are the possible causes of surge damages to electrical installations or equipment. It was the motivation for Chapter 2.

Very often in the past, experience was the main way to define the need in surge protection: determine which equipment or part of installation to protect was not an easy task for non-specialists. This led sometimes to overprotection and when the related budget was too high to no protection at all. When you are not able to localize the sources of damages or to classify them, you are tempted to protect everything and at the highest possible level. However, getting experience takes time with successes and mistakes. Hopefully, a method is used in the lightning and surge protection field since 1995 that helps the designer to define what should be protected and at which level, thanks to calculations. The protection is not over-designed that would be creating an economic problem or under designed that could create a safety issue, but designed to an acceptable level. This method, its applications to SPDs and to specific cases such as PV installations or street lighting, is the scope of Chapter 3.

The surge protection field is covered by many standards. There are standards addressing specifically SPDs and their application, other standard dealing with electrical installations, other dealing with specific applications, all of them showing a part of the SPD domain. It was necessary to provide the reader with a tool to find his way between the various standards, understand when they apply and how they coordinate. Chapter 4 fills that task.

An SPD is based on Surge-Protective Components (active part of the SPD) and other components (wiring, enclosure, terminals, resin, additional circuits for monitoring, etc.). The SPCs play of course an important role on the SPD operation and also on the SPD end of life. It was then necessary to describe these components in Chapter 5.

To correctly use SPDs, it is important to understand how they are made, what their main characteristics are and how these characteristics are proved by tests. It is also necessary to get familiar with the SPD characteristics given in SPD's technical brochures, their symbol and meaning in the selection process. There are also specificities for a few SPDs such as telecom and signal SPDs or two ports power SPDs. They are all described in Chapter 6.

To be efficient an SPD should be first well selected, then its appropriate location in the electrical installation should be defined and finally the SPD should be installed. Poorly selected SPDs will provide poor protection but best SPDs badly installed may also provide poor protection. There are easy application cases and more specific ones that are described in Chapter 7. This chapter is organized in order to start from simple cases and go further in detail for more complex ones. To get a first idea of SPD selection process, it is possible to concentrate on Section 7.1. Notes in italics are provided to avoid losing the reader who wants to concentrate on main items. They may be skipped and still get a clear picture of how to select an SPD. But to enlarge the vision, it is possible to read the notes. Generic cases described in Section 7.1 will cover more than 80% of the cases but the notes will help to clarify specific aspects of the 20% missing cases. Following sub-chapters will concentrate on aspects that are needed to consider to cover these 20% of cases: selection of the current rating of SPD, consideration on the protective distance that will generally conclude to the need of more than one SPD on a circuit, need of coordination in energy and in voltage protection between these SPDs located in different places of installation circuit, need of coordination with equipment to be protected, including the immunity levels of equipment to be protected, the rather tricky selection of the SPD disconnector that will lead generally to compromise between protection and power continuity. Then a six-step selection process for power SPDs is presented that summarizes the main points to take into account for an easy selection of SPDs. Due to the fact that systems and SPDs are different for telecom, data and signal SPD-specific rules are finally presented. Very often, it seems non-understandable when SPDs are apparently selected and installed according to the SPD standards and equipment to be protected fails. It is a common thinking that standards may solve all problems. It is important, at this stage, to notice that a standard will probably never cover 100% of the cases. A standard is not a scientific document that would describe physical law applicable wherever and whenever but only based on scientific facts. A standard is a compromise between the desire of exhaustivity and the need of simplicity. A rule that would be covering 99% of the cases but would be so complex to formulate that only 50% of readers would understand would not make a good rule for a standard. It would be better to simplify the wording, forget voluntarily exotic cases and get a text that at least 80% of readers will understand and apply. It is also important to notice that international standards are written in English that is not the mother language of most of the standard writers and readers and thus text should remain understandable by all. Sometimes rules are simple to write but difficult to apply in practice. For example, the 50-cm rule for SPDs (connecting conductors should not exceed a total length of 50 cm) is known to be difficult to apply, especially on existing panel boards. This rule is simple but it is very important to explain how to compensate when this rule cannot be applied in practice. For all these reasons, general standards cover only general cases when specific standards, such as IEC 61643-12 for power SPDs selection and application principles, can cover more cases, approaching probably an exhaustive approach of the problem. To help understand the limitation of simple rules, case studies are also (indented text surrounded by a red line) in clause 7. This will help to understand how to apply specific rules based on real cases (very often adapted to simplify them) and enlarge the point of view.

There are many cases where applications require specific rules such as PV, wind turbine or DC installations. This is covered by Chapter 8. It was thought interesting as well to apply the rules to specific applications such as electrical vehicle or boat charging stations, pipeline cathodic protection, nuclear installations or data centres. Finally, due to the fact rules may apply in a different way in various places in the world, a sub-chapter is dedicated to a few country specificities.

Finally, it was necessary to give information on new topics presently discussed regarding SPDs to be sure that the book is up to date and valid for a few years. This is covering smart SPDs (SPD with additional functions, for example surge counting or power quality), specific SPD disconnectors that are supposed to help solving the SPD disconnector selection or multi pulses (most of the lightning surges are multi pulses and it is important to know if this must be considered during type tests or if tests with high magnitude surge current cover the need). This is the aim of Chapter 9.

Instead of writing a usual conclusion that would be a simple summary of the main aspects of SPDs, we have tried to discuss about the future of SPDs as we see it, dealing with what is still to know and what are the possible solutions to solve unsolved problems.

The second reason to write this book was to present a comprehensive and up-to-date book on SPDs. Very often, books are dealing with the lightning protection of structure, and SPDs are just a part of it. As can be seen, from the present book, there is a lot to stay on SPDs that was justifying a complete book on the topic. There are excellent guides or texts from SPD manufacturers but, of course, they not only present general rules but also their product line specificities and a few aspects may be missing from other manufacturers. The present book, written by many hands and with most of the writers having responsibilities in the development of standards and thus used to aggregate information from many sources, was able to present an independent and complete picture of the SPD business.

During three decades of training sessions all over the world, it appeared that a few items were regularly requesting a clarification and especially regarding practical problems. The rules were clear but their application to real cases and the need of adaptation were causing problems. What are we allowed to do when it is not possible to apply fully the rules? What is the influence of changing the parameters? Which solutions can be practically applied? A typical question was the application of the so-called 50-cm rule. But the impact of the protective distance was also often requesting clarification and more recently the selection of SPD disconnector. Data SPDs seemed also difficult to select due to the varieties of data systems. This book is intended to answer to all these questions and many more as well as to help understanding the impact of a partial application of the rules and when necessary to find an alternative way to meet the goal: protecting equipment and installation.

Not only this book is intended to facilitate the task of various types of users for defining what should be protected and for selecting and installing SPDs, but also to help them finding their way in the specific vocabulary used in standards and in the SPD fields. Standards are made by people who are devoting a large part of their time to this voluntary activity and due to the rules necessary to build standards they mainly speak the same language that is sometimes different from the user way of speaking. Texts produced by standard committees are also often referring to other standards and without a clear and complete view of these other standards, it may be difficult to catch all the subtilities. When the standard reader is a laboratory officer or an SPD designer, it is quite easy to read the standards because it is your usual business. But when the reader is someone who wants to apply SPDs and especially not on a regular basis, it may be difficult to read the standards and especially differentiate what is fundamental and what could be of minor importance provided an alternative way to meet the goal is used. On the reverse, as discussed before, a few standards are voluntarily presenting simplified rules, but it may not be clear to the reader up to which point the simplification is acceptable to his own case. Finally, the market language is somewhat different from the language used in standards. A Type 1 SPD exists only in Europe. Outside Europe, it is generally known as an SPD tested according to class I with the symbol T1 in a square. But they are the same type of SPDs and reading brochures may be confusing. Type 1+2 SPD, an SPD that is tested both with class I tests and class II tests, does not exist in the standards but in many catalogues or sometimes presented as Type 1+Type 2. Type 1 SPDs do not exist for telecom and signal SPDs, which are known by the test category (D1), but due to an equivalent use, they are often cited in catalogues as Type 1 SPDs. This book is intended to help avoiding these pitfalls and also to find their way in the various standards and their detailed text. By reading this book, even if concentrating only on the main sub-chapters, the reader should be able to get more information from reading the standards.

It should be noticed that the case studies or examples presented in the book cannot be extrapolated to other cases without a careful analysis of the similarities. However, they should be able to demonstrate how the methods apply or to expand the use of usual methods to other cases.

Figures have been detailed to try illustrating most of the concepts.

Note: To illustrate a few concepts, case studies have been presented. For clarity sake and also for non-disclosure reasons, the real situations encountered in field have been generally simplified.

To make the reading easier, the following glossary is provided that includes the main terms and symbol used throughout the book. More specific definitions can be found in the relevant standards given in the bibliography.

1.5   Glossary and acronyms*

Terms	Abbreviated terms/ acronym	Definition
1,2/50 µs–8/20 µs combination wave (for Type 3 SPDs)		Wave characterized by voltage amplitude (U_{oc}) under open-circuit conditions (1,2/50 µs waveshape) and a current amplitude (I_{cw}) under short-circuit conditions (8/20 µs waveshape)
1,2/50 µs voltage impulse		Voltage impulse with a virtual front time of 1,2 µs and a time to half-value of 50 µs

(Continues)

*IEC 61643-12 ed.3.0. Copyright © 2020 IEC Geneva, Switzerland. www.iec.ch.

(*Continued*)

Terms	Abbreviated terms/ acronym	Definition
8/20 μs current impulse		Current impulse with a virtual front time of 8 μs and a time to half-value of 20 μs
Alternative current	AC	
Active part of an SPD		The component that provides the surge protection by opposition to the other components of an SPD used for safety or operation (e.g. a GDT or a MOV)
Bonding bar		Metal bar on which metal installations, external conductive parts, electric power and telecommunication lines, and other cables can be bonded to an LPS
Class I tests		Tests mainly carried out with the impulse discharge current I_{imp}
Class II tests		Tests carried out with the nominal discharge current I_n
Class III tests		Tests carried out with a 1,2/50 μs voltage and 8/20 μs current combination wave generator
Combination SPD		SPD that incorporates both voltage-switching components and voltage-limiting components (typical combination SPDs are SPDs using both GDT and MOV in series as main active part)
Direct current	DC	
Effective voltage protection level	$U_{p/f}$	Voltage at the connection point of the SPD that includes the voltage protection level of the SPD and the voltage drop across a disconnector (when known) and connecting leads
Follow current	I_f	Peak current supplied by the power system and flowing through the SPD after a discharge current impulse (typical of spark gaps)
Impulse discharge current (for Type 1 SPD)	I_{imp}	Crest value of a discharge current through the SPD having a current waveshape 10/350 μs
Isolated LPS		LPS with an air-termination system and down conductor system installed in such a way that the path of the lightning current has no contact with the structure to be protected
Junction box	JB	Connexion box often used for PV modules
Lightning equipotential bonding	EB	Bonding to the LPS of separated conductive parts, by direct connections or via SPDs, to reduce potential differences caused by lightning current
Lightning protection system	LPS	Complete system used to reduce physical damage due to lightning flashes to a structure that consists of both external and

(Continues)

(*Continued*)

Terms	Abbreviated terms/ acronym	Definition
		internal lightning protection systems (lightning rod or alike, down-conductors, earthing system and equipotentiality, including SPDs)
Low current interrupt rating	I_{fi}	Prospective short-circuit current that an SPD is able to interrupt without operation of a disconnector
Main panel board	MPB	
Maximum continuous operating voltage	U_c	Maximum voltage which may be continuously applied to the SPD
Maximum continuous operating voltage for PV SPDs	U_{CPV}	Maximum DC voltage which may be continuously applied to the SPD
Maximum discharge current	I_{max}	Crest value of a current through the SPD having an 8/20 μs waveshape (is supposed to be the maximum value a Type 2 SPD can survive without acceptable degradation but this parameter is optional)
Mode of protection		Intended current path between SPD terminals that contains protective components (e.g. between line and neutral)
Nominal discharge current (for Type 2 SPD)	I_n	Crest value of the current through the SPD having a current waveshape 8/20 μs
Nominal voltage of the system		Voltage by which a system or equipment is designated, for example 230/400 V. The nominal voltage of the system phase to earth is called U_n. The line-to-neutral voltage of the system is called U_0 and the line-to-line voltage of the system is called U
One-port SPD		SPD having no intended series impedance
Open-circuit failure mode	O_{cfm}	Failure behaviour when an SPD changes to a permanent open-circuit state (or high impedance)
Open-circuit maximum voltage	U_{oc} max	Maximum voltage under standard test conditions across an open PV installation
Origin of the electrical installation		Point at which the electric energy is delivered to the electrical installation
Overcurrent protective device	OCPD	Device provided to interrupt an electric circuit in case the current in the electric circuit exceeds a determined value for a specified duration
Overvoltage category		Number from *I* to *IV* defining fixed overvoltage conditions
Prospective short-circuit current of a power supply		Current which would flow at a given location in a circuit if it were short-circuited at that location by a link of negligible impedance
Photovoltaic	PV	

(*Continues*)

(Continued)

Terms	Abbreviated terms/ acronym	Definition
PV array		Assembly of electrically interconnected PV modules or PV strings
PV module		Smallest complete assembly of interconnected PV cells
PV string		Circuit of one or more series-connected PV modules
Rated impulse voltage	U_w	Voltage value between live conductors and earth characterizing the withstand capability of insulation against transient overvoltages
Reference test voltage	U_{ref}	RMS value of voltage used for testing which depends on the nominal system voltage and the system configuration
Residual current device	RCD	Mechanical switching device designed to carry current under normal service conditions and to open the contacts when the residual current attains a given value (e.g. used in TT systems to be able to detect a fault to earth)
Residual voltage	U_{res}	Crest value of voltage that appears between the terminals of an SPD due to the passage of discharge current
Separation distance	s	Distance between two conductive parts at which no dangerous sparking can occur
Short-circuit current rating	I_{SCCR}	Maximum short-circuit current from the power system for which the SPD is rated (may be in conjunction with the specified SPD disconnector)
Short-circuit current rating for PV SPDs	I_{SCPV}	Maximum prospective short-circuit current for which the SPD is rated (may be with an SPD disconnector)
Short-circuit failure mode	S_{cfm}	Failure behaviour when an SPD changes to a permanent short-circuit state (of low impedance)
Short-circuit maximum current	I_{sc} max	Maximum short-circuit current of a PV installation
SPD disconnector		Device for disconnecting an SPD from the power system to prevent a persistent fault on the system and is used to give an indication of an SPD's failure. Disconnectors can be internal (built in) or external (required by the manufacturer)
Status indicator		Device that indicates the operational status of an SPD or of its SPD disconnector
Surge-protective device	SPD	Device that contains at least one nonlinear component that is intended to limit surge voltages and divert surge currents

(Continues)

(*Continued*)

Terms	Abbreviated terms/ acronym	Definition
Temporary overvoltage test value	U_T	Test voltage applied to the SPD to simulate temporary overvoltage (TOV) conditions
Total discharge current	I_{total}	Total current which flows through the PE or PEN conductor of an SPD during the total discharge tests
Two-port SPD		SPD having a specific series impedance connected between separate input and output terminals
Type 1 SPD	Type 1 or T1	SPD tested with class I tests
Type 2 SPD	Type 2 or T2	SPD tested with class II tests
Type 3 SPD	Type 3 or T3	SPD tested with class III tests
Voltage for clearance determination	U_{max}	Highest measured voltage during surge tests
Voltage protection level	U_p	Maximum voltage at the SPD terminals under specific impulse conditions
Voltage-limiting SPD		SPD that has a high impedance when no surge is present but will reduce it continuously with increased surge current and voltage (typical voltage limiting SPDs are MOV SPDs using metal oxide varistors being a non-linear resistor as main active part)
Voltage-switching SPD		SPD that has a high impedance when no surge is present but will decrease suddenly to a low value in response to a voltage surge (typical voltage switching SPDs are gapped SPDs using gas discharge tubes or other spark gaps as main active part)

Note: The IEC way of describing abbreviated terms is very specific with italics for first letter and subscripts for the others (acronyms are using regular letters). Outside the context of standards there is no reason to use these rules especially because a single letter in italics is not always very readable and the IEC subscript is a specific one and not the usual one used by word processors. For that reason, the symbols used throughout the text may differ in shape from the symbols used in the IEC standards but the letters remain of course the same. SPD catalogues sometimes use subscript after the first letter but rarely italics.

Chapter 2

Lightning and surges

Manu Haddad[1]

2.1 Introduction

It is estimated that, around the world, there are some 100 lightning strikes to ground every second. As was observed from various lightning monitoring systems, the lightning strike density is highest around the tropics due to the all-year round hot weather. A typical example of recently recorded map of lightning density is shown in Figure 2.1. Strike's densities in excess of 120 strikes/km^2/year are measured around many areas of the world. In Figure 2.1, it can be seen that central Europe, the Alps and the Mediterranean region as well as large areas of the United States are also subjected to high lightning strike density. This clearly indicates the hazard of lightning to ground as well as airborne structures. For aviation, it is now well established that, on average, each commercial aircraft is struck by lightning once a year. Such frequency puts a significant extra cost on structures, design and construction to ensure safety of people and integrity of stuck objects, including reduced damage to materials, electrical circuits and components and where necessary provision of electromagnetic shielding and lightning protection systems to structures and electrical circuit components. In the United Kingdom, more modest lightning densities have been observed, ranging between 1 and 2 strikes/km^2/year. This is mainly due to the lower lightning activity in the area, as can be observed in Figure 2.2 showing a lower number of thunderstorm days per year, with the highest being around 21 thunder days in very small areas of the south east of England.

In addition to lightning strikes to ground, much more frequent lightning discharges occur within the thunderclouds and even more energetic discharges are fired from the top of the thunder cloud into space. These are known as Elves, Sprites and Blue Jets and they have been observed above cloud levels stretching deep into outer space up to altitude of more 100 km. Figure 2.3 shows an example picture captured by the International Space Station (ISS), depicting intense lightning intra-cloud activity (white-yellowish lights areas) as well as Blue Jets and Sprites highlighted in the detail box.

Recent detailed studies of such transient luminous events (TLE) phenomena allowed further understanding of the mechanisms involved in such high-energy

[1]Department of Electrical and Electronic Engineering, School of Engineering, Cardiff University, Cardiff, Wales, UK

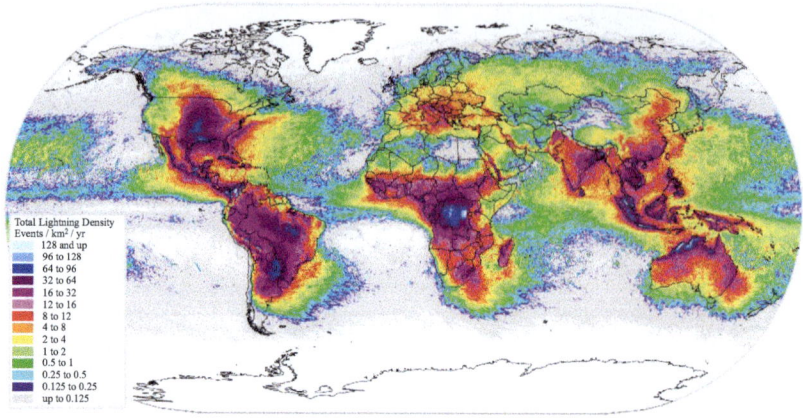

Figure 2.1   Measured total (ground and in-cloud) lightning density around the Globe (see footnote *)

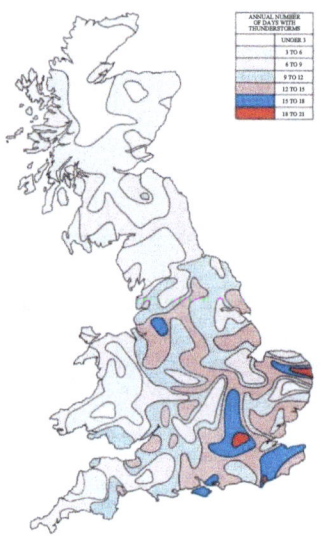

Figure 2.2   Distribution of the number of thunderstorm days in Great Britain (see footnote †)

discharges into space. Figure 2.4 illustrates the current understanding of such phenomena. The red sprites are thought to be due to the electrical discharges in the high nitrogen content (of almost 80%) of the earth's atmosphere.

---

* Global lightning density map from Vaisala, Inc., GLD360. https://www.vaisala.com/en/products/systems/lightning/gld360.
† Sourced from National Grid, United Kingdom.

Figure 2.3   ISS picture of lightning activity and Sprites (Picture Credit: NASA/JSC, NOAA/NSSL)

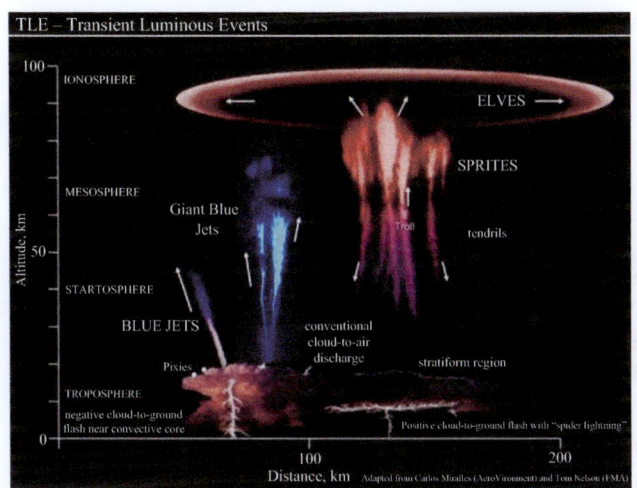

Figure 2.4   TLE events – Lightning, Blue jets, Sprites and Elves (Image credit: NOAA/NSSL, adapted from Carlos Miralles, AeroVironment and Tom Nelson, FMA) (see footnote ‡)

In this chapter, focus is on lightning strikes to ground, i.e., the electrical discharges between the thunder cloud and the ground. Special attention to ground structures, such as buildings and electrical networks will be considered. Lightning strikes can affect electrical circuits in two ways: by direct strikes and by induced effects due to the electromagnetic fields generated by the high magnitude fast lightning current.

‡https://image2.slideserve.com/3707261/slide16-l.jpgI

Following lightning strikes, high current and high-voltage surges travel on electrical networks on both low-voltage (LV) and high-voltage systems. If and when these travelling surges reach sensitive electrical and/or electronic equipment, significant damage can occur if no adequate lightning protection or shielding is put in place.

For completeness, switching surges on networks will be reviewed and their impact on the network will be addressed.

2.2   The lightning phenomenon

2.2.1   The charging of clouds

It is now fairly well understood that lightning is initiated from highly bipolar charged clouds of the type Cumulonimbus (Figure 2.5(a)). Such clouds form following evaporation and the rise of hot moist air into the high-altitude colder atmosphere (Figure 2.5(b)). The fast rise of the air turns into supercooled small ice crystals which become positively charged after collision with bigger denser crystals, known as graupel, accumulating down the body of the cloud. The graupel crystals get a negative charge and tend to be located at the base of the cloud or suspended over around the middle of the cloud. Updrafts and downdrafts within the cloud can spread the charges horizontally and over a wider area. When negative lightning strike events occur, they tend to be of lower magnitudes but are characterised by multiple strikes, up to 40 strikes, separated by few milliseconds. This allows the discharge of deeper charge pockets within the cloud.

Although most clouds tend to have a predominantly negative base, there is, on occasions, an important formation of positive charge at the base of the cloud, mainly due to precipitation and higher temperatures. Such cases generate much higher current magnitudes characterised by single-stroke lightning events.

The Cumulonimbus clouds formation spreads over areas of tens of kilometres wide and can have heights between 2 and 10 km. Their base is usually located at altitudes between 1 and 5 km.

2.2.2   The lightning discharge mechanism

The charged bipolar Cumulonimbus cloud induces the opposite charge on the ground surface. Such a system of high voltage electrodes is formed between the cloud base and the ground. With the charge involved generating potential of the order of gigavolts combined with rugged special distribution both at the base of the cloud and on the ground, extremely large field can be generated as a result. The ionisation threshold for air under normal atmospheric conditions (absolute pressure $P = 1.013$ bar, temperature $T = 20°C$ and absolute humidity $H = 11$ g/m^3) is initiated at an electric field of 30 kV/cm. This threshold value can change significantly at the base of the cloud where both pressure and temperature are lower, and the humidity is higher.

Once the ionisation threshold is exceeded at the local atmospheric conditions, an electrical discharge is initiated growing into streamers. The mechanism of such discharge can be described by the classical Townsend discharge, where a free electron is accelerated by the high electric field, causing collisions with other

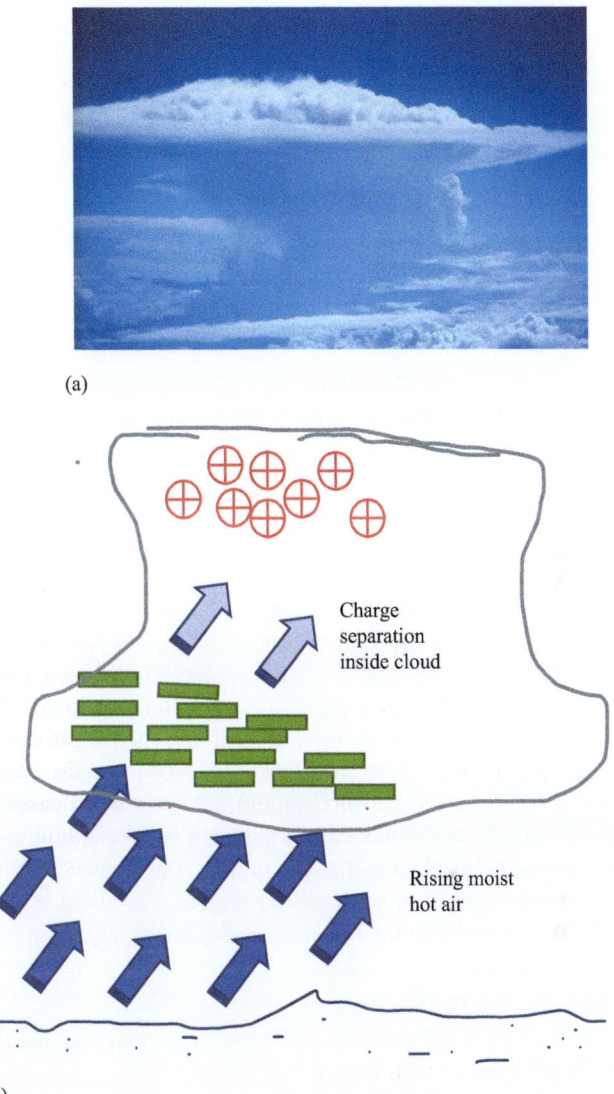

(a)

(b)

*Figure 2.5   The thundercloud – moist air rising due to hot weather causing charge
separation: (a) Cumulonimbus cloud§ and (b) charge formation in thunder
cloud*

molecules and freeing electrons by impact ionisation (Figure 2.6). Eventually, an
electron avalanche is created, and streamers are formed.

§Originally from U.S. National Oceanic and Atmospheric Administration. https://commons.wikimedia.
org/wiki/File:Fly00890_-_Flickr_-_NOAA_Photo_Library.jpg

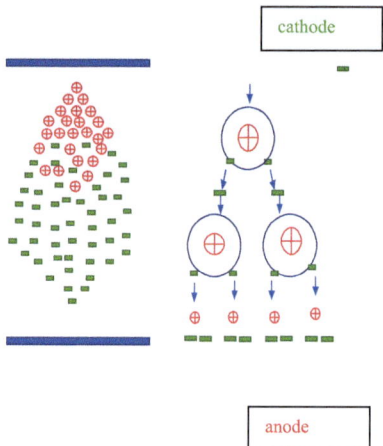

Figure 2.6   Ionisation and electron avalanche formation: (a) townsend discharge process, (b)detail of discharge initiation

These streamers are initially highly branched, which helps a leader to form and advance towards the ground. The leader discharge propagates in steps of approximately 50 m, creating an ionised path known as the stepped leader. After each step, the leader extinguishes and reignites to follow the same path channel to advance further towards the ground. These steps were clearly captured with fast photography and streak camera imaging as shown in Figure 2.7 (Rakov and Uman, 2003). Further ionised branches are formed during this process, giving the lightning strikes its well-known branched shape, as shown in Figure 2.8, depicting a typical lightning strike channel.

As the stepped leader approaches the ground, the electric field at ground level is significantly enhanced. When a high-rise structure or sharp metallic and grounded structure, such tower masts and tall tower buildings, are within the reach of the descending stepped leader area and reach, upward leaders may form at the tip of the ground structure advancing to meet the descending leader (Figure 2.9(a)). Competing leaders to connect to the ground can also form if more than one structure exists (Figure 2.9(a)). The striking distance, D, which is the last step before the ground impact point of the lightning strike is determined, is dependent on the lightning current magnitude and the height of the ground structure. Empirical and analytical formulae have been derived (IEC 62305, Waters, 2004).

The nearest point on the ground with distance D will be likely to be hit first, and one of the first equations to be developed was by Wagner:

$$D = A(I)^b \tag{2.1}$$

With constants $A = 10$ and $b = 0.67$, and I is lightning return stroke current.

Once the stepped leader channel connects the cloud and ground, a very highly ionised branch forms and the lightning current flows causing the discharge of the cloud area where the leader is initiated, and this channel is known as the return stroke.

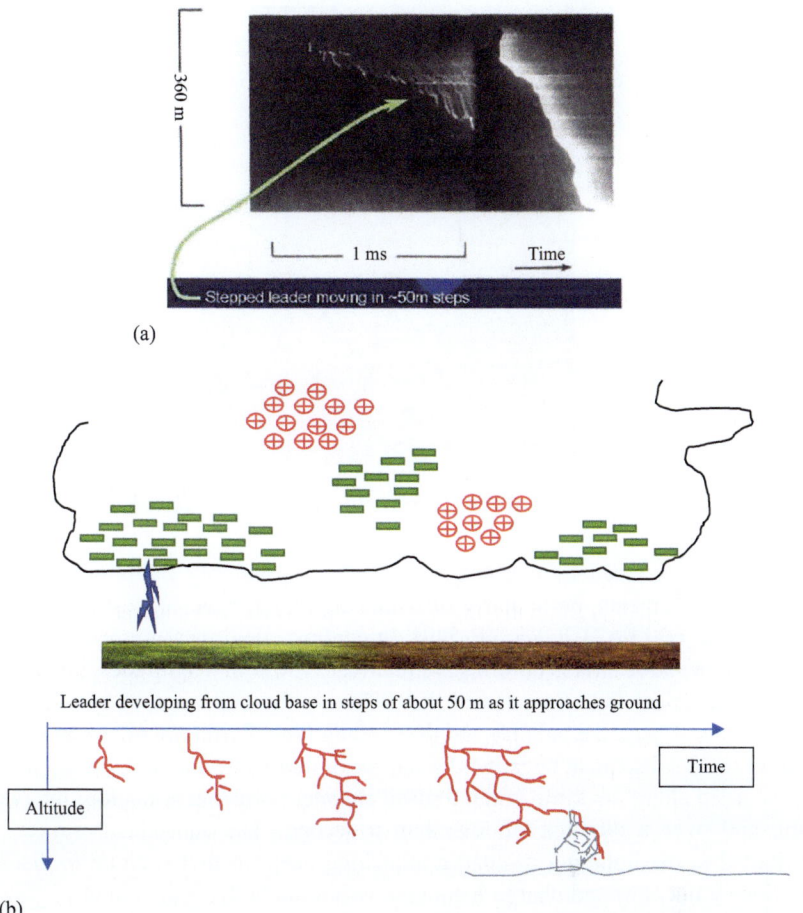

Figure 2.7 Stepped leader propagation: (a) streak camera photograph (Rakov and Uman, 2003: Lightning Physics and Effects ISBN/ISSN: 9780521035415 Copyright material reproduced with permission of Cambridge University Press through PLSclear) and (b) sketch of lightning stepped leader

Lightning strikes are classified as of ascending or descending type, owing to where the initial discharge starts and in which direction it advances; either the leader starts from the base of the cloud and descends down to the ground, or the leader is initiated at ground level and propagates upwards towards the cloud.

They can also be of positive or negative polarity, depending on the polarity of the cloud region that is being discharged, and hence the polarity of the stepped leader tips. Field measurements have revealed that positive strikes, in general, consist of one long discharge and are characterised with very high magnitude currents compared with negative lightning. It was thought that current magnitudes

Figure 2.8   Typical highly branched lightning strike channel[¶]

of up to 250 kA can be measured under positive polarity lightning strikes, but recent measurements, particularly of winter lightning, have suggested magnitudes in excess of 600 kA [CIGRE TB 549]. In contrast, field measurements have indicated that negative strikes currents are relatively lower in magnitudes, i.e., less than 80 kA. Statistical field data reveal that 90% of all cloud-to-ground lightning discharges are of negative polarity and descending type. Furthermore, it was observed that up to 40 subsequent negative discharges of dart leaders and return strokes may be recorded along the same initial ionised channel occurring at random intervals of time and over a duration of less than a second. The subsequent strikes were attributed to the lightning discharge branching deep inside the cloud to discharge previously unconnected charge locations within the body of the cloud.

It is worth highlighting that for many lightning protection designs, the mean value of lightning current of 32 kA is utilised.

Given the high values of current and energy present in lightning strikes, the temperature in the lightning channel is estimated to be 30,000°C which is about five times hotter than the surface of the sun.

2.2.3   Lightning parameters

Since the 1960s, there have been several significant studies that have summarised the aspects of measured lightning parameters such as in Berger *et al.* (1975), Golde (1997a, 1997b), Anderson *et al.* (1980), CIGRE TB063 (1991), Rakov and Uman (2003), Waters (2004), and a recent comprehensive update on CIGRE TB 549 (2013) which includes further new lightning strike measurement data from around the world.

The measurements have established the foundations for the now well-known impulse shapes of the lightning current commonly used for testing and

[¶]https://commons.wikimedia.org/wiki/File:Staccoto_Lightning.jpg. The Author of this image is Griffinstorm.

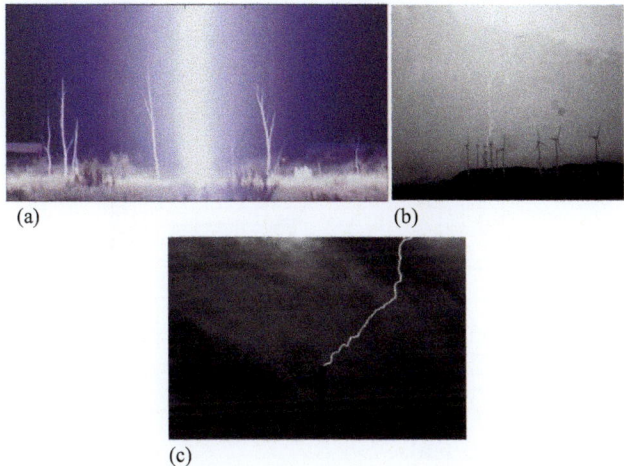

(a) (b)

(c)

Figure 2.9   Upward leaders starting at ground level: (a) formation of several
upward leaders from ground advancing to connect with descending
lightning leader with one successfully connecting to descending
lightning leader (Cummins et al. 2017: Reprinted with permission from
Elsevier), (b) upward leaders from wind turbines (Montanyà et al.,
2014), (c) intercepting upward leader from tall tower, Sky Tree in Tokyo,
Japan (CIGRE TB 633), reprinted with permission from CIGRE ©

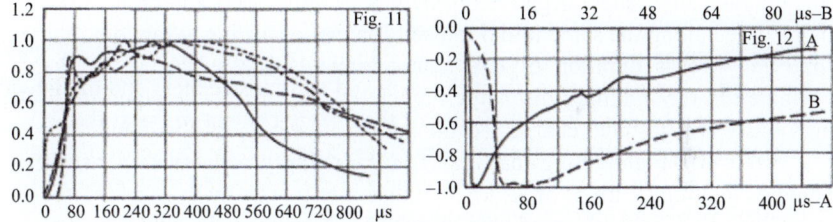

Figure 2.10   Typical measured lightning current shapes (Berger et al., 1975)

simulation. Figure 2.10 reproduces typical positive and negative waveforms
measured by Berger *et al*. It indicates fast fronts and slower tails times of the
impulse shape; all in the microsecond time range. However, it is worth noting
that, compared to positive polarity currents, negative stokes are characterised by
faster time scales, as can be observed in the figure. Moreover, small shape parameter variability from strike to strike is also indicated.

The cumulative statistical current distribution of measured lightning currents,
as reproduced in Figure 2.10, shows that, for the first return stroke of negative
lightning, a median value of 30 kA is obtained and up to 5% of all strikes exceed
80 kA in magnitude (curve (1) in Figure 2.11). In contrast, for the positive lightning
strikes, a mean value of 35 kA is obtained and 5% of the measured strikes have a

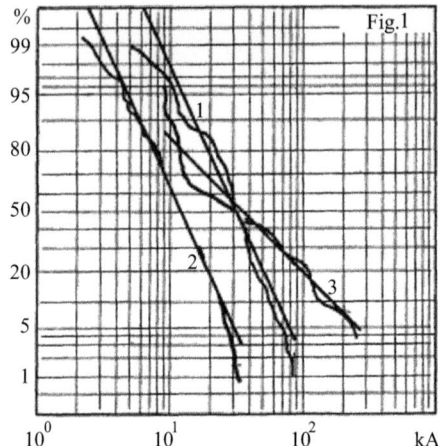

Figure 2.11   Probability distributions of measured lightning strikes to exceed currents magnitudes: (1) negative first strokes, (2) positive first strokes, (3) negative subsequent strokes (Berger et al., 1975)

magnitude in excess of 250 kA (curve (2) in Figure 2.11). The corresponding values for subsequent negative strikes are much lower.

Further analysis of the previous data has led to the development of empirical equations to express the measured current magnitude distributions. Figure 2.12 shows the curves proposed by IEEE 2005 and CIGRE 2013 technical committees. These are now being adopted in various standards.

Other lightning parameters were also evaluated from the measurements. These include charge, impulse charge, front duration and current steepness and the integral of $i^2 dt$ which can be used as an indicator of the energy content in the strikes.

Table 2.1 reports some of the key parameters measured by Berger *et al.* where it is easily observed that the positive strokes have higher magnitudes (in excess of 250 kA), charge (more than 350 C), longer durations (more than 2 ms) and energy content (higher than 1.5×10^7 A^2 s). However, the short front times seen on the negative subsequent strikes give them the highest current rate of change, up to four times the first stroke, which could be hazardous in the presence of inductive paths.

Subsequent measurements and lightning data analysis, in addition to the published work by Berger *et al.* (1975), allowed some useful comparison of peak current values measured around the world. Published work in CIGRE TB 549 gives a useful comparison of published data from various sources. Table 2.2 reports the most useful data for this work. It is noteworthy to emphasise that the 50% value is around 30 kA for most of these data.

2.3   Lightning strikes to ground structures and lines

2.3.1   Effects of lightning strikes on ground structures

Lightning strikes to ground structures are known to result in severe damage to the impacted structure/s. The key physical threats to various structures are summarised in

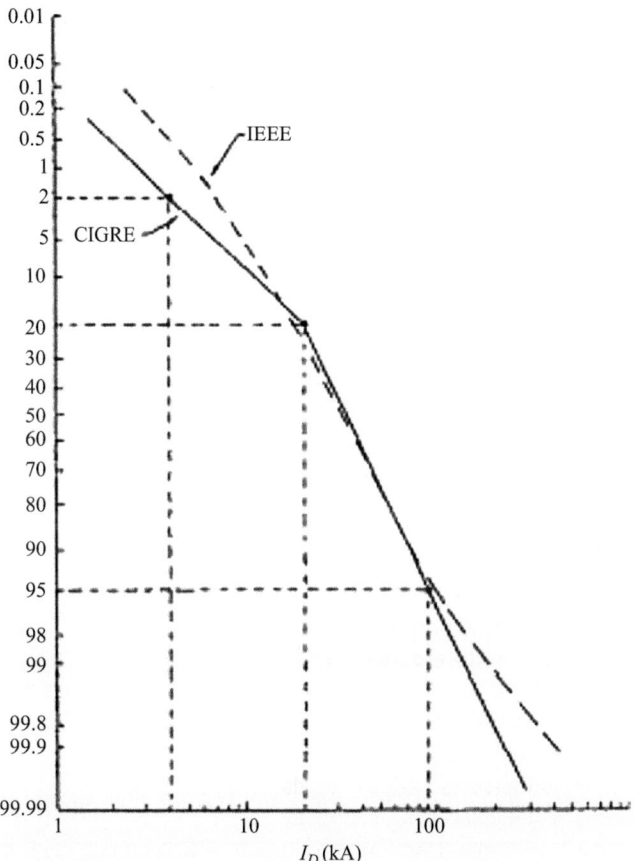

*Figure 2.12   Comparison of lightning parameter distribution curves as adopted by
IEEE (2005) and CIGRE (1991) and adopted in a number of
lightning protection standards (CIGRE TB063, 1991), reprinted with
permission from CIGRE ©*

Table 2.3 (IEC 62305-1:2011, 2011). The possible impact on living human and
animals in such structures is not included in this table. Ground structures can be of
domestic or industrial type. When a structure includes a metallic frame, the impact
of lightning strike may be reduced, as such metallic structures can act as a protective layer against lightning or can form part of and will help provide further
support to any specially installed lightning protection system.

2.3.2   Lightning surge scenarios on ground structures

Lightning strikes can impact ground structures in several ways. Table 2.4 summarises
the various scenarios of lightning strikes to building and gives some general
requirements for mitigation. Detailed description of the scenarios will be given.

Table 2.1   Measured lightning parameters

Parameter (unit)	Stroke component	Portion exceeding magnitude (kA)		
		95%	50%	5%
Peak current (kA)	Negative first return stroke	14	30	80
	Negative subsequent strokes	4.6	12	30
	Positive stroke	4.6	35	250
Charge (C)	Negative first return stroke	1.1	5.2	24
	Negative subsequent strokes	0.2	1.4	11
	Positive stroke	20	80	350
Front duration (μs)	Negative first return stroke	1.8	5.5	18
	Negative subsequent strokes	0.22	1.1	4.5
	Positive stroke	3.5	22	200
Stroke duration (μs)	Negative first return stroke	30	75	200
	Negative subsequent strokes	6.5	32	140
	Positive stroke	25	230	2,000
Maximum di/dt (kA/μs)	Negative first return stroke	5.5	12	32
	Negative subsequent strokes	12	40	120
	Positive stroke	0.20	2.4	32
Integral of ($i^2 \, dt$) (A^2 s)	Negative first return stroke	6.0×10^3	5.5×10^4	5.5×10^5
	Negative subsequent strokes	5.5×10^2	6.5×10^3	5.2×10^4
	Positive stroke	2.5×10^4	6.5×10^5	1.5×10^7

Values taken from Berger et al. (1975).

Table 2.2   Measured lightning parameters from various sources and countries

Reference	Country of source data (no. of samples)	Portion exceeding magnitude (kA)		
		95%	50%	5%
Berger *et al.* (1975)	Switzerland (101)	14	30	80
Anderson *et al.* (1980)	Switzerland (80)	14	31	69
Dellera *et al.* (1985)	Italy (42)	–	33	–
Geldenhuys *et al.* (1989)	South Africa (29)	7	33	162
Takami and Okabe (2007)	Japan (120)	10	29	85
Visacro *et al.* (2012)	Brazil (38)	21	45	94
Anderson *et al.* (1980)	Switzerland (125), Australia (18), Czechoslovakia (123), Poland (3), South Africa (11), Sweden (14), USA (44)	9	30	101
CIGRE TB063 (1991)	Switzerland (125), Australia (18), Czechoslovakia (123), Poland (3), South Africa (81), Sweden (14), USA (44)	–	31	–

Values taken from CIGRE Technical Brochure TB 549, reprinted with permission from CIGRE ©.

Table 2.3 Some key lightning strike damages to impacted structures (IEC 62305-1:2010, 2010)[//]

Type of structure according to function and/or contents	Effects of lightning
Dwelling house	Puncture of electrical installations, fire and material damage
	Damage normally limited to structures exposed to the point of strike or to the lightning current path
	Failure of electrical and electronic equipment and systems installed (e.g., TV sets, computers, modems, telephones, etc.)
Farm building	Primary risk of fire and hazardous step voltages as well as material damage
	Secondary risk due to loss of electric power, and life hazard to livestock due to failure of electronic control of ventilation and food supply systems, etc.
Theatre	Damage to the electrical installations (e.g., electric lighting) likely to cause panic
Hotel	Failure of fire alarms resulting in delayed fire fighting
School	measures
Department store	
Sports area	
Bank	As mentioned previously, plus problems resulting from
Insurance company	loss of communication, and failure of computers and
Commercial company, etc.	loss of data
Hospital	As mentioned previously, plus problems of people in
Nursing home	intensive care, and the difficulties of rescuing
Prison	immobile people
Industry	Additional effects depending on the contents of factories, ranging from minor to unacceptable damage and loss of production
Museums and archaeological sites	Loss of irreplaceable cultural heritage
Churches	
Telecommunication	Unacceptable loss of services to the public
Power plants	
Firework factories	Consequences of fire and explosion to the plant and its
Munitions works	surroundings
Chemical plants	Fire and malfunction of the plant with detrimental
Refineries	consequences to the local and global environment
Nuclear plants	
Biochemical laboratories and plants	

2.3.2.1 Lightning strikes to the structure

By direct strike to unprotected structures

In this case, the lightning strike hits a point location on the structure. This can result in severe damage to the structure, including explosive effects due to a combination

IEC 62305-1 ed.2.0. Copyright © 2010 IEC Geneva, Switzerland. www.iec.ch

Table 2.4 *Overview of sources and mechanisms of damaging lightning surges in ground structures*

Lightning strike scenario	Damage mechanism	Areas affected	Mitigation
Direct impact on unprotected structure	• Joule effect, high temperature, electro-magnetic forces, shock waves • Induced effects may be strong, depending on lightning current path from point of impact to earth • Rise of earth potential	• Mostly external parts of building affected • Internal circuits and components affected if lightning current finds its way to electricity and/or tele-communication circuits	• If risk is high, adequate protection system should be considered • Good earthing and bonding

Direct impact on protected structure	• Induced effects due to capacitive and inductive couplings between lightning carrying current down-conductor and internal structure circuits in building
	• Rise of earth potential at point of current entry into structure's earthing system
		• Cabling and equipment connected to
		  – loops formed by electrical circuits
		  – loops formed by IT/comms circuits
		  – loops formed by combination of above circuits
			• Minimise loops and mutual inductance/capacitance couplings
			• Reduce earth impedance/resistance
			• Earthing and bonding of electrodes and wiring

Strike to the vicinity of the structure	• Inductive effect due to high di/dt of strike current (radiated interference) • Capacitive effect due to high potential and fast rise of lightning generated potentials • Transferred earth potential through ground wires and pipes, due to the rise of earth potential at the point of ground of strike impact and distribution of earth potential around it • Coupling of strike into incoming circuit wiring and or pipes	• Loops of electrical, IT and comms wiring and attached equipment • Unshielded electrical/electronic equipment	• Minimise loops and routes to inductive and capacitive couplings • Improve shielding of sensitive equipment (or disconnect sensitive equipment during nearby lightning storms) • Earthing, bonding and protective devices
Remote or close strike to electrical lines or pipes supplying the structure	• Lightning voltage and current surges will propagate and enter into the building via the mains or earth routes, with risk to affect all internal circuits and equipment	• Circuits and equipment	• Protection against incoming surges earthing and bonding

of sudden temperature increase to extremely high values generated by Joule effect, lightning arc temperature, electromagnetic forces due to high magnitude and rate of rise of lightning current, accompanied by the formation of a strong shock pressure wave.

If the lightning current finds its way to the electrical network within the structure by conduction or induction mechanisms, this direct impact damage is expected not only at the point of attachment of the lightning channel to the structure but also all along its overheated dissipation path to ground which, in turn, can result in fire, destroy electrical wiring and electrical/electronic equipment connected to the wiring.

The lightning surge (voltage and current) will split and propagate throughout the circuit creating high-voltage arcing and high current forces and heating.

By direct strike to protected structures

The lightning protection rod, air termination system or mesh and the down-conductors significantly reduce the risk of lightning striking the structure directly. If adequately protected, the lightning strike should be safely diverted to ground with minimum effect to wiring and equipment. However, induced effects need to be considered carefully.

Induced surges

Inductive and capacitive couplings can be experienced by the wiring in the building not only due to a direct strike to the building, whether protected or unprotected, but also if the strike is in the vicinity of the structure and strong enough to generate hazardous effects.

Inductive coupling: If there are loops formed in the wiring, it gives rise to an inductive coupling, which will cause voltage, $V_i(t)$, to appear at the open end of the loop, and this is determined by the product of the mutual inductance, M, and the rate of rise of the lightning strike current, $dI_s(t)/dt$,

$$V_i(t) = M \frac{dI_s(t)}{dt} \tag{2.2}$$

This voltage can be several hundreds of kilovolts, which may be large enough to cause breakdown at the open end of the loop. If the loop is closed, then a very large current will flow in the loop, destroying any connected equipment, and could lead to overheating of the affected wiring. The current can be determined from the total induced voltage in the loop, $V_i(t)$, as before and the resulting current flow, $i_c(t)$, into the circuit self-inductance, L_c, circuit resistance, R_c, and the connected equipment total impedance, Z_{eq}.

$$V_i(t) = M \frac{dI_s(t)}{dt} = i_c(t) + L_c \frac{di_c(t)}{dt} + Z_{eq} i_c(t) \tag{2.3}$$

The loops formed in buildings can be very large and can be formed by electrical wiring as well as communication/IT cables, or their combination when they have a common earth. It is, therefore, important that such loops are studied and minimised at the design stage. The mutual inductances can be analytically calculated as presented in Hasse (1998). The key figures for the various loop scenarios are reproduced here in Figure 2.13 for completeness and for the reader's benefit.

As expected, the mutual inductance increases with the size of the loop but decreases with the increasing cross-section area, q, of the down-conductor. Moreover, the mutual inductance falls rapidly when the separation distance, s, between the down-conductor and the loop increases.

Capacitive coupling: Surges can also be induced capacitively. Considering a simplified case of a floating conductor or metal surface located close to a conductor carrying a surge voltage, $V_s(t)$, and away from ground, two capacitors are formed: one, C_1, between the energised conductor and the floating electrode and the other, C_2, between the floating electrode and ground, as shown in Figure 2.14. A capacitive current, I_{cap}, will flow through the two capacitors due to the voltage $V_s(t)$,

$$I_{cap} = \frac{C_{eq}dV_s(t)}{dt} \tag{2.4}$$

With

$$C_{eq} = \frac{C_1 \cdot C_2}{(C_1 + C_2)} \tag{2.5}$$

In this way, this configuration acts as a capacitive divider and, hence, a significant voltage, $V_f(t)$, is induced on the floating electrode according to the following equation:

$$V_f(t) = \frac{C_1 \cdot V_s(t)}{(C_1 + C_2)} \tag{2.6}$$

When $C_1 = C_2$, the voltage on the floating conductor is half that of the lightning surge. However, if the floating electrode is closer to the energised conductor and further away from the ground electrode, C_1 becomes much larger than C_2 and, hence, the floating potential increases significantly. Given their highly elevated magnitude, such induced voltages can cause side flashes between components within the building.

Earth potential rise and backflashover

A further source of lightning surge overvoltage for the structure for all scenarios can be generated by the rise of earth potential, $V_e(t)$, when the lightning current, $I_s(t)$, flowing in the down-conductor enters the ground impedance, Z_g, via the buried earth electrode system, where

$$V_e(t) = Z_g I_s(t) \tag{2.7}$$

If this earth potential rise is in the hundreds of kilovolts, a backflashover similar to that seen on transmission and distribution overhead lines may take place or cause the puncture of the electrical insulation on cables and equipment inside the building.

Backflashover voltages can generate voltages, V_b, of very steep rates of change, and hence, the dielectric current, I_d, in an insulation of circuits and

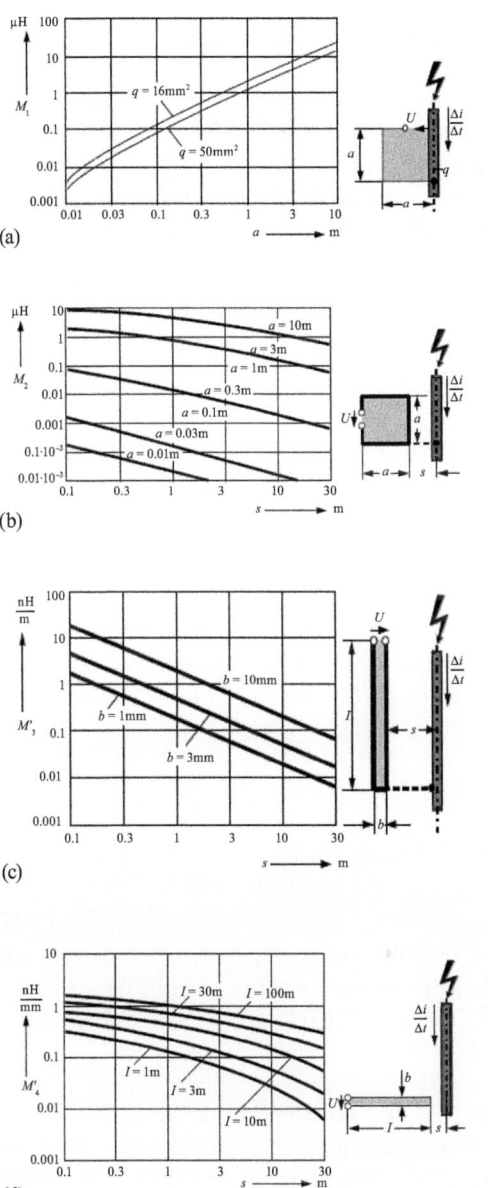

Figure 2.13 *Determination of mutual inductance for various scenarios of circuit loop and lightning down-conductor configurations (Hasse, 1998). (a) Case 1: mutual inductance for a square loop of side, a, for two down-conductor cross-section areas; (b) Case 2: mutual inductance for a square loop of side, a, as a function of separation distance, s, between the down-conductor and the loop; (c) Case 3: mutual inductance for a vertical rectangular loop of length, l, for three widths, b, as a function of separation distance, s, between the down-conductor and the loop; (d) Case 4: mutual inductance for a horizontal rectangular loop of width, b, for three lengths, l, as a function of separation distance, s, between the down-conductor and the loop*

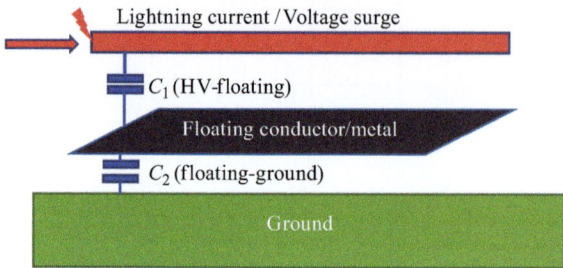

Figure 2.14   Schematic of capacitive coupling for a floating conductor

equipment that are characterised by capacitance C_i, will be stressed more than for less steep voltages.

$$I_d = \frac{C_i dV_b(t)}{dt} \tag{2.8}$$

In electrical distribution and transmission networks, steep-front backflashover voltages are regarded as one of the most stressful voltages on high-voltage insulation systems.

2.3.2.2   Lightning strikes near the structure

When lightning strikes a structure or the ground nearby to another structure, it is expected that surge voltages will appear in other buildings. The mechanisms by which these surges appear in a building can be mainly due to radiated and conducted effects, in addition to possible earth potentials being transferred to the earthing system of the building due to the ground potential distribution around the point of lightning strike impact. Figure 2.15 depicts such scenarios where both radiated and conducted surges can be transferred to Building 1.

The radiated effects are generated by the intense electromagnetic field generated by the lightning strike current which induces voltages and/or currents in loops within the building and loops formed by conductors entering the building, e.g., feeder power cables and telecommunication, IT and control wiring. These radiated fields can also be picked up by antennae like objects in the building.

The conducted effects are transferred into the building through direct connection between two points between the building, including the earthing systems and any metal pipework.

Radiated/inductively induced voltages

For both nearby and remote strikes, if the lightning strike is close enough to the building or any wiring entering the building, there is a risk of inducing significant voltages in any formed loops. The loops can be made up along the incoming power cables, both phase to phase and phase to ground, the communication wiring or their combination. Also, metallic pipes entering the building, such as gas and water, can be part of such loops. Figure 2.15 illustrates the inductive coupling to power and signal cables and pipes.

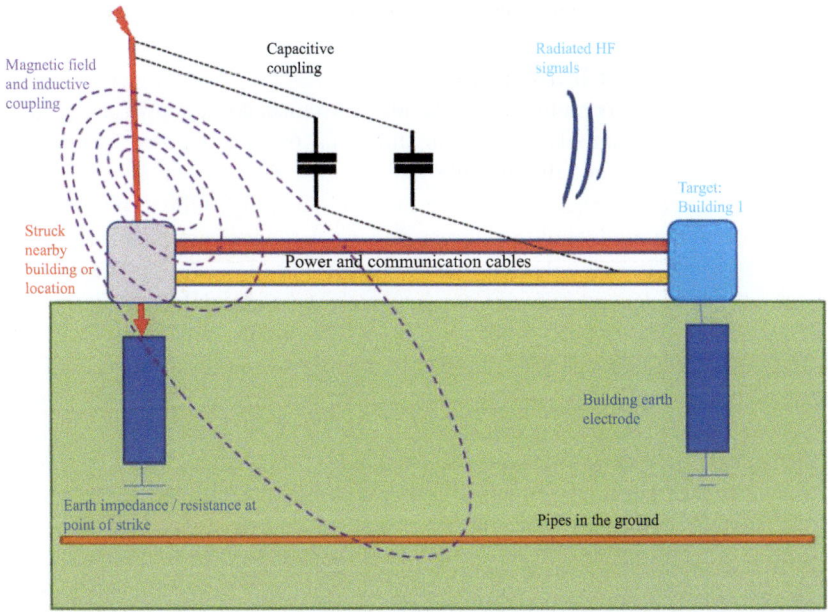

Figure 2.15 Scenarios for surge transfer between nearby strikes and buildings

When the so-generated voltage is high enough, failure of connected terminal equipment may fail, which will be the source for a high current to flow in the loop. Currents in the kiloampere range can result from such scenarios.

A typical scenario for such induction processes is seen in two wire lines (power or other signals) with connected terminal equipment when high current surges can flow in such circuits, as seen on telephone lines failures following nearby lightning strikes. A further common scenario can be generated between a power/signal wire and the earth; in this case, the insulation near the earth end of the equipment/building can break down due to the high voltage surge and, thus, allow large currents to flow through terminals and or equipment.

Capacitively induced voltages
As the power, telecommunications and IT cables are insulated from ground, and a capacitance to ground is thus formed between the cable/s and earth, in a similar way as in power systems overhead lines. Moreover, a further stray capacitance from the cable/s to the lightning channel is developed from these cables. Such a network of stray capacitances provides a route for voltages to be induced capacitively on the cables through the voltage divider formed by the stray capacitance networks, as detailed in the "Capacitive coupling" section.

Conducted currents and transferred voltages
Both longitudinal and transverse voltages are involved in such incidents; the longitudinal voltage will appear along the connecting cable between the point of

strike and the entry into the building. The magnitude of such voltage is governed by the series impulse resistance/impedance of the cable and the lightning surge current that results following a breakdown of terminal insulation/equipment. Table 2.5 reports values of the per unit length series resistance, also known as coupling resistance, R_k, for commonly used cable and metal pipes in the building industry (Hasse, 1998). The values in the table indicate that the per unit length impulse resistance is lower than that under direct current. The longitudinal voltage can be calculated from the product of the current flowing in the cable or pipe and the total impulse resistance (taking into account its length). For most practical cases, longitudinal voltage of several kilovolts can be developed as a result.

Furthermore, the rise of earth potential at the strike point, with the ground potential distribution developing around it, is another source for the pickup of high magnitude potentials by pipes and bare conductors buried in the 'energised ground'. The earth potential can easily reach tens or hundreds of kilovolts when it dissipates high magnitude lightning strike currents. The picked-up potential, which depends on proximity and location along the ground potential distribution, will then be transferred into the building with a high risk of causing backflashover onto connected equipment and wiring. Also, if the strike is close enough to the building, the transferred voltage will directly increase the earth potential rise of the building, causing the breakdown of insulation and flow of current. In all cases, after breakdown, a high magnitude current in the kiloampere range may flow through equipment and circuitry.

It is important to note that LV equipment and installations in office and domestic buildings have a relatively low lightning withstand levels, which makes

Table 2.5   Series (coupling) impulse resistance for selected cables and metal pipes

Shielding	dc resistance	Impulse coupling resistance
	R_G* in mΩ/m	R_k* in mΩ/m
Cable with aluminium sheath 35 mm Ø	0.24	0.17
Cable with lead sheath 35 mm Ø	1.00	0.95
Cable NYCY with shield 2.5 mm²	7	7
Cable NYCY with shield 6 mm²	3	3
Cable NYCY with shield 10 mm²	1.70	1.70
Telecommunication cable with copper braiding	2	2
	0.9	0.90
Wires 1.2 Ø 6 paired		
20 paired		
Steel tube 64 mm Ø × 1.3 mm	0.5	0.08
Copper tube 64 mm Ø × 1.3 mm	0.07	0.04
Shield conductor made out of steel tube 600 mm Ø/8 mm at distance of 10 cm from the cable	0.01	30
Wire 10 mm Ø made out of steel at a distance of 10 cm from the cable	1.50	300

Adapted from Hasse (1998).

Table 2.6   Recommended rated impulse voltage for low-voltage components[**]

		Required rated impulse voltage of equipment (kV)[c]			
Nominal voltage of the installation[a] (V)	Voltage line to neutral derived from nominal voltages ac or dc up to and including (V)	Overvoltage category IV (equipment with very high rated voltage); e.g., energy meter, tele-control systems	Overvoltage category III (equipment with high rated voltage); e.g., distribution boards, switches, socket outlets	Overvoltage category II (equipment with normal rated voltage); e.g., distribution, domestic appliances, tools	Overvoltage category I (equipment with reduced rated voltage); e.g., sensitive electronic equipment
120/208	150	4	2.5	1.5	0.8
230/400[b,d] 277/480[b]	300	6	4	2.5	1.5
400/690	600	8	6	4	2.5
1,000	1,000	12	8	6	4
1,500 dc	1,500 dc			8	6

[a]According to IEC 60038.
[b]In Canada and the USA, for voltages to earth higher than 300 V, the rated impulse voltage corresponding to the next highest voltage in this column.
[c]This rated impulse voltage is applied between live conductors and the protective earth PE conductor.
[d]For IT systems operations at 220–240 V, the 230/400 row shall be used, due to the voltage to earth at the earth fault on one line.

them vulnerable to lightning strikes surges in buildings. Table 2.6 gives typical lightning voltage withstand levels for building cables and integrated circuits while Table 2.7 reports the withstand levels for domestic appliances.

The lightning insulation level of equipment in electrical power applications is usually defined with the standard lightning impulse shape of 1.2/50 (IEC 60060-1:2010, 2010). Figure 2.16 depicts the standard impulse shapes and indicates the tolerances for the front time, $T1$, tail time, $T2$, and the peak value.

2.4   Overvoltages on electrical networks

This section introduces the main overvoltages encountered on medium- and high-voltage networks. Their mechanism and the basic electromagnetic principles that govern their magnitudes and shapes can be replicated in all circuits that have inductances and capacitances, including in LV networks. Furthermore, the overvoltage generated on high-voltage networks can be coupled and/or directly transferred to LV networks. Under each of the following subsections, reference to interaction with LV networks is indicated in each relevant case.

[**]IEC 60364-4-44 ed.2.0. Copyright © 2007 IEC Geneva, Switzerland. www.iec.ch.

Table 2.7 Lightning impulse withstand level of domestic appliances (CIGRE TB 550, 2013), reprinted with permission from CIGRE ©

Appliance	Surge withstand voltage (kV)	
	Common mode	Differential mode
VCR[a]	2[d]	1
DVD player[b]	≥ 6[d]	2
Television[b]	≥ 6[d]	2
PC[a]	2	3.5
Multifunctional printer[b]	≥ 6	2
Microwave oven[a]	4	
Audio micro system[a]	≥ 6[d]	2
Fax machine[a]	2	4
Air conditioner[a]	≥ 6	2
Refrigerator[c]	≥ 6	≥ 6
UPS[a]	2	≥ 6

[a]Fixed voltage input.
[b]Automatically regulated voltage input.
[c]Refrigerator without electronic control.
[d]Equipment without earthing terminal (voltage applied from active conductors to a metallic plate under the equipment).

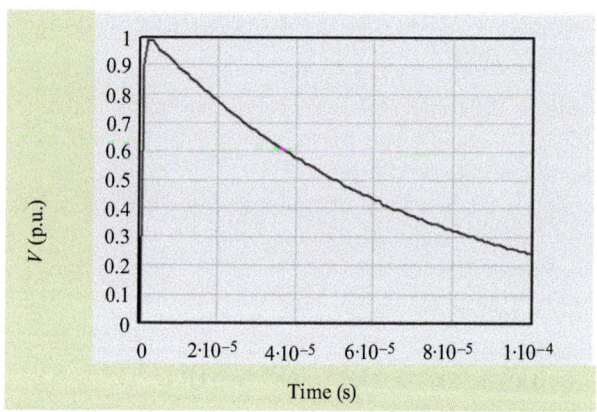

Figure 2.16 IEC 60060 Standard definition of lightning voltage impulse for testing. Shape: T1 – front time: 1.2 μs; T2 – time to half-value: 50 μs, tolerances: peak value 3%, front time±30%, time to half-value±20%

Electrical networks are subjected to overvoltages that are generated internally following switching operations and faults on the system, and to overvoltages generated by external sources to the network, such as lightning and surges that can emanate from nuclear accidents and geomagnetic surges due to solar storms. Figure 2.17 shows

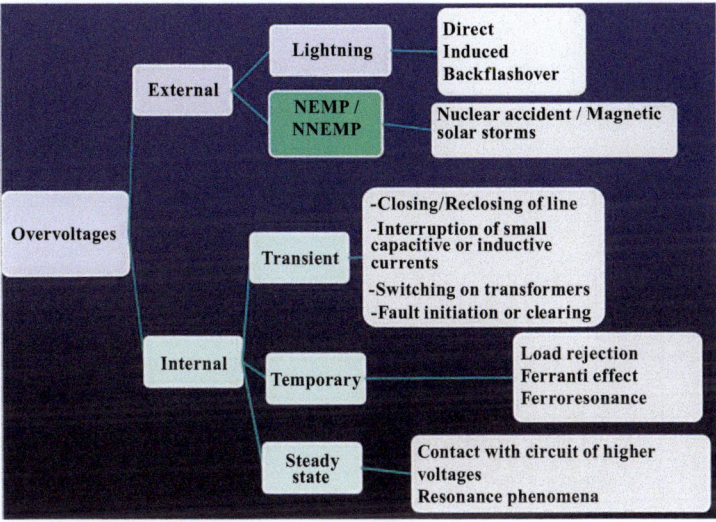

Figure 2.17   Overvoltages on power networks (adapted from German and Haddad, 2004)

a classification of the overvoltages and gives some of the key causes that have been observed on transmission and distribution networks (German and Haddad, 2004).

Such overvoltages can have different shapes and magnitudes depending on the type and source that has initiated the overvoltage. Figure 2.18 gives the IEC 60071-1 summary of the shapes and their categorisation for use in insulation coordination studies of the power networks. A clear distinction is made between slow and fast surge overvoltages. Switching surges are classified as slow-front surge voltages, as their rise time is the 10s to 100s microseconds, while lightning surges are considered fast-front surges given their faster rise times and the order of less than 10 µs. The surges generated in gas-insulated systems (GISs) are known to be much faster, with fronts of the order of few nanoseconds.

The magnitude and shape of switching surges are dependent on system parameters and, in most cases, can be between 2 and 4 p.u. However, they can be limited with appropriate mitigation techniques. Again, the magnitudes of lightning voltage surges are comparatively much higher but contain less energy than switching surges.

In order to test the withstand capability of equipment insulation systems under each category of these overvoltages, the international/national standards specify voltage test waveforms as indicated in the figure.

For system voltages up to 300 kV, lightning surges are the most onerous because the air clearances and insulation thickness are not strong enough to withstand the impinging excessive lightning surges. In contrast, for higher voltages, the switching surge is more important given the dielectric properties of long air gaps

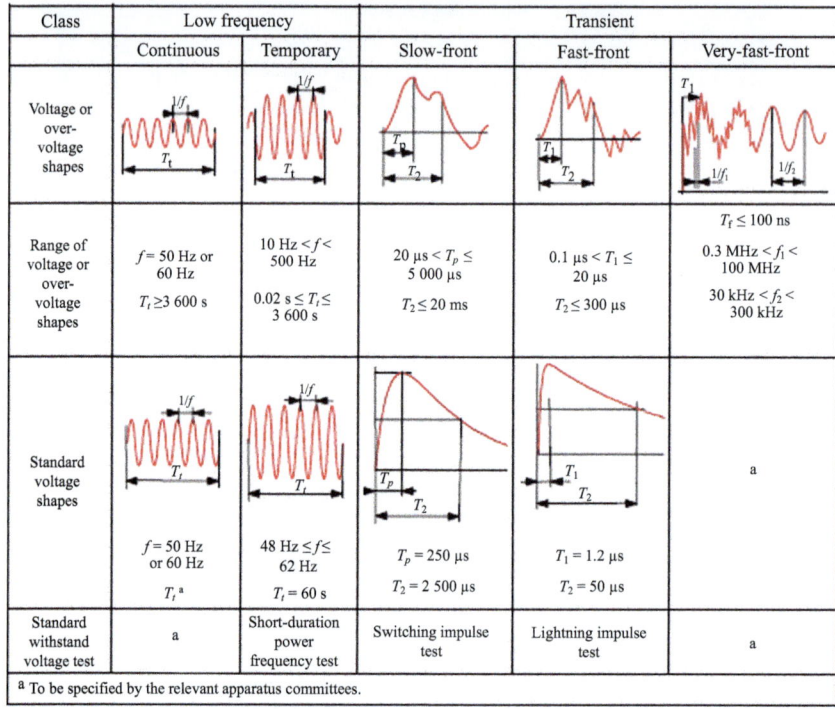

Class	Low frequency		Transient		
	Continuous	Temporary	Slow-front	Fast-front	Very-fast-front
Voltage or over-voltage shapes					
Range of voltage or over-voltage shapes	f = 50 Hz or 60 Hz $T_t \geq 3\ 600$ s	10 Hz < f < 500 Hz 0.02 s ≤ T_t ≤ 3 600 s	20 μs < T_p ≤ 5 000 μs T_2 ≤ 20 ms	0.1 μs < T_1 ≤ 20 μs T_2 ≤ 300 μs	T_f ≤ 100 ns 0.3 MHz < f_1 < 100 MHz 30 kHz < f_2 < 300 kHz
Standard voltage shapes	f = 50 Hz or 60 Hz T_t [a]	48 Hz ≤ f ≤ 62 Hz T_t = 60 s	T_p = 250 μs T_2 = 2 500 μs	T_1 = 1.2 μs T_2 = 50 μs	a
Standard withstand voltage test	a	Short-duration power frequency test	Switching impulse test	Lightning impulse test	a
[a] To be specified by the relevant apparatus committees.					

Figure 2.18   Shapes of overvoltages on high voltage electrical networks and standardised voltage shapes for withstand testing of equipment[††]

under switching surges. Therefore, such systems are first dimensioned for switching surge withstand.

2.4.1   Lightning overvoltages on lines

In the United Kingdom, the average lightning strike density is the range 1–2 strikes/km^2/year. The strikes to overhead lines are typically around 10 strikes/100 km/year. Although these figures are relatively low compared to other regions of the world, they still represent a serious threat to electrical networks and infrastructure. Shielding and surge protection are therefore required to ensure reliability and integrity of systems. Utilities adopt surge overvoltage protective measures and have access to monitoring systems that track lightning storms as they move on land, which can help to take extra measures to protect key infrastructure. Figure 2.19 shows an example of a visual real-time lightning tracking. Such tracking may also be useful for key critical installations to avoid major outages and damage.

Lightning can impact electrical networks in three ways: (i) direct hit to phase conductors, (ii) direct hit to shield/earth wire and backflashover and (iii) induced

[††]IEC 60071-1 ed.9.0. Copyright © 2019 IEC Geneva, Switzerland. www.iec.ch.

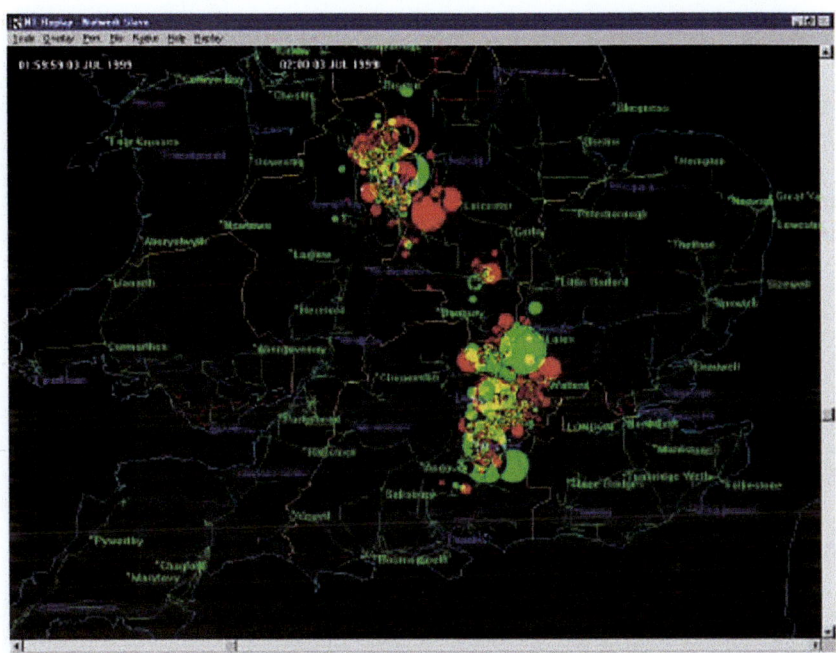

*Figure 2.19   Live tracking of lightning strikes by network companies (the example
shows intense activity in south east and the midlands, England,
sourced from National Grid, United Kingdom)*

voltage when lightning strikes in the vicinity of the overhead line. Figure 2.20
shows captured images of lightning strikes to lines, featuring scenarios (i) and (ii),
as reported in CIGRE TB 633 (2015), TB 704 (2017) and Takami and Okabe
(2007).

2.4.1.1   Direct impact on phase conductor and shielding failure

High-voltage lines with system voltages above 132 kV usually have a lightning
shielding wire, also known as the earth wire. If the three-phase conductor line has a
horizontal layout in an area of lightning risk, two earth wires are used to help shield
the phase conductors. The earth wire is positioned at a sufficient height above the
top phase conductor and forms a shielding angle of 30–45 degrees, depending on
the minimum lightning current that needs to be intercepted by the shielding wire.
The median value of lightning magnitude current of 30–32 kA is usually used for
such shielding. Such practice guarantees that any approaching lightning strikes that
have a magnitude higher than the design value above will hit the shielding wire (see
Figure 2.20(a)). However, for lower current magnitudes than the design value, the
shielding wire may fail to intercept the lightning strike and hence will not protect
the phase conductors. In this case, the lightning strike hits the phase conductor, and
this scenario is known as 'shielding failure'. Such failure can happen as the distance of the approaching strike has a striking distance that reaches the phase

	Lightning strokes (Non-outage)	Outage
Tower / Ground wire	(a)	(b)
Phase conductor	(c)	(d)

Figure 2.20   Examples of lightning strikes to earth wires and to phase conductors (shielding failure): (a) Lightning stroke to the top of the tower, (b) backflashover, (c) direct lightning stroke to the upper phase conductor, (d) flashover (direct lightning stroke to the middle phase conductor) (from CIGRE TB 633 (2015), TB 704 (2017) and Takami and Okabe (2007)), reprinted with permission from CIGRE ©

conductor before the earth wire. Failure can also happen when the approaching strike is from the side of the line or approaching the phase conductor horizontally (see Figure 2.20(c) and (d)).

Following shielding failure, the full lightning current, $I_f(t)$, will split into two surges travelling in the opposite directions towards the next terminal substations. A surge voltage, $V_s(t)$, will then be impressed on the phase conductor which will depend on the characteristic/surge impedance, Z_0, of the line:

$$V_s(t) = Z_0 \frac{I_s(t)}{2} \tag{2.9}$$

Generally, the characteristic impedance of overhead lines ranges between 200 and 500 Ω going from medium voltage (MV) lines to ultra-high voltage lines. For a lightning strike of 20 kA, the lightning surge on the line will be according to (2.9) between 2 MV and 5 MV. Such voltage level is well above the insulation levels of all lines and particularly for MV lines. Therefore, it is expected that a flashover will happen along the route of the surge (see Figure 2.20(d)) or at the terminal substation if the spark gaps and other protection systems are not present or fail to operate. It is well known that such surges can propagate through transformers into the LV network and hence reach commercial and domestic buildings. CIGRE Technical Brochures TB 287 (2006), TB 441 (2010), TB 550 (2013) give a comprehensive overview of lightning surges and their evaluation on LV networks. In particular, the transformer model is crucial for the accuracy of the computations.

When such failures occur on the distribution and/or transmission networks, and on the LV network, other faults may occur as a consequence. If the fault is not cleared quickly, follow-on ac currents will flow which injects high energy into the faulted circuit.

2.4.1.2 Backflashover due to direct impact to shield/earth wire or tower

When the lightning strike current is above the design value, it is expected that the shielding wire will intercept the lightning strike and protect the phase conductors. The strike is likely to terminate at the top of the tower (Figure 2.20(a)) or on the shielding wire in the close vicinity of the tower top. As sketched in Figure 2.21,

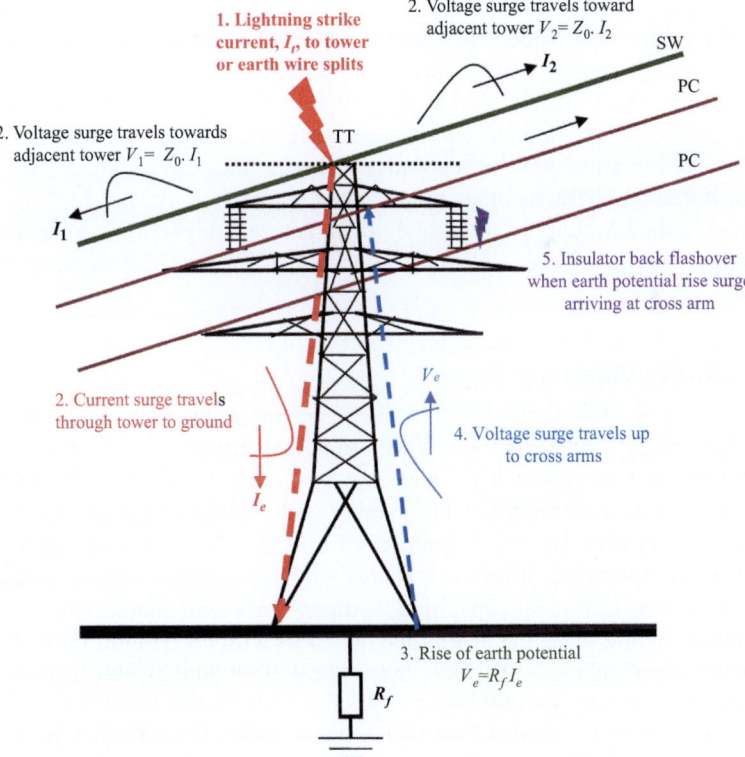

Figure 2.21 Steps of backflashover of insulators on overhead lines

the lightning current will then split into three paths, according to the characteristic impedances of the earth/shielding wire and the body of the tower (which is labelled as Step 2 in Figure 2.21). Two surges will travel in the opposite directions on the earth wire towards the adjacent towers, and the third component will flow to ground through the steel body of the tower. When this third surge reaches the base of the tower, the earth potential will rise based on the incoming current and the impedance/resistance of the tower base (Step 3 on Figure 2.21). This rise of earth potential surge will travel back up the body of the tower stressing the insulator strings (Step 4 in Figure 2.21). If the magnitude of the surge is higher than the withstand level of the insulator, a flashover from the grounded steel cross arm (or attachment point of the insulator) to the phase conductor will happen (Step 5 in Figure 2.21), and this is known as the backflashover due to lightning strike to towers (see Figure 2.20(b)). Although these events are rare (1%), they have very steep fronts which are very stressful on high-voltage insulation, particularly on transformers at terminal substations.

2.4.1.3   Induced lightning overvoltages

Overhead lines and their associated electrical networks run over long distances, usually of several kilometres for MV and approaching hundreds of kilometres for high-voltage lines. Such long distances make them vulnerable to induced surge voltages from lightning striking in their vicinity given the large loops formed by the conductors. Similar principles and analyses to induced effects in infrastructure can be adopted to evaluate the induced voltages in the case of lines as demonstrated in the published literature. MV lines are denser; hence, they have a higher risk to be close to lightning strike to other points close to the line. As the MV lines are connected to the LV supplies of ground infrastructure, a higher risk of transferring such surges to the LV network, hence the buildings, has been experienced in the past.

2.4.2   Switching overvoltages

Switching surges have slower fronts, lower magnitudes but much longer durations and higher energy contents compared with lightning surges.

As indicated in Figure 2.17, transient switching surges on electrical networks are mainly generated by the opening and closing of circuits containing inductances and capacitances. Examples of circuits elements involved in such transients are overhead lines and cables with distributed capacitances and inductances, transformers, capacitive compensators and shunt reactors. For transients involving lines, the magnitude increases with line length. Furthermore, if there is trapped charge on the line, much higher switching overvoltages can be expected.

Switching surges generated on high-voltage and MV networks can transfer onto the LV system through transformers as described in CIGRE TB 287 (2006), TB 441 (2010) and TB 550 (2013).

The main causes of switching surges that have been extensively documented in the literature (Haddad, 1990; Hileman, 1999; Weedy and Cory, 1998; Greenwood, 1991; Gary, 1984 and EPRI) are as follows.

2.4.2.1   Energising and re-energising of lines

The closing and reclosing of a line, with its series inductance, capacitance and relatively low series resistance forming an oscillatory circuit, generates transient voltages, capacitance. When a line open at the other end is switched, the reflected travelling wave at the open end can double the voltage, producing a 2 p.u. surge. The surge magnitude is even higher if the circuit breaker recloses the line with trapped charge remaining on the line. Trapped charge is due to the capacitance of the line that remains charged after removing the voltage on an open line, as the only way for this charge to leak is through shunt conductance of insulators and other insulation connected to the line. On cable circuits, the trapped charge was found to remain at high levels even after 12 h of de-energising the cable (Robson *et al.*, 2020).

Switching of capacitive–inductive circuits in LV systems can produce similar effects, depending on the values of capacitance and inductance. Such phenomenon is exploited to a good use in igniting neon tubes for domestic lightning where an inductor (choke) combined with a capacitor (starter) are used to generate a high switching surge to ignite the gas in the tube.

2.4.2.2   Interruption of capacitive currents

Following the interruption at peak voltage of relatively small capacitive currents feeding shunt capacitor banks, the capacitance of unloaded long lines or cables, a resulting surge voltage across the open circuit breaker contacts will develop and can reach up to 1.5 p.u. Within a half cycle, when the system voltage on the energised terminal of the circuit breaker reaches its maximum, the surge voltage can reach 2.5 p.u., which may then lead to a restrike of the circuit breaker, i.e., restoring the supply to the capacitive load. Several restrikes can happen with a short period of time before the voltage across the circuit breaker contacts settles to 1 p.u.

2.4.2.3   Interruption of small inductive currents

For a successful interruption of current, the circuit breaker needs to operate near the instant of zero magnitude of the current flowing in the circuit. Without this zero-current magnitude, the circuit breaker will be unable to extinguish the arc following the opening of its contacts. It is common for the ac arc inside the breaker chamber to be successfully extinguished when the current is below 1 A. However, around the zero-current instant of the alternating current, the rate of change of current, $dI(t)/dt$, reaches its maximum, and if there is complete interruption before the zero-current magnitude, the current will be chopped resulting in a very high di/dt. When the switching is on inductive loads, this rate of change of chopped current combined with the inductance value can give rise to overvoltages of 2–3 p.u. in modern transformers. If the transformer is connected to a shunt reactor, this switching surge overvoltage can reach up to 5 p.u.

2.4.2.4   Fault clearance and initiation

Following a single-phase short-circuit to earth fault, which is one of the most common faults on electrical networks, short-circuiting the phase to neutral, the voltage at the neutral will suddenly increase the system voltage. Up to 2.7 p.u. transient surge overvoltage was observed when the fault occurs at maximum system voltage.

Clearing three-phase faults to ground using circuit breakers has been reported to produce up to 2 p.u. surge overvoltages.

To reduce switching surge magnitudes to less than 2 p.u., various mitigation techniques have been adopted, e.g., circuit breakers with point-on-wave switching, circuit breakers with resistor switching and draining of trapped charge before circuit breaker closing/reclosing. Where necessary, surge arresters are used for the overvoltage protection.

It is important to note that the overvoltage generation scenarios described before can couple onto LV systems through conduction, capacitive and inductive couplings and earth potential rise. For the latter, tripping of generator circuits in power stations and faults in control rooms have been experienced during air or GIS disconnector switching in nearby high voltage substations. The disconnector operation is known to generate very high frequency/fast transients surges which both radiate an interference surge, and fast currents are injected into the ground developing high-frequency earth potential rises. LV circuits pick up such transients which may result in faults.

Furthermore, similar scenarios of surge generation, as described before, can be generated within the LV system too.

2.4.3   Temporary overvoltages

A temporary overvoltage is defined as 'an oscillatory phase-to-ground or phase-to-phase overvoltage of relatively long duration at a given location which is undamped or weakly damped in contrast to switching and lightning overvoltages which are usually highly or very highly dumped and of short times'. Temporary overvoltages are characterised by low-frequency voltage shapes and relatively long durations of up to 60 s. Diesendorf (1974), Glavitsch (1980) and CIGRE/IEEE joint working group (1990) give detailed analysis for all important scenarios of temporary overvoltages. Temporary overvoltages magnitudes range from 1.2 to 1.5 p.u., although, in most severe cases, they can reach 2 p.u.

Amongst the reported causes of temporary overvoltages are the following most common cases: ferroresonance, saturation effects, ground faults, load rejection and Ferranti effect. The latter ones occur on long lines and are known to produce high voltages at the open end of a long line, with a nonlinear voltage increase along the line from the energised end to the open end. An overview of important TOV causes and their characteristics is given in German and Haddad (2004).

Acknowledgement

The author thanks the International Electrotechnical Commission (IEC) for permission to reproduce Information from its International Standards. All such extracts are copyright of IEC, Geneva, Switzerland. All rights reserved. Further information on the IEC is available from www.iec.ch. IEC has no responsibility for the placement and context in which the extracts and contents are reproduced by the author, nor is IEC in any way responsible for the other content or accuracy therein.

The author would like to Thank Ms Kate Osbaldeston for support in obtaining copyright permissions and for sourcing some of the key images. The author thanks Vaisala for permission to reproduce a GLD360 picture. Bibliography lists two papers describing some of the principles of GLD360. The author thanks CIGRE, Cambridge University Press and Elsevier for permission to reproduce published material.

Bibliography

Anderson RB, Eriksson AJ, Kroninger H, and Meal DV, "Lightning parameters for engineering applications", *Electra Cigre*, Vol. 68, pp. 65–102, 1980.

Berger K, Anderson RB, and Kroninger H, "Parameters of lightning flashes", *Electra Cigre*, Vol. 80, pp. 23–37, 1975.

Chisholm WA and Anderson JG, "Lightning and grounding" Chapter 6 in *"EPRI transmission line reference book – 200 kV and above"*, 1011974, Edited by R Lings, Electric Power Research Institute (EPRI, Palo Alto, USA), 3rd edition, 2005.

CIGRE WG 33.10 and IEEE TF on TOV, "Temporary overvoltages: Causes, effects and evaluation", CIGRE 1990 Session, paper 33-210, pp. 1–15, 1990.

Cummins KL, Krider EP, Olbinski O, and Holle RL, "A case study of lightning attachment to flat ground showing multiple unconnected upward leaders", *Atmospheric Research*, Vol. 202, pp. 169–174, 2018.

Dellera L, Garbagnati E, Lo Piparo G, Ronchetti P, and Solbiati G, "Lightning protection of structures. Part IV: Lightning current parameters", *L'ENERGIA ELETTRICA*, Vol. 11, pp. 447–461, 1985.

Diesendorf W, "Temporary overvoltages", in *"High-voltage Electric Power Systems"*, Butterworths, London, 1974.

Furse, Thomas and Betts, "A guide to BSEN 62305: 2006 Protection against lightning", 2nd edition, catalogue, 2007.

Gary C, "La foudre" in *"Les proprietes dieletriques de l'air et les tres hautes tensions"*, EDF, Eyrolles, Paris, 1984.

Geldenhys HT, Eriksson AJ, and Bourn GW, "Fifteen years of data of lightning current test-line", in *"Lightning and Power Systems"*, IEE Conf. Publ. No. 236, London, pp. 62–66, 1989.

Glavitsch H, "Temporary overvoltages", In Ragaller K (Ed.), *"Surges in High Voltage Networks"*, Plenum Press, London, pp. 131–163, 1980.

Golde RH, *"Lightning, Volume 1: Physics of lightning"*, Academic Press, London, 1997a, ISBN 0-12-287802-7.

Golde RH, *"Lightning, Volume 2: Lightning protection"*, Academic Press, London, 1997b, ISBN 0-12-287802-7.

Greenwood A, *"Electrical Transients in Power Systems"*, John Wiley and Sons, 2nd edition, 1991, ISBN0-471-62058-0.

German DM and Haddad A, "Overvoltages and insulation coordination in transmission systems", Chapter 7 in *"Advances in High Voltage Engineering"*, Edited by A Haddad and D Warne, IEE Power & Energy Series 40, London, 2004, ISBN 0-85296-158-8.

Haddad A, "Attenuation and limitation of transient overvoltages on transmission systems", PhD thesis, University of Wales Cardiff (Cardiff University), 1990.

Harris Semiconductor, "Transient suppression devices", catalogue for commercial and military applications, 1994.

Hasse P, "Overvoltage protection of low voltage systems", in *"IET Power Series"* 12, 1998, ISBN 0-86341-213-0.

Hileman AR, *"Insulation Coordination for Power Systems"*, Taylors and Francis group, 1999, ISBN 0-8247-9957-7.

IEC 60060-1:2010, "High-voltage test techniques – Part 1: General definitions and test requirements", IEC Standard, 2010.

IEC 60071-1:2019, "Insulation co-ordination – Part 1: Definitions, principles and rules", IEC Standard, 2019.

IEC 62305-1:2011, "Protection against lightning – Part 1: General principles", IEC Standard 2011.

IEEE Members of the Task Force 15.09 on Parameters of Lightning Strokes: Chowdhuri P, Anderson JG, Chisholm WA, *et al.*, "Parameters of lightning strokes: A review", *IEEE Transactions on Power Delivery*, Vol. 20, no. 1, pp. 346–358, 2005.

Montanyà J, van der Velde O, and Williams ER, "Lightning discharges produced by wind turbines", *Journal of Geophysical Research: Atmospheres*, Vol. 119, 2014, doi:10.1002/2013JD020225.

Rakov V and Uman MA, *"Lightning: Physics and effects"*, Cambridge University Press, Cambridge, third printing, 2003, ISBN-13 978-0-521-03541-5.

Robson S, Haddad A, Dennis S, and Ghassemi F, "Non-contact measurement and analysis of trapped charge decay rates for cable line switching transients", *Energies*, Vol. 13, No. 5, article number: 1142. 2020, doi:10.3390/en13051142.

Said RK, Inan US, and Cummins KL, "Long-range lightning geolocation using a VLF radio atmospheric waveform bank", *Journal of Geophysical Research*, Vol. 115, D23108. 2010, doi:10.1029/2010JD013863.

Said R, "Towards a global lightning locating system", *Weather*, Vol. 72, No. 2, 2017.

Takami J and Okabe S, "Characteristics of direct lightning strokes to phase conductors of UHV transmission lines" *IEEE Transactions on Power Delivery*, Vol. 22, No. 1, pp. 537–546, 2007.

TB 063, "Guide to procedures for estimating the lightning performance of transmission lines", CIGRE WG33.01, 1991.

TB 094, "Lightning characteristics relevant for electrical engineering: Assessment of sensing, recording and mapping requirements in the light of present technological advancements", CIGRE Task Force 33.01.02, 1995.

TB 118, "Lightning exposure of structures and interception efficiency of air terminals", CIGRE Task Force 33.01.03, 1997.

TB 287, "Protection of MV and LV networks against lightning – Part 1: Common topics", CIGRE C4.402, Technical Brochure, 2006.

TB 376, "Cloud-to-ground lightning parameters derived from lightning location systems – The effects of system performance", CIGRE C4.404, Technical Brochure, 2009.

TB 441, "Protection of medium voltage and low voltage networks against lightning Part 2: Lightning protection of medium voltage networks", CIGRE C4.402, Technical Brochure, 2010.

TB 549, "Lightning parameters for engineering applications", CIGRE C4.407, Technical Brochure, 2013.

TB 550, "Lightning protection of low-voltage networks", CIGRE C4.408, Technical Brochure, 2013.

TB 600, "Protection of high voltage power network control electronics against Intentional Electromagnetic Interference (IEMI)", CIGRE C4.206, Technical Brochure, 2014.

TB 633, "Striking characteristics to very high structures", CIGRE C4.410, Technical Brochure, 2015.

TB 704, "Evaluation of lightning shielding analysis methods for EHV and UHV dc and ac transmission lines", CIGRE C4.426, Technical Brochure, 2017.

TB 785, "Electromagnetic computation methods for lightning surge studies with emphasis on the FDTD method", CIGRE C4.37, Technical Brochure, 2019.

TB 172, "Characterization of lightning for applications in electric power systems", CIGRE, Technical Brochure, Task Force 33.01.02, 2000.

Uman MA, *"The Art and Science of Lightning Protection"*, Cambridge University Press, Cambridge, 1st edition, 2010, ISBN 978-0-521-15825-1.

Visacro S, Mesquita CR, De Conti A and Silveira FH, "Updated statistics of lightning currents measured at Morro do Cachimbo station", *Atmos. Res*, Vol. 117, pp. 55–63, 2012.

Wagner CF, "The relationship between stroke current and the velocity of the return stroke", *IEEE Transactions on Power Systems and Apparatus*, Vol. PAS-68, pp. 609–617, 1963.

Waters RT, "Lightning phenomena and protection systems", Chapter 3 in *"Advances in High Voltage Engineering"*, Edited by A Haddad and D Warne, IEE Power & Energy Series 40, London, 2004, ISBN 0-85296-158-8.

Weedy BM and Cory BJ, *"Electric Power Systems"*, John Wiley and Sons, Chichester, 4th edition, 1998, ISBN 0-471-97677-6.

Whitehead, "Cigre survey of the lightning performance of extra-high voltage transmission lines", CIGRE WG no. 33.01, *Electra*, Vol. 33, pp. 63–89, 1974.

Chapter 3

Risk assessment

Alain Rousseau[1]

The possible types of damages caused by lightning and surges (overvoltages or impulse current) to a structure and its connected lines should be evaluated as well as their frequency of occurrence and the extent of the associated damages. The method to perform this evaluation is a lightning risk assessment (LRA) as defined in IEC Standard 62305-2. This analysis aims to determine what are the most efficient protection means for the studied structure (lightning rods, mesh system, surge-protective devices (SPDs), Thunderstorm Warning System (TWS), etc.), where to locate the selected protection components and with which efficiency (level of protection).

Lightning and surge protection risk mitigation solutions generally include the following:

- Lightning protection system (LPS): air-termination (lightning rod, mast, overhead wire, mesh), downconductors and lightning earthing system as well as equipotential bonding, separation distance and protection of human beings (including at ground level near downconductors and earthing system).
- Surge-protective devices (SPDs): either Type 1 (able to withstand a partial direct lightning current), Type 2 (only able to withstand induced surges but providing a good voltage protection), Type 1+2 (able to protect against induced surges and partial direct lightning current) or even Type 3 (similar to Type 2 but tests performed with a different generator and supposed to bring a much lower voltage protection level U_p than a Type 2). Data SPDs according to 61643-21 standards use a different classification (see Chapter 6). In the market, the concept of Type 1 and Type 2 is also largely used for data SPDs and we will also use the generic terms of Type 1 and Type 2 (respectively, T1 and T2) for all types of SPDs with the assumption that a T1 SPD can handle a partial lightning current (associated with current testing waveshape 10/350 µs) when a T2 SPD can only handle induced surges (associated with current testing waveshape 8/20 µs).

Note: A Type 3 SPD is not necessarily an SPD with a lower U_p than a Type 2 SPD. It is an SPD tested with a combination wave generator when Type 2 (and Type 1) are tested with a current impulse generator. A combination wave generator

[1]SEFTIM, Vincennes, France

(CWG) is a generator that will deliver a 1.2/50 µs voltage waveshape when con-nected to an open-circuit and an 8/20 µs surge current waveshape when connected to a short circuit. The voltage amplitude, current amplitude and waveform that is delivered to the SPD are determined by the CWG impedance (generally 2 Ω) and the impedance of the SPD. The short circuit current is called I_{cw} and the open circuit voltage is called U_{oc}. $U_{oc} = 2 \times I_{cw}$. U_{oc} is the main parameter and should be indicated on the SPD nameplate. I_{cw} is also often indicated but it is not what was circulating in the SPD during the tests but what the CWG was able to generate if the SPD was a pure short circuit. In practice, the current really flowing through the SPD is much lower than I_{cw}, especially for MOV-based SPDs. A Type 3 SPD is neither an SPD for data/signal applications nor necessarily a two port SPD (or SPD including a filter). It is an SPD tested with another method than Type 2 SPDs.

Note: By definition, a waveshape x/y either for current or voltage is a bi-exponential waveshape with a front time x µs and a decay time to half-value y µs. However, many people use the following way for referring to lightning wave-shapes: x/y µs. Even, if this redundant, µs being assumed by definition, we will also use this practice as it is clearer for the readers.

Note: There is generally more than one SPD in an installation. SPDs on the same power lines should be coordinated (see Chapter 7). For convenience's sake, we will mention SPDs on the same power line as a coordinated set of SPDs and the complete list of SPDs and their associated devices such as SPD disconnectors, remote alarm and so on will be mentioned as an SPD system in short SPS.

- Thunderstorm Warning System (TWS): a device, group of devices or networks of devices that can detect and inform about a possible impending lightning-related event in the vicinity (e.g. from 10 to 30 min), allowing people to go to a safe place, stop a dangerous process or enforce emergency services.
- Shielding (structure, room, zone or for lines) and line routing: this will decrease the surges created on the structure's internal circuits or even directly inside equipment. However, SPDs may be used to protect against induced surges if they are near enough the equipment not to generate an additional voltage drop in the circuit between the SPD and equipment. Shielding will only protect against induced surges and not against surges that are propagating along the lines.

The primary object considered in risk studies is a structure (a building, a shelter, an industrial installation, etc.). Connected lines are considered because they may bring lightning stress from far away to the structure. The lines themselves are rarely studied by this method, even if in principle, the method is fully applicable from a physical point of view. However, the power utilities and telecom service providers have their own rules for calculating the risk based on models and experience and the lines are outside the scope of the IEC 62305-2 standard.

Note: Connected lines are considered in the calculation because they can bring surges inside the studied structure and they are often called 'incoming' lines for that reason. This may be misleading because a utility power line brings energy to the structure and the line supplying an external pump are both 'incoming' lines even if the flow of energy is going in two opposite directions (to the structure for the utility line, from the structure to the pump for the other line). Even if experience has shown that the term 'incoming line' may be confusing, we will use it in that book due to the fact it is often used in practice or more generally 'connected lines'.

Lightning may cause many types of damages to a structure: injury to human beings, physical damage of the structure and its contents (fire, explosion, mechanical damage, etc.) or failure of internal systems (electrical and electronics).

Four sources of damage are considered in the calculation of the risk (Figure 3.1):

- lightning flash to the structure,
- lightning flash near the structure (generally a few hundred metres from the building limits),
- lightning flash to the connected lines,
- lightning flash near the connected lines (generally a few thousand metres both sides from the line).

SPDs can be used to mitigate the risk due to the last three sources of damages and are then important components in the risk reduction strategy. Very often, there are more surges coming from the lines, even if underground, than direct strike to the structure itself.

Risk component are related to the sources of damage.* For a structure connected to a single line, there is a maximum of eight risk components to calculate, the global risk being the sum of the considered components. There are three components dedicated to a direct strike to the structure and three other components dedicated to direct strike to the incoming lines. From the risk perspective, damages caused by a direct strike either to the structure or to the incoming service are similar: human risk – component R_A or R_U, respectively; risk of physical damage to

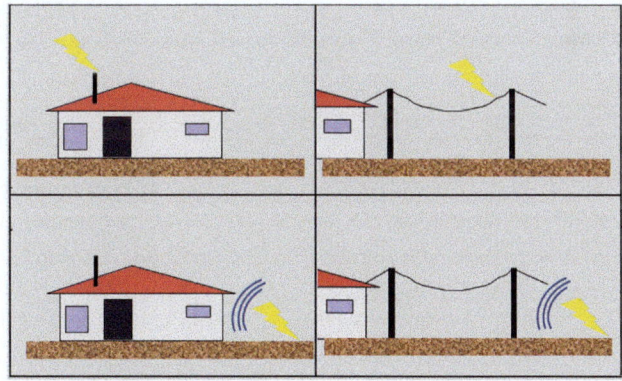

Figure 3.1　The four sources of damage considered

*IEC 62305-2 ed.2.0. Copyright © 2010 IEC Geneva, Switzerland. www.iec.ch.

the structure – component R_B or R_V, respectively; and damages to equipment – component R_C or R_W, respectively. When the connected line is short (less than 1 km, when line length is longer it is assumed the incoming surges to the structures are less damaging), the direct strike to the structure connected at the other end of the line – called an adjacent structure – needs to be considered as well (Figure 3.3). This consideration is because a direct strike to that structure will inject a significant surge into the line that will propagate and reach the studied structure. There is another component dedicated to induced surges inside the structure – component R_M and another one dedicated to induced surges to the line – component R_Z. Both components have a similar effect: damage to equipment. The total risk R for the considered structure is then

$$R = R_A + R_B + R_C + R_M + R_U + R_V + R_W + R_Z \tag{3.1}$$

The components can be grouped into the effect of a direct strike to the structure ($R_A + R_B + R_C$) and an indirect effect (all the others, $R_M+R_U+R_V+R_W+R_Z$) (Figure 3.2). From the perspective of effects, we can also group the components into effect to human beings ($R_A + R_U$), effect to the structure ($R_B + R_V$) and effects to its content ($R_C + R_M + R_W + R_Z$).

Note: It is important to understand that what is (relatively) simple for a structure connected to a single line becomes more difficult to calculate when the structure is connected to many lines. For example, for a building connected to a power line and to a telecom line the total risk will be

$R = R_A+R_B+R_{CPWR}+R_{MPWR}+R_{UPWR}+R_{VPWR}+R_{WPWR}+R_{ZPWR}+R_{CTELCO}+R_{MTELCO}+R_{UTELCO}+R_{VTELCO}+R_{WTELCO}+R_{ZTELCO}$ *where* $_{PWR}$ *stands for power line and* $_{TELCO}$ *stands for telecom line.*

It is easily understandable that for an industrial building connected to many lines for power, telecom, sensors, motors and alike, the calculation becomes very bulky and a dedicated software is needed (Figure 3.4). Fortunately, when many lines connect the studied structure to another one and run following the same path (e.g. inside a common metal tray), only one path needs to be considered for calculation (generally the weakest in terms of surge withstand).

Each risk component is calculated in the same way: a multiplication of the number of events (N) by the probability of this event will create a damage (P) and by the amount of damages generated (L for 'losses'): $R_i = N \times P \times L$.

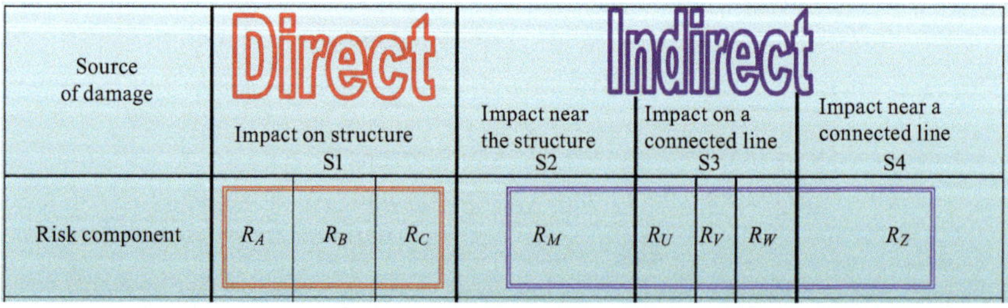

Source of damage	Direct			Indirect				
	Impact on structure S1			Impact near the structure S2	Impact on a connected line S3			Impact near a connected line S4
Risk component	R_A	R_B	R_C	R_M	R_U	R_V	R_W	R_Z

Figure 3.2   Sharing of risk components between direct and indirect

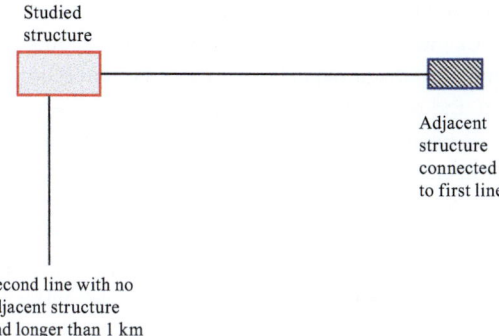

Figure 3.3   *Example of a structure connected to two lines, one connected to an adjacent building and one longer than 1 km where adjacent building either does not exist or is not considered because too far away*

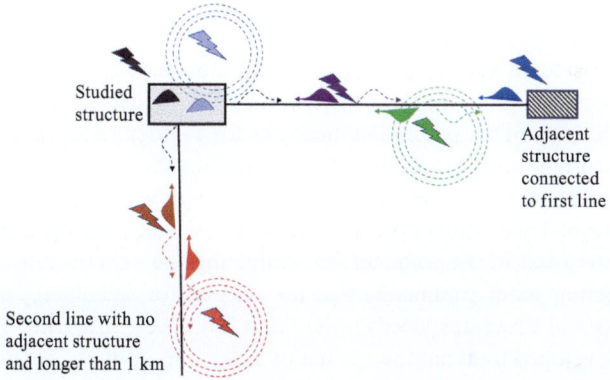

Figure 3.4   *Various surges to be considered. Dotted surges are just for information as they do not create a danger for the studied structure (but can for adjacent structures). In red are induced surges to second line, in green induced surges to first line, in dark blue direct impact on the adjacent structure, in purple is direct impact on first line, in brown is direct impact on second line, in light blue are induced surges into the structure and in dark grey is direct impact on the structure (this will create a huge partial direct impulse current for the structure and connected lines)*

It is considered that in the absence of any means of protection, any lightning-related event will lead to a damage (100% probability of damage, so $P = 1$). Protection means will reduce this probability of damage under a minimum level that is called the residual risk.

The amount of damage may be small (a broken device) or large (structure destroyed by fire) and this amount of damage needs to be estimated. It is important to notice that both N and P may be calculated with enough accuracy when L is related to a level of uncertainty. This is why, in the proposed new version of the standard, the frequency of damage ($F = N{\times}P$) is generally preferred to the risk ($R = N{\times}P{\times}L$).

The number of events N depends mainly on the structure or lines dimensions and on the lightning exposure of the area (in terms of lightning flashes per year and per km^2).

The probability that this event creates a damage (P) is related to the natural resilience of the structure, its contents and connected lines to a lightning event, as well as existing lightning protection measures such as lightning rods or SPDs.

The generated amount of damage (L) can be calculated on the basis of the existing mitigating measures (e.g. fire detection system) and to the time of the presence of people in a dangerous zone inside or outside the structure (the later can be mitigated thanks to a TWS).

As soon as the risk components are calculated, the risk is evaluated as the sum of these components and is compared to a tolerable risk level (RT) that is proposed by the standard. If the calculated risk is lower than this tolerable level, no additional lightning protection measures are needed and the level of risk can be accepted (structure considered self-protected). If this is not the case, lightning protection measures should be provided until the risk falls below the tolerable level.

The efficiency of the protection measures for a structure is related to a level of protection ranging from IV to I, with I being the best level of protection. This level defines the percentage of cases for which the structure will be protected. Normally, an LPS at level I will cost more than at level IV and a specific economic risk method is proposed in the standard for evaluating the cost/benefit of these measures. Collecting input parameters that are needed for calculating the risk level, especially for old structures, needs time. For a few cases, simplified methods have then been developed to avoid the burden of collecting all these parameters.

IEC 62305-2 is the reference standard for lightning risk calculation. Its formulas are then used by many other standards, including electrical installations (IEC 60364 Part 4-44 Section 443), photovoltaic systems (IEC 60364 Part 7-712) and wind turbines (IEC 61400-24). The first 62305-2 standard was published in 2006 but it was based on an IEC technical report dated 1995, so it has a long experience. The current edition is edition 2 (published in 2010), and edition 3 is in preparation (expected by the end of 2022).

It should be noted that the LRA is only a first step of a protection plan.

The second step is to define the protection components (LPS, SPDs, TWS, etc.) that will allow the risk to be reduced to the level defined by the risk analysis and their associated installation rules (the right product improperly installed fails to protect efficiently. For SPD this will be detailed in Chapter 7). This second part of the protection plan is sometimes called a Technical Study.

The third step is to produce a Protection Specification that will allow contractors to quote the appropriate protection measures. While the Technical Study may include variants, very often the Protection Specification is very specific and detailed. This specification does not include all the justifications that are normally given in the Technical Study but adds detailed information on location, quantities and even a bill of material. In many cases, the Technical Study is not performed and only a Protection Specification is produced. This is particularly the case when it is not necessary to demonstrate that the proposed protection measures will effectively reduce the risk to the expected level.

The fourth step is to install the protection measures and produce a Protection Technical File (as-built) that details what has been done, what protection components have been used and their compliance with standards. This step is as important as the studies because inappropriate products or products improperly installed will fail to protect at the expected level.

Then the fifth step is to perform an inspection to prove that what has been installed is conform to what was intended and defined in the Protection Specification.

After this point, the maintenance and inspection program can start. A protection system needs to be inspected regularly, typically on a yearly basis and, when needed, maintenance on the protection system is performed (e.g. to replace a disconnected damaged SPD or add additional SPDs when new sensitive equipment is installed).

3.1   What needs to be protected: industrial needs

When an industrial plant is studied, it is necessary to first determine which structures need to be studied. The risk process will study each of these structures separately but their interconnection will be considered as connected lines.

For an industrial facility, there are at least three targets for the lightning protection plan:

- People protection: This includes the workers, the maintenance team, the visitors, the suppliers and all the people who need to be on site for a limited time or for an extended period. In the case of damages that can expand outside the facility fence, for example due to an explosion or a fire ball, the protection of people included in the potential damaged area outside of the site also needs to be considered (the risk that covers this is called $R1$).
- Assets protection: This includes the plant itself as well as all the machines used for the production, including the incoming electrical power supply lines, distribution lines, data lines and other services such as gas and water. In the absence of one or more of these elements, the production could be reduced, slowed down or even interrupted. In this case, protection needs to be considered on an economic basis, including the cost of lightning protection and surge protection, the cost of key elements of the production line, the economic loss until the production restarts and the cost to repair (risk that covers this is called $R4$). A specific risk exists for utilities that are providing gas, water, power, TV and telecommunications (called $R2$). Damages related to $R2$ affect customers outside of the facility, as the service provided by the utility will be interrupted. Normally the risk method concentrates on a structure and its content but for $R2$ its takes into account customers outside the studied structure.

Note: If an industrial plant that produces, for example, fertilizer is stopped because of a lightning damage, the people directly related to this facility will be impacted (no more work except for maintenance and safety teams, no sales if stock is damaged as well, etc.). If a power utility experiences lightning damage one of its substations, a few consumers may be without power supply if there is no other substation that can supply them. This is the purpose of risk R2. But if an industrial plant that produces oil is the only one in the vicinity (e.g. in the case of islands), many people may be short of oil when the stock is depleted, so this will also have an impact on people not directly related to this facility. This is why some people expand the use of R2 to such cases even if this is not exactly the scope of the method according to the standard (we will call it R2 extended).

- Environment protection: An industrial site differs generally from other sites by the fact that they handle products that may be dangerous for the environment. The quantity of such products may be small (a few hundreds of litres of oil for an emergency power generator or a few bottles of cleaning products e.g.) or large (oil tanks, refineries, chemical product bags stored on pellets, etc.). For small quantities, the risk can generally be neglected but for large quantities, there may be damage to the environment resulting from a lightning-related event. The damage to the environment can be direct (such as leakage of an oil tank or oil pipe punctured by lightning) or indirect (e.g. toxic fumes due to a fire created by lightning or polluted water due to the mixing of chemical products with water used by the fire brigade to stop the fire that may escape from the site). Damage to the environment can remain local, spoiling the soil or the local water table or propagate into rivers and seas.

Protection of people is the first thing to study. This is covered by the specific risk calculation called R1 in the IEC standard 62305-2 ed. 2 (see Section 3.4).

Note: If animals can be impacted by lightning, for example in an industrial farm, this is covered by the economic risk (R4) and not the risk R1.

Note: There is another risk called R3 that covers specifically historical buildings and national heritage sites that are not considered here because it primarily takes into account the risk of fire, which is already covered by the analysis given regarding R1, as far as surge-protective devices are concerned.

What are the mechanisms that can impact the people in the case of a lightning event?

The first possibility is a direct strike for people working on the structure roof (e.g. a tank roof). They can also be impacted by a direct lightning strike when walking from a building to another one, even if this is rather rare (this can happen if the site is large and buildings separated from each other for safety reasons, for example to avoid the domino effect). Protection for people on a roof may be provided by an LPS but it is then necessary to stay away from the lightning conductors and lightning rods. This risk associated with the vicinity of the LPS could be partially mitigated by an isolated LPS but people on roof should not use any metallic ladders that protrude above the isolated part. It is much more efficient to prevent people from entering the roof area in stormy periods, by using a reliable TWS. It is more complex to protect people on a pathway between buildings and quite difficult to ask all the people in the site to stop moving from one building to another so once again, the best way or even the only way is to incorporate a TWS into a standard operating procedure detailing the actions to be taken (visitors included) in the event of a TWS alarm. A TWS can be used for reducing the risk for people in the 62305-2 statistical calculation and the key parameter is called 'Failure to Warn Ratio', the ratio of lightning events that occurred in area surrounding the site with no prior warning from the TWS, with respect to the total number of lightning events affecting this area. A FTWR may be equal to 0 (then the risk disappears) but this generally has an impact on other parameters of the TWS such as the time delay between alarm and the first lightning event. It is then necessary to check whether this time delay is compatible with the time needed to have all people locate to a secure place.

The second possible mechanism that can impact people relates to people on the façade of the building (open windows, balconies or even scaffolding). Generally, the rolling sphere (electro-geometric model used to determine how to install the LPS) can show that a lightning strike on the façade is not possible. For tall buildings, the rolling sphere can impact the façade but it is considered that for building height lower than 60 m, this risk is negligible. For structure above 60 m, the façade is at risk but situation is almost the same as previous and only the top 20% of the structure needs a specific consideration. When the structure cannot be struck by lightning on the façade, the risk comes from the downconductors. People should stay away from the LPS downconductors. Situation can be managed thanks to isolated cables as mentioned earlier or non-conductive cable guards. There is a risk for people located at the ground floor near (less than 3 m) a non-isolated downconductor. This risk (touch voltage) is tricky because it is existing for anyone passing in the vicinity, and downconductors may be unnoticed by people passing by. One solution is then a warning panel that can be seen by pedestrians at a minimum distance of 3 m, but a better solution is to isolate the last 3 m of the downconductor by a tube made of 3-mm thick reticulated in polyethylene.

A third mechanism is the step voltage for people walking above the lightning earthing system. It is generally considered that if people stay at 3 m from the point where the downconductor enters the soil, the risk is limited. If the earthing system must be installed on a walking path, specific measures such as equipotentialization or isolation should be applied.

It must be noticed that when a TWS is available and applicable, this is the best way, associated with specific procedures, to protect people from all these risks.

The last mechanism, that is probably the trickiest one, is for people working in contact with panel boards and electrical conduits. A surge on an incoming line can propagate over a large distance, typically up to 1,000 m; thus, a lightning event can happen on a power line far away from the building and be unnoticed by the people working in a building. A surge can then propagate to the building electrical installations when a maintenance team is working on panel board, electrical wiring or inside machines. This can create a severe life hazard.

All these risks are covered by the risk $R1$ of standard IEC 62305-2. However, SPDs can only be used to reduce the last risk mechanism. As a matter of fact, provided one or more SPDs are used to avoid equipment sparkover and maintain systems insulation, the human risk is reduced. It is however still necessary to avoid touching or approaching live conductors of any type.

The best way to consider the protection of assets is to calculate the economic risk, called $R4$ in IEC 62305-2 standard. However, for this tool to be powerful, a lot of information must be collected, such as the price of pieces of equipment, price of the structure itself, cost for repair for damaged equipment, cost due to the loss of production, among others. Unfortunately, these data are rarely available in full and generally many assumptions have to be made. For example, for a piece of equipment connected both to power line and to data line, it is necessary to know what is the source of damages when this piece of equipment fails due to surges. It may be a surge coming from only one of the lines: power line or data line. In that case, the cost of damage will easily be associated with the line causing the damages. But when the damage can be caused by either of the two lines, is the price of equipment equally shared between the two lines? The cause may also be related to both lines at the same time due to a differential ingress (a typical example is a telecom box connected both to power and telecom line that fails due to a voltage between the two ports even if each port seems to be protected, we will come back on this in Chapter 7). The global loss cost should be compared to the yearly cost of protection (including the price of installation and the maintenance cost) and in this respect, variants should be evaluated: should a mesh system be used or a group of lightning rods? Which SPD should you consider in the economic risk calculation: the brand you are used to or the cheapest, when they both comply with the SPD standard? At this stage, the protection scheme is not defined and the cost of protection is largely estimated limiting the benefit of the economic risk study. For the economic risk to be of value, a lot of time is needed to study the parameters and the different protection options to get an accurate estimation of the protection system. The industrial site economic details should be known. For this reason, a simplified economic risk exists that is much less demanding in terms of economic accuracy and details. At the end, the main purpose of the economic risk is not to decide to protect assets or not, but to help selecting the most appropriate and economically efficient protection methods. SPDs can be efficiently used to reduce this risk. A coordinated set of SPDs can provide the protection of piece of equipment inside the electrical and electronics installation and thus keep production lines operational as well as protecting key elements such as computers or PLCs. An SPD protection scheme is usually based on installing powerful SPDs at an installation line entrance to stop the most damaging surges and installing downstream SPDs coordinated with entrance ones that will protect specific cabinets, panels or pieces of equipment. One of the targets of the risk management is to identify these key elements that deserve surge protection.

When the industry facility provides services to a lot of connected customers such as a power station or radio station, a specific risk method can be used called $R2$ in IEC 62305-2 standard. For this risk calculation, the method will be based on a determination of how many people are connected to the service and how many users will lose this connection in the case of damages due to lightning. As mentioned previously, it should be noted that the method described in IEC 62305-2 standard does not cover the utility network itself (power lines, telecom lines, gas pipes, etc.) but only the structure that is providing the service. SPD can be used effectively to reduce that type of risk for gas, water, power supply, TV and tele-communications structures.

Finally, for environment protection, the risk calculation method is derived from risk $R1$ where the number of people and time of presence in a dangerous area outside of the structure, or even outside of the site fence, are key parameters. This environmental risk is an additional part of the $R1$ risk calculation (we will call it $R1$ extended).

When a surge damages safety equipment (fire alarm, smoke detectors, pollution detectors, etc.), this damage can create a hazard for human beings or for the environment (e.g. pollution of rivers by releasing chemical products or stored polluted water, due to electromagnetic disturbances on control circuit). The basis for this study is a scenario that will determine what damage can happen and the extent this damage scenario may have. SPDs can be used to reduce this risk by protecting electrical appliances and avoiding dangerous sparkover that can trigger fire or explosion. They can also be used to protect fire detection and fire alarm systems as well as protecting smoke, fire or gas detectors throughout the facility. Finally, the protection of specific equipment that are used to fight against environment ingress can also be efficiently provided to reduce that risk. For example, a pipe that is carrying polluted water could be monitored to detect leakage and avoid unnoticed pollution ingress and in case pollution is detected, the monitoring system can trigger the shutdown of the pipe. Equipment providing this detection, monitoring and launching the shutdown sequence needs to be surge protected to finally protect environment.

Industrial site protection will generally rely on a mix between LPS, SPDs (Type 1 and Type 2), TWS and shielding/routing of cable to cover their risks.

3.2   What needs to be protected: domestic needs

Regarding the protection of dwellings, two cases should be considered: houses and buildings.

For buildings, lightning protection could be provided but generally, in cities, only high-rise buildings deserve lightning protection. Building of smaller dimensions and especially smaller heights are easily protected by surrounding tall buildings.

If an LPS is provided, Type 1 SPDs must be provided at line entrance at ground level. However, for such buildings, the power supply is generally provided by underground HV lines and Type 1 HV Surge Arresters do not exist. Type 2 HV Surge Arresters may be provided by the power utility but it is not a general practice when the line is underground especially in urban environment. Type 2 SPDs are then necessary at the secondary side of the building transformer HV/LV mainly to protect the LV side of the transformer and the main LV panel board. This will not protect each of the flats inside the building, and generally, Type 2 SPDs are also necessary at each flat entrance or at least at each floor.

Note: A Surge Arrester (SA) and a surge-protective device (SPD) are almost the same devices. SA is used by high-voltage industry when SPD is used for low-voltage and data/telecom industry. An SA is generally based on varistors (see Chapter 6) with possibly a spark gap in series (air gap or enclosed gap). An SPD is generally based on varistors and spark gaps (either Gas Discharge Tube, enclosed gap and rarely air gap) and sometimes silicon avalanche diodes. Many SPDs are made for indoor use but a few of them are also applicable for outdoor use and then there is not so much difference between an SA and an SPD except the voltage. We can however mention that SPDs are used in low-voltage environments and as such it is possible for inexperienced people to be near SPDs or even to manipulate it when such a thing cannot exist with SA. Safety tests and especially end of life tests are then more numerous for SPD than for SA and, for example in Europe, SPDs fall under the Low-Voltage Directive that imposes a safe behaviour.

When an LPS is installed on dwellings, the distance between the LPS and the system on the roof (or on the facades near downconductors), such as aircraft warning lights, antennas and HVAC, must be considered. If this distance is lower than the separation distance calculated for the LPS, Type 1 SPDs need to be provided at the roof entering point inside the building of the lines connected to these systems. The purpose is not to protect these systems but to avoid a damaging partial lightning current to enter the structure. To protect these systems, other SPDs near the system also need to be used. Type 2 SPDs are also needed at each floor and especially for the panels located at the highest floor. The purpose of the Type 2 SPDs is to decrease the surge that propagates down to the building entrance. When an isolated LPS is used, T1 SPDs are not needed on the roof and only T1 SPDs at the line entrance at ground level are needed. It is possible that induced surges on long cables on the roof may need to be considered but generally cable shielding is enough to avoid the stress. If shielding is not possible, then T2 SPDs are also necessary on the roof circuit to avoid surges to penetrate inside the structure.

If no LPS is provided on a building, this can be because direct strikes to the building are very rare (low flash ground density or protected effect of neighbouring structures) or because the direct risk to that building is not considered. In such a case, due to the fact that surges from an underground HV cable are also very rare in a city, there is generally no Surge Arrester provided. But Type 2 SPDs at the secondary of the HV/LV transformer remain a good option, even if the windings of the transformer will provide a first level of attenuation of any surge propagating from the HV underground system. In this case, installing Type 2 at each building floor is not critically needed.

In both cases, a building protected or not protected by an LPS needs to protect sensitive installations such a fire detection system, lift or alarms with a T2 SPDs. In addition, systems that are important for people's health (such as breathing aid systems or remote health alarms) should also be protected by a T2 SPD.

For houses, the situation is different because the power supply is of low voltage without a transformer attenuating the incoming surges. Even if a house is smaller in size and in height than a building, they are often located at the edge of the cities or even in rural areas, with limited neighbouring structures that may reduce the risk. In addition, the power supply is often based on overhead lines, and the number of surges coming from the lines is also much greater than in cities. For all these reasons, surge protection is often more used for houses than for buildings.

If the house is protected by an LPS, a Type 1 SPD is needed at the power installation entrance. Generally, if there is no LPS, Type 2 SPDs at the installation entrance are enough except if the line is overhead on the last meters before entering the building.

It is needed to protect sensitive and expensive equipment with T2 SPDs and especially systems that are important for people health (such as breathing aid systems or remote health alarms).

3.3   What needs to be protected: specific needs, including nuclear facility and data centres

It is often falsely considered that application of standard 62305-2 is enough for determining the risk level of a facility. In a few cases, it is not necessary to perform such an analysis because the level of protection is determined by the user or by a law from an authority having jurisdiction. But, in many cases, application of risk methods is helpful because it provides the level of protection for the structure to avoid over protection or under protection and addresses the lines, especially when lines are more prone to surges than the structure. However, it is often necessary to add a simple analysis to determine, for example which equipment needs a specific protection. This additional analysis is generally mentioned as a deterministic analysis by opposition to the statistical analysis using 62305-2 standard.

A few structures require a more in-depth analysis of the risk because consequences can be important in the case of failures either for environment or for economics.

Many oil and gas plant or chemical plants can present a risk to the environment resulting from direct lightning strikes or surges resulting from nearby strikes or strikes to lines. The resolution of this risk is discussed in Section 3.1.

Nuclear facilities are considered here a specific example. Regulation for these facilities is of the highest level in terms of safety demonstration requirements. Nuclear facilities are not just nuclear power plants but also include laboratories, production plants, storage and treatment facilities. In many places in the world, the nuclear activities are covered by specific national or international agencies that define their own rules for lightning and surge protection. For nuclear facilities, the risk is specifically related to nuclear safety, including all sensors that monitor the radioactive level as well as all equipment that provide basic safety: fire, intrusion, process monitoring and communication. The risk is mainly related to the quantity and the process used (a local radioactive source used for imagery does not generally require a long analysis where a large process, including many sensors and safety measures, likely needs a specific analysis).

Data centres have other types of requirements because losing data, not being able to supply data, or store data on request, can cost a fortune. For these structures, the key risks are related to continuity of power, cooling of mainframes and fire detection and prevention. Intrusion prevention is also an important parameter because data should be protected against illegal access.

These two examples will show how the risk can be estimated on the basis of scenarios that are to be considered in addition to the regular risk analysis in IEC 62305-2.

For nuclear facilities, the statistical risk methods of IEC 62305-2 will allow a level of protection for the structures to be provided based on the human risk ($R1$) and the environmental risk ($R1$ extended) with the opportunity to revise the generic loss parameters to cover the specific range of losses associated with the facility. For a nuclear facility, all the risk components R_i are to be considered, each component being equal to $N \times P \times L$ with N number of dangerous events, P probability that a dangerous event creates a damage and L amount of losses due to damages.

For each risk component, N can be assessed precisely using a local N_{sg} value (lightning ground strike point density) that is provided by a Lightning Location System that meets the requirements of the IEC 62858 standard. But this is not enough, N_{sg} is by definition an average over a 10-year period so this must be analysed in detail, taking into account the worst year instead of the average.

Probability P can also be assessed with a reasonable accuracy, especially for the structures. However, a few structures in the facility may need to be studied with different stresses than considered in IEC 62305-2. For example, a higher current than 200 kA (above level I) and at the other end a lower current than the minimum current considered for level I (3 kA) should be considered where applicable. For a 'normal' industrial structure, the maximum current generally leads to the most severe stress, but for a few structures inside a nuclear facility, a low current not intercepted by an LPS can damage an important outdoor sensor or allow disturbing surges to penetrate inside very low voltage monitoring circuit. These two extreme cases, higher current than considered by the standards and lower current than considered by the standards, should be addressed.

Probability associated with SPDs (P_{SPD}) is definitely a weak point. This is the probability that an SPD fails to protect. In today's approach, this probability is only related to the SPD being destroyed, assuming only that a failed SPD is not able to protect anymore. As all SPDs are coordinated together, none will fail before the T1 SPD upstream will fail and thus, P_{SPD} relates to the probability that the highest current associated with the LPS is exceeded. But an SPD that does not fail may not be able to perform as intended when the coordination between SPDs leads to a too high current in the downstream T2 SPD (too high to correctly protect but low enough not to fail). Another possibility of protection failure is because the SPD is disconnected due to maintenance or due to another cause such as an internal thermal runaway. In such a case, the SPD does not protect even though the lightning current does not exceed the capability of the T1 SPD. Conversely, an SPD may protect much better than expected thanks to an appropriate coordination or due to a stress lower than considered (e.g. when an induced surge occurs it is likely that protection will be better than for a direct strike on the line, but probability P_{SPD} is considered the same in all cases in IEC 62305-2). P_{SPD} should be better and more specifically studied to cover the nuclear facility needs.

The weakest point is L (losses) as mentioned previously because it fails to cover possible scenarios that only a specific and detailed analysis can provide. For example, malfunction of sensors or damage of a specific pipe needs to be considered even if the structure remains protected against typical major risks such as fire or mechanical damage.

For all these reasons, the basis for a nuclear facility risk calculation is not the risk ($R = N \times P \times L$) but the frequency of damage ($F = N \times P$). The frequency of damages calculation should be the basis for the risk level for specific equipment important for the nuclear safety and for establishing priorities between scenarios. Scenarios refer to components of the safety chain and it is necessary to allocate a frequency of damage to each of these components to determine the best protection strategy. The general analysis defined in 62305-2 standard cannot reach this level of details but is a good basis for allowing such complex calculations. As a result, a list of safety components with their needed level of protection can be established. A further study will determine how this protection can be achieved on the basis of specific SPD data (e.g. voltage versus current curves, mean time between failure,) and accurate SPD installation rules (choice of SPD disconnector, lead lengths, location, etc.).

The key point for nuclear facilities is the need to demonstrate that the reliability of the installations is always guaranteed in the case of a lightning-related event. For most of the installations, the protection plan is defined using the LRA, lists protection measures and identifies generic installation rules. Sometimes these generic rules are not fully applicable in practice for extremes and rare cases. The lightning current characteristics are such that the protection will not be completely adequate. These cases are so rare that for most of the facilities this is acceptable, but for nuclear facility these rare cases are already too large and should be evaluated. For example, the case of longer lead length (see Chapter 7) that will limit SPD protection efficiency should be evaluated (see Section 8.8).

Data centres are another case of a specific need where the LRA may require tailoring to the specific application. Once again, a global protection plan can be established using IEC 62305-2 rules, but without tailoring it will not fully cover the specific risk related to the loss of power or loss of the cooling system. Risk $R1$ covers nicely the fire risk as well as the risk to human beings. Risk $R2$ may be used to determine protection measures needed to avoid power supply surge damages, but the losses proposed by the standard are not fully adequate because it is designed to consider many external users and not a single one. In addition, the tolerable risk is far too high for a data centre. There is no factor identified to address the calculation of the surge damage rate for the cooling facilities, when it is one of the main objectives of a data centre risk assessment. A level of protection I may be applied to address the problem, but a 98% efficiency is generally too low for data centres and in addition, it will not help to define the SPDs required to achieve that goal as the probability P_{SPD} is presently inadequate to cover that need. It is then necessary to develop an additional deterministic approach supported by specific calculations and stringent installation rule requirements. Determine all the sources of surges, including those coming from roof equipment and external systems (such as light pole in the parking lot) and assess whether their path is crucial. Surges generated on the internal cabling due to lightning strike to the LPS or due to coupling between disturbed services and sensitive equipment installations should also be considered. The protection plan will then be mainly based on SPDs located in various places (panel boards, UPS, in front of equipment, etc.) and specific installation rules (mainly the choice of SPD disconnectors and cabling rules).

3.4   Method to address the protection needs: the general rules for risk assessment

When a lightning strikes a building, it may create a mechanical damage (hole on the roof, destruction of concrete walls, etc.) or trigger a fire or an explosion where such an environment may exist.

When a lightning strikes a building or its connected systems (power, telecom, data, metal pipes, covered by the generic name of 'services'), the surges due to the lightning can also trigger a fire or an explosion but can also damage electrical and electronic devices. Even when lightning only occurs near the building or near the services, the associated induced surges can damage equipment and especially sensitive equipment. If this piece of equipment can affect a person's health or welfare (hospitals, medical centre, etc.), such a surge can also create a hazard for human beings. Lightning can, at the same time, be the origin of a fire and cause the destruction of fire detectors, creating cumulative hazards.

The methodology to determine the lightning risk should then cover all these possible stresses. It is called a 'lightning risk assessment' or sometimes 'lightning risk management'. This is described in international standard IEC 62305-2.[†] It is important that the risk is correctly estimated because the outcome of the risk calculation is a required level of protection that determines the efficiency of the protection measures. Overprotecting a structure would cause an economical issue, and under protecting may create a safety hazard.

Lightning protection is in fact a generic term that also includes surge protection against overvoltages of atmospheric origin. Surge protection should be provided any time an LPS is installed. Furthermore, it is frequent that there are more possible damages coming from the connected lines (power or telecom/data) than from direct lightning on the structure. Thus, it is necessary to analyse the possible sources of damages and evaluate their frequency of occurrence as well as the extent of the associated damages.

The risk calculation process is as follows:

- determine the possible sources of damages;
- determine their frequency of occurrence;

[†]IEC 62305-2 ed.2.0. Copyright © 2010 IEC Geneva, Switzerland. www.iec.ch.

- determine the associated amount of damages;
- determine if the risk is acceptable and if not propose solutions;
- determine the cost efficiency of proposed solutions, especially when there are more than one protection solution.

The lightning risk is evaluated for a 1-year period. The method is typically applicable to a single structure. When there are many structures at a site that need to be studied (e.g. in industrial sites as explained earlier), the calculations should be performed for each structure, taking into account the other structures that may reduce the number of strikes to the studied structure or that are connected to the studied structure by lines (called generally adjacent building).

The risk due to lightning according to 62305-2 is the sum of different risk components, differing in their source of damage (S1, S2, S3, S4) and type of damage (D1, D2, D3) defined as follows:

- S1: lightning flash to the structure;
- S2: lightning flash near the structure (flash at a distance up to a few hundred metres but too far to be captured by the structure itself or its LPS);
- S3: lightning flash to the lines connected to the structure (including lightning flash to the adjacent structures: in that case, a part of this flash will flow through the line to the studied structure). This also includes underground cables even if generally it is falsely believed that an underground service does not require protection;
- S4: lightning flash near the lines connected to the structure (flash at a distance up to a few hundred metres but too far to be captured by the line itself).

and

- D1: injury to living beings by electric shock (mainly human being but in a few occasions, livestock may be concerned as well, especially for economic calculations);
- D2: physical damage (fire, explosion, mechanical destruction, etc.) due to lightning current effects, including sparking;
- D3: failure of internal systems due to lightning electromagnetic pulse (LEMP), i.e. the magnetic field generated by lightning and associated conducted or induced surges.

Based on these sources of damage and types of damage, the user can calculate up to eight risk components R_A, R_B, R_C, R_M, R_U, R_V, R_W and R_Z. The total risk R is defined as the sum of risk components applicable to the assessment (the number of risk component to calculate is depending on the type of risk the user wants to cover and the structure parameters).

If the structure is partitioned in individual zones having specific characteristics (e.g. a building having an explosive area when the remaining part of the building is a low risk of fire), each risk component shall be evaluated for each zone. Assuming the whole building is an explosive area would overestimate the risk and ignoring the small explosive area would underestimate the risk. The total risk R of the structure is the sum of all the considered risk components over all the zones of the structure (including outdoor zones if needed).

Each of the risk components R_i ($i = A, B, C, M, U, V, W$ or Z) is expressed by the following equation:

$$R_i = N_i \times P_i \times L_i \tag{3.2}$$

where N_i is the number of dangerous events per year; P_i is the probability of damage to a structure; L_i is the consequent loss.

The number N_i of dangerous events is influenced by the lightning ground flash density (N_g) or more precisely by the lightning strike ground point density (N_{sg}), by the dimensions of the studied structure, its surroundings, the connected lines and adjacent buildings connected to the studied structure.

The probability of damage P_i is influenced by the characteristics of the studied structure, the connected lines and the protection measures provided.

The process generally starts assuming that there are no lightning protection measures (thus $P = 1$: each lightning flash to the structure or to the services will create a damage). If the calculated risk is below what is considered acceptable, the structure is considered self-protected and no additional protection measures are needed. This does not mean that no lighting damage will ever occur, but the statistical risk is low enough to allow the user to ignore it. If the calculated risk exceeds what is considered acceptable, it is common to add protection means from the easiest to the more complex one until the risk decreases below the tolerable level. Protection means are related to a protection level ranging generally from IV (lowest level, weaker protection) to I (highest level, stronger protection).

What is considered tolerable may be dependent from a country, from an authority having jurisdiction or from the user. Tolerable risk values are proposed by the standard when needed but it is possible to fix other values.

The consequent loss L_i is influenced by the use to which the structure is assigned, the attendance of people, the type of service provided to the public, the value of goods involved in the damage and the measures provided to limit the amount of loss.

The third draft edition of 62305-2 (currently under discussion) has recently introduced the concept of frequency of damage.

$$F_i = N_i \times P_i \tag{3.3}$$

This frequency of damage defines how many times per year a damage related to a lightning event will occur. The user or the designer performing the risk calculation can then assess this frequency of damage, without introducing loss parameters (as it has been explained previously, the concept of losses is too simplified and too broad. It is not able to cover specific cases such as nuclear facilities because of the complexity of some applications) and determine the need of protection accordingly. The extent of damages due to a lightning strike (similar to the concept of losses) can then be much better evaluated by a deterministic approach, including various scenarios, without calculating L_i.

Each type of damage may produce a different consequential loss in the studied structure. The type of loss that may appear depends on the characteristics of the structure itself and its content. The following types of loss are considered:

- L1: Loss of human life that includes permanent injury and more generally any injury due to lightning in, on or near a structure.
- L2: Loss of service to the public that addresses cases where damages to a structure would create an inconvenience to or burden to the public. This includes power and telecommunication stations as well as water and gas distribution. For example, damage to a power station will result in a loss to many people located outside the structure. Radio and TV could be included in this list as well as train stations, airports or traffic controls. What is considered service to the public is often different from one country to another as indicated previously.
- L3: Loss of cultural heritage that applies to museum, old buildings and any national heritage building for which partial destruction would be a cultural loss. At the moment, this applies only to the structure itself and not to the content that may be as important as the structure itself or even more in the case of a museum, for example. In that case, the whole structure, including its content, should be considered national heritage.
- L4: Loss of economic value that applies to the structure, its content, and loss of activity. In general, L4 is applied to determine which protection means are most cost efficient. It is also possible to use this method to determine whether protection is economical even if not required due to one of the criteria given earlier. Obviously, most of the structures are occupied, even if not permanently, by human being (employees, customers, safety or maintenance team) and the key risk to calculate is related to L1. In such a case, a few structures may deserve an additional analysis, based on L4, but with the target to provide more protection that issued from the analysis based only on L1. It may be the case, for example, for a shopping centre. Another example is a family owned business or homes where the cost to replace the structure and contents that cannot be replaced (such as family photos or heirlooms) exceed the cost of protection.

Risks are evaluated on the basis of the type of loss that may be concerned for the studied structure. Risk $R1$ is associated with the type of loss L1 and so on. A maximum of four risks may be calculated for a given structure: $R1$, $R2$, $R3$ and $R4$. When calculating a risk $R1$–$R4$, the risk components may be grouped according to the source of damage and the type of damage.

Risk components associated with lightning flashes to the structure:

R_A: Component related to injury to living beings caused by electric shock due to touch and step voltages inside the structure and outside in the zones up to 3 m around downconductors. It applies also to people that may be located on the roof and terraces but obviously it is safer to stay inside when thunder roars. Early warning may be given by a TWS to avoid staying in an exposed place. R_A is associated with loss L1. But in the case of structures with livestock such as farms, barns and factory farming, loss L4 may also be considered. Generally, SPDs cannot be used to reduce this risk component.

R_B: Component related to physical damage caused by dangerous sparking inside the structure triggering fire or explosion which may also endanger the environment. This applies to physical damage to the structure (hole in the roof, head wall damages, concrete spalling, etc.). R_B is associated with all losses: L1, L2, L3 and L4. SPDs are not used to avoid direct lightning damage but they are part of the LPS because they provide equipotential bonding between the incoming services and the structure.

R_C: Component related to the failure of internal systems caused by LEMP. R_C is normally associated with losses L2 and L4. This can also apply to L1 when there is a risk of explosion or to hospitals or other structures when the failure of internal systems immediately endangers human life (retirement home, medical assistance, etc.). SPDs are the main source of mitigation for reducing this risk component.

Risk component associated with lightning flashes near the structure:

R_M: Component related to the failure of internal systems caused by LEMP (i.e. the electromagnetic field generated by a lightning channel that creates surges in structure's circuits and connected services). R_M is normally associated with losses L2 and L4. This can also apply to L1 when there is a risk of explosion or for hospitals or other structures where the failure of internal systems immediately endangers human life (retirement home, medical assistance, etc.). SPDs can be used in conjunction with shielding (shielding of the structure, a zone, a room or a circuit) for reducing this risk component.

Risk components associated with lightning flashes to a line connected to the structure:

The lines that are considered are the lines connected to the structure (reminder: the generic term is 'incoming service' even if the flow of energy is from the studied structure to a neighbouring structure or equipment). A line that would connect two parts of the same structure would be ignored in that process (but will deserve a specific analysis to determine if SPDs are needed for this line).

R_U: Component related to injury to living beings caused by electric shock due to touch voltage inside the structure. This can happen when a spark is created near people between an electrical appliance or power socket and a metal grounded part. This can also happen, when maintenance is performed on electrical or tele-communication system inside the structure during stormy weather. People inside structure may not be aware that lightning can occur soon if they are located far from a window. R_U is associated with loss L1. In the case of structures with livestock such as farms, barns and factory farming, loss L4 may also be considered. SPDs may be used to mitigate this risk component in conjunction with preventive measures based on TWS to avoid people working on services or equipment during stormy periods.

R_V: Component related to physical damage (fire or explosion triggered by dangerous sparking between external lines and metallic parts generally at the entrance point of a line into the structure) due to lightning current flowing in the lines. R_V is associated with all losses: L1, L2, L3 and L4. SPDs are the primary source of mitigation for reducing this risk component.

R_W: Component related to the failure of internal systems caused by over-voltages generated on incoming lines and transmitted to the structure. R_W is normally associated with losses L2 and L4. This can also apply for L1 when there is a risk of explosion or for hospitals or other structures where the failure of internal systems immediately endangers human life (retirement home, medical assistance, etc.). SPDs are the primary source of mitigation for reducing this risk component.

Risk components associated with flashes near a line connected to the structure are as follows:

As previously, the lines that are considered are the incoming lines.

R_Z: Component related to the failure of internal systems caused by over-voltages induced on incoming lines and transmitted to the structure. R_Z is normally associated with losses L2 and L4. This can also apply to L1 when there is a risk of explosion or for hospitals or other structures where the failure of internal systems immediately endangers human life (retirement home, medical assistance, etc.). SPDs are the primary source of mitigation for reducing this risk component. In general, induced surges on incoming services are more frequent than direct strikes to them but the extent of damages is generally smaller with induced surges. To protect against induced surges, Type 2 SPDs are enough but to protect against partial lightning current Type 1 SPDs are needed.

Summary of risk components associated with each type of loss:

- $R1$: Risk of loss of human life:

$$R1 \text{ (general case)} = R_A + R_B + R_U + R_V \tag{3.4}$$

  $R1$ (structures with the risk of explosion and hospitals with life-saving

  electrical equipment or other structures when the failure of internal

  systems immediately endangers human life)

$$= R_A + R_B + R_C + R_M + R_U + R_V + R_W + R_Z \tag{3.5}$$

- $R2$: Risk of loss of service to the public:

$$R2 = R_B + R_C + R_M + R_V + R_W + R_Z \tag{3.6}$$

- $R3$: Risk of loss of cultural heritage:

$$R3 = R_B + R_V \tag{3.7}$$

- $R4$: Risk of loss of economic value:

$$R4 \text{ (general case)} = R_B + R_C + R_M + R_V + R_W + R_Z \tag{3.8}$$

  $R4$ (properties where animals may be lost)

$$= R_A + R_B + R_C + R_M + R_U + R_V + R_W + R_Z \tag{3.9}$$

Formulas for risk $R1$–$R4$ are similar but values may be quite different. N_i keeps the same value for all risks $R1$–$R4$ because it is only related to physical parameters (structure dimensions, N_{sg}, etc.). However, losses L_i will take a different value for each case. Furthermore, tolerable values are different for $R1$, $R2$ and $R3$ (a given saving instead of a tolerable risk is associated to $R4$, see later). Thus, probabilities P_i may also take different values for each case and the protection means to be applied to a structure will be the combination of the protection means necessary for each of the evaluated risk $R1–R4$.

The risk components corresponding to each type of loss are also combined in Table 3.1.

Tolerable risks RT:

Typical tolerable risk values per year are given in Table 3.2. As explained previously, these values could be changed by site owner or authority having jurisdiction.

Table 3.1   Risk components associated with each type of loss[†]

Source of damage	Flash to a structure	Flash near a structure	Flash to a line connected to a structure	Flash near a line connected to a structure
	S1	**S2**	**S3**	**S4**
Risk component	R_A R_B R_C	R_M	R_U R_V R_W	R_Z
$R1$ (general case)	■ ■		■ ■	
$R1$ (structures with the risk of explosion, and hospitals or other structures where the failure of internal systems immediately endangers human life)	■ ■ ■	■	■ ■ ■	■
$R2$	■ ■	■	■ ■	■
$R3$	■		■	
$R4$ (general case)	■ ■	■	■ ■	■
$R4$ (properties where animals may be lost)	■ ■ ■	■	■ ■ ■	■

Table 3.2   Typical values of tolerable risk RT[†]

	Risk considered	RT (per year)
$R1$	Risk of loss of human life	10^{-5}
$R2$	Risk of loss of service to the public	10^{-3}
$R3$	Risk of loss of cultural heritage	10^{-4}
$R4$	Risk of loss of economic value	10^{-3}

In principle, for the risk of loss of economic value associated with risk $R4$, a cost/benefit comparison should be performed. A tolerable risk value is often not needed. As discussed earlier, sufficient data for this analysis are often not available; therefore, it may be easier to use a representative value of tolerable economic loss.

Procedure to determine the protection means to be used for risk $R1$–$R3$:

For each risk $R1$, $R2$ and/or $R3$ to be considered, the following steps shall be followed:

- calculation of the appropriate risk components R_i ($i = A, B, C, M, U, V, W$ or Z);
- calculation of the risk R as the sum of these risk components;
- comparison of the risk R with the tolerable value RT.

If $R \leq$ RT, lightning protection means are not necessary.

If $R >$ RT, lightning protection means shall be adopted in order to reduce the risk $R \leq$ RT.

In cases where the risk cannot be reduced below the tolerable level in spite of using the highest levels of protection (I), better protection means (more efficient than LPL I) or complimentary solutions such as TWS should be used.

Reduction of the risk may also be achieved by dividing the structure into zones. This allows the designer to consider the characteristics of each zone of the structure in the evaluation of risk components and to select the most suitable protection measures tailored zone by zone. A more accurate evaluation of input parameters may also be useful.

If all these methods fail to reduce the risk to a tolerable level, the site owner should be informed and the highest level of protection (i.e. I) should be used.

Procedure to determine the protection means to be used for economic risk $R4$[§]:

Whether or not there is a need to evaluate protection to reduce risks $R1$, $R2$ and $R3$, it is useful to evaluate the economic justification of protection measures by calculating risk $R4$.

The cost of loss may be calculated by the following equation:

$$\text{CL} = R4 \times C_t \tag{3.10}$$

where C_t is the total value of the structure (animals, building, content and internal systems, including their activities), and $R4$ is the risk related to the loss of value in the structure, without protection measures.

The cost CRLO of residual loss in spite of protection means may be calculated by the equation:

$$\text{CRLO} = R'4 \times C_t \tag{3.11}$$

where $R'4$ is the risk related to the loss of value in the structure, with protection measures (this risk is of course neither equal to 0 and is called residual risk).

[§]IEC 62305-2 ed.2.0. Copyright © 2010 IEC Geneva, Switzerland. www.iec.ch.

The annual cost CPM of protection measures may be calculated by means of the equation:

$$\text{CPM} = \text{CP} \times (i + a + m) \tag{3.12}$$

where CP is the cost of protection measures, including raw protection material and installation cost; i is the interest rate (to be obtained from local financial authorities); a is the amortization rate (to be based on how many years are considered until lightning protection is amortized. With a linear amortization rate over 20 years the rate is 0.05); m is the maintenance rate (to be obtained from lightning protection installer or designer).

When i, a and/or m are not known, the values given in Table 3.3 can be used instead.

The annual saving SM in local currency is then

$$\text{SM} = \text{CL} - (\text{CPM} + \text{CRLO}) \tag{3.13}$$

Protection is economically justified if the annual saving SM > 0.

It should be noted that when all parameters are not known to perform such a complete economic risk calculation, a simpler calculation of risk based on the assessment of $R4$ only and comparing it to the tolerable risk defined earlier (see Table 3.2) can be performed instead.

Formulas used to calculate risk components:

Table 3.4 summarizes the various equations used to calculate the risk components.

In Table 3.4, N_D is given for the average annual number of dangerous events due to lightning flashes to the structure; N_M is for the average annual number of dangerous events due to lightning flashes near the structure; N_L is for the average annual number of dangerous events due to lightning flashes to a line entering the structure; N_i is for the average annual number of dangerous events due to lightning flashes near a line entering the structure; N_{DJ} is for the average annual number of dangerous events due to lightning flashes to the adjacent structure; P_A is for the probability that a flash to the structure will cause injury to living beings by electric shock; P_B is for the probability that a flash to the structure will cause physical damage; P_C is for the probability that a flash to the structure will cause failure of internal systems; P_M is for the probability that a flash near the structure will cause failure of internal systems; P_U is for the probability that a flash to a line will cause injury to living beings by electric shock; P_V is for the probability that a flash to a line will cause physical damage; P_W is for the probability that a flash to a line will cause failure of internal systems; P_Z is for the probability that a flash near a line

Table 3.3   Typical values of i, a and m

Rate	Symbol	Typical value
Interest	i	0.04
Amortization	a	0.05
Maintenance	m	0.01

Table 3.4   Risk components equations[¶]

Damage	Source of damage			
	S1	**S2**	**S3**	**S4**
	Lightning flash to a structure	**Lightning flash near a structure**	**Lightning flash to an incoming line**	**Lightning flash near a line**
D1 (injury to living beings by electric shock)	$R_A = N_D \times P_A \times L_A$		$R_U = (N_L + N_{DJ}) \times P_U \times L_U$	
D2 (physical damage)	$R_B = N_D \times P_B \times L_B$		$R_V = (N_L + N_{DJ}) \times P_V \times L_V$	
D3 (failure of electrical and electronic systems)	$R_C = N_D \times P_C \times L_C$	$R_M = N_M \times P_M \times L_M$	$R_W = (N_L + N_{DJ}) \times P_W \times L_W$	$R_Z = N_i \times P_Z \times L_Z$

will cause failure of internal systems; and $L_A = L_U$ is the loss due to injury to living beings by electric shock; $L_B = L_V$ is the loss due to physical damage; $L_C = L_M = L_W = L_Z$ is the loss due to the failure of internal systems.

3.5   Probabilities related to SPDs

Most of the probabilities described in IEC 62305-2 are easy to evaluate and sound (many publications at international conferences explain and justify these probabilities). However, the probabilities related to SPDs are presently incomplete. Probabilities related to SPDs refer in fact to two probabilities:

- P_{EB}: Reducing the probability of injury to living beings by electric shock and probability of physical damage to a structure related to flashes to a connected line, depending on line characteristics;
- P_{SPD}: Reducing the probability of failure of internal systems related to flashes to a structure, the probability of failure of internal systems related to flashes near a structure, the probability of failure of internal systems related to flashes to connected line and the probability of failure of internal systems related to flashes near a connected line, when a coordinated SPD system is installed.

[¶]IEC 62305-2 ed.2.0. Copyright © 2010 IEC Geneva, Switzerland. www.iec.ch.

Table 3.5   Values of probability P_{SPD} for coordinated SPDs[//]

LPL	P_{SPD}
No coordinated SPD system	1
III–IV	0.05
II	0.02
I	0.01
SPD having better protection characteristics (lower protective level U_P) compared with the requirements defined for LPL I	0.005–0.001

The probability P_{EB} is mainly related to the SPDs located at line entrance (equipotential bonding) when the probability P_{SPD} is related to a coordinated set of SPDs (SPDs coordinated and installed to form a system intended to reduce failures of electrical and electronic systems).

P_{EB} is related to the surge withstand of the SPD located at the entrance of electrical installation (generally Type 1 SPD) that depends on the level of protection of the LPS. The probability P_{EB} mainly depends on the current I_{imp} that the SPD can withstand. This current I_{imp} can easily be determined for simple cases but may need the use of software for more complex cases.

IEC 62305-2 standard indicates that probability P_{EB} can be reduced using SPDs with better protection characteristics (higher nominal current I_n, lower voltage protective level U_p, etc.) compared to the requirements defined for lightning protection level I at the relevant installation locations. Unfortunately, no formula is proposed to calculate these lower probabilities. In addition, the key parameter for better protection is only related to the voltage protective level U_p. Using an SPD with a higher nominal current I_n will not necessarily result in a better voltage protection level.

P_{SPD} is defined in the IEC 623505-2 standard with exactly the same table as for P_{EB} and the same indication related to the possibilities to reduce the probabilities below the value defined for lightning protection I. The difference is that P_{SPD} is related to a coordinated set of SPDs (Table 3.5). SPD standards originally defined a coordinated set of SPDs as a system of SPDs that work together on a power line in such a way the lightning energy injected in the system is shared between all the SPDs so none of the SPD's energy withstand is exceeded. It concentrates on SPD survival not on equipment protection. However, an SPD may not fail but may not be able to protect either. Let us take the example of an SPD based on MOVs that is a typical case for most of Type 2 SPDs. Two currents may characterize such an SPD:

- its nominal discharge current I_n, used to define the voltage protection level U_p;
- its maximum discharge capability (usually called I_{max}).

When the current flowing through the SPD has a magnitude between these two values, the voltage protection level will be exceeded. Thus, the SPD will survive but not the piece of equipment supposed to be protected.

Presently, a better definition for SPD coordination exists that relates to the SPD voltage protection level. The voltage at the terminals of each SPD of the coordinated system should remain below its voltage protective level. It is much more stringent because the current should then remain below the nominal discharge current I_n for Type 2 SPDs when, for energy coordination, the current in each SPD need only remain below its maximum discharge current.

The present description of P_{SPD} (or P_{EB} because it has an identical table in IEC 62305-2 standard) is insufficient as it relates only to energy sharing. It mentions the possibility of reducing P_{SPD} for an SPD providing a better protection level, but only for a few cases. The SPD should first have the surge withstand capability defined by level of protection I. Unfortunately, the standard fails to provide a formula to calculate this lower probability. The problem is due to the definition of U_p, the voltage protection level of the SPD, U_p. This value of U_p is based on the measurement of two parameters (when applicable, depending on internal protective components inside the SPD): the residual voltage and the sparkover voltage. For a spark gap, the appropriate parameter is the sparkover voltage when for a MOV the appropriate parameter is the residual voltage. If the SPD installed at the entrance of the installation is a MOV type, the protective level is only defined by the residual voltage U_{res} at its nominal discharge current I_n. However, the curve U vs I of a varistor is non-linear (see Chapter 5). For a typical varistor, the residual voltage at I_n (5 kA) is 1.5 kV when the residual voltage at a higher current (e.g. 12.5 kA) may be up to 2 kV.

We will use the example of an LPS with lightning protection level III (i.e. max current of 100 kA) to protect a building. Equipment inside the building deserves protection because it serves an important role for building safety. It has a surge withstand of 1.5 kV and is supposed to be protected by a MOV-type SPD with four internal varistors connected between active conductors and the earth terminal (three phases plus neutral, TT system) (Figure 3.5).

Note: Active conductors are phase and neutral conductors by opposition with earth (PE conductor). Live conductors refer to phases by opposition to neutral.

In case the maximum lightning current occurs (100 kA), the current inside the power system will be 50 kA (50% being dispersed in the earthing system), which results in 12.5 kA in each varistor. The voltage at SPD terminals will be 2 kV,

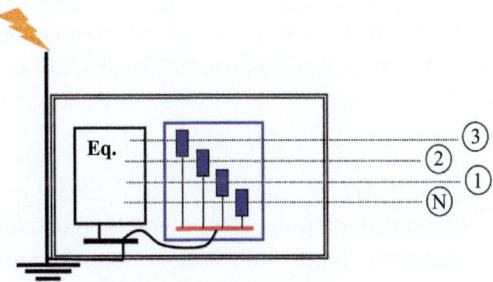

Figure 3.5   Three phases and neutral SPD Type 1 MOV-based protecting an equipment (Eq.)

which is higher than the withstand of the equipment. In that case, the equipment would fail.

The current inside each varistor must remain below or equal to 5 kA to provide adequate equipment protection. This means a total current inside the SPD of $4 \times 5 = 20$ kA and a current inside the LPS of $2 \times 20 = 40$ kA (still assuming 50% of the lightning current flows through the earthing system). Any time the lightning current to the LPS exceeds 40 kA, there is a probability of failure of the sensitive equipment.

According to IEC 62305-1, the probability of having a positive lightning current (related to Class 1 test, i.e. 10/350) of 40 kA is 45%. As there are 10% of positive flashes, this relates to a probability of 4.5%, rounded to 5%. For simplicity, we consider that equipment to be protected will fail due to insulation breakdown and for that reason we consider only long duration surges (10/350), but it must be considered that the failure rate of equipment may be much higher than evaluated if we also consider steep front and short duration surges. Any current that exceeds 5 kA may lead to an equipment failure because, for electronic components, the surge voltage applied is as important as the surge duration.

The probability P_{SPD} should then be based on the combination of the two following events:

- either the SPD fails (it is the probability P_{SPD} existing in present IEC 62305-2 standard, i.e. in our case 5% associated to LPL III)
- or the SPD does not fail but does not protect the equipment (that probability is also 5%, as determined just earlier)

The probability P_{SPD} depends on these two events:

- the probability $P1$ that the value of the expected current, associated with the current flowing through the SPD at its point of installation, exceeds the current tolerated by the SPD;
- the probability $P2$ that the value of residual voltage at SPD terminals exceeds the required protection level U_p.

Then the probability P_{SPD} is given by:

$$P_{SPD} = 1 - (1 - P1)(1 - P2) \tag{3.14}$$

P_{SPD} should then be equal to $9.75\% = 1 - 95\% \times 95\%$, rounded to 10%. P_{SPD} should then be 10% instead of the 5% proposed by the standard.

Of course, this is an extreme case based on a single SPD performing both the equipotential bonding function and protection function. In most of the cases, the presence of many cascaded SPDs will help reduce this problem.

However, the tables currently proposed in IEC 62305-2 are incomplete as they provide no formula for calculating the probability P_{SPD} and they are confusing by restricting the use of a better probability only to the case where the current withstand is greater than the one defined for level I. It also infers that only two SPDs that are energy coordinated will solve the problem, when in general there could be many SPDs on a single line (e.g. one SPD at installation entrance – main panel board, one SPD in one or more subsidiary panels and finally an SPD in front of sensitive equipment or in front of UPS). All these SPDs should be coordinated in such a way that the current injected in the downstream SPD in front of sensitive equipment will be much further reduced, providing a much better voltage protection than the U_p written on the SPD nameplate. At the moment, there is no way in the IEC 62305-2 standard to benefit from such a frequent installation. However, this could be allowed in Edition 3 of the standard.

3.6 Specific case of LV electrical installation

The international standards committee in charge of electrical installations of buildings has developed a simplified version of the risk assessment for its own needs based on IEC 62305-2 standard. This simplified method is described in IEC 60364-4-44 standard and IEC 61643-12 standard.

For this simplified method, only the economic risk ($R4$ in its simplified form, i.e. with a tolerable risk value without calculating of the savings) is calculated and only surges on the lines (induced and direct) are considered. In this method, the calculated risk level (CRL) is assessed to determine whether SPDs are required at the entrance of the installation or not.**

Note: The simplified method also includes a list of cases where SPDs are necessary (without calculation) and a few cases where the simplified method cannot be used and only 62305-2 is applicable. Only the complete method described in IEC 62305-2 standard is to be used in the next cases and as a consequence, there is no need to apply the simplified method for

- *structures with the risk of explosion,*
- *structures where the damage may also involve surrounding structures or the environment.*

Protection against transient overvoltage shall be provided for buildings where surges affect

- *care of human life, e.g. safety services, medical care facilities;*
- *public services and cultural heritage, e.g. loss of public services, IT centres, museums;*
- *commercial or industrial activity, e.g. hotels, banks, industries, commercial markets, farms.*

Note: This simplified method uses the term 'overvoltage' and not 'surge' but the stresses considered by this method cover both the overvoltages induced on lines by a nearby lightning strike and a direct lightning strike to the line. In the case of induced overvoltage, the stress is mainly a voltage generated along the line that creates a mild surge current due to the line impedance. In the case of a direct strike to the line, the generated stress is mainly a surge current that will create an

---

**IEC 61643-12 ed.3.0 "Copyright © 2020 IEC Geneva, Switzerland. www.iec.ch"

overvoltage due to the line impedance. In both cases, the stress taken into account for the SPDs design will be a surge current with waveshape 8/20 μs for induced surges and 10/350 μs for partial direct lightning current; both of them being current waveshapes and not voltage waveshapes. This is why, in general, the term 'surge' is more appropriate and more general.

Secondary SPDs and coordinated sets of SPDs that are important for protecting sensitive equipment are not defined by this simplified calculation. It is assumed that if an SPD is needed at the entrance of the installation, other SPDs downstream, called coordinated SPDs, will be used to protect specific equipment.

This method has been defined mainly to allow an easy determination of the need of an SPD, providing mainly a safety function. There is no need to perform this simplified risk calculation if the complete method defined in IEC 62305-2 is applied. As it is aimed to be simple to use, many assumptions have been made and parameters fixed. The number of remaining parameters to determine in performing this simplified calculation is then limited: mainly ground flash density and length of lines.

Note: Ground flash density is now often replaced by lightning ground strike point density that corrects for actual ground strike points but this new concept will take some time to be introduced in new editions.

CRL is given by the following formula:

$$\text{CRL} = \frac{f_{\text{env}}}{(\text{LP} \times N_g)} \tag{3.15}$$

where f_{env} is an environmental factor and LP is given by the following formula:

$$\text{LP} = 2\,\text{LPAL} + \text{LPCL} + 0.4\,\text{LPAH} + 0.2\,\text{LPCH} \tag{3.16}$$

where LPAL is the length in km of low-voltage overhead line; LPCL is the length in km of low-voltage underground cable; LPAH is the length in km of high-voltage overhead line and LPCH is the length in km of high-voltage underground cable (Figure 3.6).

The total length (LPAL+LPCL+LPAH+LPCH) is limited to 1 km or by the distance from the first surge protection point installed in the power network (either

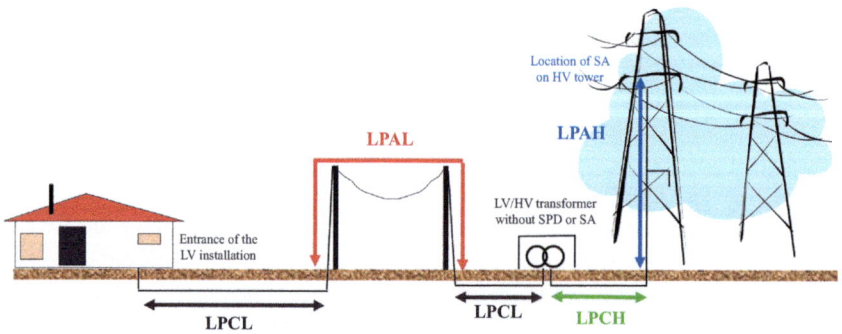

Figure 3.6   Illustration of an installation showing the lengths to consider

Table 3.6   Environmental factor f_{env}

Environment	f_{env}
Rural and suburban environment	85
Urban environment	850

SPD for LV networks or Surge Arrester for HV networks) to the entrance of the installation whichever is the smaller. If the distribution networks' lengths are totally or partially unknown, LPAL shall be taken to be equal to the remaining distance to reach a total length of 1 km (e.g. if the building is supplied by an underground cable connected to an overhead LV line and then to an HV underground cable and the only length that is known is the length of the underground cable, being 100 m, then LPAL shall be considered equal to be 900 m). It is assumed that when the line length is longer than 1 km, the surge will be reduced and not harmful any longer. LPAL is the most important parameter because overhead LV lines will collect more direct strikes and induced surges than underground LV cables. In addition, HV overhead lines are also prone to numerous lightning impacts but when associated with an HV/LV transformer, the stress is reduced by 80% on the LV lines. This explains the coefficient used for each line length in the LP formula.

Rural and urban environments are in general clearly understood (Table 3.6). In fact, it is related to the density of houses and buildings. A suburban environment can be considered neither a city nor a rural area, and suburbs with mainly houses can fit in that category. When there is a significant number of tall buildings, the environment can be considered a city. When the distance is large between houses, environment can be considered rural. But a dense village, for example, would fit into the suburban category. Of course, in the case of doubt, it is wiser to use the most severe case that is rural and suburban environment.

As soon as CRL is calculated, it must be compared to 1,000. If CRL $\geq$ 1,000, no SPD is needed but if CRL $<$ 1,000, an SPD is required at the installation entrance.

Let us apply the method to a farm in a place where $N_g = 1$, with the following hypothesis: the farm is supplied by an LV underground cable 400-m long to the limit of the property and from that point an LV overhead line 600-m long is connected to an LV/HV transformer protected by an SA.

Then it comes
LPAL $= 0.4$ km
LPCL $= 0.6$ km
LPAH $=$ LPCH $= 0$ km
and LP $= 2\times0.4+0.6 = 1.4$
CRL $= 85/(1.4\times1) = 60.7$
and SPDs should be installed at the LV line entrance in the farm.

But if we consider a building in a city in a place where $N_g = 1$ and supplied by an HV underground cable with an HV/LV transformer directly inside the building.

Then it comes

LPAL = LPCL = 0 km

LPAH = 0 km

LPCH = 1 km

and LP = $0.2 \times 1 = 0.2$

CRL = $850/(0.2 \times 1) = 4{,}250$

SPDs are not necessary to provide building safety.

An interesting case is an electric vehicle supply installation made of many charging pedestals powered by a main charging cabinet. This cabinet includes an HV/LV transformer. The installation is located in a suburb of a city where $N_g = 1$. The length of the HV underground cables is 1 km. There is no surge protection on either the HV side or the LV side of the transformer. The charging stations are supplied by LV underground cables for a total of 100 m.

Then it comes

LPAL = 0 km

LPCL = 0.1 km

LPAH = 0.9 (because the total length is limited to 1 km)

LPCH = 0 km

and LP = $0.1 + 0.2 \times 0.9 = 0.28$

CRL = $85/(0.28 \times 1) = 303.5$

SPDs should be installed at the LV line entrance of the charging station (at the charging main cabinet).

If protection of any pedestal is important, the SPD installed at the main cabinet will not be enough because its main task is to ensure safety of the installation. It is then advised to install additional SPDs at the pedestals.

IEC standard 61643-12 presents an interesting comparison between the simplified version presented earlier and the complete 62305-2 method (economic risk $R4$) for three buildings inside a chemical facility.

The buildings are the industrial power station that supplies the site, the chemical unit itself (assuming there is a single unit) and the administrative building. The result is summarized next.

An HV overheard line 500-m long distributes power to the power station through a 100-m underground HV cable. The power station is supplying LV power to an administrative building and to the chemical unit. The area is rural with $N_g = 1$.

As the site is an industrial site, LV SPD should be provided at the LV side of power station because it is one of the cases where SPDs at the entrance of an installation is mandatory. The power station is the entrance point for that installation. As the chemical unit may have impact on the environment, IEC 62305-2 should also be used for the chemical unit instead of the simplified version. The simplified risk method applies in fact only to the administrative building.

Table 3.7 is the result of the comparison (for more details on the related calculation please refer to IEC 61643-12 standard Annex H).[††]

[††]IEC 61643-12 ed.3.0 "Copyright © 2020 IEC Geneva, Switzerland. www.iec.ch"

Table 3.7   Comparison for a chemical plant between simplified version and 62305-2 complete method

	Power station	**Chemical unit**	**Administrative building**
		Simplified method	
SPD	Required	Required	Not required
Note	SPD is required because of industrial activity	SPD is required because of industrial activity with possible risk for the environment	SPD to protect specific equipment such as UPS, computer or telecom exchange may still be needed for production
		Complete method (62305-2) risk $R4$	
SPD	Required	Required	Not required

In spite of being based on different parameters and calculations, in this particular case, the complete and the simplified methods are consistent for the economic risk. This is not surprising because the simplified version is derived from the complete one. However, applying the complete method for the human and environmental risk ($R1$) or for the loss of service ($R2$ extended) may lead to different results and especially may lead to the implementation of an LPS on the chemical unit and therefore different types of SPDs plus possibly additional SPDs.

3.7   Specific case of photovoltaic systems

The committee in charge of electrical installations for PV plants has also developed a simplified method derived from 62305-2 to suit its needs.[‡‡] For this simplified method, only the loss of service to the public risk ($R2$) is concerned because it is the first goal of a PV installation to produce electricity. Only induced surges on the DC lines are calculated. Inducted surges are the predominant stress for PV installations. For AC circuits, the simplified method presented earlier applies when necessary.

In this method, the risk assessment is based on the evaluation of the so-called critical length L_{crit} and its comparison with the length L. As before, the method relies on N_g and line length.

SPDs shall be installed on the DC side of the installation when $L \geq L_{crit}$. Where:

- L is the cumulative length between inverters and the furthest point of PV modules in a chain, considering each of the paths. In the case of many inverters, the length to consider is the sum of the length L for each inverter (Figure 3.7).
- L_{crit} (m) depends on the type of PV installation (Table 3.8).

[‡‡]IEC 60364-7-712 ed.2.0 "Copyright © 2017 IEC Geneva, Switzerland. www.iec.ch"

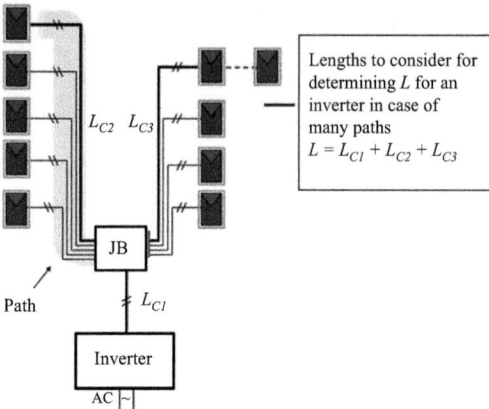

Figure 3.7   Method for determining length L. JB, junction box

Table 3.8   Critical length L_{crit}

	PV installation on the roof of a building	**PV farm**
L_{crit} (m)	$115/N_g$	$200/N_g$

The original study listed in bibliography considers more cases. Especially a difference is made between a house and an office building.

Table 3.9   Length of line L above which a coordinated set of DC SPDs is needed

Case	**House**	**Office building**	**Farm**
		Length L in m	
$N_g = 1$	115	450	200
$N_g = 2$	58	225	100
$N_g = 2.5$	46	180	80
$N_g = 4$	29	113	50
$N_g = 10$	12	45	20

It is possible to determine for each N_g, the minimum length that leads to the need of a DC SPD, based on the length derived from typical cases (see bibliography for more details). This minimum length is reported in Table 3.9.

Example of application of this simplified method: a PV farm supplying a large building (Figure 3.8).

The PV farm is crucial as it is supplying a large portion of the power needed by the building, with only a small part provided by the power utility network. The PV panels are not motorized to follow the sun (no tracking) and the only part to be protected against surges is then the DC circuit. The panel producer has already installed T2 SPDs in the Junction Boxes. There is a single but large inverter. The

Figure 3.8   PV farm installed near the parking lot used in the example

loss of a panel is easy to manage but the loss of the inverter means the total loss of production. The question is whether an SPD should also be installed on the DC side of the inverter. The inverter is located in a nearby building. The distance between PV panels and the inverter is not substantial, and the critical length will be obtained from each PV chain to the closest junction box. There are six chains and some of them are very long. The cumulated line length $L = 550$ m. The ground flash density $N_g = 1$ flash/year/km^2. The total length L is greater than the 200 m identified in the previous table for a PV farm when $N_g = 1$. The simplified method recommends the use of a coordinated set of Type 2 SPDs. The installation of Type 2 DC SPDs specific for PV applications (PV SPDs) at the DC input of the inverter is justified and the decision by the manufacturer to install Type 2 PV SPDs in JBs is also confirmed.

3.8   Specific case of street lighting

Street lighting systems can be damaged by lightning occurring in the vicinity. This damage can occur as a result of:

- direct strike to a pole or luminaire,
- surges resulting from a strike to power lines feeding the circuits powering the lighting system,
- induced surges on underground power lines (that is the most likely scenario).

A simplified risk assessment method has been developed to determine the benefit of using SPDs on lighting circuits based on the lightning ground flash density, type of lighting technology used and length of cabling of the lighting system. This case will serve as an example of how it is possible to determine a simplified risk method derived from the general requirements of IEC 62305-2 method, when a particular risk is studied. This is especially true when SPDs are concerned because a limited number of parameters are needed, mainly the line length and the ground flash density.

In many cases, the pole and associated lighting circuits struck by lightning are lost and a lot of problems will then be generated on other poles. Thus, the main concerns are the surges generated on the circuit (induced surges) and the risk of loss of service to the public ($R2$) defined in IEC 62305-2. However, as already discussed the assessment of loss of service to the public is based on the number of users not served. This does not work well for the street lighting as the number of users is less relevant than the amount of time the service is not available. As $R2$ is not fully relevant, the assessment of loss of service related to overvoltages in street lighting lends itself well to the use of the frequency of damage concept proposed for introduction in edition 3 of IEC 62305-2 and already presented earlier in this chapter. In this concept, the frequency of damage is equal to the number of dangerous events multiplied by the probability of damage ($F = N{\times}P$). When no protection is provided, P is equal to 1. The calculated frequency of damage should be simply compared to the tolerable frequency of damage that in this case will be set by the authority having jurisdiction.

A reasonable simplification formula for calculating the number of damages per year generated by induced surges on lightning circuits has been presented (see bibliography) and is given by the following formula:

$$N_{\text{damages}} = N_g \times 0.015 \times L_T \times P_{\text{li}} \tag{3.17}$$

where N_g is lightning flash ground density in strikes/year/km^2, L_T is total cumulated length in m of the power supply lines from the Main Lighting Panel Box (MLPB) (see Figure 3.9), and P_{li} depends on surge withstand of lighting circuit as given in Table 3.10.

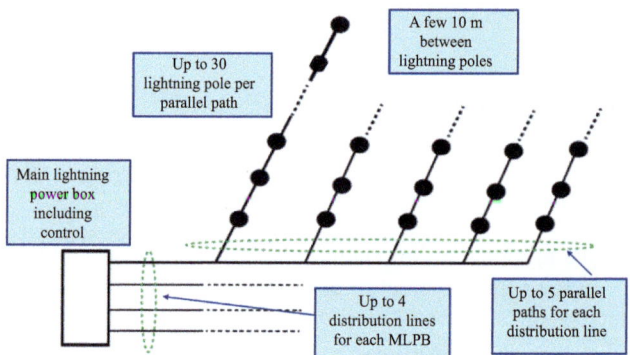

Figure 3.9   Typical layout for lightning poles power supply

Table 3.10 P_{li} derived from IEC 62305-2

Withstand voltage U_w in kV	Typical case	P_{li}
1.5	Sensitive device (sensitive electronic driver or LED circuit)	0.6
2.5	Standard device (electronic driver or LED circuit)	0.3
4	Strong device (discharge lamp, magnetic ballast)	0.16
6	Self-protected (driver with built-in surge-protective device)	0.1

The justification for using Type 2 SPDs to protect lighting circuits and the MLPB is obtained by comparing N_{damages} (estimated number of damages per year) to the tolerable frequency of damage. The tolerable frequency of damage may differ from one user to another.

Bibliography

IEC 62305-2 (currently edition 2), 2010. "Protection against lightning – Part 2: Risk management".

IEC 60364-4-44:2007+AMD1:2015+AMD2:2018, January 2018. Electrical installations of buildings – Part 4-44: Protection for safety – Protection against voltage disturbances and electromagnetic disturbances.

IEC 60364 Part 7-712, April 2017. Low voltage electrical installations – Part 7-712: Requirements for special installations or locations – Solar photovoltaic (PV) power supply systems.

IEC 61400-24, July 2019. Wind energy generation systems – Part 24: Lightning protection.

Alain Rousseau, Fernanda Cruz, Sébastien Sarramegna, Semi Taofifenua, and Wacapo Taine "Lightning risk evaluation – Field experience" 33rd ICLP 2016.

Alain Rousseau, Fernanda Cruz, Stéphane Pedeboy, and Stephane Schmitt "Lightning risk: how to improve the calculation?" CIGRE International Colloquium on Lightning and Power Systems ICLPS 2019.

Alain Rousseau and Mitchell Guthrie "Lightning risk assessment for street lighting systems" 34th ICLP 2018.

Alain Rousseau and Mitchell Guthrie "Direct lightning protection risk assessment on PV Systems" 10th APL 2017.

Alain Rousseau and Mitchell Guthrie "Risk assessment of lightning-induced surges on PV Systems" CIGRE ICLPS 2017.

Alain Rousseau, Céline Sainte-Rose-Fanchine and Mitchell Guthrie "Application of environmental risk according to IEC 62305-2 Edition 2" XIII SIPDA (International Symposium on Lightning Protection) 2015.

Christian Bouquegneau and Alain Rousseau "Lightning strike-point density for risk assessment" CIGRE ICLPS 2014.

Mitchell Guthrie, Alain Rousseau, Jacob Struck and Josephine Covino "A proposed methodology for risk assessments for temporary lightning exposures" DDESB 2010.

Alain Rousseau and Alexander Kern "How to deal with environmental risk in IEC 62305-2" 32nd ICLP 2014.

Alain Rousseau "Surge protective devices: risk or not ?" 26th ICLP 2002.

Chapter 4

Standard environment

Alain Rousseau[1]

Standards are often late compared to the market needs and manufacturers development of new products to meet those needs. It is typical that 3–5 years are needed to develop and publish a standard. The result is that technology will generally lead standards in terms of what is available in the marketplace. On the other hand, standards are a good indication of accepted norms for the technology that are currently established. Surge-protective devices (SPDs) are used in a number of increasing applications as new technologies are developed. An example was the rapid expansion of photovoltaic (PV) applications that led to an urgent demand for SPDs tailored to the application. It was urgent that standards be developed to address safety and performance tests for SPDs designed to be installed in the PV installations. This led to the development of product and application standards now published as IEC 61643-31 and IEC 61643-32, respectively. In the same way, the growing need for SPDs connected to DC power circuits and equipment for protection against indirect and direct effects of lightning or other transient overvoltages justified the current development of IEC 61643-41.

There are many International Electrotechnical Commission (IEC) standards that incorporate the use of SPDs into their standards. In addition to the series of SPD standards (IEC 61643), IEC standards that incorporate SPDs are lightning protection standards series (IEC 62305) and the electrical installation standards series (IEC 60364). Specific standards also address and specify SPDs for PV installations (IEC TR 63227) and wind turbines (IEC 61400-24).

It is sometimes difficult for a user to know which standard should apply. These standards are not always aligned because they are not all published at the same time. They do not target either the same type of applications. There are procedures at various levels to harmonize these standards:

- At the IEC level: Technical Committees (TCs) identify those other TCs that are users or providers with interest in the content of each standard, Technical Report, or Technical Specification that are produced by the TC. The interested TCs have full access to all documents during the development process and can submit comments for consideration.

[1]SEFTIM, Vincennes, France

- At the TC level: SC37A, TC64 and TC81 have established an advisory joint working group (AJWG), to specifically address coordination of efforts and resolve potential conflicts. The AJWG meets only when necessary and provides liaison reports during plenary meetings. This AJWG does not meet often, good proof of a solid common basis for the standards.
- At the member level: Both National Committees and their working group members know all these standards or even participate to the creation of these standards and perform a smooth alignment of all the standards.

When a standard is updated, all the standards that reference the original version or specific clauses of that standard may no longer be aligned, at least for a certain period until which time the referencing standard is updated. It is then important to identify the current standard and which is the most applicable for a type of user.

Note: User, in that chapter, refers to the user of a standard. In the specific case of SPDs, it may be a standard technical committee, a laboratory, a design office, an industrial designer, an electrical contractor or anybody who uses a standard to correctly select, install, use or control an SPD.

It is important for users that all standards dealing with surge and lightning protection are consistent. If we consider AC SPDs, they are described primarily in IEC 61643-11, but also in IEC 62305-4, in general standards such as IEC 60364 for electrical installation or specific standards such as 61400-24 for wind turbines and IEC TR 63227 for PV applications.

When electrical contractor designs the electrical distribution system for a structure or a panel builder designs his panel, they will likely begin the design based on the electrical loads anticipated and the source of power to be used. They should also consider the potential electromagnetic environment the structure may experience, especially in the design of surge protection. When SPDs are used, it will generally be driven by the requirements of IEC 60364 or a national equivalent unless other factors are identified. If the installation is based on IEC 60364 applications, the SPD installed will probably be to protect against induced surges (Type 2). If the electrical distribution system is designed and installed before a risk assessment to determine whether a lightning protection system (LPS) should be installed and it is later determined lightning protection is required, a Type 1 SPD will be required. The result will be additional costs to either preplace the Type 2 SPD in the main electrical panel for the building or install the Type 1 SPD in a separate panel should the necessary space be unavailable in the existing panel. If the SPD is not changed, this can create a building loss when a lightning strike occurs because Type 2 SPDs are not able to handle the stress and this may cause a significant damage or even a fire. The lightning risk assessment will also determine the surge withstand of the required Type 1 SPD. Consistency is required between standards but it is also necessary to understand bridges between standards to draw the attention of the users on the fact other standards may apply as well.

A good example of coordination between TCs is the risk assessment for SPDs. In that context, the question to be answered is 'do I need an SPD?'. The pilot for risk assessments for structures is the IEC lightning protection committee (TC81). However, the committees that deal primarily with SPDs (Electrical installations and protection against electric shock (TC64) and low voltage SPDs (SC37A)) needed a specific simplified risk assessment only for SPDs and not for the complete structure. IEC SC37A developed a basis for such a simplified risk version, and IEC TC64 started to develop a simplified method. It was then quickly agreed that TC64 would base its simplified risk method on the TC81 assessment method and refer to the TC81 method when appropriate. SC37A stopped its own work and referenced, when necessary, the TC64 simplified method and TC81 complete method where a more complex assessment is necessary.

Note: TC stands for Technical Committee and SC for Sub-Committee. A Sub-Committee is related to another TC. For SC37A, it is related to TC37 in charge of high-voltage surge arresters. There is another Sub-Committee under TC37 umbrella: SC37B dealing with components for SPDs.

The IEC TC81 risk method changed from edition 1 to edition 2 and will further evolve in a 3rd edition. There is a difference in the revision schedules for the simplified method developed by TC64 and the more detailed complete method developed by TC81. In some applications, different revision periods could create a conflict but this is not the case here because the use and scope of the assessments are different.

Following is an explanation of the organization of the three main series of standards for user clarity's sake:

- TC81 addresses lightning protection of structures. These structures generally include electrical installations. Lightning threats to a structure includes the following:
  - direct strike to the structure: the main protection means is the LPS;
  - induces surges inside the structure due to a nearby strike: the main protection means include SPDs and shielding;
  - direct strike to an incoming line that propagates to the building installation: the main protection means are SPDs;
  - induced surges on an incoming line that propagates to the building installation: the main protection means are SPDs.
- TC64 addresses electrical installations and SPDs are a component of this installation. SPDs provide a safety function avoiding fire, sparkover and dangerous voltages and provide protection of assets by avoiding electrical damages.
- SC37A produces standards for testing SPDs and for SPD selection and application.

Focusing only on SPDs

TC81: Allows risk calculation (IEC 62305-2 standard) to determine when SPDs are needed in the complete protection plan (when an LPS is installed, SPDs are always necessary at the installation entrance for equipotential bonding between the electrical installation and LPS earthing system, but other SPDs may be needed to protect specific parts of the installation or specific equipment). When SPDs are needed for an equipotential function or for a protection function, the rating of the SPDs can be determined using IEC 62305-3 and IEC 62305-1 standards for equipotential function and using IEC 62305-4 standard for protection. A few application rules are summarized in an annex of 62305-4 but when more details are needed, it refers to 61643-12 or 61643-22 standards.

TC64: Provides a simplified risk calculation (60364-4-44 clause 443) for an electrical contractor to assess the use of SPDs in the electrical installation. In a few cases, the complexity does not allow the use of the simplified method and the standard refers to IEC 62305-2. When an SPD is needed, installation rules can be found in IEC 60364-5-53 clause 534 and when more detailed information is needed on SPD application rules, this standard refers to IEC 61643-12.

SC37A: Provides a series of SPD standards for AC SPDs (IEC 61643-11), data/signal SPDs (IEC 61643-21), PhotoVoltaic SPDs (IEC 61643-31) and DC SPDs (future IEC 61643-41). Common requirements and tests for SPDs will soon be published as standard IEC 61643-01, to consolidate general requirements and test methods and avoid repetition between various standards of the 61643 series. SC37A also produces a series of application standards: application of AC SPDs (61643-12), application of data/signal SPDs (IEC 61643-22), application of PV SPDs (IEC 61643-32) and possibly in future application of DC SPDs (IEC 61643-42) or even general application clauses grouped in a single standard (IEC 61643-02). SC37A refers to the IEC 62305 series of standards when direct lightning protection is considered and refers to the IEC 60364 series for installation rules of AC SPDs but brings more details on physical rules and phenomena as well as detailed application principles that are then used by TC64 and TC81 to produce their own standards.

A few typical users and their main and secondary needs in terms of standards are listed in Table 4.1:

Standards in the field of electricity are generally developed at international level IEC or for telecom needs ITU (International Telecommunication Union).

Other entities develop regional standards such as CENELEC for the European market or ETSI (European Telecommunications Standards Institute) for telecommunications. They may be the adaptation of IEC standards to the CENELEC needs or standards developed specifically for CENELEC.

IEC standards are numbered 6xxxx. Standards that are existing only at CENELEC level are numbered 5xxxx. When regional or national standards development organizations (SDO) adopt the IECs as their own, they are given designations that include the SDO. Examples are EN IEC XXXXX for CENELEC and ANSI/IEC XXXXX for the American National Standards Institute in the United States. In some countries, the IEC standards may be incorporated into existing national standards rather than adopting the entire IEC standards as published.

For example, the standard for AC power SPDs testing is numbered as IEC 61643-11 and the CENELEC version is numbered EN IEC 61643-11. CENELEC versions may contain changes compared to the IEC version. While CENELEC countries are required to apply the CENELEC standards, IEC members can decide to adopt the published standard (generally translating the text in their own language) or not or to incorporate a part of the IEC requirements in their national standards.

Other organizations also developed standards for a typical market such as UL.

In Europe, it is very rare that a standard is developed only at national level but, of course, outside Europe there are many specific national standards.

IEC also produces Technical Specifications and Technical Reports. A Technical Report is for information and a Technical Specification is generally

Table 4.1   Application of standards for typical users

Type of user	62305 series	60364 series	61643 series
Laboratory			61643-01/11/21/31 or 41 for requirements and tests
Building designer	For LPS need and design	For electrical installation rules	
Electrical contractor	Only if simplified risk method does not apply or if SPDs are not mandatory	For SPD need and installation rules	When more detailed installation rules make it necessary
Inspection office	For LPS compliance	For SPD installation rules compliance	For compliance with SPD standard 61643-x1 by marking on product or data sheet
Design office	For lightning protection need and design	For electrical installation rules and basic SPD installation rules	For detailed SPDs application rules
Standard committee	For lightning protection	For electrical installation	For SPDs detailed requirement and application rules
Insurance company	For lightning protection need and compliance	For SPD installation rules compliance	

produced for products or technology that is still under development and not mature enough to be a standard. Technical Specifications are subject to review not later than 3 years after publication and will become a standard or disappear after a sufficient period of review.

CENELEC follows the same path with standards (EN) and Harmonized Documents (HD) or Technical Specifications (CLC TS). This is mainly the case for installation rules that vary from one country to another and a standard is then not possible. Application of an HD in a European country is less stringent than a standard and a country may decide to adapt the HD when adaptation of an EN standard is not accepted. HDs are often incorporated in national existing standards and do not exist at the national level with the same number. CLC TS have almost the same meaning at the CENELEC level as then TS at the IEC level. For that reason, standard IEC 60364 exists in Europe as HD 60364 and the application guide for 61643-11 is a standard at IEC level when it is a CLC TS at the CENELEC level that countries may adopt or not.

This book mainly concentrates on the IEC rules because of the wider audience. On a case-by-case basis, specific standards from other organizations will also be discussed.

Note: CENELEC is a non-profit organization, which is the association of the National Electrotechnical Committees of European countries. There are presently 34 countries member of CENELEC. IEC is a non-profit organization that brings together more than 170 countries worldwide.

4.1   Surge-protective devices

The IEC 61643 series currently provides six detailed standards for power, data/signal SPDs and PV SPDs. The product standards deal primarily with safety but also include performance requirements and tests. The application guides provide the basic rules for the selection of these SPDs that are then used by IEC 62305-4 and 60364. The application guides also provide many annexes for detailed studies by engineers looking for specific applications of SPDs or a better understanding of the key parameters such as SPD coordination (for which 61643-12 has a pilot function) and the selection of SPD disconnectors.

The list of published standards regarding SPDs is as follows:

IEC 61643-11: 'Surge protective devices connected to low-voltage power systems – Requirements and test methods'

IEC 61643-12: 'Surge protective devices connected to low-voltage power distribution systems – Selection and application principles'

IEC 61643-21: 'Surge protective devices connected to telecommunications and signaling networks – Performance requirements and testing methods'

IEC 61643-22: 'Surge protective devices connected to telecommunications and signaling networks – Selection and application principles'

IEC 61643-31: 'Requirements and test methods for SPDs for photovoltaic installations'

IEC 61643-32: 'Surge protective devices connected to the d.c. side of photovoltaic installations – Selection and application principles'

The 61643-11 has been recently completely redrafted to combine tests and pass criteria as much as possible. The purpose was to limit the number of tests and thus the cost of testing while providing a high level of safety, reliability and efficiency for the SPDs.

The reference test voltage (U_{ref}) has been introduced depending on the type of network (TT, TN, IT, etc.) and how the SPD is connected (phase-earth, phase-neutral, etc.). Specific values have been introduced for temporary overvoltages applicable in a few countries (e.g. the United States, but also Japan). I_{max} that was an important SPD parameter in the past is now an optional test. The main technical modification deals with safety and primarily the short circuit withstand rating. Previous tests concentrated mainly on thermal runaway and short circuit current (the test focused on the internal wiring and especially high short circuit current, replacing the active part by a metal block). In many cases, in practice, the short circuit current will be low and appear slowly as a degradation process. It is possible that previous tests did not cover all the possible failure modes, even though some of these cases may be rare. A new test has then been introduced to try damaging the

SPD in a repeatable way and use this damaged SPD for the new short circuit current test.

This concern is especially important for the PV SPDs where the short circuit current may be quite low when the sun radiation is low on the PV arrays. IEC 61643-31 has included such a test.

A new test has been proposed in CENELEC to address possible short impulses generated by the electronic control super-imposed on the AC voltage for wind turbines.

There are few modifications anticipated for the telecommunications and data/signal SPD standards because these standards are mature and meet the need of the stakeholders (users, manufacturers, laboratories, etc.).

A very helpful table associated with the application of AC SPDs has been introduced in IEC 61643-12. This table gives the surge withstand of fuse (both with 10/350 μs impulse and 8/20 μs impulse) facilitating the user's task for coordination between the SPD, its disconnector (when a fuse) and upstream short circuit protection. This is one of the most critical SPD applications even if damages to SPDs are very rare thanks to robust test standards. The same table should be developed for circuit breakers to complete the disconnector picture.

To address the challenge of selecting the SPD disconnector, a new SPD Specific Disconnector project was launched. It is not clear whether this will lead to a standard, but the technical discussions around this topic provide a rather good background to understand the SPD disconnector specification and selection (see Chapter 7). Another topic receiving interest is 'smart' SPDs. As products of this type already exist in the market, it is necessary to investigate whether this may lead to a new standard and, if so, provide a definition of what constitutes a smart SPD. Some oppose the use of the term 'smart' SPD arguing that all SPDs are smart and the present term is SPD plus additional functions or SPD monitoring device (see Chapter 9). There have been discussions regarding the need to standardize multi-impulses surge current stress on SPDs without a decision (see Chapter 9).

As discussed previously, most of the SPD standards are developed at the IEC level first and then adapted for the CENELEC market. On a few occasions, specific standards are developed directly at the CENELEC level. This was the case for the EN 50539-11 standard that appeared in Europe for PV SPDs before a standard was developed at IEC level under number 61643-31 (EN 50539-11 is now replaced by EN IEC 61643-31). Normally, there are few modifications between the CENELEC and IEC versions of standards but in a few circumstances, such as for the application rules, the two versions may differ. In practice, the phenomena described in IEC version are valid for the CENELEC version. There could be national or regional differences that justify a few modifications. There can also be cases such as IEC 61643-11 where an amendment that has been developed in Europe for EN 61643-11 has additional requirements for portable (pluggable) SPDs that do not exist at IEC level.

Note: The main changes provided by this amendment to EN 61643-11 relate to the fact that in such SPDs, phase and neutral should be considered in the same way. In addition, for SPDs with a protection mode connected to PE, this protection mode shall consist of at least one voltage limiting component (e.g. MOV) and one voltage switching component (e.g. GDT) connected in series.

Note: A specificity exists for the guide dedicated to SPDs for PV applications. The product standard is 61643-31 and exists in both IEC and CENELEC versions. The application guide is an IEC standard (IEC 61643-32) and, as previously explained, is a CENELEC CLC TS. The CENELEC document differs from the IEC version and the amount of changes was considered too large to keep the same number. For consistency, the European version has been numbered CLC TS 51643-32.

There currently are two CENELEC SPD standards without IEC equivalent (others have been either already incorporated at IEC level or transferred to another commission):

- CLC TS 50539-22 for wind turbine specificities
- CLC TR 50656 for SPD application in conjunction with Class II equipment

CLC TS 50539-22: Low-voltage surge protective devices – Surge protective devices for specific application, including d.c. – Part 22: Selection and application principles – Wind turbine applications.

This European Technical Specification is especially interesting because it assists in selecting SPDs for wind turbine circuits in conjunction with IEC 61400-24 standard.

It describes the following stresses used to define SPDs to protect the generator alternator excitation circuit to cover the repetitive transients that may exist.

- Maximum operating voltage of the system phase to phase: 750 V
- Repetitive transients superimposed on the voltages up to 2.95 kV

These spikes are much higher than the maximum operating voltage and require SPDs with a very high operating voltage and as a consequence, a very high protective level.

This CLC TS also gives typical values of vibration withstand for SPDs used in wind turbines.

CLC TR 50656: SPD application in conjunction with Class II equipment

This Technical Report provides guidance for applying SPDs that may bridge double or reinforced insulation (such as between the primary and secondary side of an isolating transformer or between active parts and touchable conductive surfaces of Class II equipment). This document specially addresses installation issues related to street lighting. The concept of Class II SPDs is introduced. A Class II SPD is an SPD in which protection against electric shock does not rely on basic insulation only, but in which additional safety precautions such as double insulation or reinforced insulation are provided. In this case, connection to earth of the Class II SPD is not permitted.

4.2   Lightning protection

There are four standards in the 62305 series.

IEC 62305-1: 'Protection against lightning – Part 1: General principles'

IEC 62305-2: 'Protection against lightning – Part 2: Risk management'

IEC 62305-3: 'Protection against lightning – Part 3: Physical damage to structures and life hazard'

IEC 62305-4: 'Protection against lightning – Part 4: Electrical and electronic systems within structures'

The first part of the series (62305-1) describes the lightning environment and gives general rules that will be developed in the other parts.

The need for protection, the selection of protection measures and the economic benefits of installing protection measures are determined thanks to risk management, according to 62305-2.

The design, installation and maintenance rules of lightning protection means are addressed in the following two parts:

62305-3 concentrates on the external part of lightning protection and provides protection measures to reduce physical damage and life hazard in a structure.

62305-4 concentrates on the internal part of lightning protection and provides protection measures to reduce failures of electrical and electronic systems in a structure.

The relationship between the four parts is represented in Figure 4.1.

62305-1 defines a key parameter of lightning protection: the lightning protection level (LPL). This number ranges from I to IV, with LPL I being the best. The LPL is related to the probability that defines maximum and minimum lightning

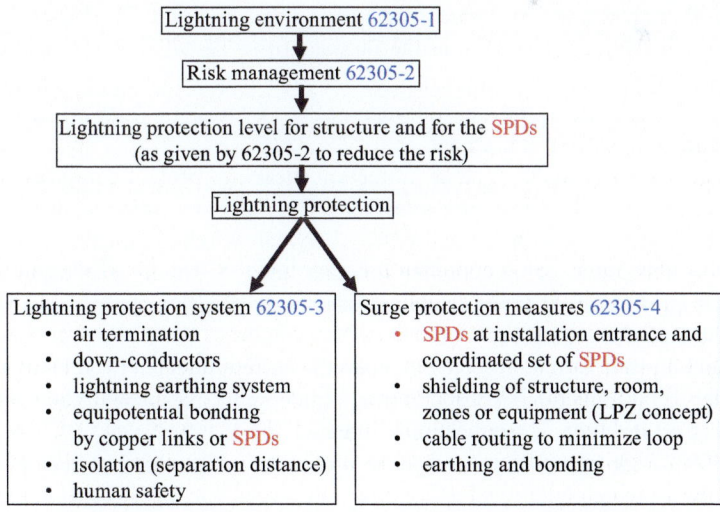

Figure 4.1   Organization of IEC 62305 series of standards

stresses (and mainly current magnitude) will not be exceeded. In a few cases, protection better than LPL I may be needed. These cases are associated with a lower probability of occurrence than LPL I.

62305-1 indicates that SPDs play a major role in the design of a Lightning Protection Zone (LPZ). An LPZ is a zone where the lightning electromagnetic environment is fixed.

Outside of a structure, two zones can be defined:

- LPZ_{0A} where a direct lightning flash and the full lightning electromagnetic field exist.
- LPZ_{0B} where a direct lightning flash is not possible (thanks to lightning protection, including possible natural components) but where the full lightning electromagnetic field exists. A typical case is the environment on a roof near a lightning rod. No direct flash is possible at a given lightning protection level but the electromagnetic field is strong and will create disturbances for equipment located on the roof.

The protected structure is generally defined as LPZ 1, where the surge current is limited by current sharing and by T1 SPDs at the zone boundary. Structure shielding (for structures with dedicated shielding or when shielding is provided by construction for metal framed and metal reinforced concrete structures) will also attenuate the lightning electromagnetic field inside the structure (the protected structure is supposed to be a clean area).

Other zones LPZ 2, 3 and more can also be defined inside the structure where the surge current may be further limited by additional SPDs (coordinated T2 SPDs) at the zone boundaries. Additional shielding may also be used to further attenuate the lightning electromagnetic field. Typical case for LPZ 2 is, for example, when a specific room is shielded in a data centre for IT purposes. In that case, the room shield will attenuate the magnetic field inside this room and T2 SPD at the room entrance will reduce the current at the output of the T2 SPD. Due to propagation rules, the surge current at the output of T2 SPD may still not be negligible for sensitive IT equipment and an additional SPD (Type 2 or Type 3) can also be installed in front of this equipment.

62305-1 also presents another protection means to protect equipment against conducted surges called isolating interfaces. Examples of isolating interfaces are isolation transformers with earthed screen between windings for power lines and metal-free fibre optic or optocoupler for data lines. SPDs are needed to provide insulation protection of these isolating interfaces if SPDs are not incorporated inside the isolating interface.

62305-1 provides a method to calculate the current injected in SPDs at the LPZ 1 boundary as well as typical currents (magnitude and waveshape) to be considered depending on the type of stress (direct, induced, etc.).

62305-2 addresses SPDs through the probability P_{SPD} that has been presented previously.

62305-3 addresses SPDs that are needed for equipotential bonding and refer either to a specific annex in cases where isolation (separation distance) is not maintained between circuits and LPS, and to 62305-1 for the calculation of the current magnitude for incoming lines.

62305-4 addresses SPDs both at LPZ boundaries and in coordination between SPDs. The rules presented in this standard primarily refer to 61643-12. A specific case is introduced, generally ignored by SPD standards: voltage generated in the loop formed by the circuit downstream of an SPD (called U_i in the standard). Let us consider an example: an SPD is installed at the main panel board and another one (coordinated with the first one) in a subsidiary panel that supplies sensitive equipment. If the PE cable runs along the floor and the active conductor (phase and neutral) runs in a different conduit, there will be a loop formed by

- the SPD (either the dynamic impedance of an SPD based on MOVs or the arc resistance for a gapped SPD);
- the active conductors;
- the PE cable;
- the internal circuit of equipment.

In that loop, the magnetic field generated by the initial lightning current will create a voltage U_i that will apply to the equipment terminals.

This can be summarized in Figure 4.2.

A direct strike to the structure will generate a surge (red impulse) on the T1 gapped SPD at the entrance that will be reduced by the secondary MOV-based SPD. At the output of this second SPD, a smaller surge (red impulse) will propagate towards the sensitive equipment to be protected. At the same time, another surge (purple surge) is directly generated in the downstream circuit due to the magnetic field (dotted blue circles). In the worst case, this additional surge will be added to the red surge and create a higher hazard level for the equipment.

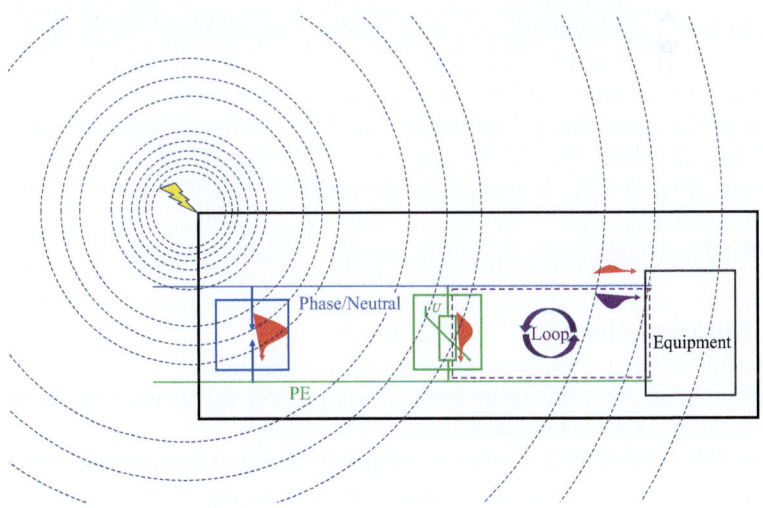

Figure 4.2   Additional voltage U_i

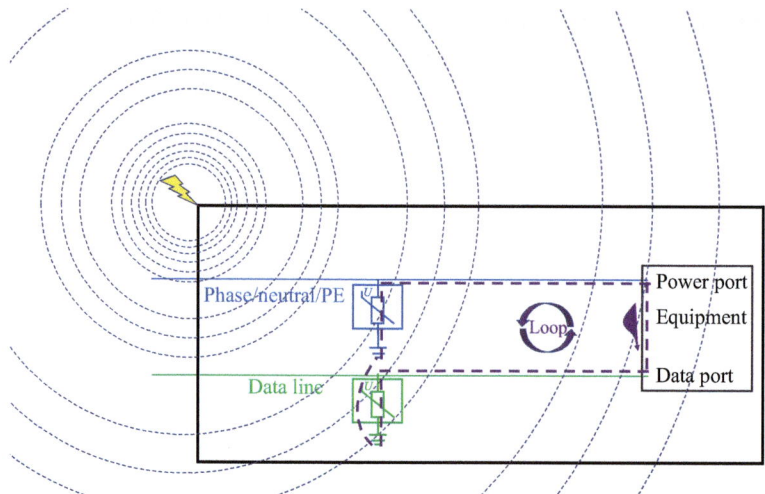

Figure 4.3   Additional voltage U_i between two systems connected to the same equipment

Very often this is not necessary to be considered because

- phase, neutral and PE circulate in the same conduit or even the same cable and thus the loop size is very small and U_i negligible;
- the structure provides shielding thanks to metal cladding or rebar mesh that will reduce the generated magnetic field and thus U_i;
- the circuits are routed in metal cable trays that will also provide shielding and reduce loop size.

However, in a few cases, this does need to be addressed. For example, it is usual that power lines and data lines are routed in different conduit for electro-magnetic compatibility (EMC) reasons. In that case, the loop between the data line and the power line needs to be considered and can be large enough to generate a significant U_i voltage that will appear between the power and data terminal of equipment (Figure 4.3). Regardless of the presence of SPDs on both systems, a surge downstream of the SPDs may be generated and should be evaluated (once again for shielded structures this phenomenon is generally negligible).

4.3   Building electrical installations

For this part, we will concentrate on the main parts of the standard that deal with SPDs: IEC 60364 parts 443 and 534.

The title of IEC 60364 is Low-voltage electrical installations – Part 4-44: Protection for safety – Protection against voltage disturbances and electromagnetic disturbances, and Part 443 is Protection against transient overvoltages of atmo-spheric origin or due to switching. It is in Part 443 that the discussion on the use of

SPDs is provided as well as the previously mentioned simplified risk assessment method.

Switching overvoltages are included in Part 443 but it is indicated that, in general, overvoltages of atmospheric origin (another way to say 'lightning surges') are bigger than switching surges and thus SPDs for protection against lightning surges also cover switching surges. There are not many details in the document explaining how the switching surges can be determined. It is indicated in a note that overvoltages due to switching can be longer in duration and can contain more energy than the transient overvoltages of atmospheric origin.

An important concept is then developed: overvoltage categories. This is a classification of rated impulse voltages (U_W where W stands for withstand) for equipment connected to the low-voltage electrical installation. Overvoltage categories are defined for insulation coordination.

What is insulation coordination? It is a way to determine the insulation withstand of the power system equipment and to coordinate them so that they can be used in the power installation. Equipment insulation is designed to withstand the highest power frequency system voltage, most of the temporary power frequency overvoltages and surges of atmospheric origin up to a certain level. Generally, the biggest part of atmospheric surges will enter the electrical installation from the power utility network. The highest level of insulation is needed at this point. Downstream of that point, surge protection measures will try to reduce the stress so that equipment with the lowest insulation withstand capability can be used. When you know the rated impulse withstand of specific equipment, you can determine in which part of the installation you can use it and if surges at this place exceed the withstand level provide SPDs to decrease the stress to an acceptable level.

This approach refers to the insulation of equipment. In the past, with limited electronic component, equipment insulation was the key point. For example, high-voltage transformers are insulated by oil, gas or paper and if a surge breaks down this insulation, the transformer is damaged and needs to be repaired and the insulation restored. Insulation is still today an important parameter because a failed insulation means repair or permanent damage. It can also create a hazard such as fire and is then important for safety. But electronic components are much more sensitive than insulation and when damaged by surges, repair is more difficult. Another concept exists in EMC standards: immunity. Part 443 indicates that where a temporary loss of function of equipment is critical, the equipment level immunity should be considered as well. The reference point for insulation is ground (active conductors: phase and neutral, to earth), whereas immunity mainly concentrates on phase to neutral circuits. Damages can occur between phase and neutral when the insulation to ground is still maintained. Most of the lightning surges occur between conductors and ground, which is why insulation withstand is critical but it is not the only point to address when developing a surge protection plan, immunity matters as well.

IEC 60364-4-44 Part 443 provides a table similar to Table 4.2 with fixed values of insulation withstand for each overvoltage category. These values depend on the nominal voltage of the low-voltage system.

Table 4.2   Rated impulse voltage of equipment related to overvoltage categories**

Nominal voltage of the installation (V)	Rated impulse voltage of equipment U_W (kV)			
	Overvoltage category			
	IV	III	II	I
	Very high insulation	High insulation	Normal insulation	Reduced insulation
Examples	Energy meter, telecontrol systems	Distribution boards, switches, socket-outlets	Domestic appliances, tools	Sensitive electronic equipment
120/208	4	2.5	1.5	0.8
230/400277/480	6	4	2.5	1.5
400/690	8	6	4	2.5
1,000	12	8	6	4
1,500 DC			8	6

Note: Table 4.2 provides a guidance (refer in particular to the line 'examples') to estimate the rated impulse withstand of equipment when it is not specified by the manufacturer or by the applicable standard for that type of equipment. In general, it may be assumed that the impulse withstand is at least five times the nominal voltage between active conductors (e.g. 1.2 kV between phase and neutral and 2 kV between phases for a 230 V/400 V system). Immunity levels may even be lower.

For information on the statistical occurrences of these surge levels, one source available is IEC TR 62066 dated 2002 'surge overvoltages and surge protection in LV AC power systems – general basic information'. This Technical Report was jointly produced in 2002 by five IEC committees, including TC81, SC37A and TC64. This report indicates that on 230/400 V system, induced voltages reaching 6 kV can be observed only every year or every 4 years depending on the type of system for a flash ground density equal to 2.2 strike/year/km^2. Higher values can also be obtained, especially in the case of direct strikes to a line, but the probabilities for these are very low (e.g. with the same hypothesis, surge above 20 kV can be observed every 22 years).

More details on switching surges can also be found in this technical report. Recordings of switching surges in industrial sites over long periods are presented in that report. One per thousand of the switching surges have an amplitude exceeding 2.5 kV with most of the surge having amplitudes below 1 kV. However, switching surges can occur much more frequently than lightning surges and can occur daily or even many times a day when related to the operation of circuit breakers and switches. This frequency of occurrence can create damages in the long term and aging of insulation or components. Generally, switching surges that occur frequently are quickly identified because they produce damages in the absence of any lightning event. In practice, it can be considered that the frequency of occurrence of switching surges is inversely proportional to the third power of its amplitude, even though the report indicates that deviation with this law may be observed because switching surges depend on local installations.

Part 534 is dedicated to SPD installations (IEC 60364 Part 5-53 Selection and erection of electrical equipment – devices for protection for safety, isolation, switching, control and monitoring with Part 534 titled Devices for protection against transient overvoltages). The first level of SPDs shall be installed as close as possible to the origin of the installation. Where the structure is equipped with an LPS (the LPS may be a natural one, i.e. protection against direct strike provided by the structure itself) Type 1 SPDs shall be used at line entrance. For protection against induced surges (or switching surges), Type 2 SPDs shall be used at line entrance. It is also indicated that in the absence of an LPS, Type 1 SPDs should also be used at the line entrance and not Type 2 SPDs if a direct strike to the incoming power line is considered. For example, when overhead incoming power lines are struck by lightning in close proximity to the entrance of the structure (between the closest pole and the service entry), most of the current will be conducted through the service entry and a Type 1 SPD is justified. It is generally assumed that after a few spans, multiple sparkover at the LV overhead line pole will reduce the stress and do not justify Type 1 SPD any longer at line entrance.

Additional Type 2 (or Type 3) SPDs may be needed to protect the downstream installation and sensitive equipment. These SPDs located in secondary panel boards or at the socket outlets shall be coordinated with SPDs located upstream.

SPDs between active conductors and PE are mandatory because the standard mainly concentrates on installation safety (insulation withstand). However, SPDs between phase and neutral are recommended to ensure equipment protection (immunity, as discussed earlier). SPDs between phases are generally not needed when SPDs between phase and neutral are provided.

This standard introduces the concept of SPD assembly. An SPD assembly consists of an SPD fitted with all SPD disconnectors required by the SPD manufacturer as discrete components. This term makes sense when an SPD disconnector is added in series with the SPD, in the SPD circuit. It is important to introduce the concept of SPD assembly because many SPD requirements apply also to the SPD disconnectors (e.g. SPD disconnectors should have the same surge withstand as the SPD itself). An SPD disconnector is a device that will automatically disconnect the SPD from the mains in the event of an SPD failure. It is used to prevent a persistent fault on the system. This disconnector should also give an indication of the SPD's failure because a failed SPD can no longer provide protection and should be changed in a timely manner. A few SPD requirements, mainly the safety ones, require the SPD disconnector be included as a part of the qualification testing. When the SPD disconnectors are included in the SPDs, there is no risk of leaving it out of the installation or using a disconnector different from those listed in the SPD data sheet, leading to lower surge performance and possible loss of power supply, or other damages. An SPD assembly is then a group of devices working together and none of these devices should be missing or inadequately selected. When all the disconnectors are included in the SPD, generally the term 'SPD assembly' is ignored and SPD is preferred.

All the rules given in IEC 60364-5-53 Part 534 are reproduced and expanded in 61643-12, see bibliography and Chapter 7 for more details.

4.4 Photovoltaic

There are many standards that refer to SPDs in the PV environment:

- standards from the SPD committee: SC37A with IEC 61643-31 for product and IEC 61643-32 for application;
- standards from the electrical installation committee: TC64;
- standards from the solar PV energy systems committee: TC82.

It is then necessary to provide details of these different standards, as far as surges are concerned.

Regarding PV SPD (surge protection of PV installations), the basic document is IEC 61643-32 edition 1 09/2017: Low-voltage surge protective devices – Part 32: Surge protective devices connected to the DC side of photovoltaic installations – Selection and application principles. It is the application guide for the IEC 61643-31 standard that deals with PV SPDs.[†]

The introduction of the standard clearly explains what can be found in the standard (the key elements are in blue):

This part of IEC 61643 provides useful information for the selection of SPDs connected to photovoltaic installations. It provides information to evaluate, with reference to the IEC 62305 series, IEC 60364 series and IEC 61643-12, the additional needs for surge protective devices (SPDs) to be installed on the DC side of a photovoltaic (PV) system, to protect against induced and direct lightning effects. It gives guidance for selection, operation and installation of SPDs, including the selection of SPD test class, surge current values and cross section of bonding conductors. Guidance for the selection of SPDs connected to the AC side is also given. The specific electrical parameters of a PV array or a PV source require specific SPDs on the DC side. This part of IEC 61643 considers SPDs used in different locations and in different kinds of PV systems. It gives examples and provides a simplified and common approach to determine impulse discharge current values for the DC side of different PV installations.

It is also indicated that for PV installations that include batteries, additional requirements may be necessary (e.g. because the short circuit current may be higher).

The key parameter for selecting an SPD for PV applications is the end of life due to low short circuit current at dawn that can create a fire hazard when using regular AC or even DC SPDs. The Y scheme is the preferred scheme ($+$ and $-$ connected to a mid-point by elementary SPDs and this mid-point is connected to ground by another elementary SPD) as it reduces the risk of failure and thus break of insulation to earth (Figure 4.4).

[†]IEC 61643-32 ed.1.0. Copyright © 2017 IEC Geneva, Switzerland. www.iec.ch.

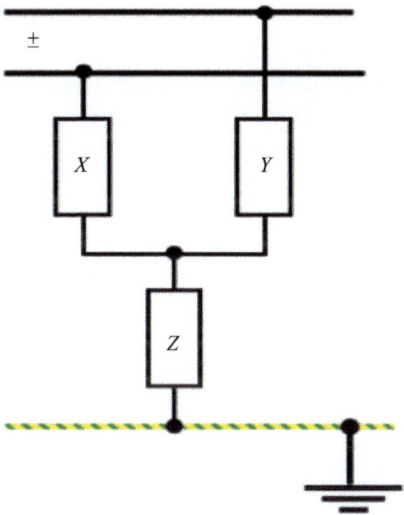

Figure 4.4   Y scheme for PV SPDs

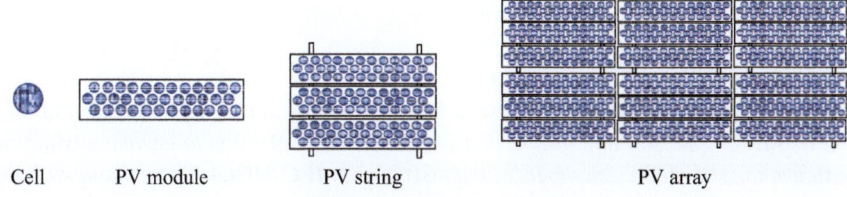

| Cell | PV module | PV string | PV array |

Figure 4.5   PV elements

For batteries, the short circuit current will be bigger and not depend much on solar activity, the short circuit current being supplied mainly by the batteries.

PV standards and applications are based on specific parameters and vocabulary.[‡] For the benefit of the reader, most of them are explained next:

- PV module or panel: smallest complete protected assembly of interconnected cells;
- PV string: circuit of series-connected modules;
- PV array: assembly of electrically interconnected PV strings;
- PV installation: erected equipment of a PV power supply installation (a PV installation starts from a PV module or a set of PV modules connected in series with their cables, provided by the PV module manufacturer, up to the user installation or the utility supply point) (Figure 4.5).

[‡]IEC 61643-32 ed.1.0. Copyright © 2017 IEC Geneva, Switzerland. www.iec.ch.

- STC (standard test conditions): Standard set of reference conditions used for the testing and rating of PV cells and modules.
- $U_{oc\ STC}$: Open-circuit voltage under standard test conditions across an unloaded PV array.
- U_{ocmax} (open-circuit maximum voltage): Maximum voltage across an unloaded PV array.
- $I_{SC\ STC}$: Short circuit current of a PV array under standard test conditions.
- $I_{SC\ max}$: Maximum short circuit current of a PV array.
- U_{CPV}: Maximum DC voltage which may be continuously applied to the SPD's ($U_{CPV} \geq U_{ocmax}$).
- I_{SCPV}: Maximum prospective short circuit current for which the SPD is rated ($I_{SCPV} \geq I_{SC\ max}$).

Equipment within a PV installation that may require protection includes

- the inverter,
- the PV array,
- the Junction Box (also called combiner box) that groups together PV arrays or strings and connects them to an inverter,
- the wiring (installation itself),
- components installed between the inverter and the PV array,
- equipment for controlling and monitoring the PV installation.

The installation of SPDs on the DC and AC sides of a PV installation is mandatory unless indicated otherwise by a risk assessment. For large PV installations, IEC 62305-2 is usually applied, but for smaller PV installations other risk evaluation methods such as those described in IEC 61643-12, IEC 60364-4-44 Clause 443 for SPDs on the AC side and IEC 60364-7-712 for SPDs on the DC side may be used.

A PV installation will generally use many SPDs. Coordination of SPDs on the same line is then essential. This means in practice that it is generally best to use SPDs from the same manufacturer following the coordination rules provided by this manufacturer.

To determine the type and location of SPDs, for the AC part of the PV installation, the solution is very simple: if there is an LPS on the structure, Type 1 SPDs should be used. If not, Type 2 SPDs are enough. Installation rules are as follows: one SPD at the entrance of the AC installation and, if needed because the line length is too long (generally more than 10 m), another SPD in front of the inverter.

For DC SPDs, it is a little more complex. First, the PV SPD should be marked 'PV' and the permanent voltage for the SPD (U_c for AC systems) is referred to as U_{CPV}. Given the specific U/I-characteristic of PV systems, only SPDs explicitly designated for use on the DC side of PV systems shall be installed. These SPDs shall comply with the requirements of IEC 61643-31.

The type of SPD depends on whether an LPS is installed (to protect the PV panels or to protect a host structure) and whether the separation distance is maintained. The separation distance (s) is the minimum distance that must be kept between the LPS circuit and any grounded or powered equipment. This separation distance does not apply to the earthing circuit, because in any case, all earthing

Table 4.3   Location and type of SPDs for PV installations according to IEC 61643-32[§]

Location of the SPD	At the entrance of the AC installations	In front of the AC side of inverter	In front of DC side of inverter and at JB location in front of PV arrays
In the case of PV installation without LPS	Type 2	Type 2	Type 2
In case of an LPS when separation distance s is maintained with PV installations	Type 1	Type 2	Type 2
In the case of an LPS when separation distance s is not maintained with PV installations	Type 1	Type 1	Type 1

circuits must be connected together. This distance is generally in the range of the m or less. For PV installations, this is the distance between the LPS circuit and any part of the PV installations. Table 4.3 indicates the location and type of SPD versus the presence of an LPS.

Any time a Type 1 SPD is used, 16 mm^2 conductors should be used and any time a Type 2 SPD is used, 6 mm^2 is enough (these sizes are minimum values based on lightning stress, short circuit current may need bigger sizes).

SPDs and their selection and installation rules shall comply with

- IEC 61643-11 for surge protective devices connected to AC low-voltage power systems;
- IEC 61643-31 for surge protective devices connected to DC photovoltaic systems;
- IEC 61643-21 for surge protective devices connected to telecommunication and signalling lines;
- IEC 61643-12 and IEC 62305-4 for the protection of AC power systems;
- IEC 61643-22 and IEC 62305-4 for the protection of the control and communication systems.

The SPD protective level (U_p) should be defined on the basis of the PV modules surge withstand U_W, with a margin of 20%, $U_p \leq 0.8\ U_W$. The standard provides a very useful table similar to Table 4.4 that gives typical values for U_W.

Another standard relevant to SPD and other safety rules related to PV installations is IEC 60364-7-712: 04/2017: Low-voltage electrical installations – Part 7-712: Requirements for special installations or locations – Solar photovoltaic (PV) power supply systems.

Table 4.4   Typical surge withstand U_w of PV equipment according to IEC 61643-32[||]

U_{ocmax} (V)	PV module	Inverter
100	800	2,500
150	1,500	2,500
300	2,500	2,500
600	4,000	4,000
800	5,000	4,000
1,000	6,000	6,000
1,500	8,000	8,000

This standard introduces a few other definitions:

- MPPT: Maximum power point tracking – PV array operation is always at the point where the product of electric current and voltage yields the maximum electrical power under specified operating conditions.
- PCE: Power conversion equipment – A system that converts the electrical power delivered by the PV array into the appropriate frequency and/or voltage values to be delivered to the load, or stored in a battery or injected into the electricity grid. An inverter is a PCE which converts DC voltage of the PV array into AC voltage.
- PV array/string combiner box: An enclosure where PV sub-arrays/strings are connected and which may also contain overcurrent protection and/or switch-disconnectors.

The fault conditions are different for PV installation depending on whether the fault is created inside the PV strings (arc) or between the PV string and earth (arc to ground). In this last case, it also depends on whether the PV installation (polarity) is grounded or not. The standard addresses three categories of arcs due to fault in PV installations:

- Series arc which may result from a faulty connection or a series break in wiring.
- Parallel arc which may result as a partial short circuit between adjacent wiring at different potentials. The standard requires cables in PV array wiring to be suitable for use with Class II equipment (double or reinforced insulation). It is important to be aware that SPDs are connected between polarity and earth break, by nature, the Class II insulation and therefore motivate a Y connection.
- Arcs to earth which result from failure of insulation.

UL rules (US-based standards) frequently impose the use of DC PV arc-fault circuit protection devices in PV circuits as for example in UL 1699B. A Ground Fault Detection and Interruption (GFDI) is a device specially designed for PV solar arrays. A ground fault at the PV generator will trigger the GFDI, interrupting the leakage and preventing damage to the system. In the case of an SPD fault, the fault current will flow through the GFDI and it is likely that the low trigger rating of the GFDI will not allow the SPD disconnector to operate before the GFDI, leading to power interruption. There is no surge protection provided by the GFDI itself.

Another SPD application document is a technical report produced by the PV energy systems committee. IEC TR 63227 edition 1 (published in 2020) is titled Lightning and surge voltage protection for PV power supply systems.

IEC TR 63227:2020 deals with the protection of PV power supply systems against the effects of lightning strikes and surge voltages. When lightning and/or surge voltage protection is required to be installed, this document describes requirements and measures for maintaining the safety, functionality, and availability of the PV power supply systems.

According to the report, magnetic coupling and resulting disturbances and current injections can be reduced considerably by increasing the separation between the PV modules and air-termination systems and associated down-conductors. This report recommends an LPL III for the LPS, but in some cases the need for an increased availability of the system (e.g. a PV farm) justifies risk calculations in accordance with IEC 62305-2 to confirm the level of protection required for the application.

Finally, a Technical Specification from the same committee also addresses lightning and surge protection IEC TS 62738 edition 1 08/2018 Ground-mounted PV plants – Design guidelines and recommendations. This standard refers to 61643 series for SPD selection and application.

Lightning risk should be based on IEC 62305-2. For typical installations, reasonable measures include the following:

- Equipotential bonding of array racking or tracker structures.
- Type 1 or 2 SPDs installed in the string combiner boxes.
- Type 1 or 2 SPDs installed on DC and AC side of inverters.
- Surge arresters at substations and feeder termination points.
- Building mounted air terminal and down conductors.
- Type D1 or C2 SPDs for array communication equipment.

All of these measures are recommended as a matter of best practice, and further study is recommended during the detailed design phase of the project.

4.5 Wind turbine

In the series of standards related to wind turbines, IEC 61400 Part 24 deals with lightning protection. This part is based on the principles of 62305 series with application-specific requirements. For example, winter lightning is addressed both regarding the risk method and the testing means for the protection of the wind turbine; especially the blades.

Regarding surges, this standard is based on the LPZ concept developed in IEC 62305-4 (Figure 4.6).

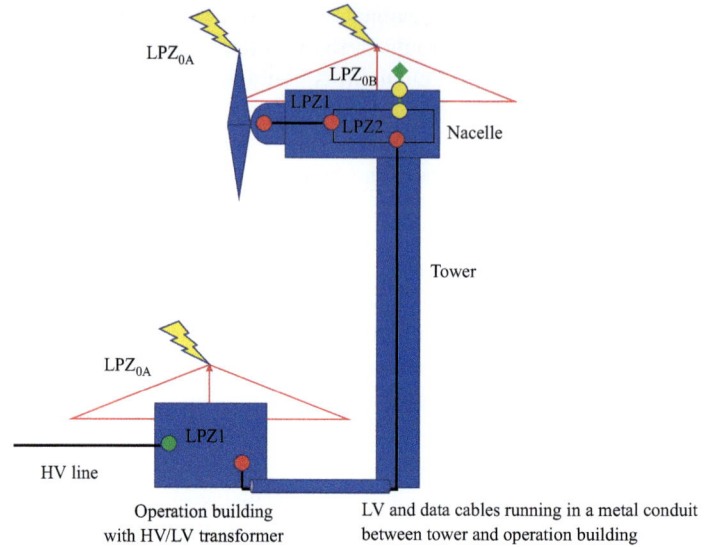

Figure 4.6   Example of application of the LPZ concept for a wind turbine. Red dot: LV SPD, yellow dot: data SPD and green dot: HV surge arrester. Green diamond: sensor

This standard gives an interesting summary of the minimum impulse withstand that is helpful for defining the suitable SPD and their protective levels[¶]:

AC ports (230 V/400 V):

- 2 kV line to ground;
- 1 kV line to line;

DC ports (50 V):

- 1.0 kV Line to ground
- 0.5 kV Line to line

Telecommunication ports connected to an external line:

- 1.5 kV port to earth
- 1.5 kV between signal conductors

Telecommunication ports connected to an unshielded internal line:

- 1 kV port to earth

Signal ports in general:

- 0.5 kV port to earth

[¶]IEC 61400-24 ed.2.0. Copyright © 2019 IEC Geneva, Switzerland. www.iec.ch.

SPDs in a wind turbine are subjected to specific environmental stresses; especially vibrations for SPDs located in the nacelle or in the turbine tower. This standard refers to the CENELEC standard EN 50539-22 when no specific values are given by the wind turbine manufacturer.

Since a nacelle is not permanently manned, SPDs protecting the critical parts of the electrical and control systems of wind turbines may require remote monitoring.

The SPD continuous operating voltage should be selected considering the repetitive transients superimposed on the operating voltages due to switching operations in the wind turbine electrical systems (power inverter cycling).

Due to the high occurrence of lightning flashes to wind turbines, SPDs shall be able to withstand multiple lightning flashes. However, there is no specific requirement in IEC 61400-24 to cover multiple pulses occurring in a single lightning event.

It should be noted that only energy coordination is listed in IEC 61400-24 even though voltage protection coordination is a more stringent rule.

Bibliography

IEC 61643-11:2011 Edition 1.0 (2011-03-09) Low-voltage surge protective devices – Part 11: Surge protective devices connected to low voltage power systems – Requirements and test methods.

IEC 61643-12:2008 Edition 2.0 (2008-11-13) Low-voltage surge protective devices – Part 12: Surge protective devices connected to low voltage power distribution systems – Selection and application principles.

IEC 61643-21:2000+AMD1:2008+AMD2:2012 Edition 1.2 (2012-07-27) Low voltage surge protective devices – Part 21: Surge protective devices connected to telecommunications and signaling networks – Performance requirements and testing methods.

IEC 61643-22:2015 Edition 2.0 (2015-06-25) Low-voltage surge protective devices – Part 22: Surge protective devices connected to telecommunications and signaling networks – Selection and application principles.

IEC 61643-31:2018 Edition 1.0 (2018-01-10) Low-voltage surge protective devices – Part 31: Requirements and test methods for SPDs for photovoltaic installations.

IEC 61643-32:2017 Edition 1.0 (2017-09-20) Low-voltage surge protective devices – Part 32: Surge protective devices connected to the d.c. side of photovoltaic installations – Selection and application principles IEC 61643-11: "Surge protective devices connected to low-voltage power systems – Requirements and test methods".

IEC 62305-1 currently edition 2 (2010); edition 3 foreseen for end of 2022 "Protection against lightning – Part 1: General principles".

IEC 62305-2 currently edition 2 (2010); edition 3 foreseen for end of 2022 "Protection against lightning – Part 2: Risk management".

IEC 62305-3 currently edition 2 (2010); edition 3 foreseen for end of 2022 "Protection against lightning – Part 3: Physical damage to structures and life hazard".

IEC 62305-4 currently edition 2 (2010); edition 3 foreseen for end of 2022 "Protection against lightning – Part 4: Electrical and electronic systems within structures".

IEC 61400-24:2019 "Wind energy generation systems – Part 24: Lightning protection".

CLC/TS 51643-32:2020 "Low-voltage surge protective devices – Surge protective devices connected to the DC side of photovoltaic installations – Selection and application principles".

IEC 60364-4-44:2007+AMD1:2015+AMD2:2018 CSV Consolidated version "Low-voltage electrical installations – Part 4-44: Protection for safety – Protection against voltage disturbances and electromagnetic disturbances".

CLC HD 60364-4-443:2016 "Low-voltage electrical installations – Part 4-44: Protection for safety – Protection against voltage disturbances and electromagnetic disturbances – Clause 443: Protection against transient overvoltages of atmospheric origin or due to switching".

IEC 60364-5-53:2019+AMD1:2020 CSV Consolidated version "Low-Voltage electrical installations – Part 5-53: Selection and erection of electrical equipment – Devices for protection for safety, isolation, switching, control and monitoring".

CLC HD 60364-5-534:2016 "Low-voltage electrical installations – Part 5-53: Selection and erection of electrical equipment – Isolation, switching and control – Clause 534: Devices for protection against transient overvoltages".

IEC TR 62066 2002 "Surge overvoltages and surge protection in LV AC power systems – General basic information".

IEC 60364-7-712 Edition 2.0 2017-04 "Low voltage electrical installations – Part 7-712: Requirements for special installations or locations – Solar photovoltaic (PV) power supply systems".

IEC TR 63227:2020 "Lightning and surge voltage protection for photovoltaic (PV) power supply systems".

IEC TS 62738:2018 "Ground-mounted photovoltaic power plants – Design guidelines and recommendations".

CLC/TS 50539-22:2010 "Low-voltage surge protective devices – Surge protective devices for specific application including d.c. – Part 22: Selection and application principles – Wind turbine applications".

CLC/TR 50656:2016 "SPD application in conjunction with Class II equipment".

Chapter 5

Surge-protective components

Vincent Crevenat[1]

5.1 Introduction

This chapter is about components that are commonly used to achieve surge protection. They can be either used in devices that are specially designed for this purpose or used in equipment's or systems that are dedicated to various applications other than specifically surge protection. They can be used alone or combined. They are commonly called SPC standing for surge-protection component.

Note: Lately a controversy about the definition of components took place under IEC world but also at other standardisation organisation levels such as ITU and IEEE. This controversy was triggered by the questioning of some and was about where stops the scope of the noun 'component' when addressing power transformers that are specifically designed to get a surge-protection function. Experts had this fight starting with the statement that a component cannot be something weighting several kilogrammes or even up to tones. In addition, some argued that a surge-protection device is something that contains at least one nonlinear component and is intended to limit surge voltages and divert surge currents (IEC 61643-11). This last point cannot be for these 'surge-protection' transformers as they are neither non-linear component nor diverting any current. This is of course an expert fight that is not of concern when the real goal is to list all components that may be used for surge protection. For instance, if someone claims that a simple inductor is not a surge-protection component, he may be right if we consider this inductor alone but it becomes a component that is inseparable from the surge-protection concept as soon it is to be combined with other components such as GDTs (see Section 5.3.3) or TSS (see Section 5.3.4) for instance. Then, the frontier to call a component 'surge-protective component' or not is not so easy to determine.

In this chapter, all components are going to be addressed whatever if they have a direct or indirect function in surge protection.

Let us start with the definition of what is a component:

Component: A part that combines with other parts to form something bigger or to achieve a function.

[1]CITEL, Reims, France

It is also to be understood that SPCs are parts of an electrical circuit. They can be connected with various techniques depending on their behaviour and the intention to use them to achieve their surge protection functions. In general, three ways are possible such as shown by Figures 5.1–5.3 (in-line, parallel or serial mounting).

The following figures do not differentiate the application neither the mode of protection (common and differential). They are simply showing the concept of using an SPC between a threat and something to protect. The arrows indicate the propagation direction of the surge wave that is typically a couple of both current and voltage waves. Real application may have several wires such as by example lines, neutral and ground for power application or pairs of wires plus a ground connection for data/signal application. This will be addressed in more specific examples all along this chapter. It does not show if the waves are, both or only one, absorbed, stopped, shorted (diverted) or simply not affected. This will depend on the used component.

Each of these mountings of the surge component may interfere with the 'passage' of a combined surge wave (current and voltage). Some components, and depending on these mountings, will only interact with the surge current; some will

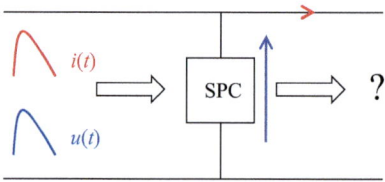

Figure 5.1   Parallel mounting

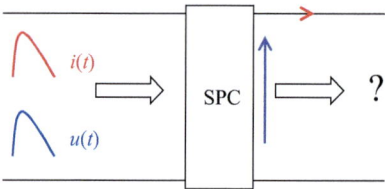

Figure 5.2   In-line mounting

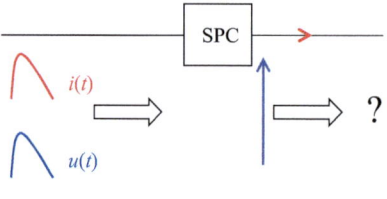

Figure 5.3   Serial mounting

interact with only the surge voltage and some will interact with both the current and the voltage waves.

In order to clarify the reading of this clause, the writer decided to categorise the components with main functions or behaviour when addressing the surge current or surge voltage (current and voltage being obviously combined when surge protection is the topic).

- Voltage-limiting components (clamping components)
- Voltage switching components (crowbar components)
- Current-limiting components
- Current switching components
- Filtering components
- Isolating components
- Component association (hybrid components)

For better understanding of this segmentation, see in Table 5.1 and the following figures that are part of the next section.

Besides the surge impulse characteristics that the SPC has to handle, one must also consider possible and specific aspects of the application. As an example, when an SPC is to be used on a telecom application that is with high frequency, low voltage and low power, the behaviour of the selected component should not interfere with the signal to be transmitted.

For each listed component in this chapter, history, different types (if any), construction, characteristics and usage (selection, normal life and end of life) will be addressed as far as possible.

Note: There is one discussion that we will try to avoid in this chapter: The surge component response time's discussion. It may happen that some people claim that some surge-protection components are faster than others. This discussion could be relevant if the criteria to measure this time response were similar to all types of SPCs. This is unfortunately not really the case. Comparing time reaction of GDTs, MOV, diodes is often a game that is played to give a very rough idea of the component's ranking about reaction time. This is really misleading the designer as each of these components has benefits and drawbacks depending on field of analysis. They are somehow all complementary to each other and are very frequently used altogether to achieve the best surge-protection functions. It is more accurate to speak about parasitic capacitance, inductance, sharpness of the response V/I curve or transition time from a state to the other state than simply time response.

Note: When figures show electrical diagrams where current or voltage is represented with arrows, in this chapter, the rule will be the one shown in Figure 5.4.

5.2   Voltage-limiting components

5.2.1   Introduction

The term voltage-limiting component is now the usual term to be used at least for IEC document's users. The term stands by itself. Clamping voltage components is

Table 5.1 Classification of SPC and usage as surge-protection components

Category	Components	Surge withstand	Protection provided	Surge protection usage	Preferred applications	SPC alone	Notes
Voltage-limiting components (clamping components)	Varistors	++	++	++	Power	Yes	Can be used as primary protection and alone
	Diodes	--	++	+	Signal	Yes/no	Cannot be used alone
	Zener diodes	-	++	-	Signal	Yes/no	Cannot be used alone
	Avalanche diode (TVS or SAD)	+	++	++	Signal (power)	Yes	Can be used as primary protection and alone
Voltage-switching components (crowbar components)	Gas discharge tube (GDT)	++	++	++	Power and signal	Yes	Can be used as primary protection and alone
	Spark gap (SG)	++	++	++	Power	Yes	Can be used as primary protection and alone
Current-limiting components	Thyristors (TSS)	-	++	++	Signal	Yes	Cannot be used alone
	Resistor	++	-	++	Signal	No	Cannot be used alone
	Inductors	++	-	++	Signal and power	No	Cannot be used alone
	Capacitors	++	-	++	Signal and power	No	Cannot be used alone
Current-switching components	PTC	-	-	--	Signal	No	Cannot be used alone
	Fuses	-	-	--	Power	No	Cannot be used alone
	TBU	-	-	--	Signal	No	Cannot be used alone
	PTC	-	-	--	Signal	No	Cannot be used alone
	Circuit breakers (CB)	-	-	--	Power	No	Cannot be used alone

Filtering components	RLC	+	+	++	Signal	Yes/no	Can be used as primary protection and alone
	Stud	++	++	++	Signal	Yes	Can be used as primary protection and alone
Isolating components isolation devices	Transformers (SIT)	+	+	++	Power and signal	Yes	Cannot be used alone
	Optical transmitter	+	++	++	Signal	Yes	Cannot be used alone
Component association (hybrid components)	VG*	++	++	++	Power and signal	Yes	Can be used as primary protection and alone
	GD*	++	++	–	Power and signal	Yes	Can be used as primary protection and alone
	VGD*	++	++	–	Power and signal	Yes	Can be used as primary protection and alone
	GTV*	++	++	–	Power and signal	Yes	Can be used as primary protection and alone

* V for varistor, G for GDT and D for diode.

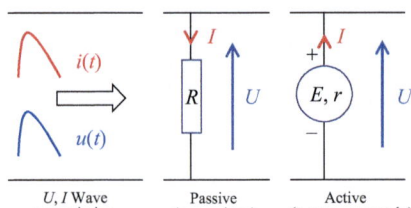

Figure 5.4   *Electrical drawing convention used in this chapter*

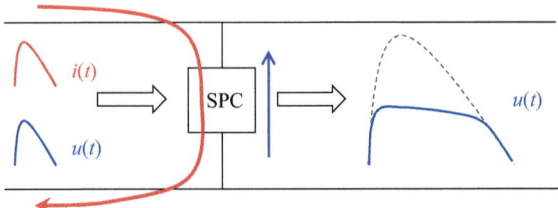

Figure 5.5   *Voltage-limiting component parallel mounting*

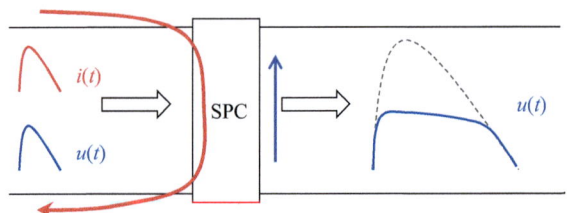

Figure 5.6   *Voltage-limiting component serial mounting*

also a common name. This type of component literally 'cuts' or 'clamps' the voltage wave at a certain voltage avoiding the voltage wave to reach high values. 'High' is a relative meaning to moderate in regards to the concept of transient surge itself and the application of course. About 1,000 V for a duration of 30 µs is not so high for an equipment connected to a permanent 230-V AC, a power system for instance.

Usually, the voltage-limiting components are parallel mounted. But some components may exist with IN and OUT leads to shorten the lead connection length avoiding the lead length voltage drop due to fast current rises.

These components limit the voltage by diverting the current. They act such as a pressure valve that releases the pressure by diverting the gas or fluid towards another direction than the rest of the circuit.

This process with the effect on current and the 'transmitted voltage wave' is shown with Figures 5.5 and 5.6.

It can be noticed that these components limit the voltage during the full surge duration and therefore, are subjected to energy directly proportional to the current, the limiting voltage and the time duration of the surge event. It is then easily understandable that the surge withstand parameters of these components are consequently linked to their ability to absorb energy and dissipate it as heat, and to their maximum operating temperature. As a consequence, a component dedicated to divert direct lightning current is usually of nice mass/volume.

The ideal static *V*/*I* and dynamic *u*(*t*) electrical characteristics of ideal voltage-limiting components can be represented such as in curve Figures 5.7 and 5.8.

By definition, these ideal curves are not realistic when using any of the technology used today and real characteristics may better look like as shown in Figures 5.9 and 5.10. In addition, dynamic behaviour of the voltage-limiting components can highly be altered by parasitic components such as parasitic capacitance, inductances or resistances. All these will be detailed further down from Sections 5.2.2 to 5.2.5.

Today, the two technologies for voltage-limiting components for surge protection are the metal oxide varistors (MOVs) (see Section 5.2.2) and semiconductors (see Section 5.2.3). MOV is the main one and used by almost all surge-protective devices (SPDs) manufacturers for power applications. Semiconductors are also used massively used for data/signal surge-protection devices but more seldom for power application. Details will be given in the following subclauses.

Note: It can be said that an MOV also uses the semiconduction electrical phenomenon but we will more focus on the manufacturing process that clearly differentiates these two technologies.

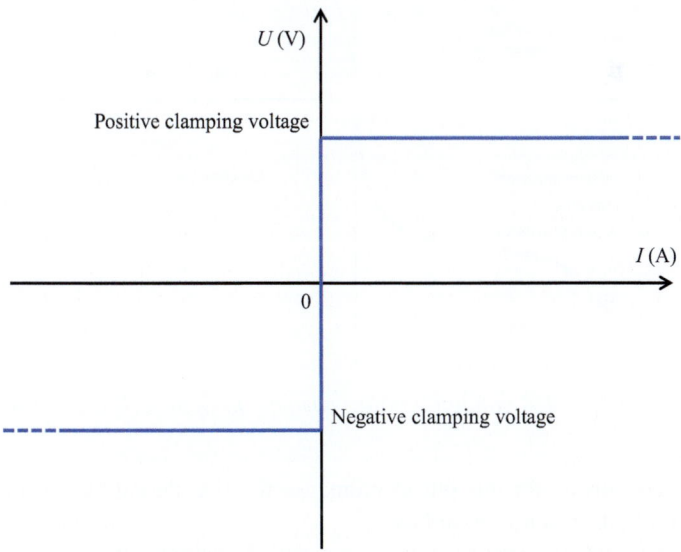

Figure 5.7   U/I voltage-limiting component ideal curve

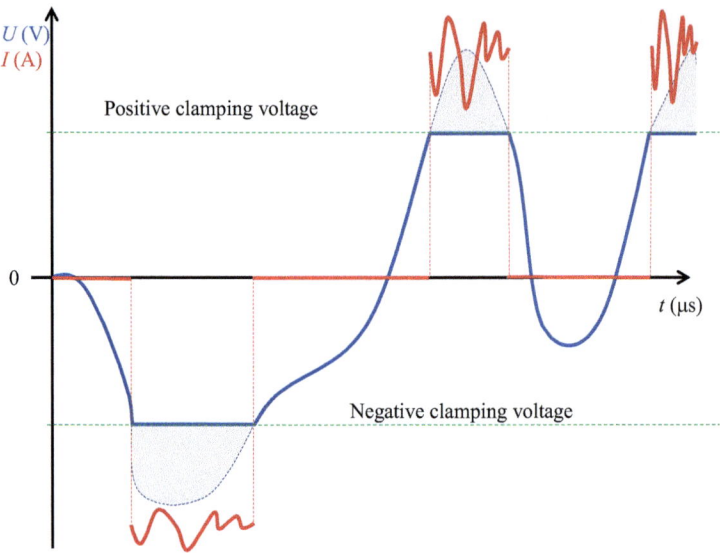

Figure 5.8   u(t) voltage-limiting component ideal curve

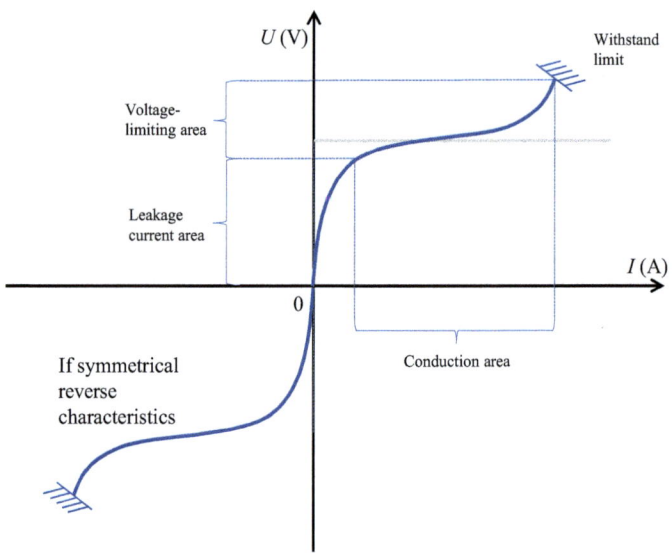

Figure 5.9   U/I voltage-limiting component approached reel curve

As a conclusion for this introduction, see the IEC definition for limiting voltage component, which is as follows:

SPC has high impedance when no surge is present but will reduce it continuously with increased surge current and voltage.

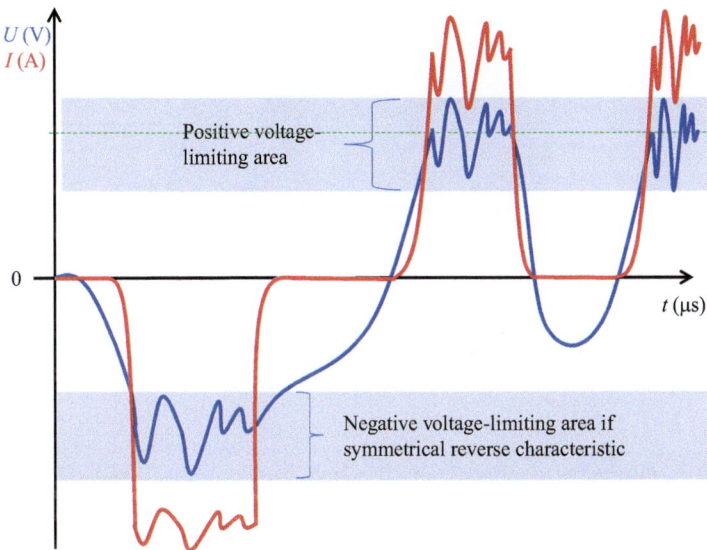

Figure 5.10 u(t) voltage-limiting component approached reel curve

NOTE 1 Common examples of components used as non-linear devices are varistors and suppressor diodes. These SPDs are sometimes called 'clamping type'.

NOTE 2 A voltage-limiting device has a continuous voltage/current characteristic.

5.2.2 Metal oxide varistors

5.2.2.1 Introduction

Varistor, SiC varistor, ZnO varistor, VDR, ZNR, varistor, etc. are common terms used since several decades to name what we now commonly call MOV (with an exception for the SiC standing for silicon carbide varistor). All these acronyms reflect either the electrical behaviour or the constituent of the component.

First a varistor is a nonohmic resistor that changes its resistance depending on the voltage applied to its connection leads. Some claims that this is also related to the current flowing through it, but ultimately, are not current and voltage related by Ohm's Law? ZNR stands for zinc oxide non-linear resistor and expresses one aspect of this electrical characteristic. This is also why voltage-dependent resistor (VDR) may be alternatively used to name this ohmic resistance that decreases with the increase of the voltage (or current). Maybe one day the term CDR standing for 'current-dependent resistor' will arise, but this will not be a good idea to add again a new name. In the past, silicon carbide (mix of silicon and carbon) had this electrical characteristic and therefore was used as surge components. But very soon MOV surpassed SiC varistors due to far better surge-protection behaviour. It is to be noticed that the concept of VDR derivates from a semiconductor mechanism. But this mechanism (by opposition to the controlled and localised semiconductor

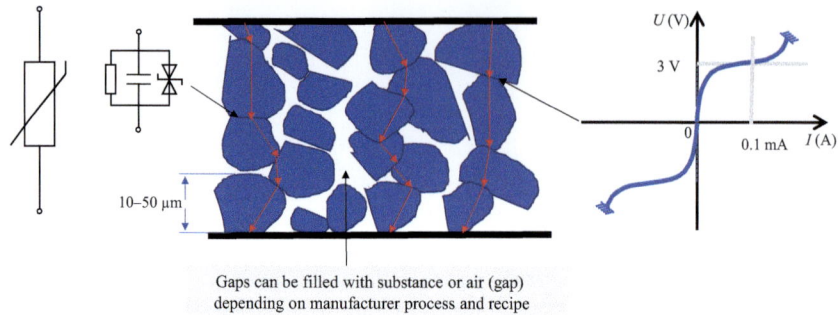

Gaps can be filled with substance or air (gap)
depending on manufacturer process and recipe

Figure 5.11   Internal structure of an MOV

effect for Si diodes for instance) is multiple millions of time and located randomly in-between grains that make the structure of the component itself. The structure of this component can be seen in Figure 5.11.

MOVs are made of several materials that are mainly metallic oxides. The major component is zinc oxide but small amounts of bismuth, cobalt and manganese or antimony oxides may be part of the mix. This is the origin where from the ZnO varistor name and metallic oxide varistors come.

Even if work (the early 1950s) on the electrical parameters of ZnO ceramics seems to have been initiated in Russia, ZnO MOV was developed by Michio Matsuoka and his research group at Matsushita Electric (now known under the brand name of Panasonic, Japan) in 1969. Nowadays, some new researches are still developing new models or theories, but most of the knowledge used today about these MOVs was published during the following decade of its discovery.

5.2.2.2   Manufacturing

The process to make an MOV is very similar to the process of making ceramics (such as dishes plates for instance). This is so close that in fact some persons call the MOV block a 'ceramic'.

Then, work starts with the preparation of the mixture of the various metal oxides to get a powder where of course percentage and size of each grain are major factors. To this powder, it is added some adjuvant that will disappear during the manufacturing process but that are necessary for the process of shaping the MOV. A mixture is made following a precise and secret recipe by the manufacturer, and it is shaped in a disc of controlled thickness using moulds, mechanically pressed and possibly dried in hoven. At this stage, the final shape is obtained but the disc is very fragile and has no possible usage. These are put in a very high temperature furnace with or without gas control (but not exceeding melting temperature) for quite long time (some hours). This process is called sintering.

Once the MOV block has become a hard 'ceramic', it is the time to add conductive means for electrical connection purpose. This is in general made of thin layers of silver that are placed on the surfaces to be connected. This is usually done by serigraphy deposit process and then the intermetallic migration between this silver layer and the MOV ceramic is realised by again a heating process (hoven or furnace).

Few millimetres of the MOV disk border material are not as uniformed as the material that constitutes the centre of the component. To limit the electrical current in this zone, usually the silver layer does not cover the whole surface.

Some specific MOV blocks are specificity designed to be used for high-voltage application and these may have connections that are not with silver but with aluminium that is pulverised at high temperature and pressure on the surface to be connected. This usually occurs due to the fact that MOVs for surge arresters (high voltage SPDs) are not connected with soldered leads but are rather piled up and mechanically connected by surface contact crushed by force. See Figure 5.12 where low-voltage MOVs are on the top half and high-voltage MOVs are at the bottom. The bottom right HV MOV that is broken into pieces to see the very thin aluminium deposit on its surface. The single round MOV in the centre of the figure sizes 20 mm in diameter.

It is also possible that sides of the MOV block are protected by a thin layer of varnish, paint or glass deposit. This may help to reduce unwanted leakage current or possible flashover, the side of the component during its usage.

It is to be considered at this stage that due to the making process (sintering), MOV discs shrink from their original shape to the final shape. This leads to have

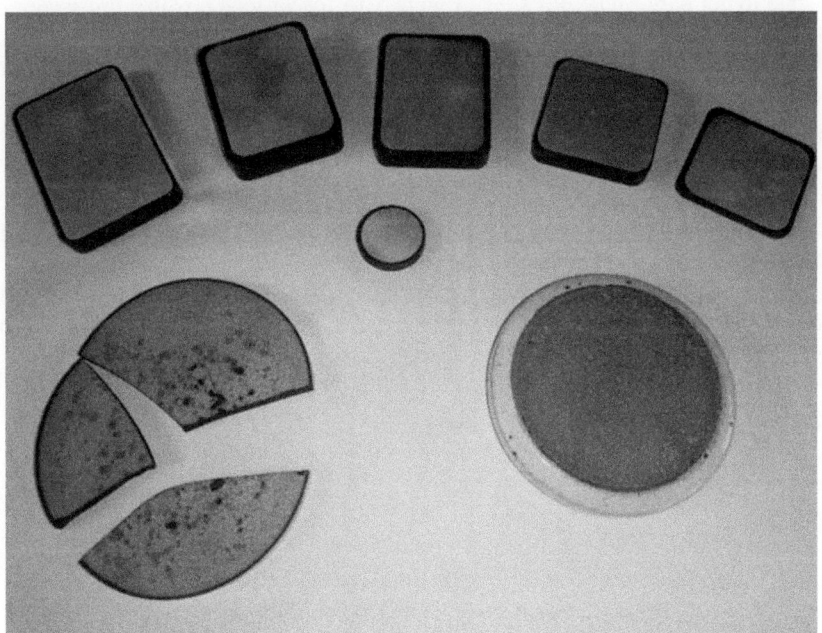

Figure 5.12   Silver electrodes compared to aluminium electrodes

some noticeable size variation between batches. It is not rare that MOV manu-facturers give quite large tolerances in sizing their MOVs as shown in Figure 5.13.

These shapes are the commonly used shapes but one can think of any shape as this is conditioned only by the mould design.

Making MOVs could be compared to cooking a cookie (recipe, mix, shape, mould, cooking) but only very good cooks can achieve to make repeatedly good MOVs.

To finalise the component, one must adapt the electrical connectivity as well as the mounting to each specific application.

It may be using standards terminals, specific terminals, and at the end of the day MOVs are often protected by one or several layers of paint or epoxy to avoid possible direct contact. Some manufacturers provide MOVs totally embedded in plastic housing equipped with mounting flanges permitting assembly on any kind of surface as soon this surface can be drilled and screwed.

Electrodes are very often made of copper with surface treatment permitting an easy soldering to the silver layer (thinned copper is mostly used). The shapes of the electrodes depend on the manufacturer, users and usage. Various electrode shapes are commonly found: Q shape, plate with holes, stars, washer, disk, etc. The cross section is of course driven by the size of the MOV that is directly linked to the surge current capability. It is easily understandable that a 44 mm by 44 mm MOV cannot be leaded with this 0.5-mm diameter wire while a 14-mm MOV disc can (see Figure 5.14).

Leaded MOV is mostly proposed by the MOV manufactures but it is also possible to get bare MOV discs. See Figure 5.14 for various possible presentations that MOV manufacturer can provide.

As for all electronic components, connecting leads for MOVs may have two or more functions. One is to electrically connect the component to a circuit; one other may be to provide mechanical means to secure the MOV on a PC board or other electric circuitry. The last one may be to conduct heat with possible two goals – one being to dissipate the heat in order to increase the MOV surge withstand capacity, and the other one to monitor the MOV over-heating status (indicating the 'soon' end of life).

Finally, MOVs intended to have electro static discharges (ESD) protection are frequently presented in surface mount devices (SMD) package. Some are simple tinny

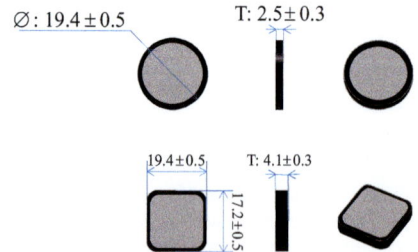

Figure 5.13   Example of lead less MOV block size

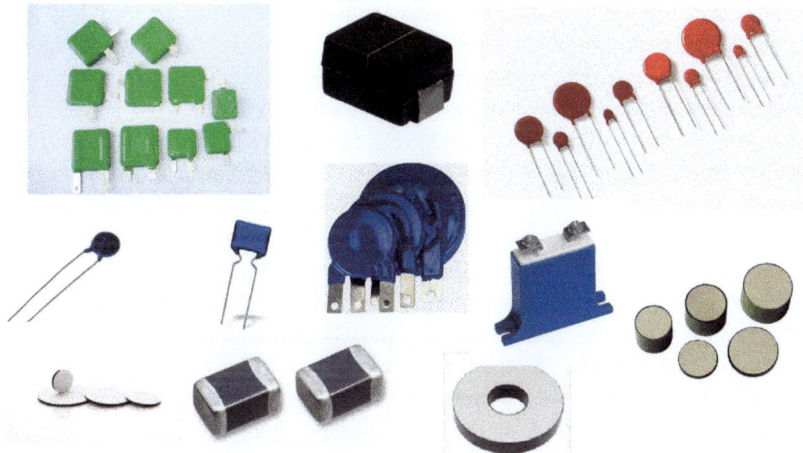

Figure 5.14   Example of MOV presentation (picture from Internet)

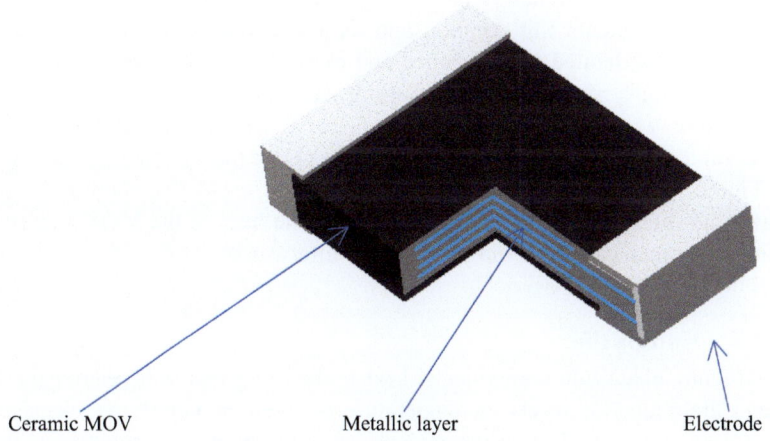

Ceramic MOV   Metallic layer   Electrode

Figure 5.15   Multi-layer SMD MOV design

bricks with two terminals, but to achieve low-voltage and relatively high-surge characteristics, the multi-layer technology is commonly used. See Figure 5.15.

5.2.2.3   Principle

Each micro-joint in between grains acts as a 'back-to-back' Zener diode, a high-ohmic resistor and a capacitor. A zinc oxide varistor consists in fact of a large number of boundaries (several millions) that create a huge network made of serial–parallel connected resistances, capacitors and somehow semiconductor junction's array.

For conventional varistors, the breakdown voltage per grain boundary is about 3 V. But depending on mix of each of the metallic oxides, and mainly the size of the grains (bigger are the grains, less is the junction number for one thickness), this can vary a lot. In consequence, MOV manufacturers are proposing various 'secrete MOV recipes' leading to very different electrical characteristics for the same size of MOV. For instance, it is not rare to find MOVs rated for 30 and 150 V/mm from the same manufacturer. This parameter can be named MOV gradient. Then for the same volume, one can imagine two MOVs with very different voltage limitation characteristics. This is crucial to understand because as already stated, mass and voltage are two of the three parameters that will condition the capability of the MOV to absorb and/or dissipate heat. And this is directly linked to the surge withstand characteristics. In other terms, a heavy and big MOV with low gradient will theoretically provide better surge withstand capability than an MOV of similar mass but with higher gradient. But this can be called into question if the MOV ceramic disc is not homogeneous enough (e.g. the grain size is not uniform and one area of the block is coarser than the rest of the block). Thus, the distribution of the current can lead to having one zone underloaded and another overloaded.

Heat is a parameter that is impacting the behaviour of the MOV. The higher is the temperature, the higher is the leakage current. It makes heat a degradation factor that will impact life duration and its associated performances. In general, MOVs are to be derated (performance and expected life duration) when the operating temperature reached 80°C and more. But this is linked to manufacturers' recipe and they usually provide specific data to address this topic. Heat has a little impact on the clamping voltage but it is more noticeable on the current.

Figure 5.16 shows the energy absorbed for a series of five chocks spaced by a minute and then letting the MOV cool down. The heat of the MOV increases for 5 min and comes back to room temperature afterwards.

5.2.2.4   End of life

As per many electronic components, heat is the thing that will destroy the MOV. When a grain junction reaches a temperature of several hundreds of Celsius degrees (500°C–800°C), it will simply melt and create a short circuit between the two grains. Once the MOV construction is understood, this is easy to guess that this will lead for the other grain junction to have to handle more voltage and thus more current. At the end of the day, this increase will also lead to increase its heat up to its melting. This, as a kind of irreversible avalanche, may end in a full short circuit of the MOV block. This heat increases of course when current passes through the MOV and it may be a current from surge or a current from abnormal voltages. In these two cases, it is due to the event itself, and the time to reach the short circuit is intrinsically dependent on the energy absorbed by the MOV and the dissipation that is not enough to control the heat increase. Time criterion may also be considered as if the heat increase of a grain junction is reached and the avalanche can process but with no time enough for the heat to diffuse in the MOV block; this, in addition to end in a short circuit, probably will break the MOV due to heat exemption

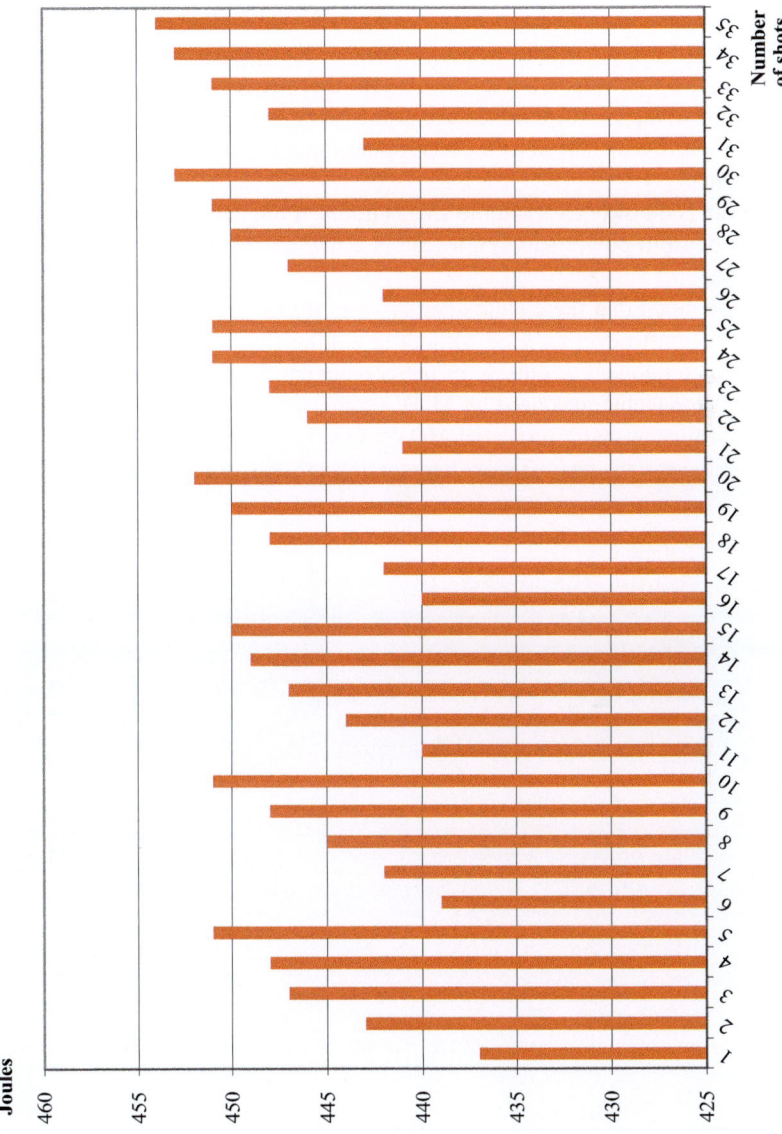

Joules

Number of shots

Figure 5.16 Variation of absorbed energy for a series of five chocks per minute and cool down between series

difference in conjunction with possible electrical arc through the MOV. This is generally called a catastrophic failure.

But failure may not be the only reason for a single event. It may also be after repetitive surges that each time degrades few grain junctions. And there is a time when the nominal supplied voltage is simply able to start an avalanche process; usually the process may be considered slow and is named thermal runaway. See Figure 5.17 for the 1-mA evolution after chocked. If these MOVs were connected to a 230-V AC power supply, it is most likely that the leakage current will start flowing and by doing so, the temperature of the MOV may increase at one point up to another point that will lead to a catastrophic failure that is not an option. This may take hours.

Whatever is the way to reach the short circuit, if the current is not interrupted (from surge or from power supply), it causes a melted channel named a puncture (Figure 5.18).

It is to be noted that there is no alternative to the failure of the MOV and it is why the MOV must never be used without any protection. These protections may be part of the component or outside of the component. Embedded thermal protection is one of the techniques used by MOV producer (Figure 5.19). A few techniques are used for thermally protected MOVs. One must understand that a thermal protection only protects against slow degradation process of the MOV.

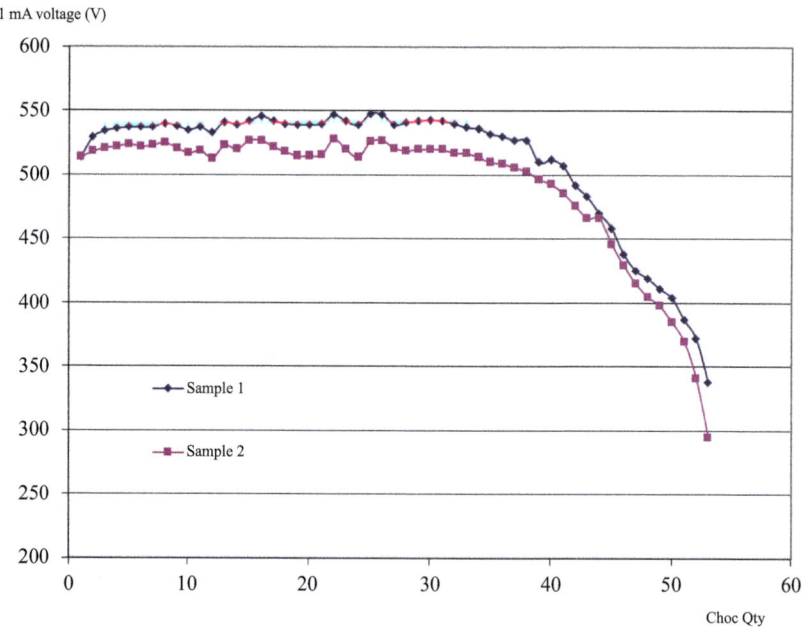

Figure 5.17   1 mA MOV voltage measured after each Choc (30 kA 8/20 and at constant temperature)

*Figure 5.18   Punctured MOV after a sudden event exceeding absorption
            energy limit*

There is no clear definition of slow and fast end of life for MOV but it can be considered that MOV failure occurring in the second's range could be the limit. It can be the temperature transfer speed to a possible thermal switch that sets this boundary. As an example, if an MOV short out in 5 s and the thermal switch had no time to operate, it was too fast, but if the MOV was disconnected in 1 s by the thermal protection, it was slow. It is also common to follow the idea that if the destruction temperature of the MOV is reached at a local point only in less than 1 ms with no chance at all for heat dissipation, this can also be another definition of the slow and fast end of life.

Another technique to deal with the catastrophic failure of the MOV is simply to permit this component to fail without any impact on the surrounding. This technique is simply used for HV surge protection as they are installed out of reach and in free air. This may be a technique on a PCB as soon as there is enough space around the MOV that when it fails, no other part of the equipment is damaged. It may also be possible to place this MOV encapsulated in a fire and/or explosion

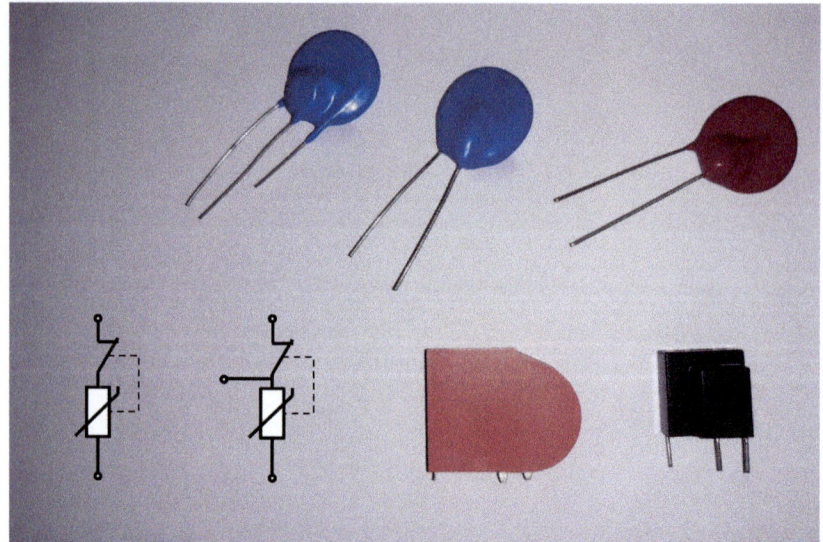

Figure 5.19   Thermally protected MOVs

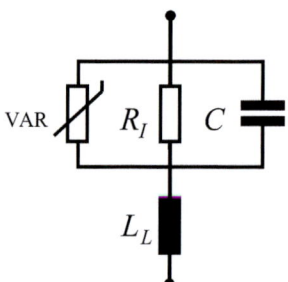

Figure 5.20   Equivalent MOV circuits

proof envelope. This implies of course to know exactly the power that will be available when the MOV turns to short circuit. (Effect of 500-A peak prospective short circuit current is drastically different than 50 kA.)

5.2.2.5   Equivalent circuit

If MOVs have to be simulated, it is possible using Electronic Spice simulation. In that case, some manufacturers provide models corresponding to their components and to be used with software using the so-called SPICE program (Simulation Program with Integrated Circuit Emphasis). However, the spice models are not always available and are not simple to use.

If one needs to have a usable electrical model, see Figure 5.20 that shows a simplified equivalent circuit.

It is simpler to split the behaviour of the MOV in region on *VI* curve as shown in Figure 5.21.

Region I is where the MOV can be almost considered a simple resistor of several mega ohms. Region II is the normal working zone of the MOV and more details will be given further down this subclause, and region III is the high current region where the maximum capacity of the MOV may be quickly reached.

Frequency response may also be addressed and parasitic capacitance and inductance are part of the equation.

In region I, the parameter that will mostly interfere will be R_I, C and L_L. If the voltage across the connection leads is DC only, it can even be only R_I that is to be considered. R_I is several gigaohms. As was expressed earlier, this resistance is impacted by temperature (if $T°C\uparrow$, $R_L\downarrow$).

In region II, R_I impact will vanish and only RVAR will be to consider and this corresponds roughly the part of the curve where the *VI* line can be considered

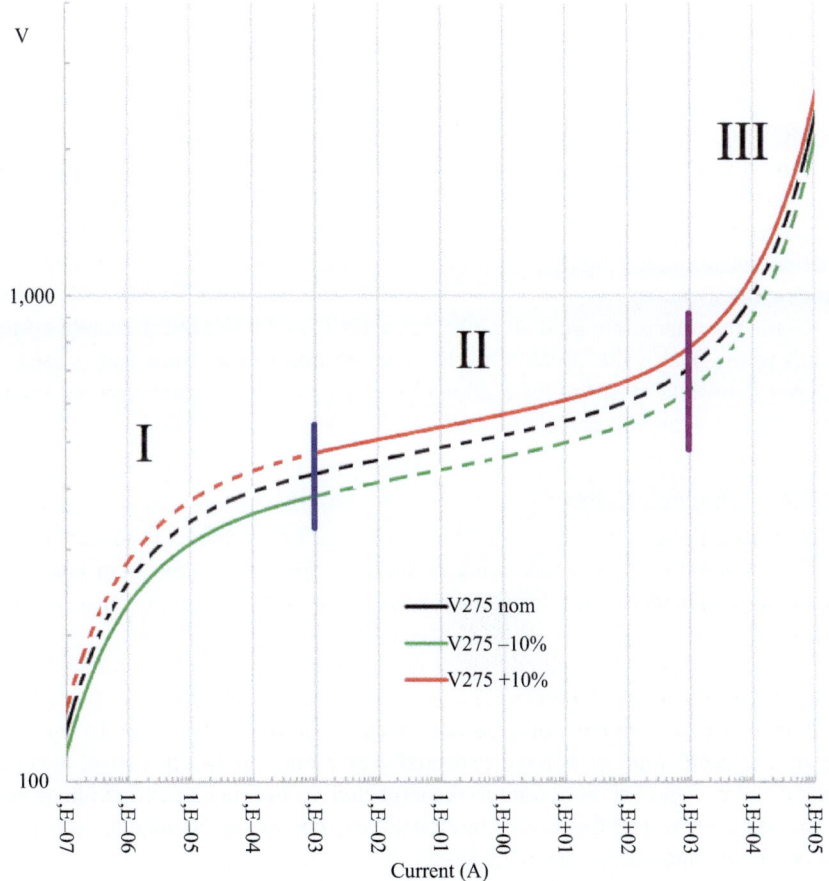

Figure 5.21   VI curve of a typical 275 MOV

straight line. This is the varistor effect. In region III, R_I is still the main impacting resistance except that on the *VI* curve one can detect that it starts to behave such as a fixed resistor and destruction is at its end.

Of course, when AC voltage is applied to the MOV, C and L_L are to be absolutely considered as they are not negligible (few nano-Henry per mm of lead length and from few pico-Farads to some nano-Farads depending on the MOV size and grains sizes). As surge is a transient, this is also to be considered that di/dt generates response time delay due to C, and additional (and unwanted) overvoltage due to L_L.

One should also know that to consider the parasitic capacitor value fixed whatever voltage is applied to the MOV is an easy technique but in reality, this capacitor can vary depending on the voltage applied as it is the case for a varicap diode. (Specific diode that is designed to have various capacitance values depending on the applied voltage.)

For the real *V/I* curve of the 'perfect' MOV, R_{VAR} must also be identified and part of the model. This can be done by using the empirical expression used since Matsuoka's researches.

$$I = \left(\frac{V}{C}\right)^{\alpha} \text{ with } \alpha = \frac{dI/I}{dV/V} = \frac{d(\log I)}{d(\log V)} \approx \frac{\log I_2 - \log I_1}{\log V_2 - \log V_1}$$

where V is the voltage across the MOV terminal, I is the current flowing through the MOV. (I_1, V_1) and (I_2, V_2) are respectively two points on the *V/I* curve. C is constant linked to the intrinsic resistance of the MOV (Figure 5.23). It appears that α is a parameter that has various values depending on where V and I are measured. Of course, if V and I are distant and are not taken on a straight portion of the curve it has no meaning. See Figure 5.23 showing a possible *V/I* characteristic of an MOV and its corresponding α.

Some manufacturers give an alpha value for their MOVs but it usually corresponds to the peak value located on the 'flat' portion of the curve that is usually between 1 mA and 10 A (on the log/log chart). This is a good indication but surely not enough to draw the complete curve.

5.2.2.6   Characteristics

As expressed in Section 5.2.1, *V–I* curve and parasitic components impacting the dynamic response of the MOV when a surge is applied are the main electrical parameter to consider predicting the electrical response of the MOV when stressed by a surge.

Figures 5.21 and 5.22 show typical curves usually provided by MOV manufactures. On the curve of Figure 5.21, it is common to only have represented the solid lines that indicate the worst case of the curve for the MOV usage. In region I, design engineers want to see the expected maximum leakage current at a given voltage and in region II, they want to see expected maximum clamping voltage at a given peak current. But this is a DC concern only. For AC and transient, capacitive and inductive parasitic components have to be taken into account.

Parameters to be considered when it comes the time to size the MOV for a design are numerous. If an MOV shows a perfect curve but can only withstand one

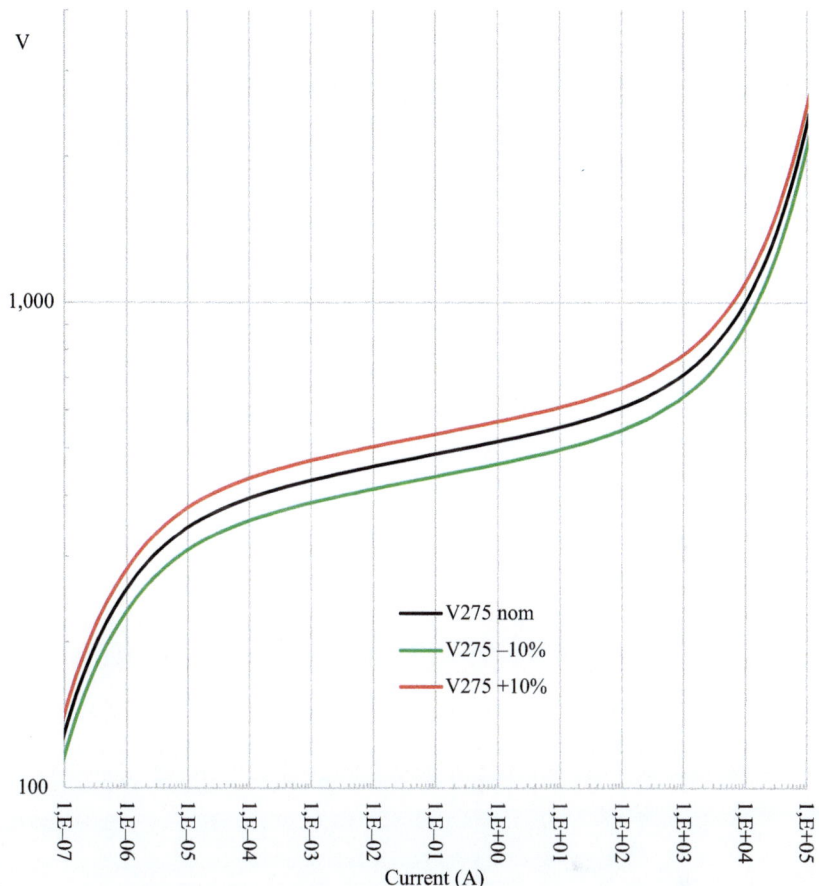

Figure 5.22   VI curves of a 275 MOV (34 mm × 34 mm)

surge, what if this MOV is supposed to see not only 1 nor 10 nor 15 but 1,000 surges? What if this MOV is supposed to protect against very short surges produced by a switching operation of some electrical devices, medium surge derivates from electromagnetic coupling or long-distance lightning bolt or even worse surge issued from a galvanic connection to a lightning strike down conductor? But this is far from being easy just by using the declared information provided by MOV manufacturers.

Of course, to select an MOV for a specific application, it is easier to compare named and well-defined characteristics instead of comparing curves point by point and knowing that it is not only one curve that is to be compared. The important point is then also to know how to pass from declared curves to declared parameters.

Depending on standards and application area, some of the parameters may have various names and their test methods may differ. The following (See Table 5.2) is an attempt to list all parameters that may be found on MOV's manufacturer's literature.

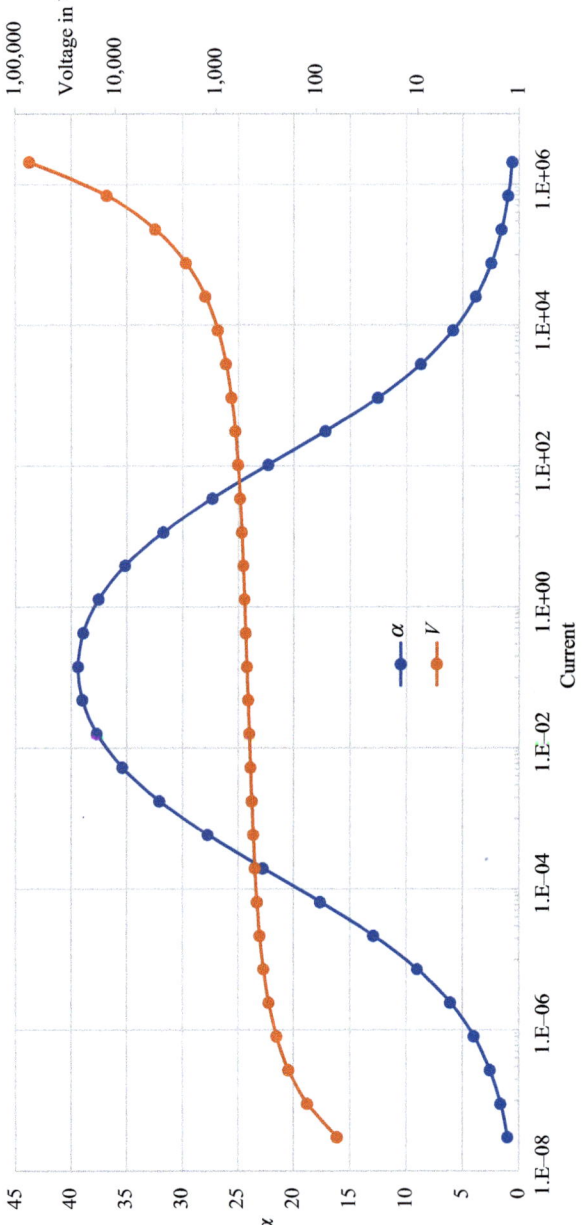

Figure 5.23   VI curve of a typical 275 MOV and its associated α parameter

Table 5.2 Typical list of parameters that can be found on MOV's manufacturer's datasheets

Acronyms	Description	Example	Note	Usage
U_c, MCOV V_{ac}, V_{dc}	Maximum operating voltage to not overpass	275 V AC	May be AC or DC	MOV selection
V_{1mA} V_1, V_0	Measured-limiting voltage at 1-mA DC current	430 V		MOV characterisation
I_n	Same as V_{1mA} Nominal surge current. This may depend on standards applied ($\times 10$, $\times 15$, $\times 20$, time in between shots, etc.)	10 kA		MOV selection
I_{max}	Max 8/20 surge current applied once in only one polarity except if over wise declared	20 kA	May be called 'non-repetitive pulse'	MOV selection
I_{imp}	10/350 impulse current either applied once or following IEC 61643 series	12.5 kA		MOV selection
V_c	Clamping voltage declared in relation to a clamping current	<600 V, @300 A	This is usually not linked to I_n, I_{max} or I_{imp}	One point on the curve
I_c	Current for clamping voltage declaration	300 A		Linked to V_c
C_{typ}	Capacitance (at 1 kHz)	1 pF		Useful when data signal may be mixed with on the power circuit (power-line communication for instance)
W W_{max}	Energy in Joules that the MOV is able to dissipate 8/ 20, 10/1,000 surges or for a current lasting 2 ms depending on manufacturer's declaration	1.1 J		Manufacturer's comparison parameter that is not really useful for users
P_{max}	Maximum power dissipation at steady state condition with ne degradation of the MOV	1.4 W		No usage. Steady state power consumption for MOV is to be avoided
I_L	Leakage current measure at a current lower than 1 mA (e.g. @75% V_{1mA})	<15 μA		MOV characterisation (only useful for control in production)
Alpha	Maximum alpha coefficient for the V/I curve	>21		MOV characterisation

Note: This table is showing parameters picked from various available datasheets on the market. In spite of standards, MOVs' manufactures keep their own way to present their MOVs.

5.2.2.7   Usage

How to select an MOV from an application? From the previous information, one should have understood that this is not an easy task. But the first questions that the user must address are the expected surge stress (whatever its source) and the system to be protected (characteristic). The reader must understand here that these are the common questions whatever is the component to be used.

Expected surges may be known from field usage feedback. For instance, if a design of equipment leads to frequent repairs and that the repair man is identifying that surges are identified to be the reason of the equipment damages. The only good solution here is based on simple try and feedback. If a designer sets a surge protection with such or such component, and the rate of return of his equipment for surge damage reason is low enough for him, he can decide that the solution is good enough. This solution is for sure showing good results but may take very long time. Only experienced companies have enough feedback experience to follow this path.

Expected surges are already defined by preset surge risk analysis. Several standards are setting surge withstand levels for general or specific equipment. For instance, IEC 61000-4-5 is a widely used standard that sets immunity requirements, test methods and range of recommended test levels for electrical and electronic equipment about surges of various sources. ITU, IEEE, MIL, ANSI, etc. also propose their own standards, setting surge stress that a dedicated product must be able to pass. For the following examples, reference will be made to IEC 61000-4-5 and ANSI C136.2 that is targeting exclusively Luminaires. (Lighting poles equipped with LED revel a huge surge sensitivity of equipment on field.)

Note: IEC 60598-1 is also about luminaire and sets different surges tests requirements.

IEC 61000-4-5 defines five levels of surge stress where the equipment is designed connected. (It may be power or signal equipment.) These levels are ranked from Level 1 (almost no stress) to Level 5 (maximum stress requesting special assessment). Let us select Level 4 for this example. It corresponds to equipment that is connected to a system that may see transient overvoltage of 2-kV line to line or 4-kV line to ground. (Electrical environment where the interconnections are running as outdoor cables along with power cables, and cables are used for both electronic and electric circuits.)

The equipment has to pass a test that consists of five positives followed by five negative 2-kV impulses (1.2/50–8/20-μs combined wave shape: max peak current 1 kA 8/20), each pulse occurring within a minute after the previous one. The outcome may be no effect at all up to total damage (but no unsafe condition). In our example, let us select that the equipment may show a temporary loss of function but with full and automatic recovery once the event is passed. In case the equipment is not able to pass this test alone, one may agree to add an SPC to the design. If we consider using an MOV between lines, it may withstand the surge test alone (worst case). In addition, the limited voltage may lead to limit the let through energy that may damage the circuitry. This may be simulated but is usually tested as, at the contrary of SPC, other components are rarely with surge parameter (just considering the normally in-line fuse that is usually a fast-acting fuse rated regarding the circuitry and not regarding surge considerations).

ANSI C136.2 has another approach of considering the surge protection. This standard gives three levels of threat (typical, enhanced and extreme). If we try to line up with the previous example, we may pick up the typical level of transient immunity that corresponds to lighting application dedicated to building entrance and building exterior. The requirement for this level is not to show malfunction but abnormal behaviour is acceptable during the event (the same as prior example). If the device has to be tested with several ring wave impulses (6 kV–0.5 A) per mode of protection and per phase angle, it has also to be shocked by 1.2/50–8/20-μs combined wave shape surges of 6 kV/3 kA (max peak current 3 kA 8/20). If an SPC is required to pass the test, and if one considers only an MOV between lines, the total number of chocs that the MOV must withstand is 14 times the ring waves and 10 times the 8/20 surge currents of possibly 3 kA (while mitigating the effect of the surge to the rest of the circuit).

Once again, simulating and testing are the only way to reach good result not under-sizing or oversizing the SPC. See Table 5.3 that summarises these two examples. It is easy to see that the need for the device is not identical depending on the selected standard. Selecting the MOV (or any kind of SPC) regarding the surge stress is not so obvious and experience mentioned earlier in this subclause may be at the end of the day part of the success in guessing the perfect SPC.

These two examples have been picked to underline that for the estimated same usage (theoretically) standards set very different surge stresses. It is also to be said that if one is more generic than the other, more and more specific application

Table 5.3   Summary of the two examples

Example 1		
Expected stress based on 61000-4-5		
L–L	L/N–PE	Ten times each modes
2 kV–1 kA	4 kV–2 kA	
MOV to select	Min size	
L–L	∅ 10 or 14 mm	
L/N–PE	∅ 14 or 15 mm	
Example 2		
Expected stress based on ANSI 136.2		
L–L	L/N–PE	Ten times each modes+ten altogether
6 kV–3 kA	6 kV–3 kA	
MOV to select	Min size	
L–L	∅ 15 or 20 mm	
L/N–PE	∅ 15 or 20 mm	

Note: The equipment and application are similar in examples 1 and 2. Only the referenced standard is different.

groups set their own requirement regarding surge and that any equipment designer must consider this parameter much more in advance in the design phases than before. Selecting the correct SPC or SPC association cannot be done after the full design is completed (see Figure 5.24 for a basic design mistake).

It is also very interesting to see that application standards are requiring surge parameters that are most likely not in-line with the surge parameters provided by manufactures. They very often provide maximum one-time surge withstand, a clamping voltage at few amps, the 1-mA value but rarely the ten time or fifteen time (time possible withstand current). Some give derating curves based on numbers of applied surge, but not in specific conditions requested by real application, and trying to guess from these curves are more gambling than real analysis.

Table 5.4 is based on experience with several MOVs provided by different producers. Be aware that this may vary depending on MOV's manufacturers but can be used as a base to start MOV selection.

It is not rare today to see equipment designed to be used in many locations of the word. Everybody who travels has experienced that his power supply of laptop is able to be plugged in several countries provided the user has the plug adaptors. But what is about the 120 V of the USA, and the 240 V of the UK? Power supply of a laptop may be equipped with an MOV. These are obviously selected to work with maximum supplied voltage. This is also true for other equipment and it is not necessary to link the MOV to the applied voltage if the system has been designed to be compatible with higher voltage. For instance, photovoltaic (PV) inverters based on the same platform and designed to work with 500, 800 and 1,200 V DC may be all equipped with the MOV set for the 1,200 V version. This points out that the selection criterion is not any more the supplied voltage (for 500 and 800 V version) but the coordination of the SPC and the rest of the circuitry. (Warning, this is true only if components for lower voltage version do not alter voltage withstand.)

Let us continue with our examples of an LED lighting system control. Let us consider that we have equipment that is interconnected to both power and signal circuitry – one to power the light and the other to control effects such as dimming for instance. Power system will be addressed at various locations in this book but to select proper SPC one has to question this topic as well. As per our example, if we still consider the line-to-line protection of the power line, the nominal voltage and its maximum voltage regulation are obviously to be part of the equation. The possible temporary over voltage is a very important parameter to consider as well. This is linked to all 'voltage characteristics' that MOVs' manufacturer usually provides. To pick the corrected voltage parameters of the MOV, there are some easy rules:

- U_c maximum continuous operating voltage (MCOV) cannot be less than the maximum voltage regulation of the system.
- Higher U_c (MCOV) reduces temporary overvoltage (TOV) consideration.
- Lower U_c (MCOV) leads to better surge protection action.
- Bigger MOV provides better protection level for high current.
- The bigger is the MOV, the bigger are the problems (size, cost, etc.).

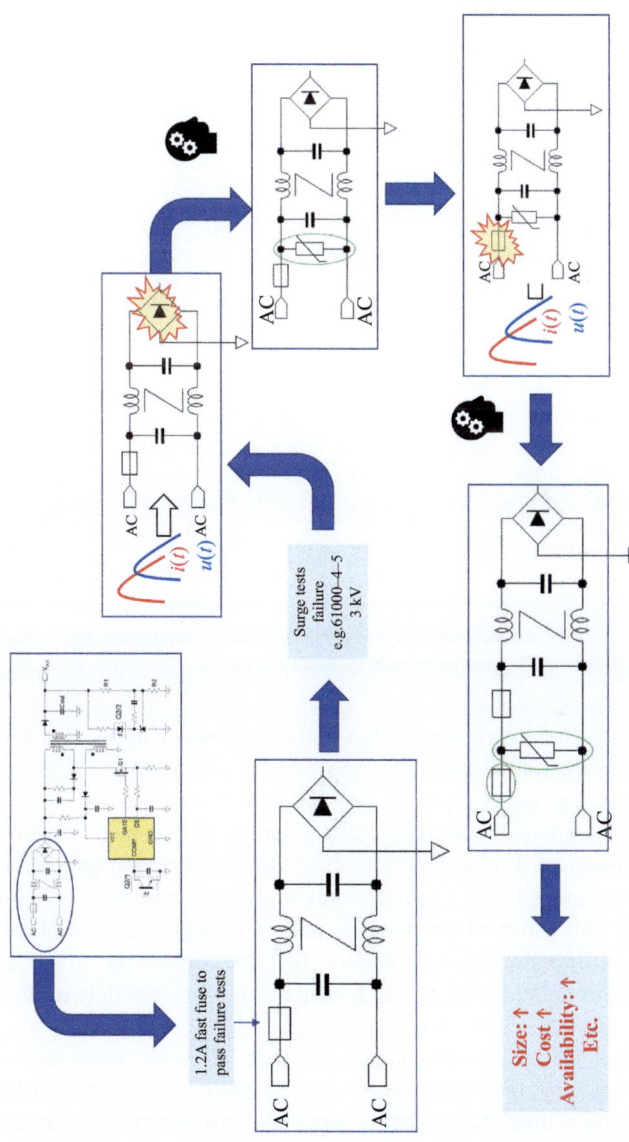

Figure 5.24 Basic mistake for design using SPC

Table 5.4   Size of MOV to pass application test

MOV shape	MOV size (mm)	61643-331 I_{imp} (10/350:kA) 0.1 to 1 (×5)	UL1449 In (8/20:kA) ×15	610051-1 I_{max} (8/20:kA) ×1	IEC 61000-4-5 Level (8/20:kA) ×10
Round	5	–	0.25	0.5	0.27
	7	–	0.6	1.2	0.65
	10	–	1.2	2.4	1.3
	14	–	2	5	2.2
	15	–	3	6	3.3
	20	–	5	10	5.5
	25	2	7	18	8.8
	34	3.5	15	30	16.5
	40	4.5	20	40	22
	50	8	30	70	33
	54	9.6	40	80	44
Square	14 × 14	0.8	3	7	3.3
	15 × 15	1	4	8	4.4
	19 × 19	1.4	6	12	6.6
	20 × 20	1.6	7	15	7.7
	25 × 25	2.4	110	22	11
	34 × 34	4.5	20	50	22
	42 × 42	7	30	65	33
Rectangular	34 × 18	2.5	10	22	11
	34 × 44	6.5	25	50	27.5
	54 × 34	8	30	65	33

Note: All values may vary from manufacturer to manufacturer and recipes may also modulate these numbers. For ANSI 136.2, shock quantity is mixing modes and cannot be summed up in this table.

This leads to reach a compromise but one should never forget that MOV will end one day in a catastrophic failure.

TOV problem, for the equipment, is in general left to the designer decision. Nevertheless, some standards are really taking care of the problem and are setting conditions. For instance, IEC 62368-1 (audio/video, information and communication technology equipment) has a specific TOV test and requirement for MOV used on these devices. This standard also requires that U_c (MCOV) be at least 1.25 times the rated voltage of the device. As a remark, this does not permit to use an MOV with $U_c = 275$ V for a 230-V AC power system even if this is widely used for other equipment.

There is also a difference if the power system is AC or DC. MOVs have a kind of reversible polarisation that makes the *V–I* curve not any more symmetrical. This occurs after a certain time the voltage is applied. This polarisation time may be shortened if this is done at hot temperature (e.g. 120°C) and this polarisation effect is recoverable vs degradation effect that is not. This polarisation leads to increase the leakage current by about 10%. This may be of importance when an SPC is dedicated to DC application (e.g. PV, traction, etc.). Some MOV users polarise in-house their MOVs before the control and the connections of the equipment on field. Then for DC power system even if the leakage current can be considered ten time less compared to AC (50–60 Hz, widely due to the capacitance of the MOV), manufactures usually apply a 10% margin on the initial U_c measurement to cover this polarisation.

To understand about the possible use of MOV for the signal line, if the surge parameters selection follows the same concept than the one for power lines SPC's selection, the nature of the transmitted information is crucial and may even forbid MOV usage. By nature, MOVs are also a non-negligible capacitor that makes their usage unwanted for frequency signal. Few hundreds of hertz will either interfere with the transmitted signal or create a very high capacitive leakage current. If the signal is digital, the transition front voltage will be smoothed and timing may be altered. Then MOV is generally used for power application with an exception permitting its usage for signal circuit or line protection that is when used in series with other SPCs such as a gas discharge tube (GDT).

It is possible to associate MOVs in series or in parallel. In series, no real care is to be applied and this is common usage to have these in series when U_c voltage cannot be reached with only one MOV. It may be due to space concerns as well. In that case, surge current capacity is kept and only the maximum power dissipation is increased (see discussion about mass–volume ratio and surge current carrying capacity). MOVs associated in parallel are also widely used. But here the association is not as simple. As already stated, MOVs have +/− non-negligible tolerances (usually 10% on the 1-mA voltage). In case 2, MOVs are not well paired, the risk is that one MOV will take big majority of the surge current as the curves on region II of Figure 5.21 are pretty flat and a small variation of the clamping voltage will lead to huge current variation. Thus, this is not possible to claim that two 'identical' MOVs when stressed by 10 kA 8/20 will equally share the current with 5 kA each. In consequence, one can use a rule of thumb that is simply rating the maximum surge-carrying capacity of MOV block reducing by 20% the sum of two MOVs (or MOV blocks). As an example, a double MOV block of 2 max 40 kA peak current each will lead to a maximum current-carrying capability of 64 kA ($40 + 40 = 80 \times 0.8 = 64$). Figure 5.25 shows a graph that can be used to follow this rule. This is pretty usable for MOVs that are not specifically sorted.

The 1-mA value is one of the important parameters and may be a good start to achieve better current sharing. It is not rare that MOV users sort their MOV to reach better current capability. It is obvious that if MOVs are with similar 1 mA values, the sharing will be much better than seen previously and one can almost claim that the current is shared equally. But it may not be precise enough and some MOV users are even sorting the MOV for parallel association under much higher current but not destructive (some pulses of 10–50 A for instance). All the tricks are to know when to stop and not to spend more money in pairing MOVs than the cost of bigger MOVs.

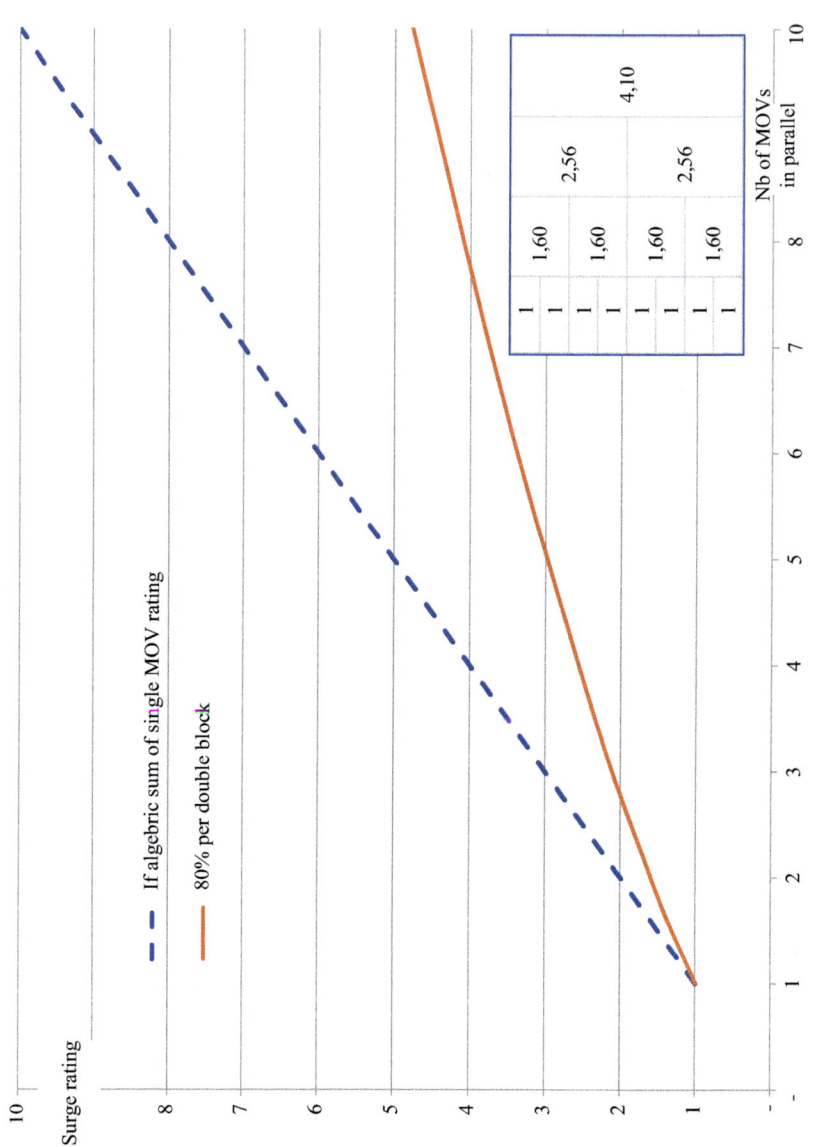

Figure 5.25 MOV's parallel association

5.2.3   Silicon PN-junction voltage limiters

5.2.3.1   **Introduction**

First trace of PN junction seems to be seen at the end of the nineteenth century and was about an AC/DC rectifier. We had to wait for almost three-fourth of century (mid-twentieth century) to see the transistor revolutionised the electrical industry. But with this first diode, it was already possible to think of an SPC. In 1934, Zener effect was discovered (named by its inventor Clarence Zener) and one among others usages is to use this PN junction as a voltage limiter component (Zener diodes). Further works lead to discover the other effect that was very similar but not totally identical and that is called the avalanche breakdown effect.

Regular diodes, Zener diodes and avalanche breakdown diodes are frequently used to achieve surge protection. An additional PN junction also exists that is considered a limiting SPC. It is known under the names punch-through or foldback diodes. These are the ones that will be discussed in the following section.

5.2.3.2   **Manufacturing**

Concept of PN junction may be one of the most explained subjects in electric schools or literatures but as a reminder, see Figure 5.26 that shows the PN junction and its associated symbol.

There exist various types of diodes. The doped regions and their shapes depend on the effect that is targeted. The depletion region (*D*) can be thinner or thicker for the same applied voltage that is for instance what varies between a Zener diode and an avalanche breakdown diode.

For an SPC, it is most likely that several PN junctions are located in the same case of the component, which will be more detailed further next. Manufacturing semiconductor is a process that is well under control these days and diodes for surge protection can be considered huge compared to the miniaturisation that can be achieved nowadays for semiconductor ships, including millions of transistors. Nevertheless, the process may be summarised as follows. A thin pre-doped silicon disc (N or P) is oxidised during a passage at very high temperature (under controlled gas). This wrapped-in insulator disc (oxide) is then photo-etched using photo-sensitive masking and acid bath. The doping of the semiconductor zone uncovered after etching is carried out by passing it at very high temperature in a gaseous atmosphere containing the doping element (P or N). Then a repetition of etching and deposits of layers (insulating or conductive) forms a structure which ends with the electrical connections. These will be connected by soldering to the

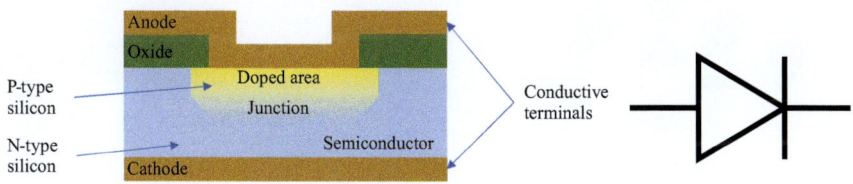

Figure 5.26   PN junction and basic electrical symbol (diode)

component connection pads or lead. The body of the component is generally formed by the resin overmoulding of the assembly (with the exception of the connection tabs or leads).

It is therefore quite obvious that manufacturers can assemble several diodes in series, in parallel, head-to-tail or even any combination in a single package. This process is common with all semiconductor SPCs. The variants will be the number of single PN junctions, their arrangement and doping characteristics (mainly the doping level, thickness of the depletion zone (D) and nature of the dopant).

Limiting surge components such as diodes (all types) are presented in various sizes and shapes. See Figure 5.27.

In general, semiconductors dedicated for surge protection limiting function are of the same principle than other PN junction but with a specific care for quick heat dissipation due to fast and high current transient. It may also have a larger depletion surface providing a larger conduction cross section to handle the huge peak current (or multiplied junctions further explained).

5.2.3.3   Principle

It is well known that a diode can be simply defined by its basic function that is to block the current in one direction and let it pass in the other direction. As shown in Figure 5.28, when the anode is positive compared to the cathode, the depletion zone is missing with electrons and the 'holes' are filled with the electrons coming for the cathode creating electron's movement: current is flowing. On the contrary, when the anode is negative the holes are naturally filled and no need for electron displacement: nor current. Of course, this is only the starting point.

Consider a simple PN junction as a forward and reverse V/I curve. The forward region is when the current is passing with very little limitation. For a silicon diode, the cost to pay is a loss of some hundreds of millivolts (the so well-known 0.7 V).

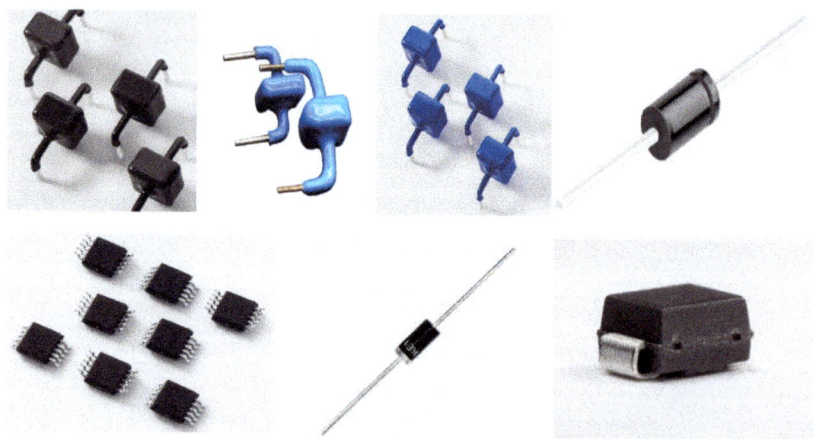

Figure 5.27   Various limiting silicon surge-protection presentations (pictures from the Internet)

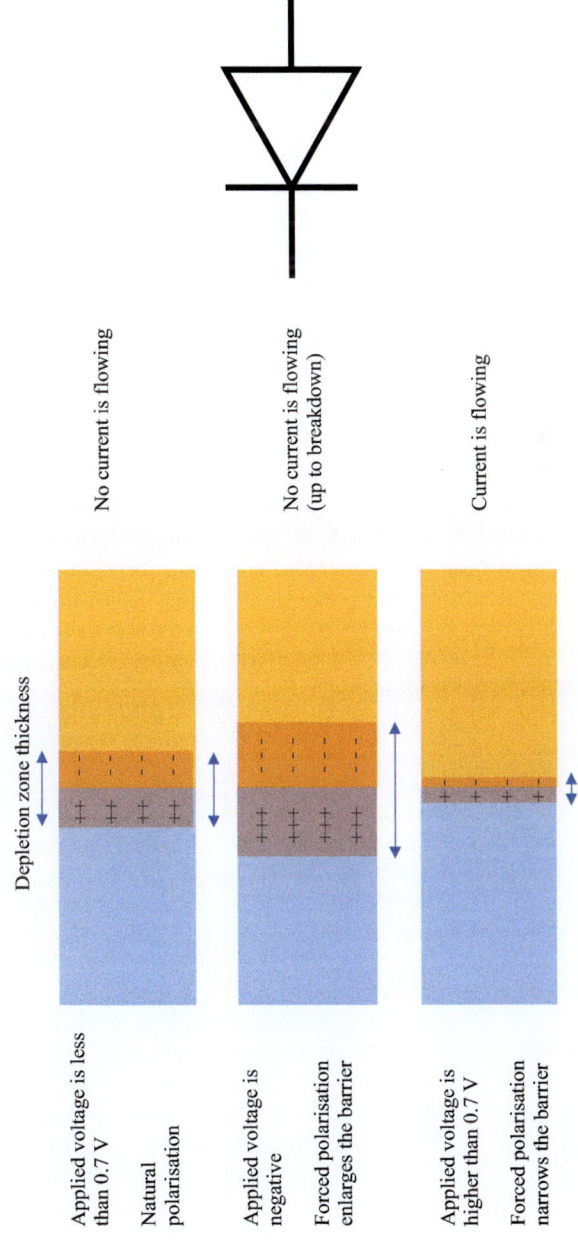

Figure 5.28   PN junction and basic electrical symbols (diode)

This voltage is not to be considered negligible as it will part of the equation that limits the maximum forward current. The maximum energy absorbed by the diode in conduction mode is limited by the maximum temperature not to reach. It is given by producer and is usually within 120°C–160°C range.

Overheating may lead to thermal runaway that is generally destructive and may lead to a short circuit.

This is the forward quadrant that is used for surge protection as it is designed to provide an insulating function when in the reverse quadrant. In case the breakdown of the reverse quadrant is reached, the standard PN junction is in a dangerous zone and may be destroyed by dielectric breakdown. One should have noticed that 0.7 V is a very low 'clamping voltage' and that applications with so low voltages are unusual. However, association techniques are commonly applied to really use widely this 'basic' PN junction for surge-protection purposes (see Section 5.2.3.4).

For Zener and avalanche breakdown diodes, the most useful quadrant is the reverse one and they are designed to withstand a certain energy while limiting the voltage as an MOV would do but with a far sharper knee curve as shown in Figure 5.29.

Zener and avalanche breakdown diodes are very similar when comparing their *V/I* curves. Nevertheless, the process when they are in their reverse quadrant is slightly different. The way to know if a diode is designed to use the Zener effect or avalanche breakdown effect is simple as if the reverse 'clamping voltage' is less than 5 V, it is a Zener diode. If the reverse clamping voltage is higher than 8 V, it is an ABD. If the reverse voltage is in between, the effects are combined.

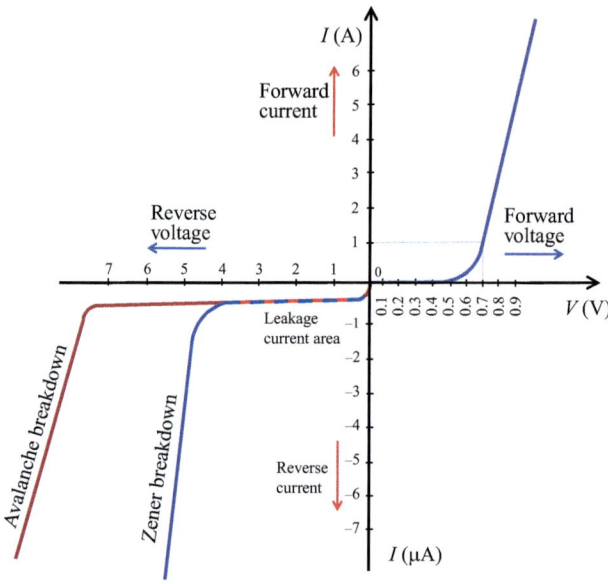

Figure 5.29   PN junction VI curve

To address the slight differences between Zener breakdown vs avalanche breakdown diodes, see Table 5.5.

Diodes are by construction with non-negligible parasitic capacitance. The depletion zone acts as the dielectric and the P and N doped zones act as the two electrodes. One can understand that this capacitance is linked with the applied voltage as the depletion thickness is dependent on it.

Constructions on Figure 5.27 show that this SPC may be equipped with terminal leads that are in our transient world means parasitic inductances. Of course, the general guidance is always shorter than connecting lead length to minimise this unwanted characteristic.

Temperature also has an effect on the forward voltage. The higher is the temperature, the lower is the forward voltage for a given forward current.

Another voltage-limiting semiconductor component is part of the family. They are in fact a multi-junction PNP or NPN such as standard transistors but the base is not used and it does not have any external connection even. This type of semiconductor PN association has a V/I curve very similar to a standard diode up to a point when it has a kind of fold back in its characteristic. This is natural that this technology is called fold-back diode. As for the forward and reverse quadrant if the doped and depletion regions are uniformed in size, one can see a symmetrical curve. But an arrangement may exist that leads to have a thinner region resulting in a non-symmetrical curve. Then forward curve is very similar to a regular diode in that case the phenomenon is called punch-through and the SCP may be called as such (Figure 5.30).

As surge event is random as soon as its polarity is questioned, it is very frequent that manufacturers assemble various PN junctions in the same case to get bidirectional SPCs. As this is a technique used to minimise the capacitance of the

Table 5.5   Zener breakdown vs avalanche breakdown diodes

Zener breakdown	Avalanche breakdown
The free electrons in depletion zone move thanks to the high electric field as it is a very narrow region	The free electrons in thick depletion zone move due to the collision with other accelerated electrons
Zener breakdown is due to high electric field and the depletion zone vanishes during conduction	Avalanche breakdown is due to collision of free electrons and the thick depletion is crossed by these during conduction
The Zener breakdown voltage is inversely proportional to the temperature	The avalanche breakdown voltage is directly proportional to the temperature
The breakdown voltage V_z is less than 5 V	The breakdown voltage V_z is greater than 8 V
When the temperature increases the breakdown, voltage decreases	When the temperature increases the breakdown, voltage increases
The VI characteristics curve shows a sharper knee	The VI characteristic curve shows a less sharp knee than Zener breakdown
Highly doped	Lightly doped
	ABDs have been named in the past SAD (silicon avalanche diodes) and Transil©

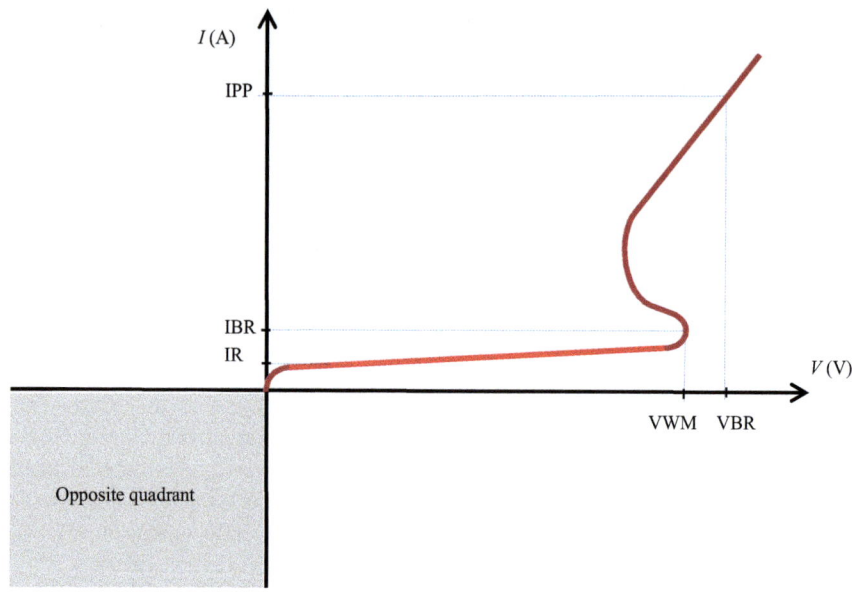

Figure 5.30   Punch-through and fold-back diodes' one-quadrant characteristics

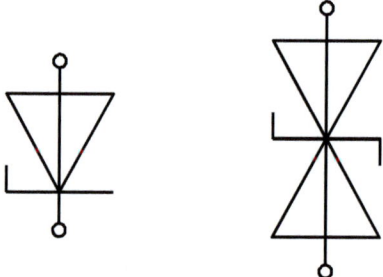

*Figure 5.31   Zener diode, ABD and fold-back diodes unidirectional and
              bidirectional symbols*

protection, this capacitive behaviour is not welcome in the application. The next chapter will give some examples of frequently used arrangements.

See Figure 5.31 that Zener and avalanche breakdown diodes have similar symbols and they frequently respond by the same acronym TVS standing for transient voltage suppressor.

5.2.3.4   Multiple PN junctions

As discussed earlier, PN junctions can have several behaviours and it is useful to assemble single-PN junction altogether in a similar way to achieve specific characteristics. Single-PN junctions have two different quadrants. If one wants to have a bidirectional characteristic, it comes naturally the idea to assemble at least two standard diodes head-to-tail or two TVS back-to-back (see Figure 5.32).

In addition, PN junctions may not be able to support reverse voltage or forward current whatever is the specific design of these. Then association of diodes and TVS may be an option, and each PN junction has its blocking or limiting function in both quadrants. It may even be designed to have no symmetrical behaviour targeted characteristics on both quadrants that are not reachable by a unique junction. See Figure 5.33 for better understanding.

One way to increase the clamping voltage is to simply have various PN junctions in series. This is a widely used technique for regular diodes to reach clamping voltage higher than 0.7 V. For instance, four times 0.7 V leads to a clamping voltage of almost 3 V that becomes interesting to protect any electronic systems using modern logic technology (such as PCs). The other advantage to assemble PN junctions in series is that it will reduce the parasitic capacitance. It is even very frequent to have a PN junction that is only intending to reduce the overall capacitance. By opposition having multiple diodes in parallel increases his capacitance but increases enormously the current-carrying capability (see Figure 5.34). If the TVS is a basic function, it is possible to find from manufacturers the same apparent component presenting identical basic characteristics but with totally different

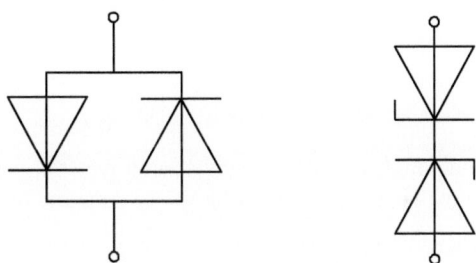

Figure 5.32   Back-to-back and head-to-tail PN junction association

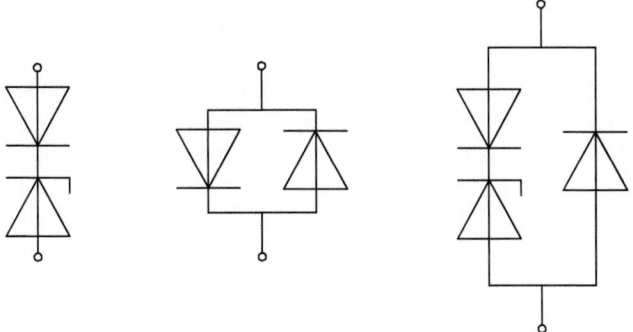

Figure 5.33   Arrangement of different types of junctions

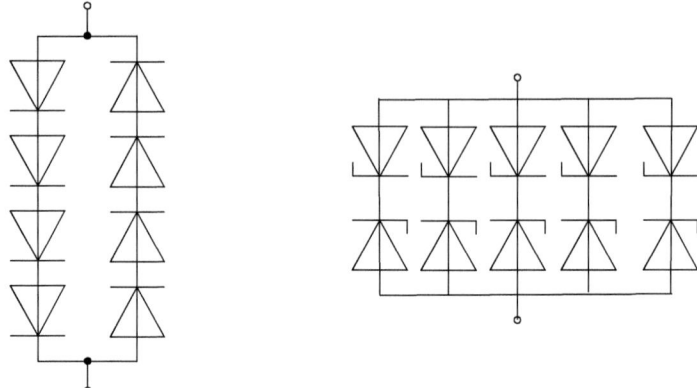

Figure 5.34   Series parallel PN junction association (array)

association. This may become a problem when a component is searched by a user to fulfil unusual and specific requirement even if still is in the field of surge protection. The detail of manufacturing is not generally given by producers.

Most of the power TVS that are available on the market are presenting *VI* curve as presented in Figure 5.30.

Once the concept is understood, it is easy to even imagine complex arrangements with multiple junctions of various nature and multiple external connections for specific application. For instance, if a TVS design is required to reach huge surge current carrying capability, it will have several TVSs in parallel to increase the overall peak pulse current. In addition, if these are mounted on a material with good heat transfer and of good mass, it will reach up to 10 or higher kiloamperes of 8/20 surge withstand.

5.2.3.5   End of life

On the contrary of the MOVs, a TVS does not degrade for each surge as soon as the maximum current or junction temperature is not reached or overpassed. This makes this technology quite robust for surge repetitions. But when this limit is reached, PN junctions have basically two end-of-life behaviours. Either they explode when the stress that they are facing is too sudden or they simply melt ending in a short circuit. Some specific designs using ABD are surrounded by a metallic tube that forces the junction to melt and, even in the case of very powerful surge, will most likely end up in a controlled short circuit that is able to handle several additional surges with no further degradation other than being a short circuit.

5.2.3.6   Equivalent circuit

Most of the diodes are available with a Spice model and *VI* characteristics are usually part of the datasheets. Of course, if they have to be studied in normal use and when they have to show not interacting with the frequency signal, they simply can be replaced by a capacitor. Just keep in mind that their capacitance varies with the applied voltage.

5.2.3.7 Characteristics

Only few manufactures are making TVS for power applications as this is a pretty new application of these components and as soon as one compares the cost of an even much more powerful MOV, the financial logic is not in favour of TVS. See in Table 5.6 usual characteristics that are provided by TVS manufactures (this table only focuses on power TVS that are usually fold-back diodes).

5.2.3.8 Usage

Diodes are widely used for low signal protection but their use is also possible for power-line protection. The sharpness of their *VI* curve is of real benefit.

The process of selecting a diode of power circuit is very similar to the one we already addressed for MOV usage (see Section 5.2.2.7). The only difference is that this is not enough common to have triggered the writing of a standard setting rules. However, users may address special care for the end of life of the diode. Short circuit protection is not an option and thermal monitoring could be imagined even if it is not really addressed by the producers. One of the best usages for these TVS is to be used as a secondary stage of a set of coordinated SPCs. If the maximum surge capability is never reached, one may expect no failure due to transient. But power supply circuits are not only subject to transient and TOV will have the same effect on TVS than on varistors.

5.2.4 Selenium rectifier

Before efficient and modern silicon diodes, a technology was used as a power rectifier and in the 1970s it was pushed for a surge protection use. This selenium diode technology behaves as an avalanche diode but with much less linearity in the voltage zone than even an MOV. Each semiconductive plate constituting this selenium element has a reverse voltage with a clamping voltage around

Table 5.6   TVS typical declared parameters

Acronyms	Name	Description	State
VBR	Breakdown voltage	Maximum repetitive breakdown voltage at IBR = 10 mA (e.g. 320 V)	From OFF to ON and ON to OFF
IBR	Breakdown current	Maximum repetitive breakdown current (e.g. 10 mA)	
IPP	Current rating	Peak current rating per 8/20 μs IEC 61000-4-5 IPPM (e.g. 10 kA)	ON
V_c	Clamping voltage	Maximum clamping voltage at IPP (e.g. 340 V)	ON
V_{WM}	Standoff voltage	Reverse stand-off voltage at IR or ID (e.g. 350 V)	OFF
IR or ID	Standby current	Standby current or leakage current or (e.g. 10 μA)	OFF
C	Capacitance	Maximum capacitance at $F = 10$ kHz, $V_d = 1$ V_{rms} (e.g. 0.3 nF)	OFF

Figure 5.35   Selenium surge suppressor component (picture from the Internet)

25 V for few amperes (forward voltage being around 1 V). They are totally obsolete in terms of performance compared to diodes and MOVs but some manufacturers (mainly in the USA) are still providing this technology claiming surge-protection functions. One advantage of this technology would be to protect equipment and installation against TOV of few seconds in association with ad hoc SPC or devices. This thought is mainly driven by the fact that these diodes are by construction able to dissipate enough energy in the case of limited power frequency overvoltage that could be due to power system fault lasting few seconds. But it is very different from transient stress due to lightning for instance and is mentioned in this section as it was one time used as an SPC.

See in Figure 5.35 a picture of such component.

5.2.5   Capacitors

Capacitors are not really considered an SPC alone. They are even frequently cited in this chapter and most likely pointed as unwanted. (They are parasitic.) But its behaviour that tends to oppose to any rapid variation of voltage may be used to help other components to act in a better way. For instance, it has been used in parallel to GDTs to limit the *dv/dt* that leads to higher triggering of a GDT. They are also frequently used in association with other components to make filters (see Section 5.6). For high-frequency application, they are frequently connected in series with the signal to transmit.

Capacitors have several technologies that are mainly the nature of the dielectric: paper, chemical electrolyte, ceramic, tantalum, plastic, air. Transient is a fast event and then big electrolyte capacitors have almost no effect on surge transmission.

The question about capacitors is more about their ability to handle overvoltage without any destruction. Studies in the 1980s have shown that standard capacitors were able to withstand transient overvoltage up to 10–15 times their rated maximum voltage. Capacitors today are widely used for EMC protection and are very often connected to protective earth (PE/Ground). This leads standard's experts in security to invite capacitor manufacturers to design capacitors with huge surge withstand and presenting the so-called self-healing characteristics. The goal is not to risk any inter-connection between active and dangerous wires and the ground (for human being in the case of direct or indirect contact to a conductive element not supposed to be hot). Then the capacitor datasheet often shows surge withstand capability properties or at least provides the classes they have been tested for. For example, the requirements for Class Y2/X1, Y3/X2 and X2 are classes of capacitors operating directly under mains voltage and are tested with surges from 2, 4, 6 or even 6 kV. For information, Class X capacitors are used between phases and between phase and earth, and those of Class Y are used between phase and earth or between neutral and the earth.

5.3   Voltage-switching components

5.3.1   Introduction

Some are using the term crowbar component to design the components that will be described in this clause. Voltage-switching component is a family of various components and this name is more linked to the electrical behaviour than to their design or principle and is now the usual term to be used (at least for IEC document's users). The term switching expresses the fact that the voltage is the trigger that literally switched the component from one state to another state (by opposition with the voltage limiting that can be considered smooth). It can be said that a voltage-switching component has a turn on level that changes suddenly its state from high-ohmic condition to very low-ohmic condition. Usually, the voltage-switching components are parallel mounted. But some components may exist with IN and OUT leads to shorten the lead connection length.

These components short the voltage output by diverting the current acting as a kind of short circuit. This process with the effect on current and the 'transmitted voltage wave' is shown in Figures 5.36 and 5.37.

It can be noticed that these components switch off the voltage during the full surge duration and, therefore, are subject to an energy that is only proportional to the current and the time duration of the surge event as the voltage may be

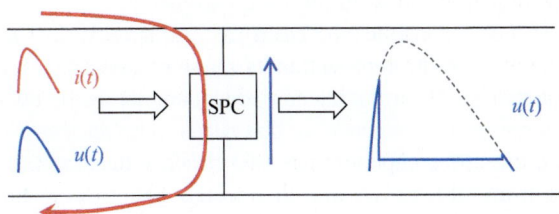

Figure 5.36   Voltage-switching component parallel mounting

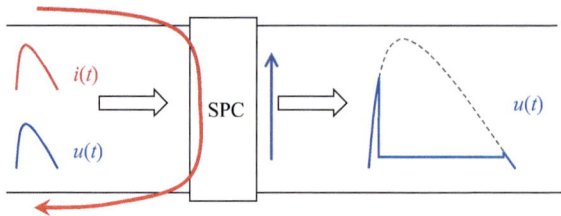

Figure 5.37　Voltage-switching component in-line mounting

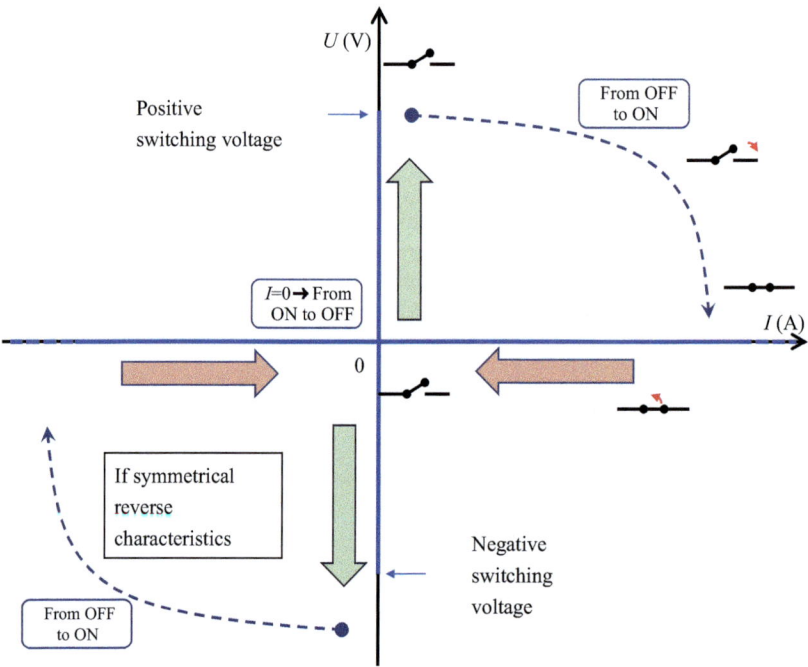

Figure 5.38　V/I voltage-switching component ideal curve

disregarded. The surge withstand parameters of these components are consequently linked to their ability to let pass through the current.

Ideal static *V/I* and dynamic *u(t)* electrical characteristics of an ideal voltage-switching components can be represented as those in curve Figures 5.37 and 5.38. Even if *V/I* characteristics for switching component are not the best way to define how they work, it is a common representation that needs to be presented (Figure 5.38).

But imagine that the component has two stable states and that there is a condition to switch from OFF to ON and then a new condition to turn back to OFF again. OFF is an open circuit or very high ohmic resistance and ON is such as a short circuit or a null resistance.

To simplify the behaviour, one can simplify the process by the following two statements:

• To turn on, the voltage must pass a certain voltage (switching or triggering voltage).
• To turn off, the current must decrease down to 0 A (Figure 5.39).

More realistic *U/I* characteristics and $u(t)$, $i(t)$ are shown in Figure 5.40.

In Figure 5.40, depending on the technology, some of the switching components may not turn from On to OFF if some conditions that are not combined are not achieved. For instance, if the current was high enough to keep heat and conduction state on, the zero crossing of the current will be with no effect. The same if large dv/dt is observed at the current zero crossing for other technology.

There are three technologies that are used for voltage-switching components today for surge protection and they are the GDTs (see Section 5.3.3), air gaps or spark gaps (see Section 5.3.2) and semiconductors (see Section 5.3.4). Air gaps or spark gaps are mainly used for power application, whereas GDTs are used for both power and data/signal application, and semiconductor are seldom used in data/signal. Details will be given in following subclauses.

As a conclusion for this introduction, see the IEC definition for switching voltage component, which is as follows:

SPC that has high impedance when no surge is present but can have a sudden change in impedance to a low value in response to a voltage surge.

NOTE 1 Common examples of components used as voltage switching devices include spark gaps, gas discharge tubes (GDT), thyristors (silicon-controlled rectifiers) and Triacs. These SPD are sometimes called 'crowbar type'.

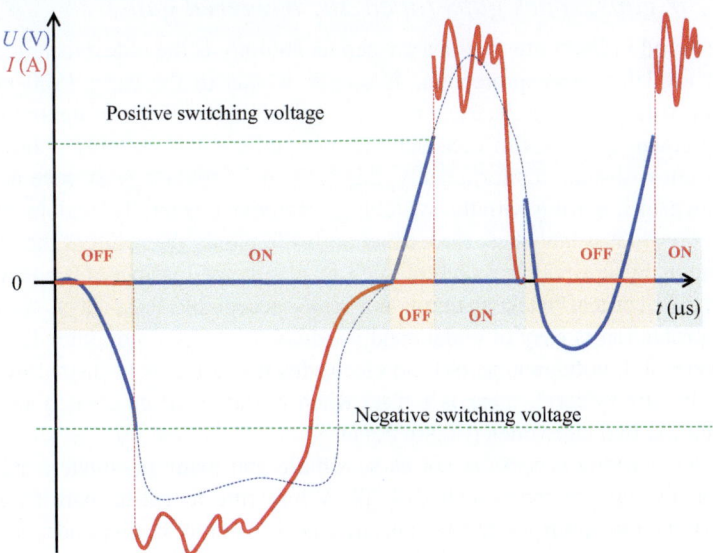

Figure 5.39   u(t) voltage-limiting component ideal curve

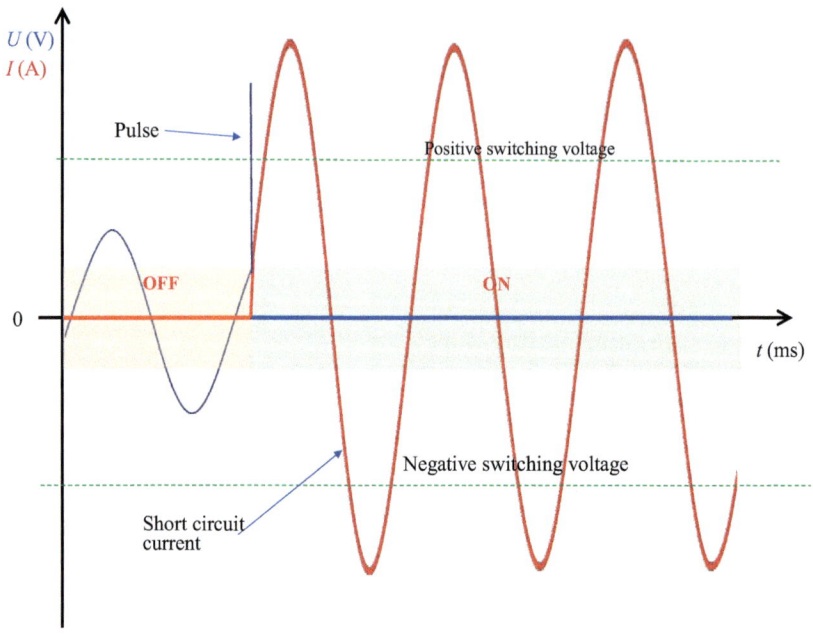

Figure 5.40   u(t) voltage-limiting component approached reel curve

NOTE 2 A voltage-switching device has a discontinuous voltage/current characteristic.

5.3.2   *Air gaps, spark gaps (open air, triggered gaps)*

It can be found in literature that this air gap technology is the oldest technique that was first used for surge protection. It seems, it was in the early 1900 and the application was to protect electrical trolleys against direct effects of lightning. This is the technique that we can consider the most efficient if we only consider the surge current withstand. Unfortunately, this is the only advantage as there is a long list of drawbacks. Staring with the switching voltage that is totally random and also that once the arc is established, the extinction is often only the result of the opening of the circuit by upstream protection such as a fuse for instance. Of course, this results in an electrical blackout that is not really acceptable these days.

The protection is easy to understand as today this is common knowledge that when a very high voltage is across too electrodes that are simply distant by a thin layer of air (atmosphere), there is a time when a spark will be created somehow connecting the two electrodes (Figure 5.41).

But the sparking process is not as so simple and many parameters are to be controlled. In spite of the so-called 3,000 V/mm rule to get a spark in air, air pressure (linked to atmospheric pressure) distance, shape of the electrodes, material of the electrode, possibly the polarity of the surge (linked with some mechanical aspects), light/obscurity, wind, humidity, air itself (other gas, pollution particles),

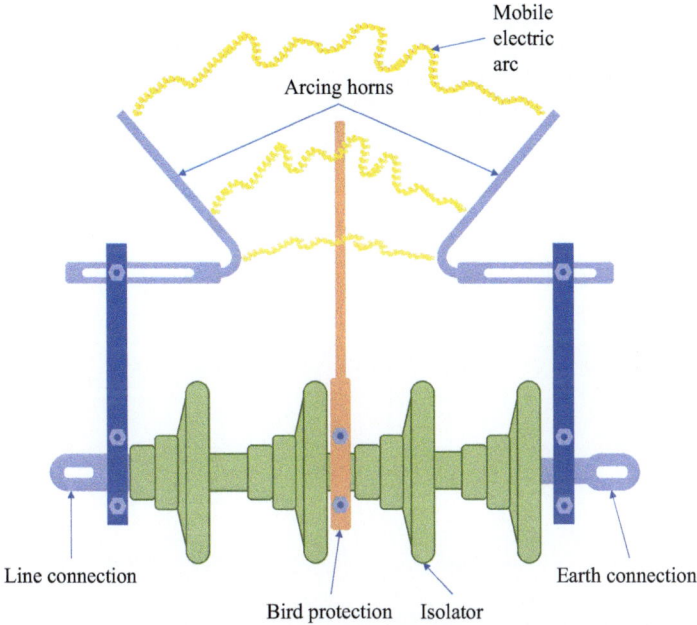

Figure 5.41   Picture of basic air gap

steepness of the surge voltage. In few words, far too many parameters that are not under control if you also consider that electrical arcs are several thousands of Celsius degrees that degrade the mechanical aspect at each surge. Then an SPC where the voltage triggering is not easy to control has a little benefit when systems to protect become sensitive. First protection for transmission line was using this technique as well. It was a component made of two electrodes spaced by few micrometres (between 70 and 150 μm). The most commonly used material for the electrode was carbon. But as the air gap for high voltage or traction lines protection, the stability of its triggering could be classified as random compared to the today's exigence. In addition, the very small air distance was subject to variation after the gap conducted some surge current waves making this component strong but not so reliable.

Some air gap designs fortunately exist successfully to limit or avoid the random triggering voltage and they are almost all components that are with side circuitry acting as a trigger system (may be an electronic circuitry itself using assembly of component where some are SPC). Then as soon as the triggering voltage is better controlled, the real benefit of air gap can be used. The high capability of this component to handle the huge energy that can create a current from direct with a lightning current conductor. This is mainly due to the fact that air expansion causes too high temperature elevation that is not so restricted and that the only degradation point will be the arc contact points (both electrodes). In that case the design must consider this hot gas expulsion.

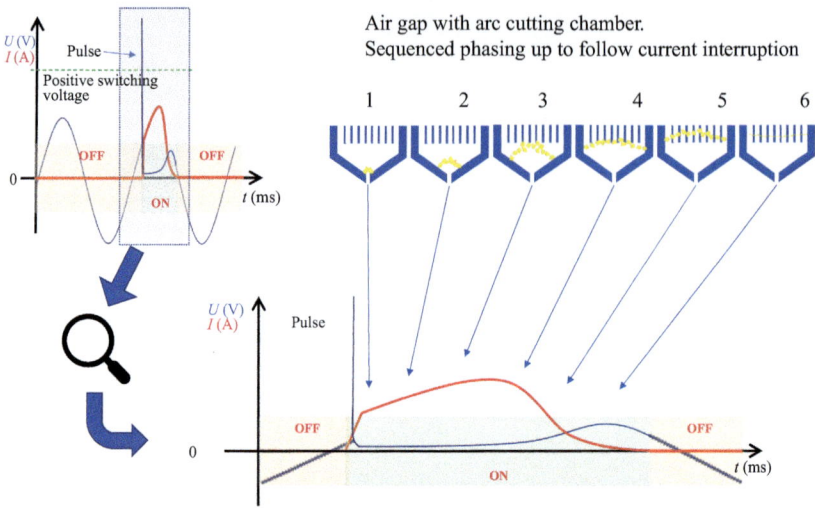

Figure 5.42   Air gaps typical behaviour depending on 'addons'

The second big problem would be that when the surge is gone, the arc may be kept in-between the two electrodes by the current provided by the installation and then this becomes a short circuit (from transformers, batteries). This current is called follow current as it is the current following the event that trigger the gap. Again, some air gaps are equipped with systems with ability to quench this follow current. A $u(t)$ and $i(t)$ curve is shown in Figure 5.42.

5.3.3   Gas discharge tubes

5.3.3.1   Introduction

Science of electrical arc in gas was already explored at the end of the nineteenth century and several laws were set at this time. In spite of that the components named GDTs are today widely used for surge protection, they are still components providing research topics for scientists. Trying to simulate the behaviour of this tinny component is still not an easy task even if using modern simulation multi-phasic software.

Its wide usage is simply due to the benefit of air gap with no disadvantage. It is the smallest volume/surge capacity ratio and the triggering voltage that can be considered uncontrolled for air gap which can be considered under control for GDTs.

5.3.3.2   Presentation

GDTs were made with emissive and radioactive material before the nuclear topic was worldwide banished for use other than military or electric production related. This was to help and better control the emission of the first electron leading to more stable and controlled breakdown.

First generations of GDT were made with either metallic enclosure or glass bulbs. They are usually made of two conductive terminals but may be with three conductive terminals. The first type is called bipolar GDTs and the second is tripolar GDTs. Since the 1970s/1980s the radioactive substances are replaced by inert materials but still help the control of the breakdown phenomenon. One must know that some industrial sites are still today protected with these radioactive GDTs. But they are gradually being replaced by their new types. See the old GDT shown in Figure 5.43.

Note: This picture includes a radioactive version (top right corner).

Today GDTs are most likely using ceramic instead of glass for their body as shown in Figure 5.44. Their body is most likely cylindrical, but rectangular shapes also exist and they are most likely dedicated to be surface mounted and with small sizes.

5.3.3.3   Construction

There exist various shapes of GDTs as shown in Figures 5.43 and 5.44. From the theory studied in the previous clause, it is also clear that depending on the usage of this component, shape is a factor that is more dependent on the application than the design. The main difference that is not related to the form factor is the number of electrodes as shown in Figure 5.45.

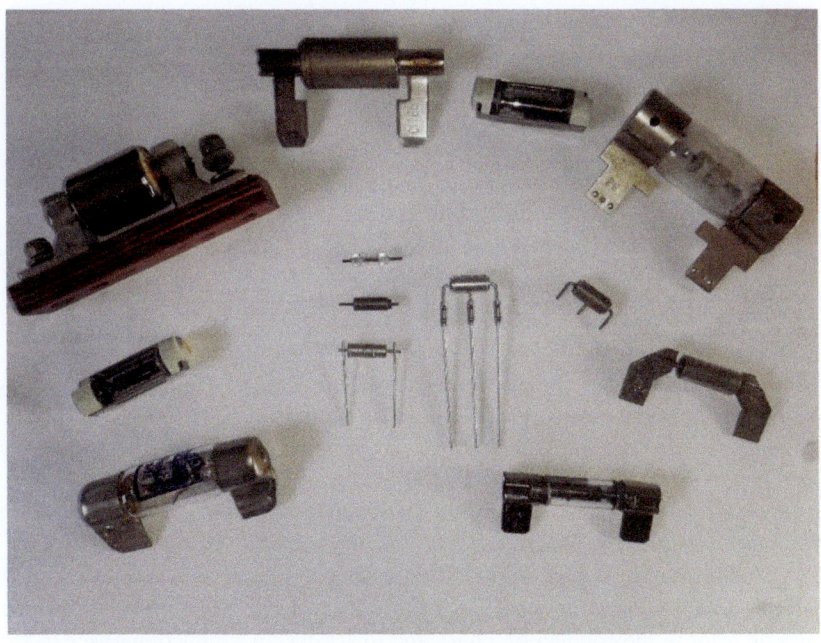

Figure 5.43   Old unipolar and tripolar GDTs

Figure 5.44   Modern unipolar and tripolar GDTs

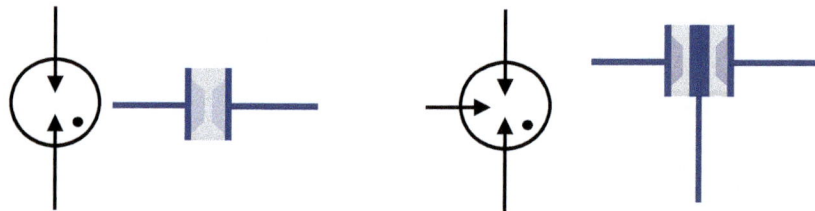

Figure 5.45   Symbols and representations for two- and three-electrode GDTs

5.3.3.4   Manufacturing

GDTs are constituted of at least two electrodes and an envelope that seals the gas inside with constant pressure. To assemble metallic conductive electrodes, the technique is to solder altogether in special oven while the heating zone is with controlled percentage and gases pressure. The gases are usually inert or noble gases such as argon, neon, Freon, but some manufacturers are also using other gases (e.g. hydrogen). With these gases, one may find additional substances that may be placed in-side as powder, or pills; In that case, the substance is added prior to the soldering process. The functions of these substances are to improve the electron emission prior to the conduction phase and to act during conduction phase to either increase or decrease the arc voltage depending on the characteristic targeted. The list of these substances is part of the secret that GDTs' manufacturers are not sharing.

The electrodes may or may not be of one part and conductive sealing caps can be assembled with the electrode before the sealing process. This is generally a solution when ceramic and electrodes assembly is difficult due to thermal expansion incompatibility, electrical behaviour of the connection, thermal withstand, thermal propagation or simply for economic reason. To illustrate this multi metallic material usage, an understandable example is when the design needs tungsten electrodes for performance reason but needs to be assemble to copper and iron for assembly.

A typical GDT construction is shown in Figure 5.46. Depending on performance requirement, various shapes for electrodes may be possible. It is even possible to find GDT that are not with symmetrical electrodes.

As already expressed, electrodes may be of various materials. In general copper, tungsten or iron are commonly used, each of which has specific characteristics that may impact the GDT performance even if with the exactly same shape. Of course, cost is a factor that is highly impacted by material choice.

To reach better dynamic characteristics, a carbon deposit may be printed inside the ceramic envelope. This artificially reduces the distance between the metalic electrodes.

As discussed, the distance of the two electrodes is of course a parameter but it is to be understood that the major factors are the combination of the gas, the pressure (and per cent if more than one gas), the material of the electrode, the possible additives, the shape of the electrode and the possible graphite lines inside the envelope. It is not rare to see two GDTs that are made of exact same parts, but the mix of the gas and pressure makes these two GDTs totally different with different characteristics.

Once the GDT is assembled, they are possibly stressed by a current of specific shape and duration. It can be AC, DC, impulse or a sequence of impulses. This phase is called 'burning' phase and most likely consists in a kind of break-in phase in either spreading the substance in all other GDTs or in a kind of break-in of the electrodes and the graphite lines.

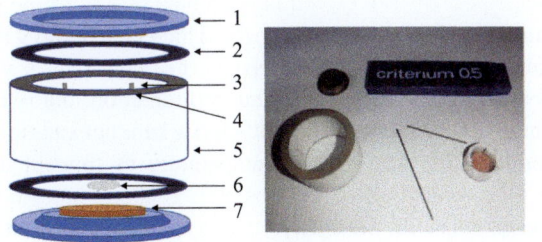

Keys:
1: Conductive metallic sealing cap
2: Solder
3: Graphite deposit
4: Solderable deposit on ceramic
5: Ceramic envelope
6: Gas and possible secret powder
7: Electrode

Figure 5.46   Modern GDT constituents

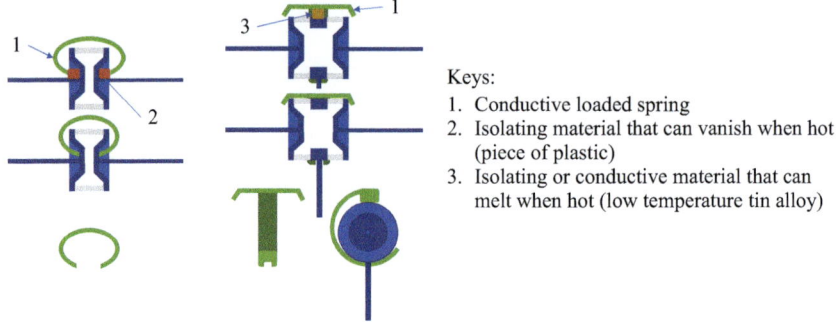

Keys:
1. Conductive loaded spring
2. Isolating material that can vanish when hot (piece of plastic)
3. Isolating or conductive material that can melt when hot (low temperature tin alloy)

Figure 5.47   Typical feature shorts out a GDT upon its temperature

GDTs can be sorted (maybe one of the only similarities with MOVs). The sorting is usually achieved by U2 measurement as there is a kind of relationship between static and dynamic breakdown measurements for GDTs of the same design.

And finally, the accessories are assembled on the body of the GDT. It is about the connection leads or other options as for instance the loaded spring permitting a GDT to end intentionally as short circuit as shown in Figure 5.47.

5.3.3.5   Principle

Among all techniques used by GDT designers, two empirical laws are the basic that one understands. These laws are more than 100 years and are the so-called Paschen's law and the Ayrton formula.

Once these two laws are understood, the first stage for understanding how GDTs are designed and work is passed.

Paschen's law gives the static conditions relative to the pressure, type of gas, gap between two electrodes and the voltage across the two flat electrodes for an electrical arc to be created. A simplified formula is given here under.

It is then one of the major parameters to characterise a GDT before conduction:

$$V_{\text{breakdown}} = \frac{A \times p \times d}{B + \ln(p \times l)}$$

where coefficients A and B are obtained by experimentation, which depend on gas. p is the pressure and d is the inter-electrode distance (gap).

It is reminded that this is a static law and that for surge protection, the phenomenon of which is rather dynamic with pretty short voltage rise time. In addition, it is also to know that this model does not consider temperature in spite of its influence.

Figure 5.48 shows the Paschen curve for air only for two-gap distances. It also shows where could be the Paschen's curve for a gap of the same dimension of 0.01 mm but with argon. It becomes clear that using various gases may have effect

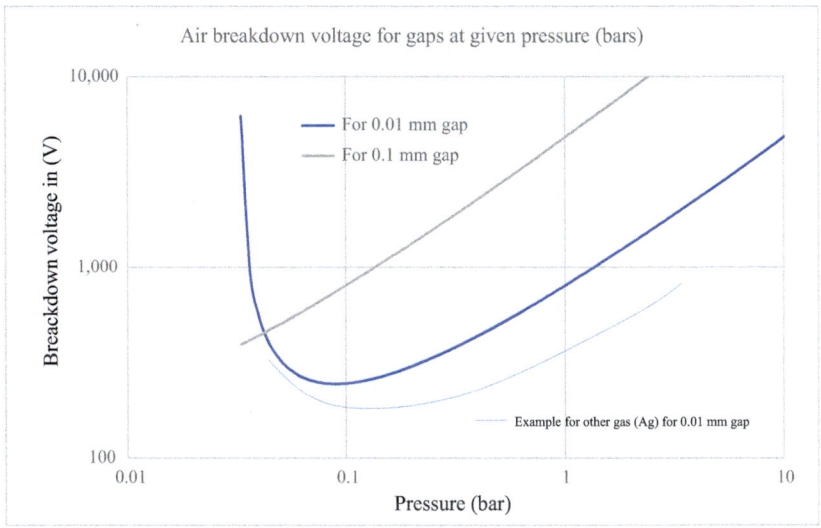

Figure 5.48   Paschen's law curves

Table 5.7   Classification of SPC and usage as surge-protection components

Electrodes	*A*	*B*	*C*	*D*
Ag	14.2	3.6	11.4	19.0
Cu	21.4	3.0	10.7	15.2
Fe	15.7	2.7	9.4	15.0

on how the GDT will operate. Some gases are higher or lower than the air curve and mixing them can lead to different tendencies (even if still in the same spirit).

It becomes obvious that gas pressure and gas mix is one of the key factors for the condition of static breakdown that manufacturers keep as their secret.

The arc formula known by the name of Mme Ayrton empirical formula expresses the relation of the voltage arc once the electrical arc is initiated. It is to be noted that this model is a static model and does not consider the voltage variation and other parameters that are related to time such as temperature and pressure rise or ionisation of the arc zone for instance. The other models can be referred to for GDT understanding but this is the oldest one and is still used nowadays:

$$V_{\text{arc}} = A + B \times l + \frac{C + D \times d}{I_{\text{arc}}}$$

where coefficients *A*, *B*, *C* and *D* are empirically obtained and depend on materials. See Table 5.7 for copper, iron and silver, and *d* is the length gap between the two electrodes (m).

Figures 5.49 and 5.50 give an idea of *U*, *I* and *l* magnitude.

The Ayrton law is then one of the major parameters to characterise a GDT during conduction.

To continue with static curves, Figure 5.51 shows the typical static *V/I* characteristics of GDT. Figures 5.52 and 5.53 are extrapolated from this theoretical static curve. It shows the two behaviours of a GDT depending on the process directions: Off to On and On to Off.

Of course, these information are only the starting points as all dynamic factors are not considered. For instance, pressure for a gas is dependent on its temperature and in the case of constant volume (GDT) pressure may reach several atmospheres. Light (photon) plays a role in the reaction time for the electron emission. This time is of major importance for the first electron to be extracted that is the first condition to get to turn to conduction. Gas, electrode material, variable pressure, electrode shape, photon, variable temperature, and of course variable voltage and current are linked and make the GDT a component that scientists are still hanger to characterise.

The switch from high impedance to low impedance is a process that takes a certain time and during this transition the arc also has various phases to pass. This process is very fast but still exists. As per physical law, it is not possible to stop a car in no time, and also it is not possible to move an electron in no time. One still needs to admit that moving an electron is still much easier and faster than to move a car. During this transition, the glow mode is there when the current is around some hundreds of milliamperes. This mode is the one used for gas discharge lamps but has no interest in surge-protection voltage (except if the system starts to oscillate

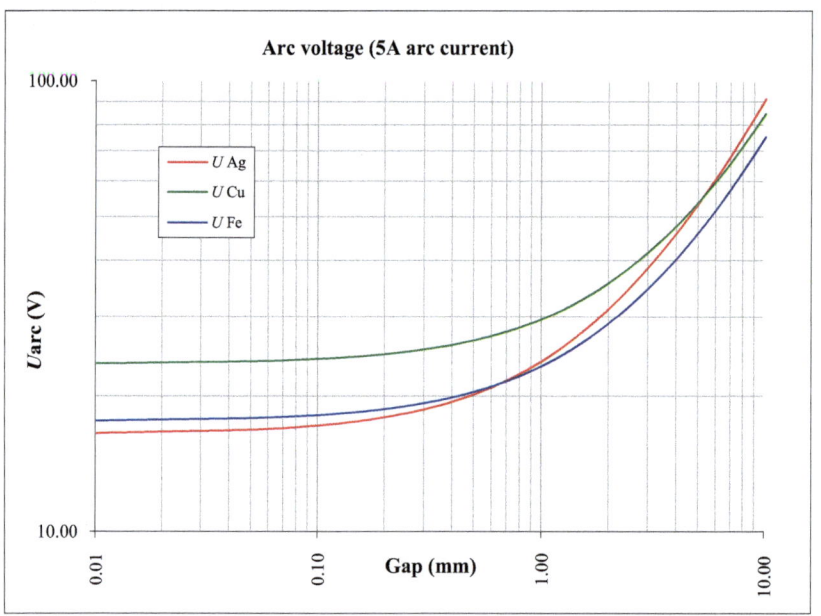

Figure 5.49   Gap/Voltage curve for an electrical arc with fixed current (from Ayrton Formula)

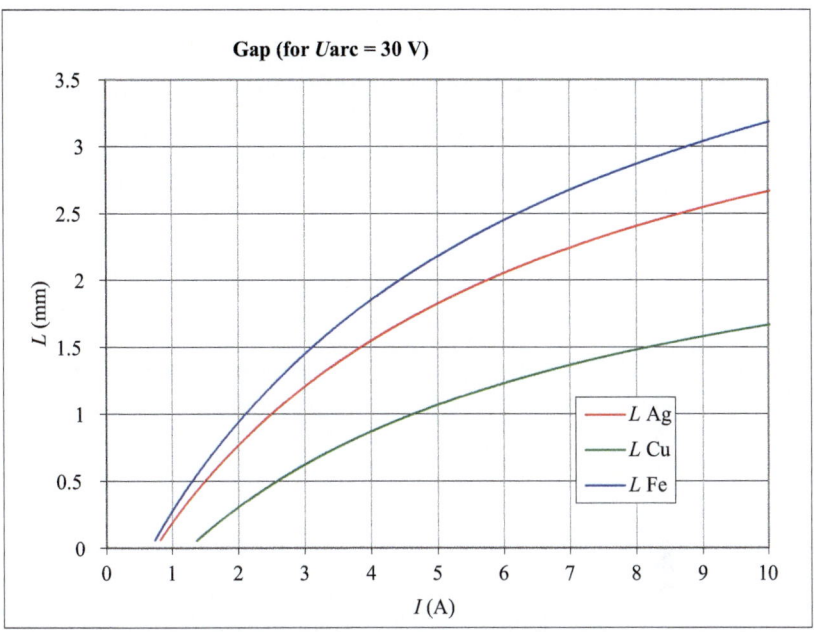

Figure 5.50   Current/Gap curve for an electrical arc with fixed voltage (from Ayrton Formula)

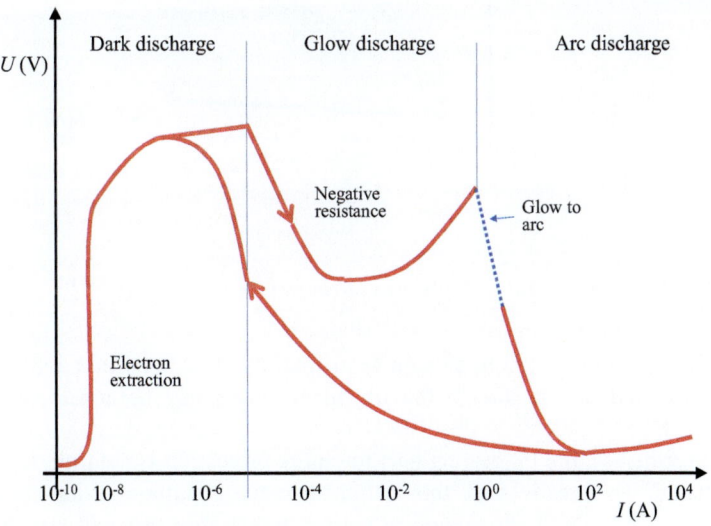

Figure 5.51   VI characteristics of a GDT

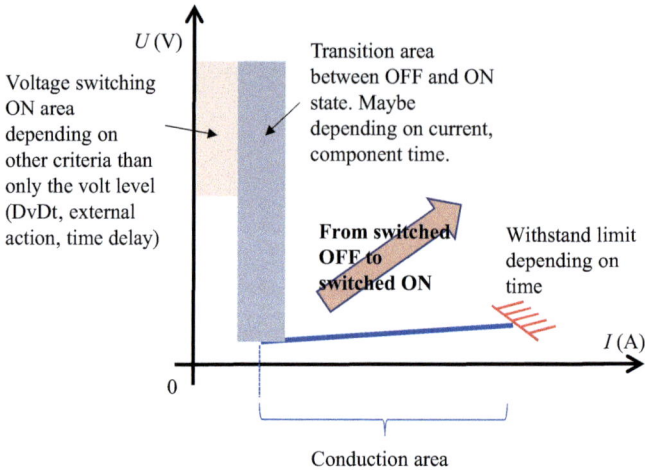

Figure 5.52   GDT VI characteristics from Off to On

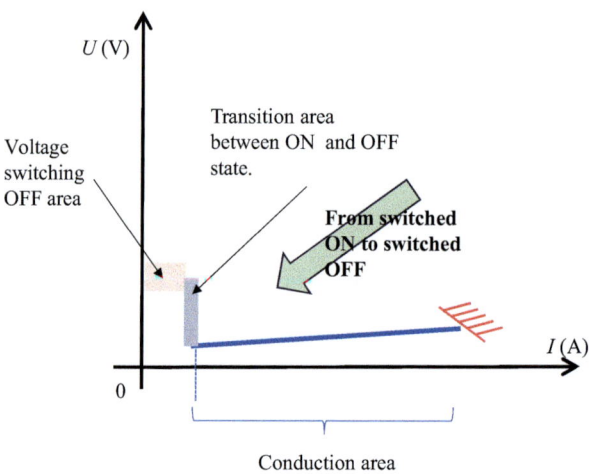

Figure 5.53   GDT VI characteristics from On to Off

when the current is not high enough to maintain the arc condition and stays in the glow zone and this is due to the impedance that may be considered negative resistance).

Nevertheless, the interesting part for surge protection is the time for the arc to be initiated. As already said, this is linked to the condition of having a voltage across the electrodes high enough to extract an electron that will turn to an electrical arc. Then this leads to admit that if the voltage rises relatively fast, the voltage condition may overcome before the breakdown of the GDT. This phenomenon is similar for air gap as well.

Figures 5.54 and 5.58 show various breakdown voltages that depend on voltage rise. It is relatively common to characterise GDTs by their spark-over voltage matched with the *dv/dt*. 1 V/s being called U1, 100 V/s called U2, 1,000 V/μs called U9, etc. For instance, the GDT4 shows that U9 is equal to 500 V. Usual tolerance for GDTs is +/−20% but some manufactures may propose better tolerances.

One can easily distinguish what can be called dynamic and static behaviour.

Oscilloscope screen shots give a lot of information to understand GDT. Figures 5.55–5.57 show the same GDT response when tested with three different surges.

The two first are typical test voltage wave shape with different voltage rises (1.2/50 6 kV and 12 kV U_{oc}). The third screen shot is the response of a GDT when stressed by a scare voltage impulse with the maximum voltage corresponding roughly to its breakdown voltage condition. This clearly shows a time to switch that is not negligible. This situation is most likely not expected for normal use in surge protection but give hints to the reader to better understand the last figure that is an attempt to explain the phenomenon.

Last, the comment on the breakdown voltage can be understood from the previous information that when testing a GDT, it is not correct to only get one measurement. The standard deviation is wide mainly for the first shots (light and temperature interaction). It is then common to measure a minimum of five breakdown voltages of each polarity for one sample (with fixed time in-between surges). And then a considerable quantity of samples from a batch is needed to get usable data.

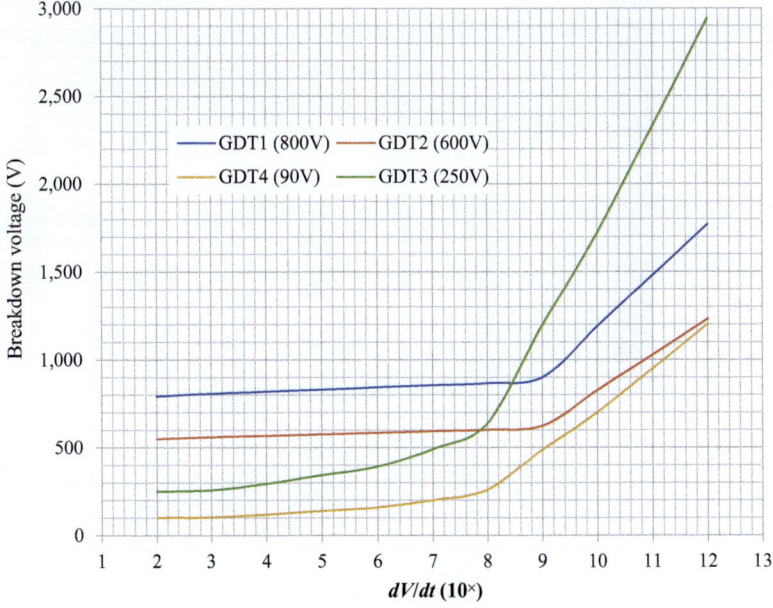

Figure 5.54 Breakdown voltage curves of various GDTs depend on voltage rise

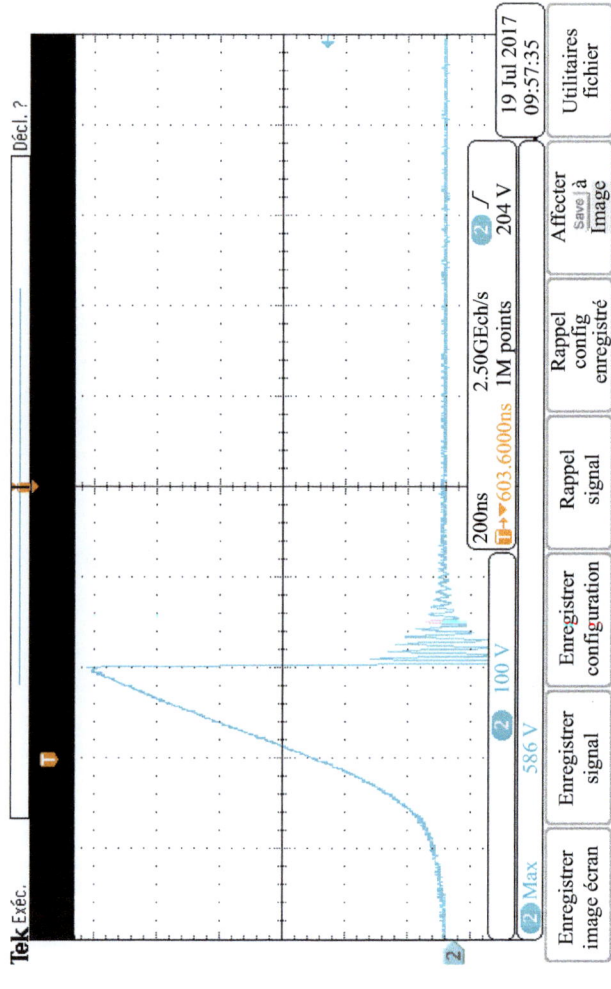

Figure 5.55   Breakdown voltage curves with about 1,000 V/μs (U9)

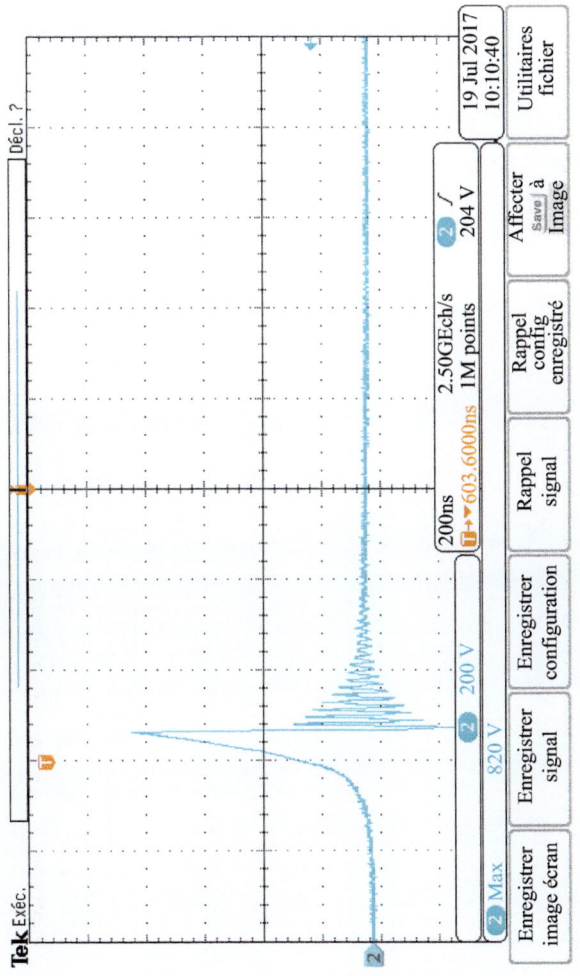

Figure 5.56 *Breakdown voltage curves with about 10,000 V/µs (U10)*

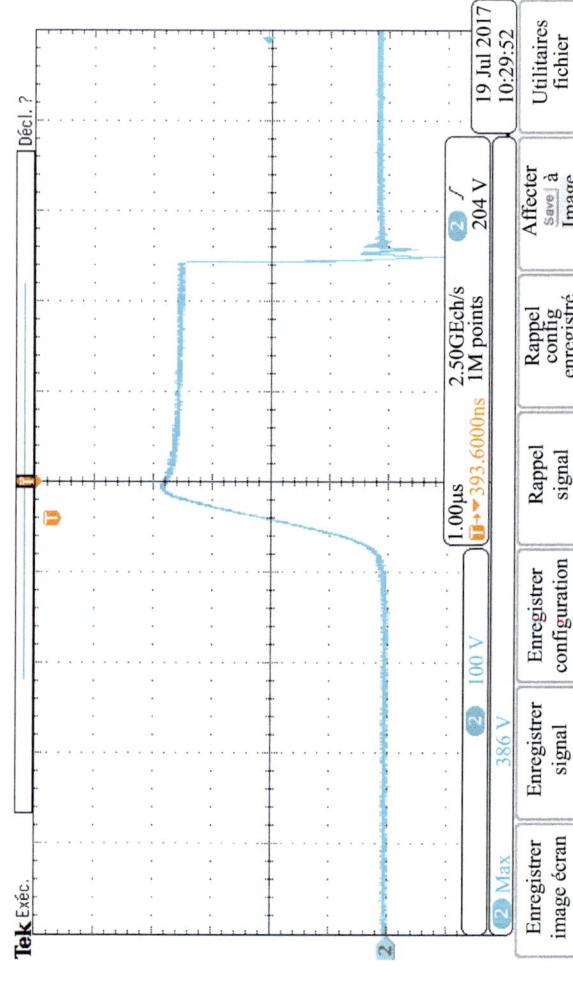

Figure 5.57 Breakdown voltage curves with a square voltage impulse

On the previous plots one can visualise the effect of the switching on the voltage that is oscillating (5.55–5.57).

Figure 5.51 shows that the *V:I* characteristics are not symmetrical from one way to another. What is happening when the surge vanishes? The GDT will extinguish and turn back to high impedance. This is what we expect and one expects that this is true if the breakdown condition is not any more valid from Figure 5.51. To be more accurate, it is usual to consider that when the current cannot maintain the arc voltage, the GDT switches back to its high impedance state and the process can start again if breakdown conditions are true again. It is very possible that the GDT still conducts the current supplied by the circuit on which the GDT is connected to if the arc condition is maintained. This is called follow current. One may consider that for sinusoidal power source, the next zero crossing will permit to reach such condition.

There is an additional condition that may cause some surprise if not well managed. When conducting the current, the GDT is absorbing thermal energy. The temperature of the gas may reach several thousands of Celsius degrees. Also, the temperature of the electrode may reach as well several thousands of Celsius degrees on surface. In case the temperature has not decreased enough and if the gas inside the GDT is still ionised, the breakdown condition is simply reduced to a minimum that will make the GDT to reconduct as soon as the voltages across the electrodes reach the arc voltage. This is far lower than the prior breakdown condition.

When a GDT is at conducting phase, the current may be from surge of power supply or both. Total energy is the one to consider. It may be possible that a GDT absorbs more energy during the follow current phase than the surge phase. The

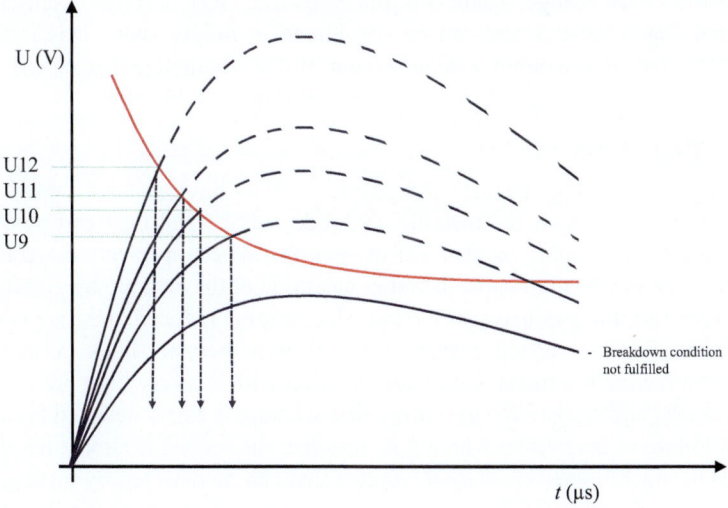

Figure 5.58   Breakdown voltage curves of various GDTs depending on voltage rise

capacity of the GDT to dissipate the energy becomes a parameter that can be important for its selection.

This is very important to consider as one may expect that a GDT turns off at the next zero crossing of the AC power supply. And that would be a mistake to consider the 'voltage-switching OFF area' shown in Figure 5.53 only set by voltage and current parameters. A possible parameter called 'Holdover voltage' gives this impression but this is only in very specific test condition and does not apply for all applications.

Pressured gases are selected for breakdown stability when (cold) and some other may have effect during the extinguishing phase (when hot).

From all the earlier information, the energy that creates the current through the component can be evaluated by calculating the integral of the instantaneous power for the duration of the event. It is interesting to see that if the arc voltage is low, the current that can handle a GDT may be very high. But when the arc voltage is low, this leads to a very poor capacity of the GDT to quench any follow current. On the other hand, when the arc voltage is intentionally high to reach this follow current interruption capability, the energy that the GDT has to absorb is much higher. It is then interesting to notice that a GDT of similar size and mass compared to another may be totally different in surge current handling capacity. As an example, similar 8 mm × 6 mm cylindrical GDT can be declared with different surge capacities with a ratio exceeding two 5–20 kA.

Each action of the GDT is a stress for the GDT (very hot temperature of the arc combined with very high pressure). This stress is proportional to the surge current and to the follow current of course. A GDT may last millions of surges if the absorbed energy is not able to melt any of the metallic material. As soon as the energy causes enough heat in the GDT, some metallic particles may be spread all other the inside envelope and may modify the behaviour for the next surge. Most likely, the static and dynamic breakdown as well as the interrupting value of the follow current may change. There is a time when the GDT becomes totally out of its original characteristics and can be considered in failure state. This is totally impossible to detect without testing except if the characteristics of the GDT become not compatible with the application and causes malfunction.

5.3.3.6   End of life

If the energy absorbed by the GDT overpasses a certain limit, various scenarios can occur. This heat can melt the metallic electrode surface and may end up in two possible failures depending on the GDT orientation. When one electrode melts, the liquid material may head towards the other electrode or the ceramic by gravity. For the first scenario, the gap between the two electrodes is reduced and may end in a short circuit. For the second scenario, it creates a thermal shock between the ceramic and the melted metal and in combination with the very high pressure, the ceramic breaks and realises the gas. If the first scenario is easily detected because it creates a failure in the system where it is installed, the second is almost not visible and leads to much higher breakdown voltage that can be detected by testing.

The probability of a GDT to terminate as a short or open circuit is not very well documented, but it may be considered that 50% may be a good compromise and this would motivate to consider both of them in applications.

Manufacturers have tried to solve this end-of-life uncertainty by proposing additional functions to their GDT. They are mainly based on thermal behaviour of the GDT. One consists of intentionally short circuiting the GDT by an extremal mean when the temperature of the body reaches a certain level. The second one is to prioritise a weak point on the envelope that will open when temperature reaches a preset value.

Both designs turn to something else than a GDT and that needs to be defined. An open circuit is defined by its isolation resistance at a defined voltage and a dielectric withstand at a given voltage.

5.3.3.7 Equivalent circuit

Few GDT manufacturers offer a GDT spice model. Even if it is very interesting to have them, it is very difficult to interpret the results that we would obtain using these models because the complexity of the behaviour of a GDT is rarely precise. If we want to simulate the follow current that a GDT will be able to turn off, the model will simulate the ideal component for the first shot but the second will not be the real behaviour and the third will be even worse (taking into account the time between shots). GDT is a living component and depending on the parameter we want to analyse, and the model would ideally be different.

It is very likely that two different GDT experts do not use the same method to simulate the same GDT. The real GDT model could possibly be obtained using the Multiphysics software which would require several hours of calculation per configuration (U, I, t), but the use with an electrical simulation software using the Spice engine, for example, leads to making it too much of simplifications.

However, from Figures 5.51–5.54 it is possible to imagine the equivalent circuit of the various phases the GDT will be 'operating'. For instance, before any operation, it is a simple capacitor. When the current starts to flow after triggering, it may be a circuit associating inductances, capacitance, resistance and electromotive force (assuming these are not variable that is not very true as all are linked with temperature and internal pressure). Once the current is finally quenched, it is back to a simple capacitance (Figure 5.59).

Note: The quenching sequence shown in Figure 5.53 is not as easy as it can be said. Finally, information provided by manufacturers is far not enough to modelise any GDT.

5.3.3.8 Characteristics

GDTs are made for several applications and they refer to various standards. The two major applications are

- for signal and data telecommunication applications and
- for power application.

These standards are ITU K12, IEC61643-331, IEEE C62.31 and some address GDTs through their application (Telcordia GR974/GR1089, UL1449, UL497B). Depending on the experts involved in the writing, the test methods are oriented

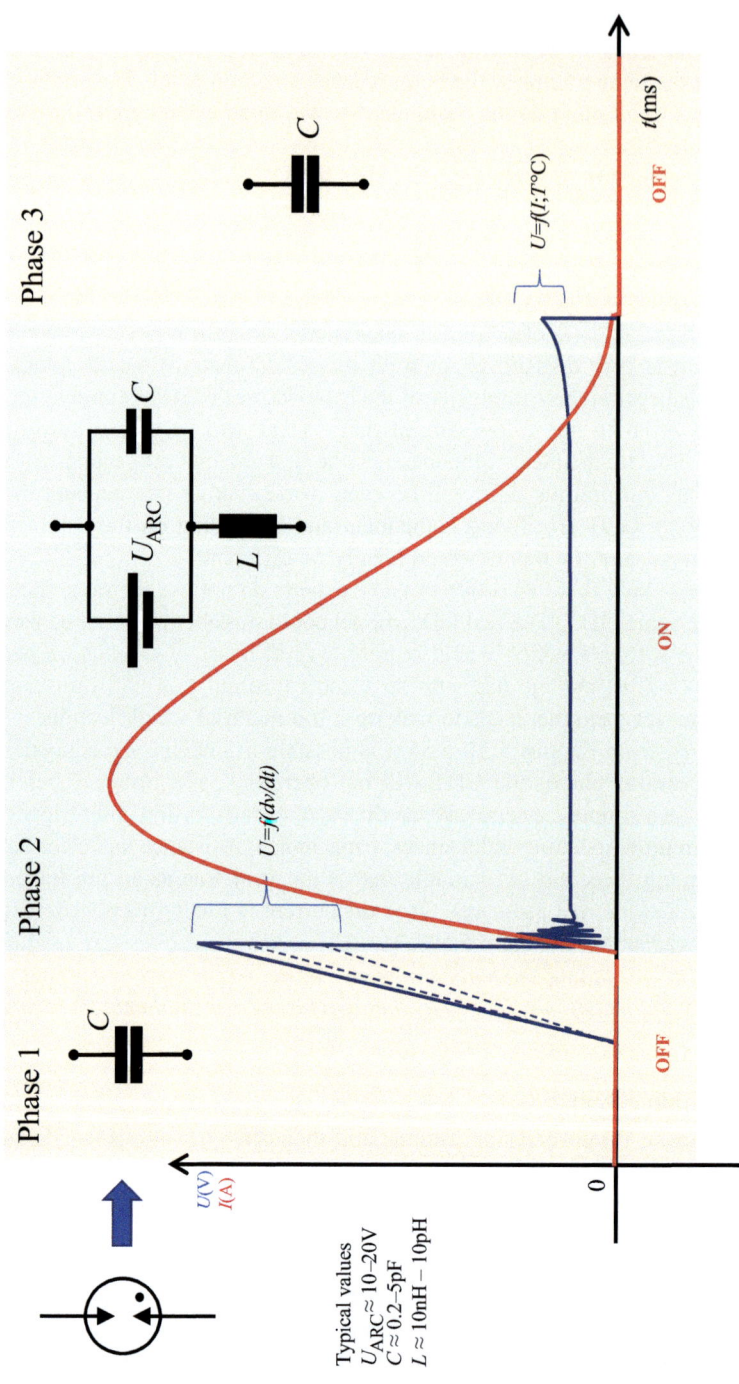

Figure 5.59   Possible simplified equivalent circuit of GDT (by phases)

either of power or telecom applications. Some standardisation groups attempted to address both topics but with little success at this time.

In addition, even for the same application, it is not rare to find information from datasheets that are not at all comparable from manufactures to manufacturers. Each parameter needs further investigation from the user. If one is given by a manufacturer without any standard test reference or precise test process and pass criteria, the user may misinterpret the parameter. For instance, even the basic surge parameter may differ depending on how the GDT is tested:

- number of impulses,
- time in between impulses,
- wave shape of the impulse,
- was the test conducted with follow current? (if yes what value, AC/DC, frequency, power factor or time constant of the power supply),
- mounting.

Table 5.8 is an attempt to list all the parameters gathered from various datasheets found on the market.

In Table 5.8 no acronyms will be given as it could be more confusing than helpful due to lack of homogeneity. This is also the reason why the reader will not be able to find the exactly same terms when comparing datasheets, which he could easily see on the Internet.

5.3.3.9 Usage

GDTs are the least capacitive SPCs. Thus, they are massively used in telecom and data transmission application. They are also the best volume–mass to current carrying an ability ratio and they are widely used for power application. If the follow current would have not been its main disadvantage, it would surpass all other surge components. The other downside which is more a marketing war than anything else is that some argue that GDT is slow and the peak voltage can be very high if the surge has a very short time rise. This is true when only looking at the peak value, but this argument becomes obsolete as soon as the stress is rather the let-through energy that can limit a GDT.

It was mentioned earlier that a GDT may create a self-oscillation if the circuit ends up in having the working point in the glow area of its *VI* characteristic. But this is simply to avoid when designing the circuit around the GDT. There is another oscillation that is due to the GDT.

Note: This is a well-known phenomenon for the electronic engineers and due to the very fast switching of the device when a surge appears. When a fast variation of energy occurs, the effect leads to oscillations. To illustrate this with a mechanical analogy, picture a long metal bar fixed on a wall and free on the other side. If a hammer of 1 kg hits the bar, it will oscillate for a certain time and the amplitude of the oscillation at it end may be much more than at the chock location. The amplitude and duration of the oscillation are more due to the speed of the hammer and to the bar characteristics than the hammer mass only. GDT are always very fast in switching from breakdown voltage to arc voltage and this may create damped

Table 5.8 Typical list of parameters that can be found on GDT's manufacturer's datasheets

Description	Example	Note	Usage
Max. continuous operating U_c	255 V		Power application
Type/class related to 61643-11	1 + 2		Power application
Application with MOV	X		Power application
Application (N-PE)	X		Power application
Impulse spark-over voltage 1.2/50 μs to 6 kV	<1.5 kV	Refer to a test from 61643-11	Power application
Nominal discharge in (8/20)	20 kA	Refer to a test from 61643-11 and UL1449	Power application
Impulse current 10/350-μs Iimp	12.5 kA	Refer to a test from 61643-11	Power application
TOV high voltage withstand	Yes	Refer to a test from 61643-11	Power application
Maximum discharge current	40 kA	Refer to a test from 61643-11	Power application
DC spark-over voltage (100 V/s)	340 V		Power application
Tolerance	20%		General
Impulse spark-over voltage@100 V/μs	7,001	Is called U2 in this chapter	General
Impulse spark-over voltage@ 1 kV/μs	1,000 V	Is called U9 in this chapter	General
Insulation resistance	10 GΩ		General
Capacitance@1 MHz	2 pF		General
Arc voltage	18 V		Optional
Glow to arc transition current	1 A		Optional
Glow voltage	70 V@10 mA		Optional
DC holdover voltages	70 V/80 V/145 V	Refer to a test from 61643-331 and ITU K12	Data
One operation 8/20 μs	10 kA	These parameters are either from data and	Data
Ten operations 8/20 μs	5 kA	telecom standards or requirements. They	Data
Hundred operations 8/20 μs	500 A	can be either claimed by current peak	Data
Three hundred operations 8/20 μs		values, wave shapes, number of operations	Data
Two hundred discharges 1.5 nF; 10 kV; 0 Ω		or time duration (for AC). Exact test	Data
Hundred operations 10/1,000 μ	50 A	reference should be found to compare GDTs	Data
Three hundred operations 10/1,000 μs	100 A		Data
One operation 10/350 μs	900 A		Data
Ten times 1 s 50–60 Hz	10 A	EIC61643-21 refers as well to surge categories	Data
1 s 50–60 Hz	55 A	(e.g. B2, C3, D1, etc.)	Data
Ten cycles 50–60 Hz	100 A		Data
...	...	...	...

oscillation in the circuit. If the circuit can oscillate because of capacitor and inductance, the GDT will trigger such damped oscillation.

Let us focus on power application. The follow current forbids the GDT to be connected alone between active lines where a prospective sort circuit current is expected. However, its usage for protection between one active wire and earth when limited short circuit current is expected is frequent. It is also very often associated with MOVs in series and this will be addressed in Section 5.8.1.1.

In general power lines with ground (PE) and neutral are designed to have no short circuit capability. However, requirement is to consider the worst scenario. In that case, it has been established that 100 A is the short circuit that the GDT has to extinguish by itself. This requirement is part of the IEC 61643-11 that is about SPDs, but other standards that address various equipment are mentioning this test when GDTs are mounted between PE and neutral. This is preferable that engineers consider the TOV tests as it may be a cause of hazards if the GDT is not well selected. GDT manufactures should provide guidance of how to handle this situation with their NPE GDTs.

5.3.4   Silicon PNPN-junction voltage switcher TSS

5.3.4.1   Introduction

TSS stands for thyristor surge suppressor. This is the term used in IEC 61643-341. In some countries, these SPCs have been named Trisil©. These are a PNPN or NPNP junction that is normally used as a thyristor. But it can also be used as surge protection. This is interesting to note that a behaviour we will use in our field is considered as a huge drawback of thyristor when used in power electronic (mainly power conversion (DC to AC). Usually, the benefit of a thyristor for this application is to be used as a controlled switch. But there is a glitch we will emphasise for our surge protection usage. This is the tendency of a thyristor to switch on alone when a too steep dv/dt is applied to its main leads. One even has a specific circuit to connect in parallel the anode and the cathode of the thyristor. This is simply an RC circuit dedicated to reduce the dv/dt and it is called a snubber circuit. Then for surge protection, this drawback is used, which makes TSS relatively efficient for surge protection. This technology is mainly used for data signal protection or small power supply of equipment. They present the same general behaviour than the GDT-switching component as they have high impedance up to their operation that turns them to low impedance up to the condition that the currant is not high enough to maintain conduction. Of course, the internal process is very different.

See Figure 5.60 for standardised TSS symbols issued from IEC 61643-341.
Note: Many producers are using the symbol shown in Figure 5.61.

5.3.4.2   Principle

First unidirectional and bidirectional TSS are available. Unidirectional TSS have a reverse *VI* characteristic that is very similar to an avalanche breakdown voltage (A) but a reversed diode may be in the package and the reverse quadrant is based on this forward diode *VI* characteristics (B). A bidirectional TSS is simply made of two

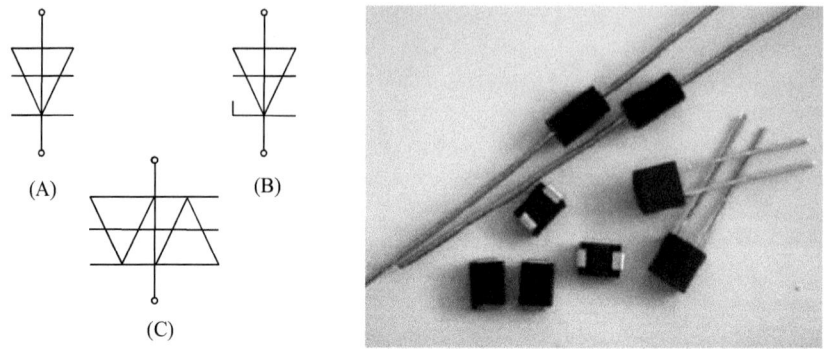

Figure 5.60   IEC symbols and possible TSS packages

Figure 5.61   Non-IEC symbol for TSS

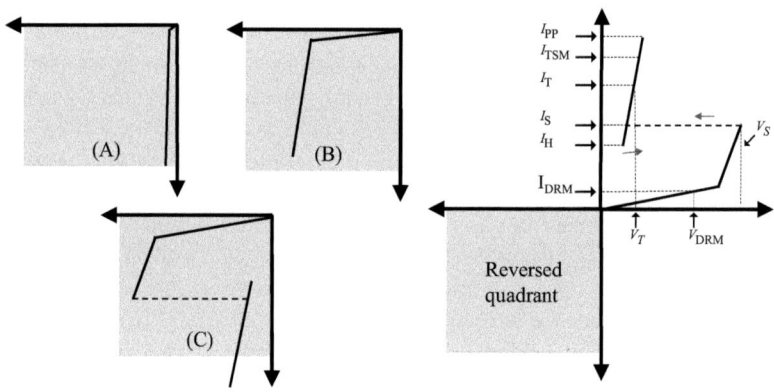

Figure 5.62   VI characteristics of the various TSS arrangements

thyristors head-to-tail using the same manufacturing technique seen for diodes (C). Two thyristors' head-to-tail are also called Triac but its usage is mainly to take benefit of the controlled switching with the help of the gate. The name of Diac has been used in some countries and this simply a Triac without any access to the gate. It is very close to TSS but with no specific ability to handle surges; it is mainly used as a switch that turns on when the voltage across its terminals is around 32 V.

Note: (A), (B) and (C) are common markers used in Figure 5.62 that shows simplified VI TSS characteristics.

Table 5.9   TSS typical declared parameters

Acronyms	Name	Description	State
V_{DRM}	Peak off-state voltage	Maximum voltage that can be applied	OFF
I_{DRM}	Leakage current	Maximum current measured at V_{DRM}	OFF
I_S	Switching current	Maximum current at breakdown	From OFF to ON
V_S	Switching voltage	Maximum breakdown voltage	From OFF to ON
I_{PP}	Peak pulse current	Maximum rated peak impulse current	ON
V_T	On-state voltage	Maximum voltage measured at IT	ON
I_T	On-state current	Maximum rated continuous	ON
I_{TSM}	Peak one-cycle surge current	Maximum rated one cycle AC current	ON
I_H	Holding current	Minimum current required to maintain on-state	From ON to OFF

From Figure 5.62 it is easy to make some parallels with GDT behaviour and one must know that TSS are much more stable. However, their current-carrying capability is much less than GDT and the typical application is then oriented towards data transmission application. On datasheets surge parameter has to be well analysed as this component is tested with various wave shapes. It is not rare to see on datasheets table giving several surge currents carrying capabilities in regard to various surge wave shapes (2/10, 8/20, 10/160, 10/560, 10/700, 10/1,000, etc.). This is due to the fact that for long time this component was only dedicated to specific application (mainly telecom), and producers were simply testing their components with the required wave shape that was customer correlated.

The parameters that are shown in Figure 5.62 are defined in Table 5.9.

As per other semiconductor SPC, the capacitance is one of the parameters to consider for its application. In general, it is a bit less than a single or triple PN junction as by construction, it starts with nor one nor two parasitic capacitors in series but three leading to a reduced overall capacitance. Typical capacitance is within the range of 10–60 pF and it is usually ensured OFF-sate.

Usual break-over voltage may exist from 25 to 700 V and on-state voltages are 4–8 V.

As all semiconductors repetitive surges are not damaging TSS as far as the declared peak impulse current and temperature limit are not overpassed.

5.4   Current-limiting components

Why not using a simple impedance in-line with a circuit that will limit the transmitted surge current? Despite that surges from lightning should be considered to be issued from a perfect current generator with no internal impedance resulting in no effect of the impedance to limit the current, this is very often used in surge protection. As said, surge current directly issued from lightning cannot be limited by impedance, but equipment is generally never directly connected to lightning conductors and thus, the surge is issued from an equivalent generator with its how impedance. This leads to saying that any impedance may limit surge current when installed in series with a circuit. But this cannot be controlled for applications and impact is then not really under control.

Nevertheless, on the same equipment, the association of various SPC technologies requires good coordination. Coordination between SPCs and other surge protection solution is one of the major techniques to know. It consists in making sure that the stress of the surge is well balanced between these SPCs and that their association provides the requested protection for the equipment. To understand, it is obvious that an SPC that has a clamping voltage (or switching) much higher than a second one connected in parallel will never see any current as it will be blinded by the lower one (e.g. 130 V MOV with V_{1mA} 200 V and a diode having a break down voltage of 100 V connected together in parallel). See Figures 5.63 and 5.64.

Then the two components can be associated but if directly connected in parallel, they are supposed to act at around the same stress level (voltage usually) and they should be roughly of the same surge-carrying capability range. What if the goal is to pick a component for its surge-carrying capability and a second one for its low and controlled protection level; a limiting current in between these can be selected to achieve this.

See the following example with the same MOV and diode picked for the previous example. In case one considers the same surge, and if we only focus on characteristic points of the MOV and the diode, see what would be the technique to select the limiting component (a resistance in that case).

Now it has become obvious that the limiting component is an essential part of the solution. Of course, this is a very simple example to express the selection mechanism and it is more complex to select inductance as rise time of the surge becomes one of the main parameters to make the math ($U = L(dv/dt)$ instead of $U = R_I$). Simulation and test are then far easier technique but it is still possible to pre-size the component using this simple technique. It must be kept in mind that if rise time and peak current are well controlled in laboratories, it is not so clear for real life where these are less under control.

The main nature of the limiting current component will depend on the application. It is not current that a resistor is used for power application as it simply generates too much unwanted power loss. In power application, inductors are preferred.

By opposition if the application is signal transmission, the frequency range of the signal must not be impacted by a frequency-dependent impedance and then if an inductor is to be selected it has to be in addition with this constraint. In consequence and for this kind of application, resistors are easier to select (this is a compromise between power loss and frequency band pass).

5.4.1   Resistors

Even if it is possible to find resistor's manufacturers providing surge parameters, it is not very common and these resistances are specific. In general, maximum power is a parameter that is given but for steady-state condition only. Some maker provides max voltage that is also linked to this steady-state situation.

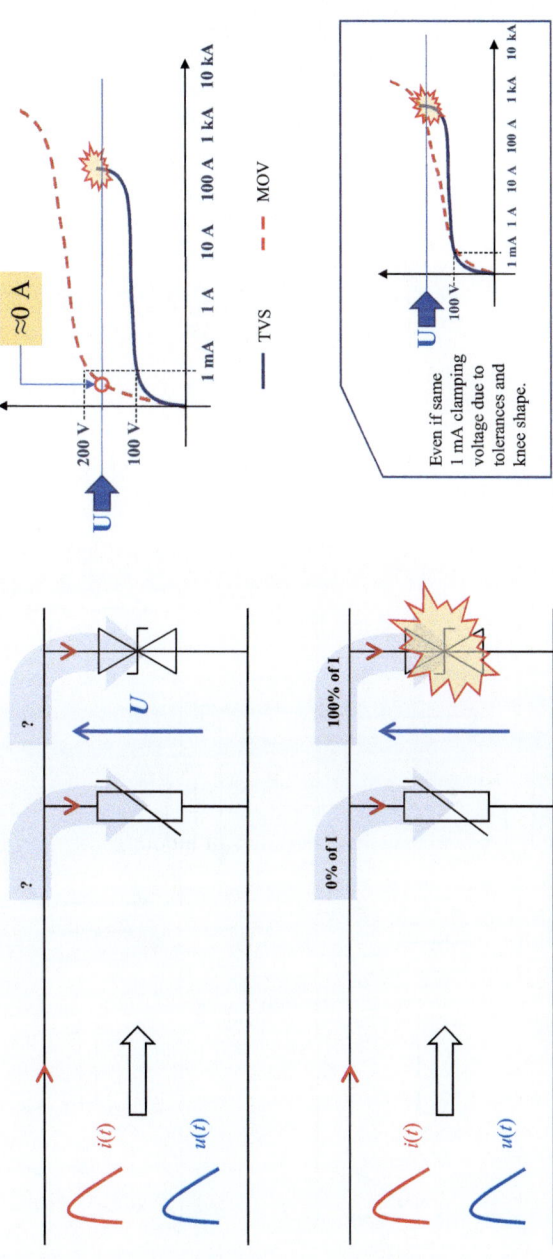

Figure 5.63   Concept of coordination between two SPCs

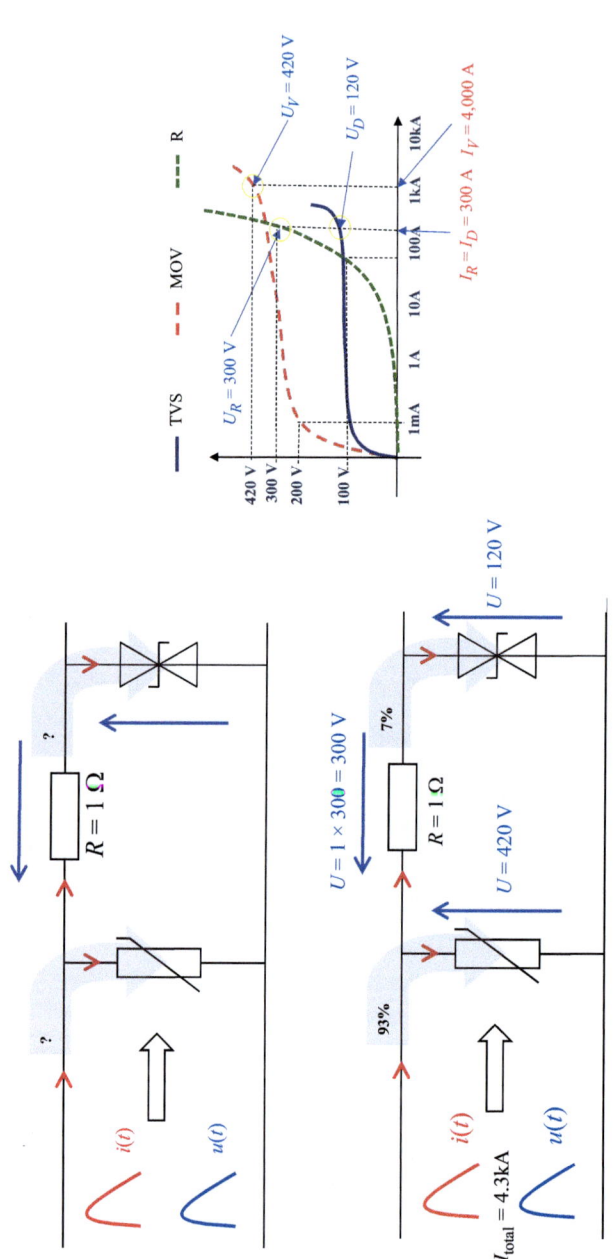

Figure 5.64   Concept of coordination between two SPCs

It seems logical to think that a resistor that can handle 5 W for hours is able to handle much more power with shorter time. Figure 5.65 shows an example of possible multiple to apply on steady-state power parameter in regards to time duration of the impulse.

Resistor technology is also important to consider. Todays, thin and thick resistor carbon films mounted on ceramic support are very common. Some resistors are using aluminium substrate as support and acting as heat sink and then present better surge withstand characteristic. But using metal itself as a resistor with or without mechanical support such as metal strip or wounded wire permits to have better energy abortion for fast event. In addition, generally, one tries to avoid wire wounded resistors as they have important parasitic impedance. This may be used at our advantage as both inductance and resistance are limiting the surge current. Carbon film resistors also have capability to handle much more power than declared rating current but far less than a metallic resistor (some 50–100 times less).

One additional limitation is the dielectric withstands of the component itself either internally or externally. A resistance of 1.1 Ω and rated for 5 W may withstand easily more than 10 kW for few µs if only power is concerned but may collapse due to electric field stress (these values are in relation to the example seen earlier).

5.4.2   Inductances

The inductor is widely used for surge protection as it tends to resist the rapid current variation of a surge and others. There are of course a few things to know with the inductor when it is used to mitigate surges. The larger the inductance is, the more efficient the opposition of the current flow will be. The lower the line resistance is, the lower the power loss for a DC or low-frequency application will be. The voltage drop generated by the current flow can reach very high values and the insulation of the wire in the coil must be made accordingly. Avoiding parasitic capacitance that is inherent to multi-layer construction is preferred as well. This usually leads to having a large coil as it is in addition almost impossible to use an inductor magnetic core. For surge protection, the overvoltage can reach several hundreds of amperes or even kiloamperes which will saturate the magnetic core leading to the complete inhibition of the inductance effect (return to the behaviour of a simple and straight wire). Regarding the positioning of these inductors on a printed circuit board, one must take into account the strong magnetic field that will be generated during the overvoltage event. This can be perturbative to another circuit or even another overvoltage protection inductor if multiple inductors are used. Some distance between these inductors may be necessary to avoid this unwanted interaction. It is not uncommon to see a 'homemade' coil with a regular insulated wire in the air. In this case, it must also be considered that the high current flowing in the coil can create enough electromagnetic forces to deform it. Then, insulated mechanical support must be part of the inductor.

Of course, all of the previous information should be deeply considered when the expected current is high, for example, several kiloamperes. In the case of lower

E

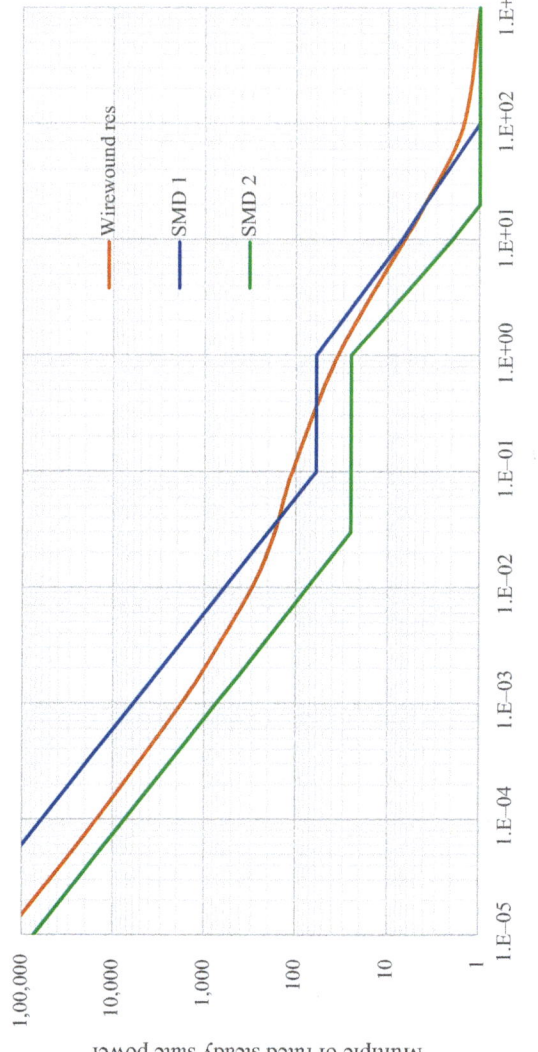

Figure 5.65   Example of overload capability of resistors

risk, everything is still to be evaluated, but a simple inductor even with a magnetic core may be sufficient.

5.5   Current-switching components

5.5.1   Introduction

Instead of diverting the surge current, why not considering simply switching off the circuitry interconnecting the lines that transmit the surge current to the line supplying the equipment to protect? Theoretically, this seems to be a nice way to consider this approach for surge protection. In fact, if the current switching was ideal it would be a really good solution (Figures 5.66 and 5.67).

See Figures 5.68 and 5.69 showing ideal curves.

Unfortunately, this is without taking care of the drawbacks that are really discrediting this kind of protection (See Figure 5.70). However, some specific technologies are claimed to be efficient. In Sections 5.5.2–5.5.5, fuses, circuit breakers, semiconductor switch and positive temperature coefficient (PTC) will be presented and their surge performance (or non-performance) characteristics will be developed. However, if some cannot be claimed to be SPC with surge-protection function, they are frequently used in solution for surge protection and this may help the reader to have some information about them.

Except the component named semiconductor switch that is claimed fast enough to be used as a surge protection, the huge gaps between time reaction and

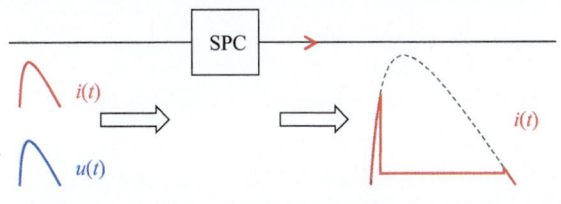

Figure 5.66   Current-switching component serial mounting

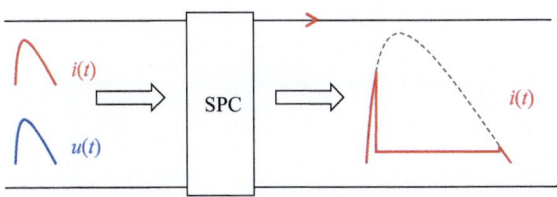

Figure 5.67   Current-switching component in-line mounting

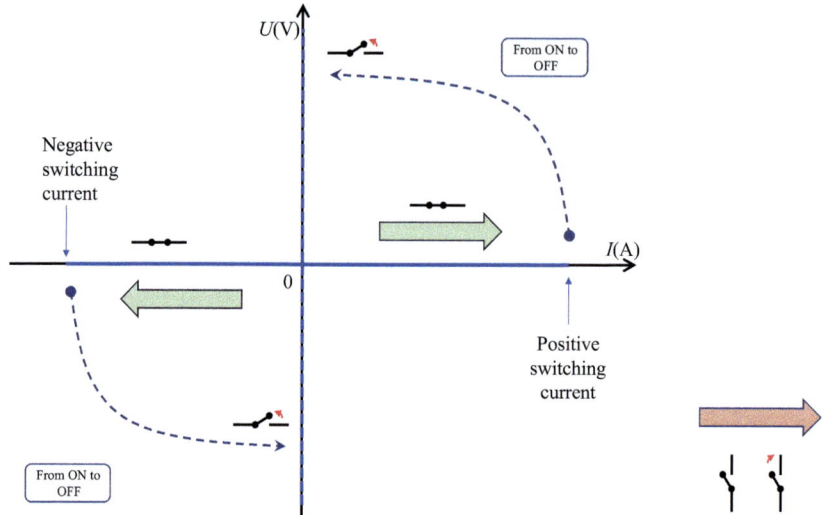

Figure 5.68   V/I current-switching component ideal curve

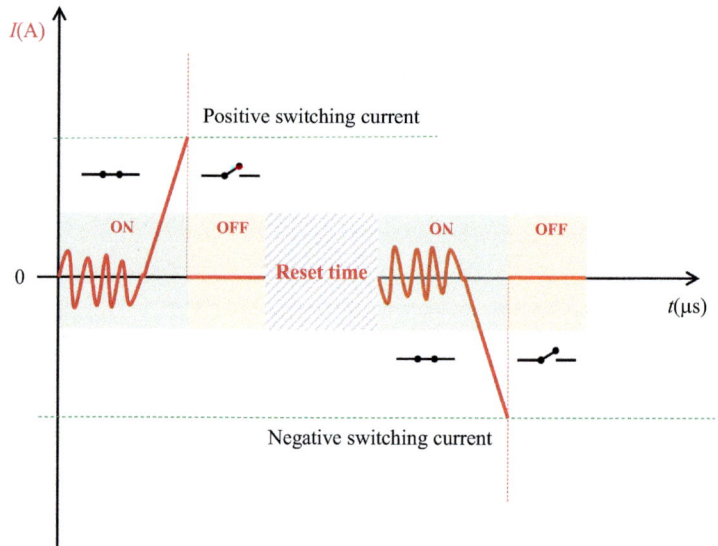

Figure 5.69   u(t) and i(t) current-switching component ideal curve

the transient aspect of a surge are totally incompatible. But if the transient also is followed by an overvoltage that is of some milliseconds and more the fuses, CB or PTC may have a play in the game.

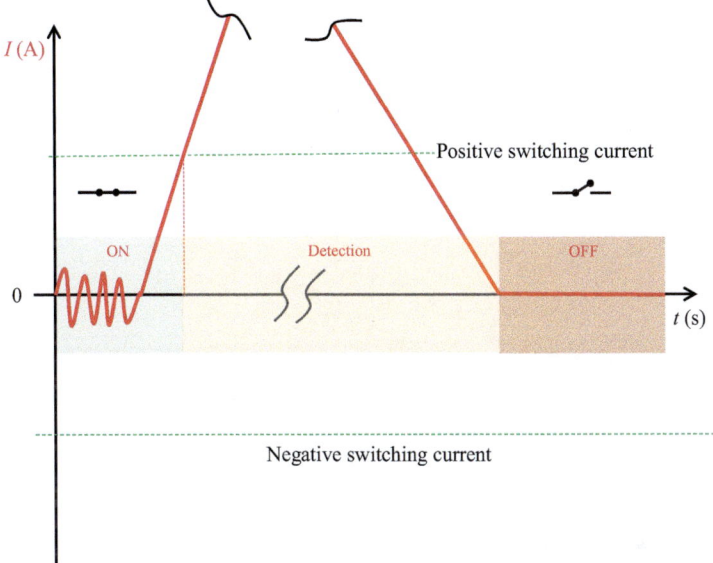

Figure 5.70   u(t) and i(t) current-switching component reel curve

5.5.2   Fuses

As already stated, fuses are not really SPCs but they are almost always to be addressed in surge protection as they are heavily used. They act mainly to avoid catastrophe failures that are very often the logical end of almost all SPCs. They are mostly used in power application but may as well be part of a data/signal surge-protection solution. Keep in mind that all electricians on earth try by using all their knowledge and techniques to avoid short circuit (between any kind of active cables or wire), and SPD acts mainly as a short circuit that lasts normally no longer than the surge but that may remain in the case of end of life of the SPC. Then fuses are part of the surge current that is diverted by the SPC (voltage limiting or switching SPCs).

How these react when some kA pass across their terminal's connections? They act as fuse and may blow if the current overpasses a certain value for a certain time. In fact, as any fuse, they are characterised with their melting energy and the surge may be of higher energy than this limit. They end up in an open circuit. The i^2t comparison of the fuse characteristic and the i^2t of current surge wave is one technique commonly used today. See Figure 5.71 for the ultimate one-shot melting limit for fuses when stressed by transient surge or lightning currents (8/20 and 10/350 wave shape).

This table gives theoretical fuse withstand and it is to be known that no negligible variation between fuse types and manufacturers can be observed. But this table gives a good trend of the real world.

Then they react to surge current; yes, but they are far too slow and most likely the surge is already gone and has destroyed the next equipment when the fuse opens. Fastest fuses may be able to melt within a millisecond. But if one considers the transient from surge current of lightning current that it acting in few µs,

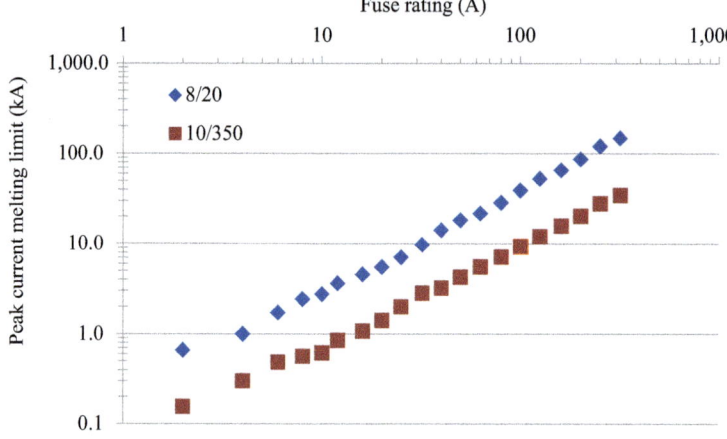

Figure 5.71　Withstand tendency of a typical fuse (gG)

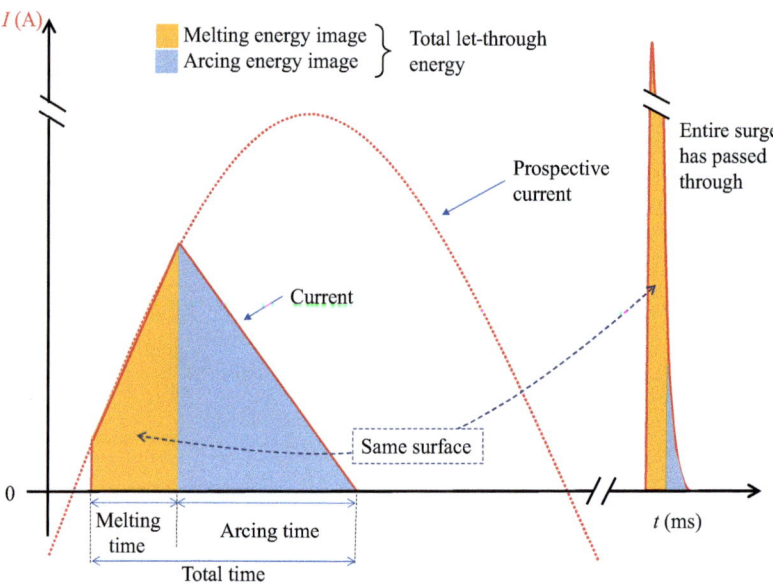

Figure 5.72　Withstand tendency of a typical fuse (gG)

dangerous voltage or current has passed without any interruption. Also consider that a fuse needs time to melt but in addition some time to really cut the current and that even if the fuse was superfast to react, and the real quenching of the current takes some more time. See Figure 5.72 to visualise this mechanism.

Note: Fuses are also a source of transient. When a fuse melts, it creates an overvoltage across its leads.

5.5.3 Circuit breaker (CB)

As fuses, circuit breakers are not used as surge protection as they are even slower than a fuse to act. In other terms, if a breaker operates due to surge current, the surge is gone far before the contacts of the switch are separated. Switches are used for the protection of the surge component but no specific surge withstand requirement is existing.

5.5.4 Semiconductor switch

This is an in-line static switch that detects in few nanoseconds a current reaching a preset value. This detection initiates the disconnection of the load in 1 µs or so. These components are fast and restable but unfortunately only dedicated to surge protection for data transmission application. Nevertheless, the concept could be extended to power applications if one day manufacturers decide to develop enough powerful components to handle load currents.

5.5.5 PTC (for positive temperature coefficient resistor)

PTCs are resistances that are reacting with temperature. They are called thermistors and even some time resettable fuse. They have several decades of resistance change when a temperature range is reached. They would be helpful for overload protection as a fuse or circuit breaker would do but their reaction time is far too slow when comes the transient protection topic. They are rated with a turn-off temperature or tripping current. In general, below the triggering temperature their resistance value is roughly 1 Ω. This conduction resistance forbids their use in-line with power circuit.

However, they are relatively and commonly used for data surge protection with the same function than a resistor or an inductor for the coordination of other SPCs (see Section 5.4). They have the additional function of line protection in the case of overload on the circuit that would possibly be due to cross-line contact with other power system of the installation.

5.6 Filtering components

5.6.1 Basic principle

Each electrical event can be analysed by its frequency spectrum. This is of course the case for surges. Even if they can be considered single events, they still have voltage and current deviation other time. Their frequency spectrum can be easily measured as shown next.

In standard bi-exponential current wave shapes being used to simulate field surges, the frequency where we can consider 90% of the total energy has been carried and can be set around 10–100 kHz depending on rise time and surge. One may consider that for application with more than 1 MHz, it would be a good value to consider filter technology for SPC solution.

Figures 5.73–5.77 show wave shapes and their frequency spectrum. Other than IEMN (Nuclear surges), these wave shapes are the ones to consider for surge protection.

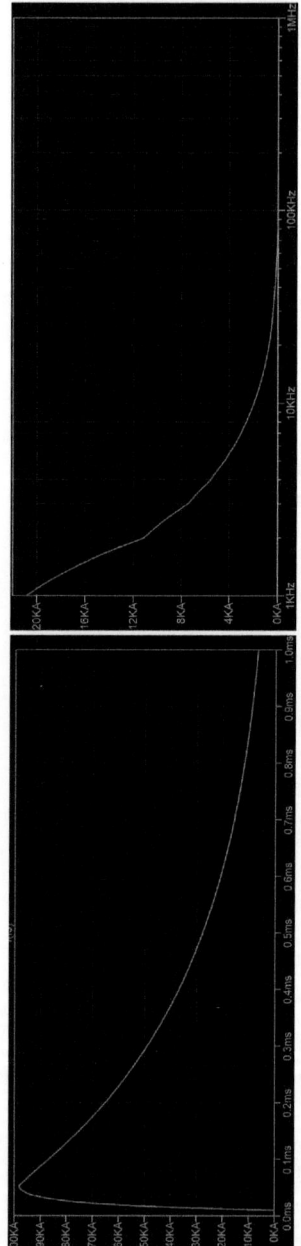

Figure 5.73   Long current wave (high peak value, rise time around 10 μs)

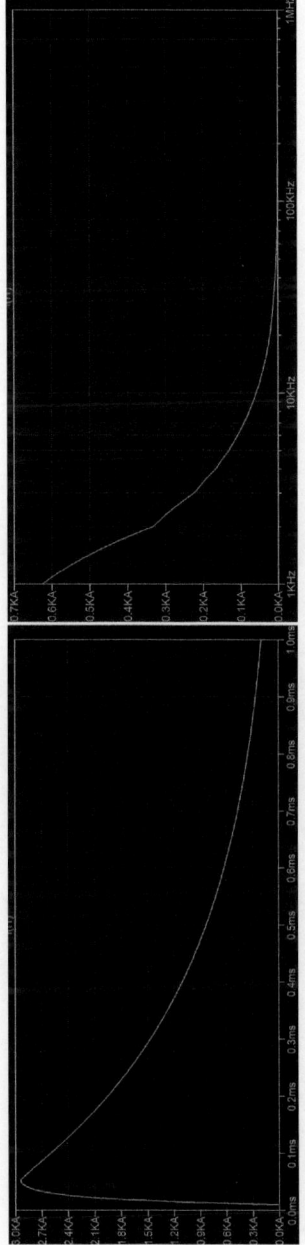

Figure 5.74 Long current wave (low peak value, rise time around 10 µs)

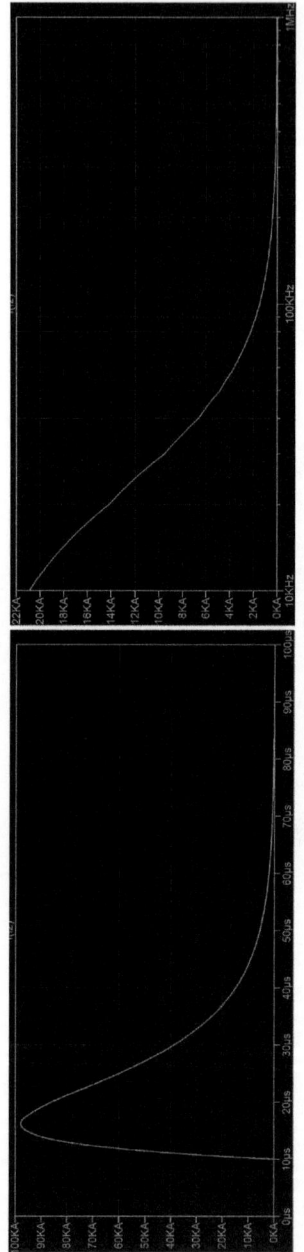

Figure 5.75 Short current wave (high peak value, rise time around 4 μs)

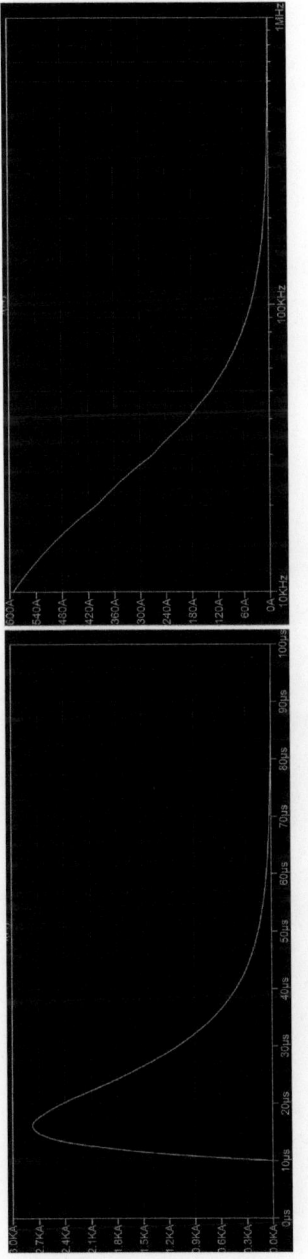

Figure 5.76   Short current wave (low peak value, rise time around 4 µs)

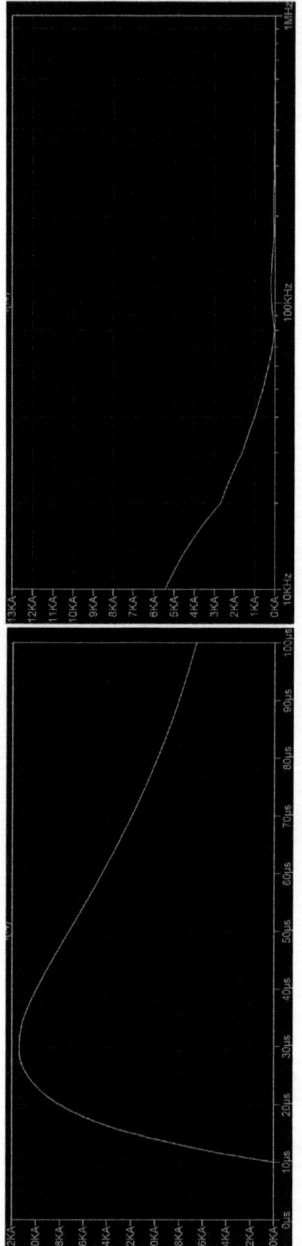

Figure 5.77   Mid current wave form (low peak value, rise time around 14 μs)

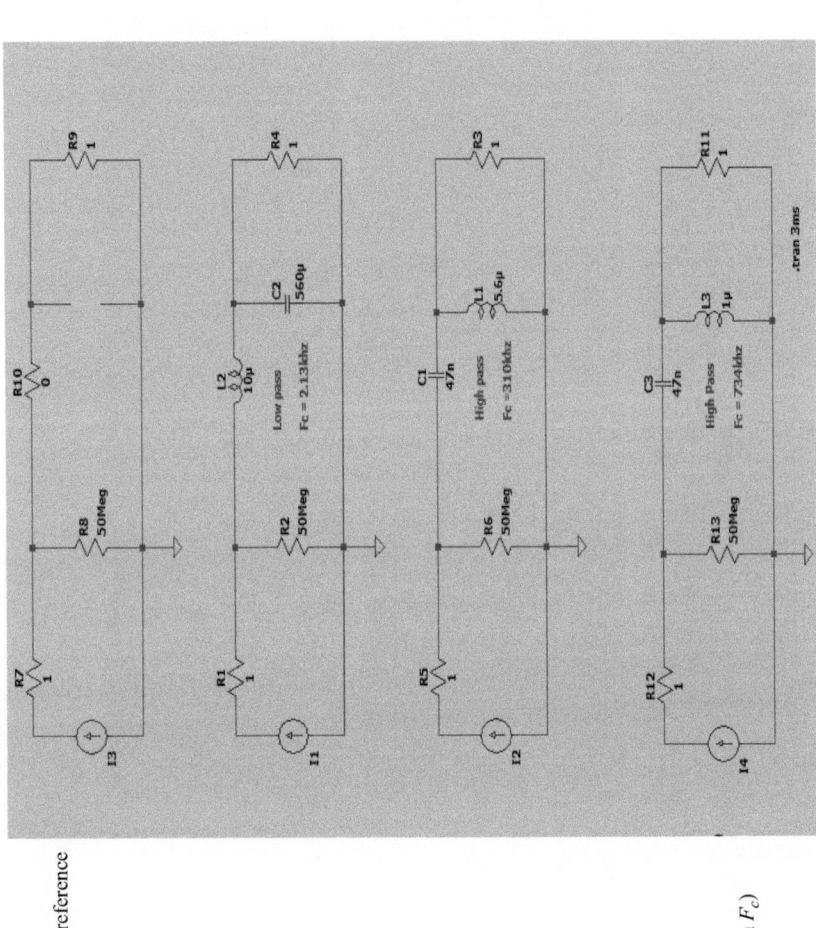

1. Circuit with no filter for reference

2. Low pass filter

3. High pass filter (low F_c)

4. High pass filter (medium F_c)

Figure 5.78   Basic circuit to express various filter responses with a transient

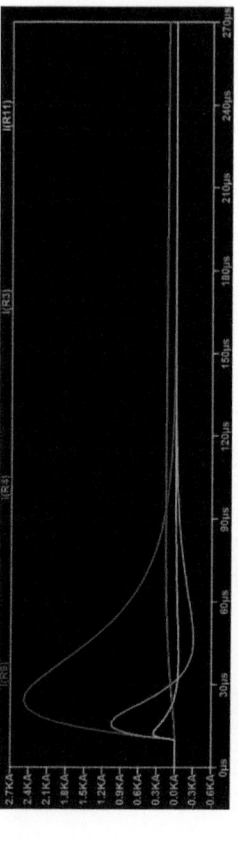

1. Current simulated in all 4 load resistors (R3, R3, R4 and R11)

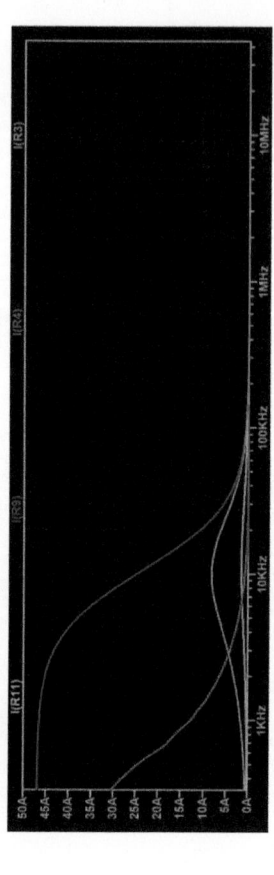

2. Spectral signature of the current in all 4 resistors

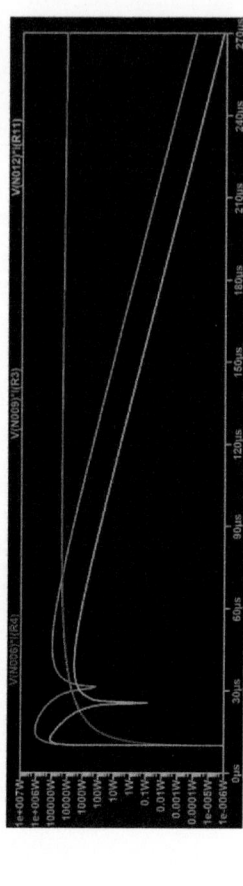

3. Instantaneous absorbed power in all 3 resistors with filter function

Total energy for a time interval of 3 ms:

$WR_9 = 164,2J$
$WR_3 = 12,7J$
$WR_4 = 8,54J$
$WR_9 = 0,71J$

Figure 5.79 Plots of various circuits shown in Figure 5.78

It may also be interesting to use a low-pass filter to smooth the fast current and voltage transition that surges are characterised by. This filter type is preferred when the working frequencies do not match the total spectrum of the surge. When the working frequency is low enough to cut substantial energy transmitted in the higher frequency range, it will reduce the peak of the passing surge current but they may last for several milliseconds.

To illustrate the effect of a filter facing a non-conventional surge current of 2.47 kA peak (non-conventional 9/39-μs wave shape, but this is not a concern for the example), see both Figures 5.78 and 5.79 that compare several circuits to really understand their effects. All filters shown here are of second order.

The energy in the load resistor for a 3-ms time period clearly shows the effect of all these filters when stressed by a transient surge.

5.6.2   Filter using LC components

In general, these filters are often presented as an electromagnetic interferences (EMI) filter rather than a surge-protection filter. Nevertheless, surge protection is part of the EMI. See Figure 5.80 for some filter. Their technology is usually using single inductors or capacitor or a combination of these components to make τ, or Π-type filters.

These filters are in general provided with information such as operating voltage, rated currents, values for the capacitors and inductance (if any) and of course frequency charts. For surge protection, the information from Section 5.6.1 should be used for its selection. Maximum peak voltage, also provided with the filter, may be used to estimate the maximum surge stress that can accept the filter. In the case of surge stress that may exceed its dielectric withstand, it is more advised to have coordination with another SPC (e.g. MOV, GDT or all other components addressed by this clause).

Of course, a filter can be realised by single capacitors and inductors. And thus, capacitor and inductors may be part of the wide family of SPCs.

5.6.3   Quarter wave stub

The quarter wave stud is a technique used for very high frequency to usually adapt transmission line impedance, but they can also be used as filters. When used as a component for surge protection, the concept is pretty easy to understand but the tuning is quite sophisticated.

It consists in a simple short circuit on a transmission line. This short circuit is realised with a piece of conductive material of a specific length that becomes an

 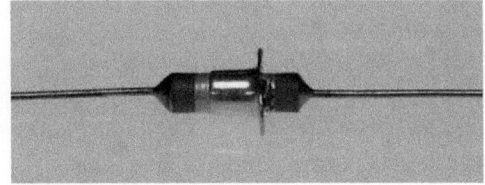

Figure 5.80   Pass-through filter examples

inductor when the frequency is very high. But for lower frequencies, the stub will simply short out all frequencies that are within the range of the usual surges. It acts as a low-pass filter. More precisely, it acts as a pass band filter but resets at each frequency that becomes a multiple of the frequency corresponding to the double of the quarter wavelength. When the frequency is corresponding to a multiple of the quarter wavelength, the impedance matching is reached.

See Figure 5.81 for the basic concept. One must consider that the chart is purely theoretical for the dotted lines as at higher frequency, coaxial line and connectors have their own limitations.

For surge protection, the used part of the frequency response is the one indicated in grey in Figure 5.81. Figure 5.82 shows various quartet wave SPD frequency responses. It is interesting to notice that the higher is the frequency, the larger is the working band width. The dotted line shows an SPD with intentional mismatch in the impedances that would permit to enlarge the bandwidth (Figure 5.83).

5.7   Isolating components

5.7.1   Basic principle

A device that is providing total isolation from a circuit to another circuit is the ultimate and easiest concept that one can imagine (Figure 5.84). But is this the easiest one to realise?

As for all topics addressing surge, this is not so obvious as isolating technologies have some limitation. First, the concept of common and differential mode of surge transmission is a very important principle for this technology as these types of components have by themselves to face both types of surge aggressions. A surge travels along conductors and the theory of transmission line is totally to be applied to understand how a surge can be transmitted and altered such as change in shape, amplifies, downsides or even vanishes due to impedance changes and specific component behaviour all along its path. This surge is to be considered as a dangerous ball of energy that may take various forms and the only way to make it acceptable is to convert this energy to be safe or simply send it away with no damage. (See Figure 5.85 to illustrate this line transmission approach.)

This surge may travel between several lines in parallel but for its treatment, it has been defined that when it travels along the line that is made of conductors connected to the Earth/ground, chassis or any part, this can be named as common between various equipment, and all other active conductors, the used term is common mode. Derivative terms are common mode protection, common mode transmission, etc. by opposition, when the propagation is between conductors that cannot be named common, and the term differential mode of propagation is used. It is reminded that one effect of reflection of a pulse in a line terminated by an open circuit is the doubling of this pulse. The overvoltage in the line is due to the reflexion of the surge (when open circuit). This possible overvoltage increase due to reflections is to be considered for power frequency transformers line length multi, point connection and possible superposition of subsequent surges (Figure 5.86).

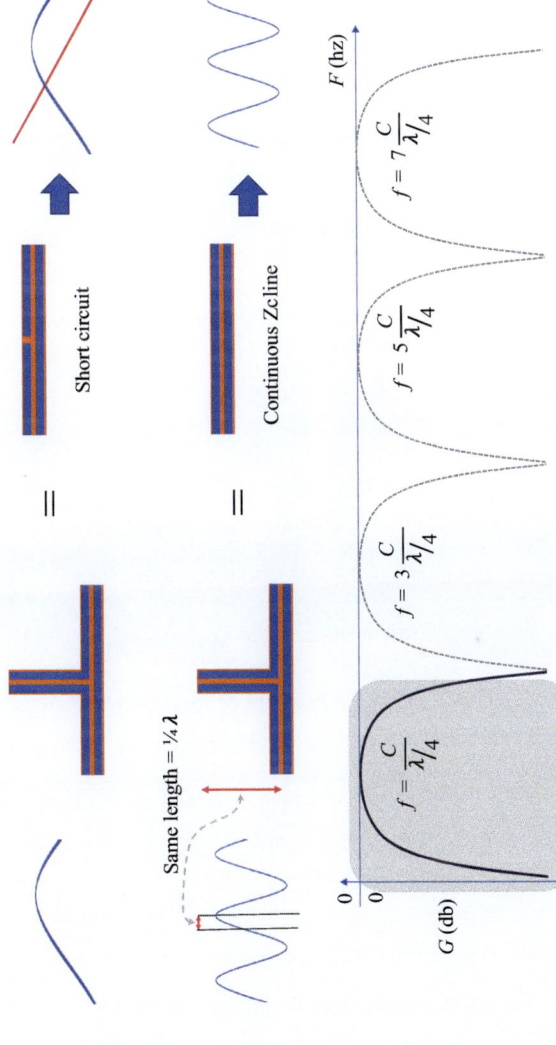

Figure 5.81 Quarter wave surge protection principle

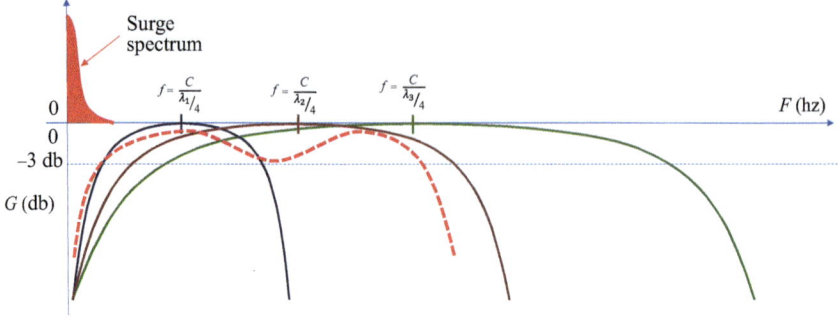

Figure 5.82   Quarter wave surge-protection comparison

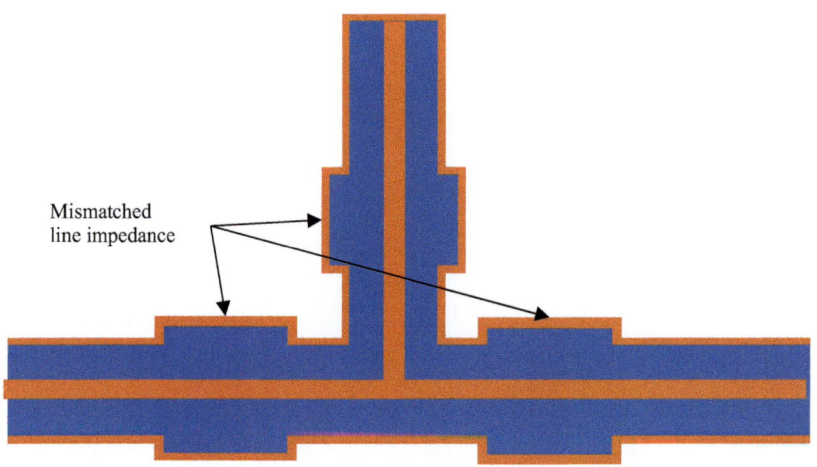

Figure 5.83   Mismatched impedance concept of quarter wave surge protection

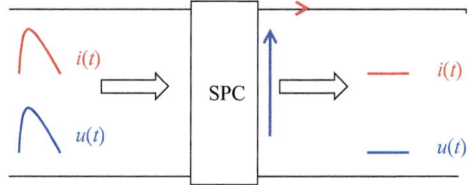

Figure 5.84   Perfect isolating component

Once the common and differential mode of propagation is defined, the transmission or blocking of the surge by the isolating component is specific for each mode.

If the SPC must transmit signal or power that is in the surge frequency spectrum, the surge itself is transferred to the rest of the circuit in differential mode.

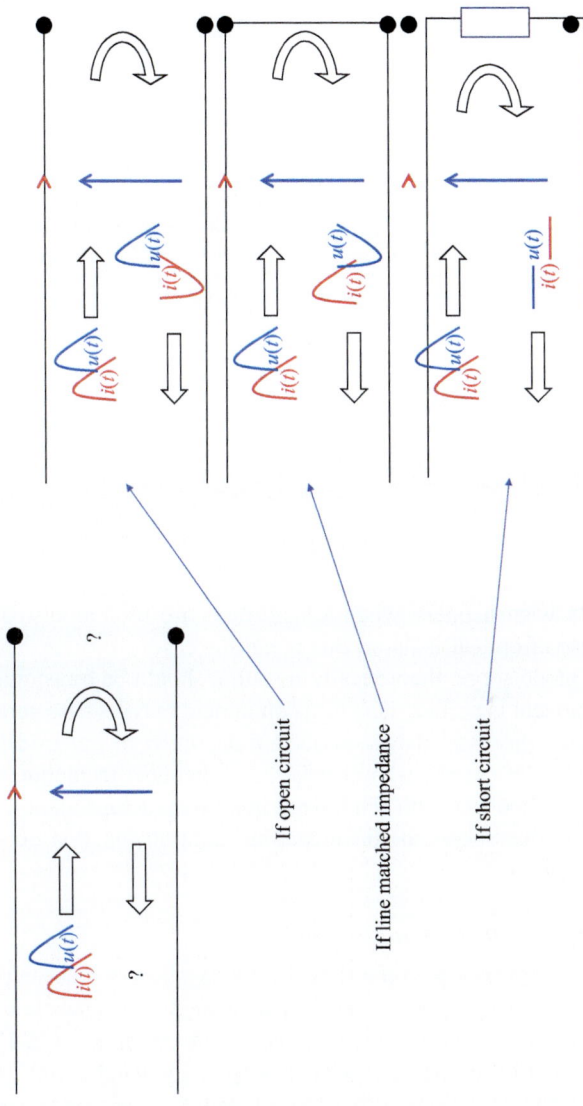

Figure 5.85   Basic concept of line transmission

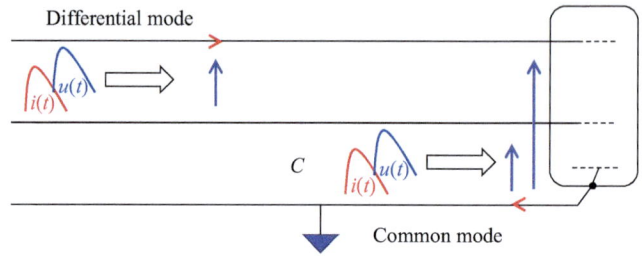

Figure 5.86   Common and differential mode of surge propagation

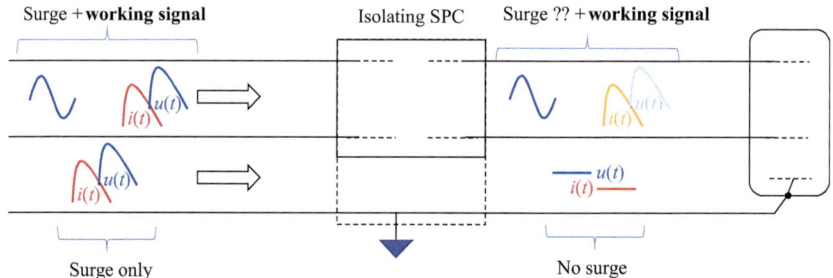

Figure 5.87   Common and differential mode of surge propagation

This is most likely what happens when transformers are used as insulating components. In the case of optical isolator, this is not the case.

For common mode surge, theoretically no surge should be transmitted. But in reality, as no component is perfect, isolating component may let pass partial energy from the surge. The only way that a surge is transmitted to the load is through parasitic capacitance. Of course, whatever is the component technique, producers tend to make their component with minimized input/output capacitance.

See Figure 5.87 that shows the difference of required function depending on the mode of propagation.

5.7.2   Surge isolation transformers

Using a transformer as a means to mitigate the propagation of a surge pretty efficient but with some care and guidance will be discussed in this subclause.

As said earlier, the transformer is intended to transmit either signal or power. When signal, the size of the transformer is usually of some cubic millimetres and can be easily mounted on a PCB. When power, these transformers may also be mountable on PCBs or have to be installed on their own as they can be of several kilograms or even tonnes. As discussed in introduction of this clause, is size or weight a parameter that defines a component? Here, one is to choose his own interpretation. In this subclause, it will not be made any difference. See Figure 5.88 showing insulation transformers.

Figure 5.88   Various shapes and transformer sizes used as isolating transformers (from Internet)

Main functions of a transformer will not be discussed here and they of course must be designed for the intended application. Voltage, frequency, primary/secondary ratio, power, etc. are all parameters that an engineer who wishes to use a transformer may identify and align with his application.

As far as the surge-protection function of these transformers is the topic, here are the elements to consider.

For differential mode, one must keep in mind that in general transformers are used in the frequency spectrum of typical expected surge and then it is admitted that transformer does not bring any benefit for this mode of protection. See Figure 5.85 for the simulation of transformers when its primary is stressed by a transient surge. The surge at the load is of course slightly reduced but still is dangerous. However, if they are used as part of a filter, this may not be accurate and surge reaching the primary of a transformer may not be transmitted totally but in that case the transformer in part of a filter.

The current in the primary may also saturate the magnetic core and inhibit the energy transfer to the secondary. This leads to surge-mitigation effect but will also block the signal or power transfer as long as the core is saturated. In addition, the current of the surge in the primary may generate heat and more important may generate high over voltage in-between each round of the winding. If the dielectric strength is broken, it may end in the destruction of the transformer and possibly catastrophic failure in the case of power transformer. It is most likely recommended to have upstream differential protection if the expected surge is foreseen to be able to overstress the transformer and also to protect the load from differential surges and others (Figure 5.89).

For common mode protection, the surge-protection benefit is real. The only limitation is the maximum breakdown voltage between primary, secondary and chassis if the transformer is grounded. See Figure 5.90 for the simulated effect of a surge applied on the primary winding of on an ideal surge isolation transformer (SIT).

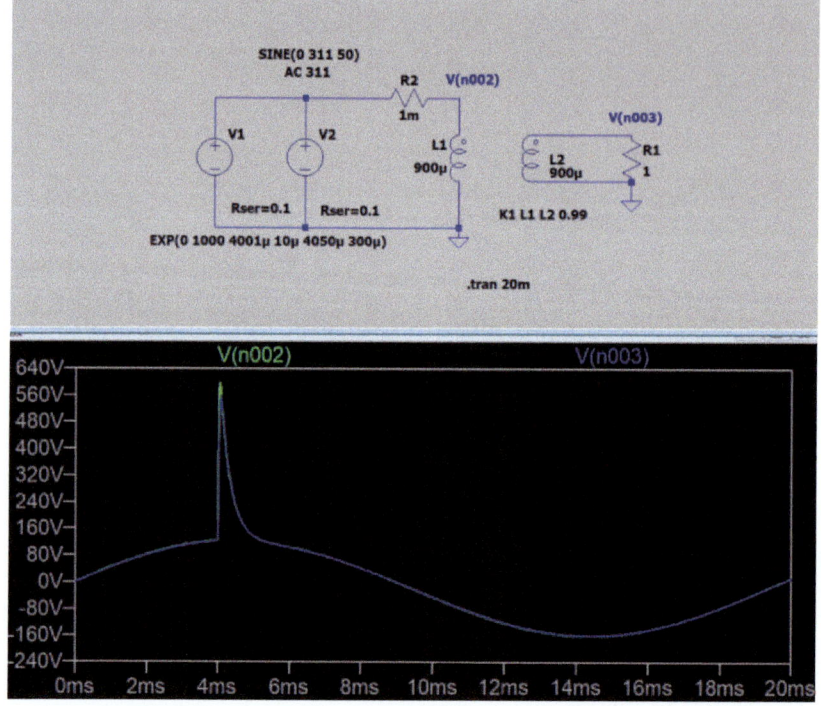

Figure 5.89   Surged transformer (differential mode)

Of course, likewise per all other components, parasitic capacitances may transmit part of the surge to the rest of the circuit. Isolating transformer may be constructed with or without metallic screens separating primary and secondary windings. Examples of simulated real isolating transformers are shown Figures 5.91 and 5.92.

It is obvious that common mode surge stress is almost annihilated thanks to the transformer behaviour. The surge stress is transferred from the rest of the circuit to the dielectric withstand between the transformer winding. It is then easy to imagine that if a transformer is given for 10 kV primary to secondary impulse withstand and that if the estimated surge exceeds this value, the option for the engineer in charge of the design are as follows: either pick a bigger transformer or limit the common mode expected stress by using SPC (or device).

It is interesting to reduce the capacitive coupling by decreasing the global capacitance between the primary and the secondary circuits. See C7 in both Figures 5.91 and 5.92 that is in series with the transformer parasitic capacitances. This is the best practice and that can be easily done on equipment but becomes tricky as soon as it concerns power distribution applications as it modifies the grounding reference from the power circuit downstream the transformer and then specific measures for protection against direct contact (human protection) are expected. But this is well known by electricians.

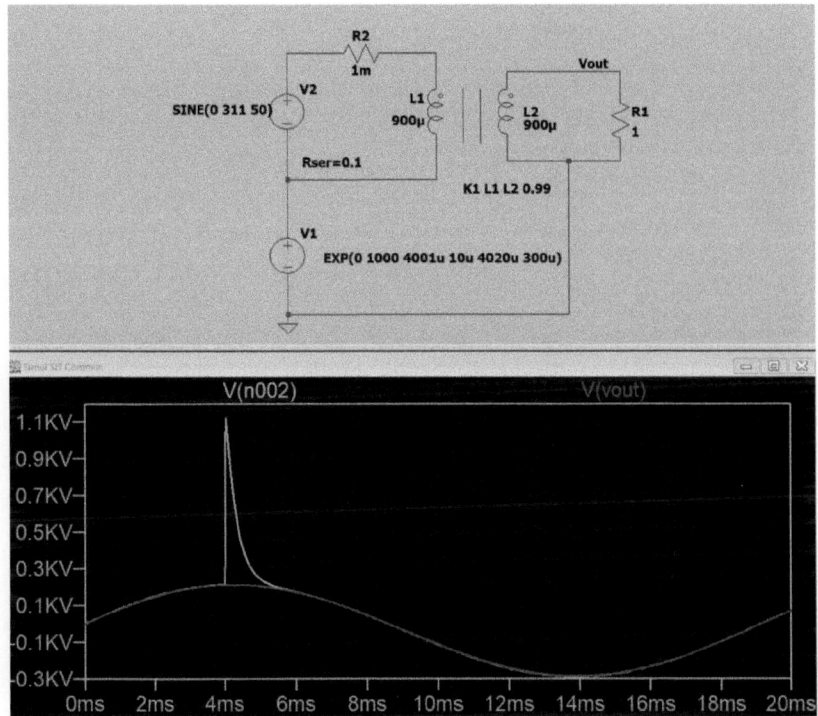

Figure 5.90 Ideal isolating transformer for common mode surge

Of course, the construction of a few grams' transformer will differ from the construction of a 500-kg transformer. But surge-protection principle remains very similar. One can easily imagine that parasitic capacitances are more diffi- cult to control for big fabrication as capacitances are mainly due to surface facing other surfaces, but distances in-between conductive parts can be increased.

Data transmission SIT are widely used in data transmission applications while power transformers are not really well known and only few manufactures claim surge-protection parameters. This is due to the fact that concerns (and budget) for power transformer are more oriented towards power parameters than towards surge-protection characteristics. However, the benefit is real and few surge pro- tection specialists even design surge-protection devices based on association with limiting, switching and filtering SPCs altogether. The only drawback is surely the weight, volume and cost.

Standards IEC 61643-351 and IEC 61643-352 address the data signal trans- former SIT (product and application) and some projects are running to produce a standard for power SITs. IEEE and ITU documents are also available but are most likely very similar to the IEC ones.

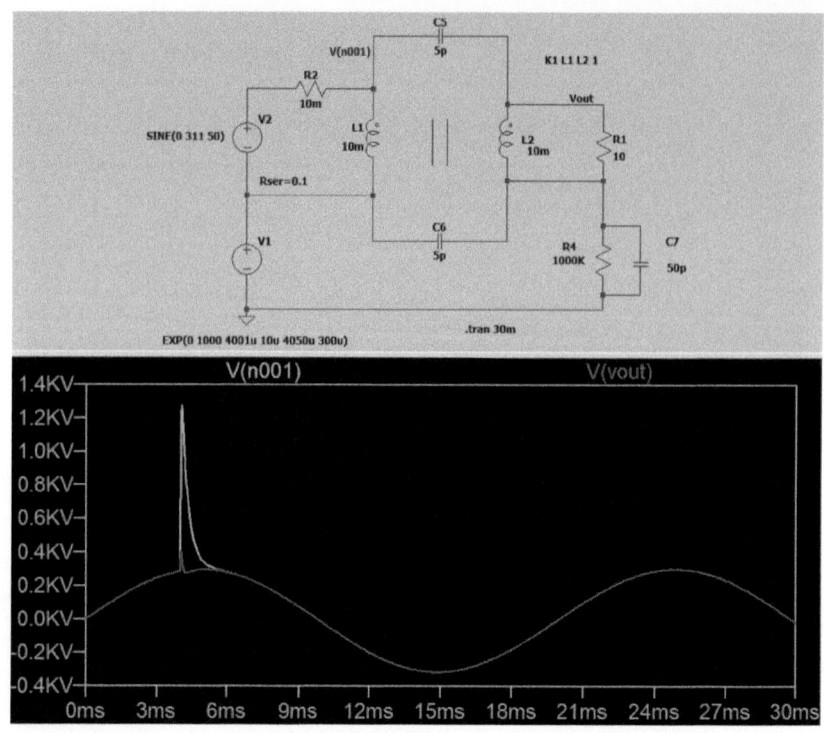

Figure 5.91   Screenless SIT and common mode surge

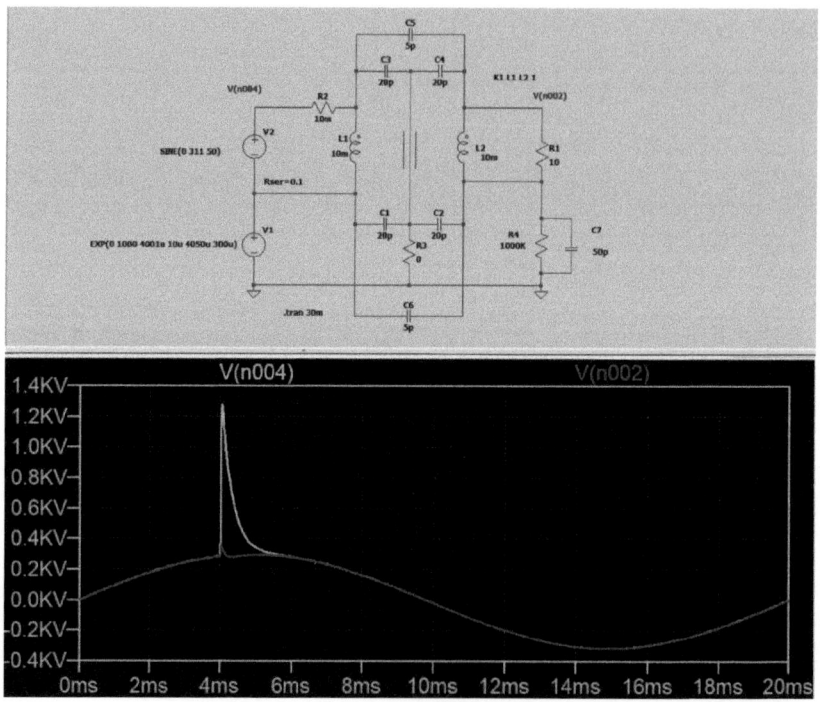

Figure 5.92   Screened SIT and common mode surge

5.7.3   Optoelectronic components

These components are well known by electronic engineers and they are widely used to isolate two circuits when no power has to be supplied. Most of the equipment dedicated to be used on rough environment are designed with an entry circuit that is not connected to the rest of the circuitry. Basic optocouplers are with two connecting leads at the entry and two or three outputs. The basic principle is that in the same package, there is an electroluminescent diode that emits a light through a transparent dielectric towards the base of a photo-transistor (usually the light is in the infra-red-light spectrum). The isolating barrier in-between input and output may be of several kilovolts (isolation is one of the most important parameters and one can find values starting from 500 V up to 20 kV). Then no electrical energy is transmitted from the input to the output and the maximum output signal is only limited to the outside power supply. Then in principle if a surge reaches input side whatever is the propagation mode, it will not be transmitted to the output side. This is to moderate due to possible parasitic capacitance coupling between input and output. If differential surge transfer is not to be expected, parasitic capacitance is a possible path for a common mode surge if the dv/dt of this surge is high. Typical coupling capacitances are part of the datasheets provided by manufacturers and may be in the range of several nano-Farads to few pico-Farads. This can easily be simulated and can be easily handled with the use of other SPCs.

Some optoisolators have the base (or grid) of the transistor available for specific use. These components are not welcomed as this terminal may act as an antenna during the surge event and may generate a wrong operation of the transistor.

The emitting diode is a very sensitive component and must be protected itself from surges possibly arising in differential mode. Then typically the input circuit is already a possible association of SPCs. The input diode may have different surge withstand depending on its polarity (forward voltage of an LED is in the range of 2 V and maximum reverse voltage is from 5 to 30 V). It is then pretty common to have a unidirectional TVS connected towards its terminals (see Section 5.2.3). An optoisolator is a good technique to consider for any signal transmission application and the only limitation is its max operating frequency that is intrinsically linked to the LED/transistor speed (Figure 5.93).

An optoisolator is a good technique to consider for any signal transmission application and the only limitation is its max operating frequency that is intrinsically linked to the LED/transistor speed.

As a conclusion for this topic, signal transmission using optic fibre is the exact same principle but due to the distance between the emitter and the receiver, all drawbacks that were due to the proximity of these (capacitive coupling) simply vanish.

5.8   Component association: hybrid

Naturally a question rises after having addressed a component by their single technology. Does a component made of several technologies exist? Of course, yes

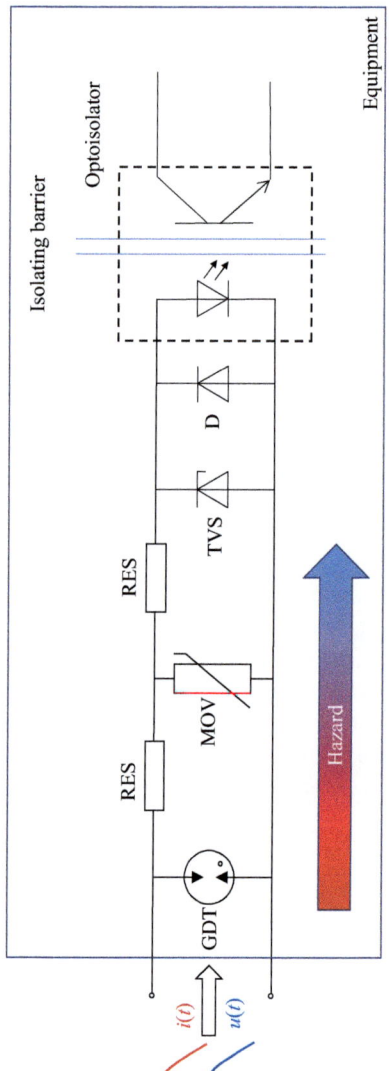

Figure 5.93 Input differential mode protection of an optoisolator

it exists. The idea behind is that each of the individual functions has advantages and disadvantages. Does their association lead to better or worse functionalities? This is obvious to say that some associations make a better component and some are already available on the market.

Then, hybrid components exist and this is also almost mandatory that surge protections are using single components and associate them to try to reach the perfect surge-protection solution.

Here again the question about where is the limit what defines a component that may be posed. The following subclause will try to answer this question but one may consider that a microcontroller is a component but contains multiples basic functions.

Usual technology association may be of voltage and current switching or limiting functions. The association of an MOV with a GDT in series is the main one, but it is not the only possible combination.

5.8.1 Varistor and gas discharge tube

5.8.1.1 Serial association (VG)

If one lists the drawbacks of the MOV, we can raise the fact that the limiting voltage lasts for the entire surge duration, and the leaking current is continuously ageing the MOV, which is catastrophic end of life when stressed by temporary overvoltages (swells or loss of neutral for AC power systems). By opposition, the GDT drives the surge current with almost no residual voltage (without considering the peak voltage of its breakdown) and the GDT is not being aged when connected to a power supply. Its biggest drawback is its limited capacity of extinguishing the follow current that comes with power application. On the contrary, the surge current stops flowing through the MOV when the voltage becomes lower than its clamping voltage.

Then if an MOV is mounted in series with a GDT, the MOV can limit the current due to the power supply to reach the GDT extinction and the GDT will block any leakage current during normal condition. In addition to this the limiting voltage of the MOV may be drastically reduced providing a much better surge protection. One question then arises, how to select the MOV and the associated GDT? The target is to get the MOV with an MCOV as low as possible to get a smallest mass–volume compatible to the expected surge current and the voltage of the power supply.

One criterion could be to set an energy that will absorbed by both the GDT and the MOV due to the surge and follow currents to reach the targeted surge characteristics.

If the component is dedicated to protect a DC-powered system, the approach may be as for these two examples addressing AC and DC power supplied circuit.

DC power system:

$U_{max} = 90$ V DC, $I_{cc} = 1$ kA, expected TOVs: 120 V for uncertain time and expected surges 15 times 2 kA 8/20.

Note: If protected by an MOV only, the MOV would be with a minimum 1-mA voltage to 140–150 V. It could be considered a 14-mm MOV disc and the expected limiting voltage at 1-kA peak current may be of −350 V.

The selection of the GDT is simply to make sure that it will never trigger for normal conditions, including TOVs; then the min breakdown DC must be higher than 120 V. Safety margin is of course to be considered. Then a GDT with a 230-V breakdown voltage suits perfectly and a size around 5 mm × 5 mm would be more than enough to handle the expected surge parameters. If we consider U_{arc} is around 20 V (this is very conservative if one considers the 1-mA condition next).

For the MOV selection, safe-side would be not to permit any follow current after the surge. At 1 mA, the condition is to maintain an arc in the GDTs that are not fulfilled. If U1mA is at least equal to 70 V (U_{max}–U_{arc}), this leads the MOV to limit the surge voltage to 200 V and at 220 for the hybrid component (compared to the 350 V of the MOV only).

AC power system:

U_{rms} = 230 DC, I_{cc} = 25 kA, expected TOVs: 440 V for possibly minutes and expected surges 15 times 20 kA 8/20.

Note: If protected by an MOV only and if the required criterion about the TOV is not to fail, the MOV would be with 700–750-V minimum 1-mA voltage. It could be considered a rectangular 34 × 34-mm² MOV block and the expected limiting voltage at 20-kA peak current may be of −2,000 V (in the case of accepted failure for the TOV, the expected limiting voltage would be of 1,300 V).

The selection of the GDT is simply to make sure that it will never trigger for normal conditions, including TOVs; then the min breakdown DC must be higher than 680 V. Safety margin is of course to be considered. Then a GDT with 800 V +/−15% breakdown voltage suits perfectly and a size around 8 mm × 6 mm would be able to handle the expected surge parameters. As per the previous example, U_{arc} is around 20 V.

Following the same rule compared to the previous example, the U1mA value for the MOV is at least equal to 360 V. This leads to an MOV that will limit the surge voltage to 200 V and a hybrid component-limiting voltage of 1,000 V (compared to the 2,000 V of the MOV only).

For these two examples, the safest criterion (1 mA) is used to select the MOV but one can consider that this can be optimised by the manufacturers. Performance of specific GDTs with better follow current interruption characteristics may leave the manufacturers with better choice. As a hint, if a possible follow current is of 10 A, for AC at the peak of the sinusoid, it is expected that the hybrid component easily passes the full surge test sequence and this would lead to have a voltage protection around 650 V.

As a conclusion, the association of MOV and GDT in series is a very good solution to get rid of the natural ageing of the MOV, to disregard the LV TOVs, to achieve a much better limiting voltage compared to MOV only, to get good surge carrying capability with limited size component not to have to be concerned about follow current (Figure 5.94).

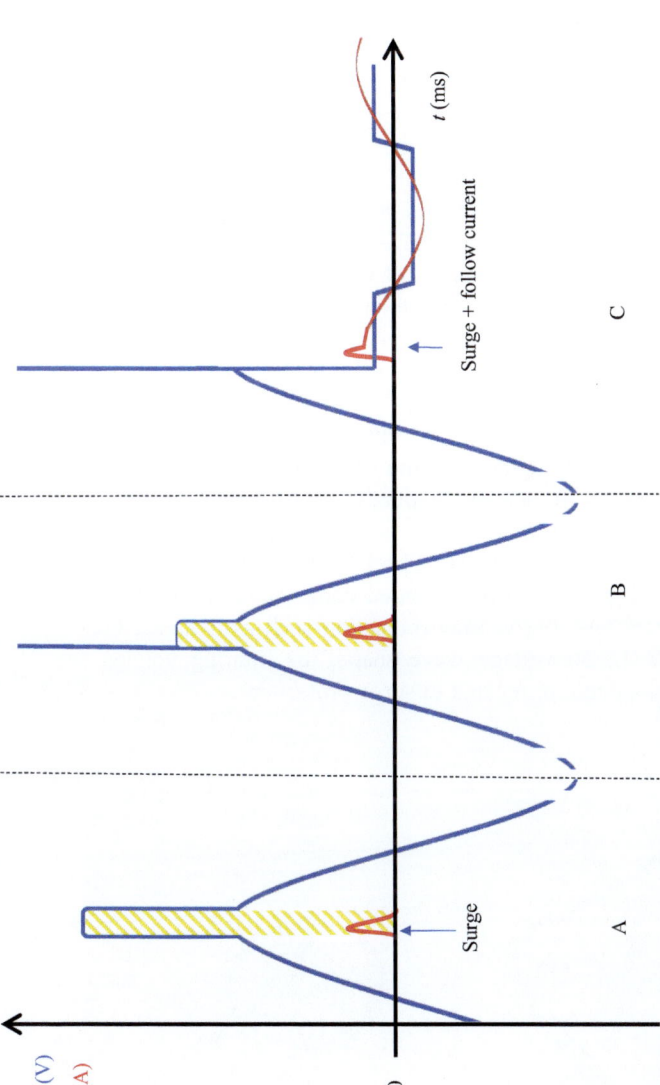

Keys:

A. *VI* characteristic when protected with an MOV. The voltage surge lasts for the surge current duration and is much higher than the peak AC voltage of the power supply.

B. *VI* characteristic when protected with a hybrid varistor GDT in series. The voltage surge lasts for the surge current duration and is only slightly higher than the peak AC voltage of the power supply.

C. *VI* characteristic when protected with GDT. The voltage surge is limited to few ns but the GDT remains in conducting modes short circuiting the circuit.

Hatched yellow zone represents the surface under the voltage trace that can be considered an image of the stress applied to the circuit.

Figure 5.94   Behaviour comparison for MOV, VG and GDT

The only point that could be still considered as a drawback is the GDT switching effect explained earlier and the peak value due to the dynamic voltage breakdown of the GDT (Figure 5.95).

Some claim that this hybrid technology avoids catastrophic failure. From Sections 5.2.2.4 and 5.3.3.5, the reading may keep in mind that both MOV and GDT may end up badly if no special care is taken. Special enclosure, fuse or thermal disconnector may still be needed depending on the system where the component is installed.

As seen, usage can be from protecting low AC or DC voltage but it can also be interesting to use one of the other great benefits of the GDT in series with the MOV. The capacitance of the assembly is lower than the GDT alone and thus high-frequency usage is permitted especially when the high-frequency signal is super-imposed with AC or DC power supply. This is widely the case for telecom RF installation. Another huge application is for the TV and Internet diffusion in cities using coaxial cables where it is very frequent to have signal and working power feeding a 1,000-W amplifier (or all other equipment such splitters for example). GDT only is totally prohibited for such application as supplied voltage may over-pass its extinguishing capabilities (e.g. 90 V–15 A) and MOV is also not possible because of its too high capacitance.

Besides its huge benefit, this type of component is not widely available. Only few manufactures are proposing complete solution. But many SPD manufacturers are using this hybrid technology in their production.

Surge current parameters such as the ones for MOVs and GDT for power application are given (surge current wave shape, peak values, repetition, etc.). This depends on standards and even type of the declared component. Associated with these surge's current parameters, the voltage protection such as dynamic break-down voltage with 6 kV 1.2/50 voltage wave shape, the limiting voltage are necessary to select the component in regards to surge protection.

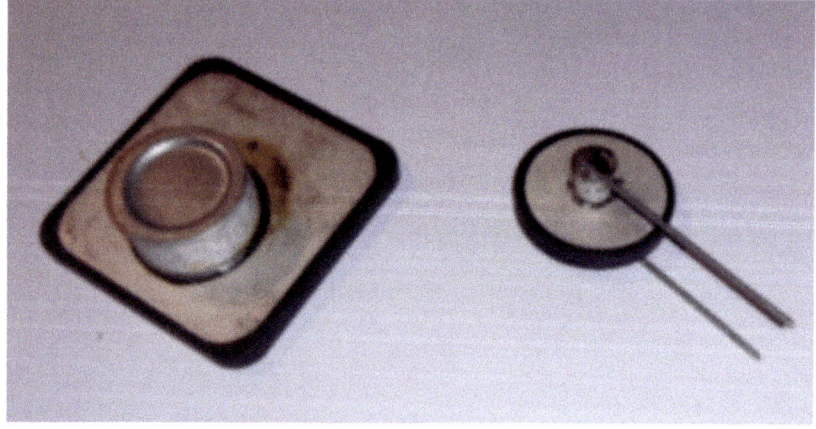

Figure 5.95   Serial assembly of MOV and GDT

The minimum breakdown voltage under AC and DC condition (if different) is necessary to select the component with respect to the power supply characteristics such as TOV, the maximum rated voltage and the possible additional function. All other characteristics only addressing the GDT or the MOV alone are not useful.

One may thing that a TVS could have the same function that the MOV in this arrangement or even a TSS may replace the GDT. This is theoretically true but in practice (in 2021) semiconductors are not as capable with the current handling, as MOV and GDT.

5.8.1.2 Parallel association (GV)

GTD and MOV are sometime associated in parallel but the limiting voltage of the MOV must be in the range that will create the GDT to operate. This association is useful when one wants to avoid a too high peak voltage when expecting that very fast voltage rises in association with a powerful surge (long or high surge current). The goal is that the MOV will limit the voltage spike up to the time that the GDT operates. Then the MOV is fully released from carrying surge current as the GDT limits the voltage to its arcing voltage. This permits to have a large current carrying capability with more controlled peak residual voltage in a controlled size. The GDT is the main component here and is totally blinding the MOV during its operation, leading this assembly to act as a pure GDT when triggered. This is to understand by this comment that the follow current is a major concern if it is expected.

The selection of the MOV-handling capability and its limiting voltage characteristics is crucial as it clamps to low the GDT will never operate. Too high clamping will not provide expected voltage limitation and its size must permit to withstand the surge current before the GDT operates or even when a slow surge is coming.

The selection of the components is explained with Figure 5.96.

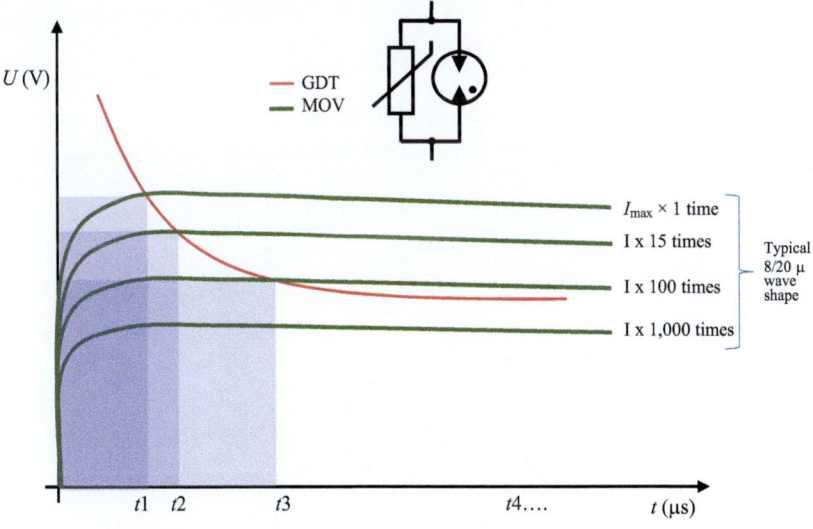

Figure 5.96 Example de GV

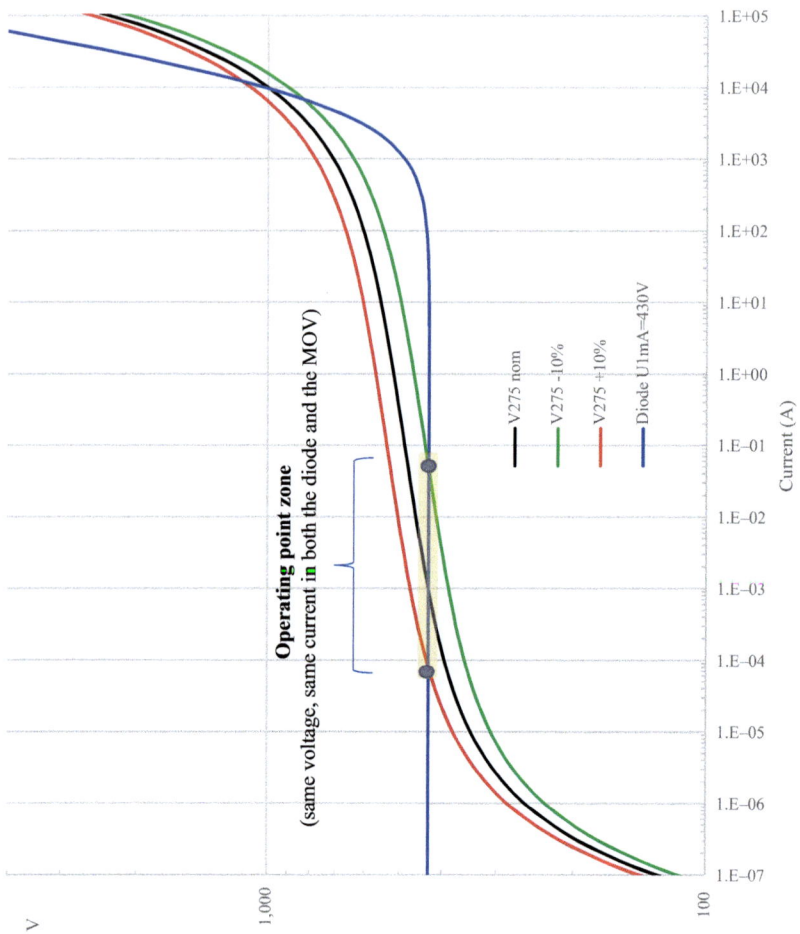

Figure 5.97 MOV and TVS VI comparison

The main difficulty is that both components have quite wide tolerances and even if the MOV can be sorted, the GDT is in addition not so stable. Imagine in Figure 5.92 all possible curves if the tolerance has to be drawn. This makes this association not easy to tune. But it may be useful for very specific application that requires this characteristic.

5.8.2 Other association with TVS, GDTs, MOVs, etc.

Some are claiming an association between MOV and TVS. This association is even more difficult to tune compared to GDT in parallel to MOV. But this is possible when a diode of almost the same surge-carrying capability compared to the parallel MOV and with very similar *VI* curves. Superposition of *VI* characteristics of an MOV and TVS shown in Figure 5.97 speaks itself. Some claims that it is useful to get this association because MOVs are slow and diodes are fast, but at the end of the day, the result has less than little benefit and is more a marketing approach than a real technical solution for cost that is most likely prohibited.

5.9 Conclusion

SPCs are various and are not perfect. They all present different characteristics. Most likely they are to be assembled to take benefit of the advantage of one technology while another technology corrects its disadvantage. Most of the components can be part of equipment and simply used in SPDs. Very often the border between component and SPD is thin and sometime SPDs are more defined by the behaviour of the component. However, MOV, GDT and TVS are the most used components and they are really linked to applications.

Bibliography

International and national standards and regulation

IEC 62368-1:2018: Audio/video, information and communication technology equipment – Part 1: Safety requirements.

IEC 61643-11:2011: Low-voltage surge protective devices – Part 11: Surge protective devices connected to low-voltage power systems – Requirements and test methods.

IEC 61643-12:2020: Low-voltage surge protective devices – Part 12: Surge protective devices connected to low-voltage power systems – Selection and application principles.

IEC 61643-21:2000+AMD1:2008+AMD2:2012.

Low voltage surge protective devices – Part 21: Surge protective devices connected to telecommunications and signalling networks – Performance requirements and testing methods.

IEC 61643-22:2015: Low-voltage surge protective devices – Part 22: Surge protective devices connected to telecommunications and signalling networks – Selection and application principles.

IEC 61643-32:2017: Low-voltage surge protective devices – Part 32: Surge protective devices connected to the d.c. side of photovoltaic installations – Selection and application principles.

IEC 61643-311:2013: Components for low-voltage surge protective devices – Part 311: Performance requirements and test circuits for gas discharge tubes (GDT).

IEC 61643-312:2013: Components for low-voltage surge protective devices – Part 312: Selection and application principles for gas discharge tubes.

IEC 61643-321:2001: Components for low-voltage surge protective devices – Part 321: Specifications for avalanche breakdown diode (ABD).

IEC 61643-331:2020: Components for low-voltage surge protection – Part 331: Performance requirements and test methods for metal oxide varistors (MOV).

IEC 61643-341:2020: Components for low-voltage surge protection – Part 341: Performance requirements and test circuits for thyristor surge suppressors (TSS).

IEC 61643-351:2016: Components for low-voltage surge protective devices – Part 351: Performance requirements and test methods for telecommunications and signalling network surge isolation transformers (SIT).

IEC 61643-352:2018: Components for low-voltage surge protection – Part 352: Selection and application principles for telecommunications and signalling network surge isolation transformers (SITs).

UL 1449, Edition 5: UL Standard for Safety Surge Protective Devices.

Standler, R. B. (2002). *Protection of Electronic Circuit From Overvoltages*. Dover publication, Mineola, New York, USA.

Standler, R. B. (1990). *Use of Metal-Oxide Varistor with Series Spark Gap Across the Mains*. University of Kentucky, USA.

Vojta, A. and Clarke, D. R. (1997). *Microstructural Origin of Current Localization and "Puncture" Failure in Varistor Ceramics*. Materials Department, College of Engineering, University of California, Santa Barbara, USA.

Crevenat, V. (2015). Surge current rating (or Imax). Citel, France.

Bartkowiak, M., Comber, M. and Mahan, G. (1999). *Failure Modes and Energy Absorption Capability of ZnO Varistors*. IEEE, USA.

He, J. (2019). *MOV, From Microstructure to Macro-Characteristics*. Tsinghua University Press, China.

Rousseau, A. and Crevenat, V. (2010). *Protective Levels at Equipment Terminals for Various SPDs*. International Conference on Grounding and Earthing, France.

IEC 62305-2 (currently edition 2) "Protection against lightning – Part 2: Risk management".

Matsuoka, M. (1970). *Nonohmic Properties of Zinc Oxide Ceramic*. Matsushita, Japan.

Chapter 6

Surge protective devices

Hubert Bachl-Hesse[1]

A surge-protective device (SPD) is defined as a device containing at least one nonlinear component intended to limit surge voltages (transients) and to divert surge currents. This implicates that such a device is designed using one or more components that either show a continuous voltage-dependent decrease in impedance, or a sudden voltage-dependent change in impedance.

An ideal SPD should

- under normal supply voltage and system conditions 'be invisible' and not influence the system characteristics in any way and
- as soon as a surge voltage appears limit the amplitude to a value low enough to avoid any damage to installations and equipment or malfunction of equipment, without any significant impact on power quality, e.g. avoiding any dips or even interruptions due to follow currents or slow recovery to a high-impedance state again.

Today, there is an expanding international standard series that covers safety-, endurance-, performance, mechanical and other aspects of SPDs. The following parts are already published:

- IEC 61643-11 SPDs connected to low-voltage AC power systems
- IEC 61643-21 SPDs connected to telecommunications and signalling networks
- IEC 61643-31 SPDs for photovoltaic installations

An additional part, IEC 61643-41 SPDs connected to low-voltage DC power systems, is in preparation.

Unfortunately, there is still no straight borderline between the standard for SPDs connected to low-voltage AC power systems and the standard for SPDs connected to telecommunications and signalling networks, except the designation related to the intended application. This is further weakened, e.g. by the fact that IEC 61643-21 covers rated currents for SPDs up to 16 A. There are ongoing discussions to improve this situation, one solution proposal being the introduction of an application-related power limit with 100 W or 100 VA, which somewhat relates to inherently limited class II power units according to UL 1310. But nothing is really decided so far.

---

[1]hbconsult, TGM, Vienna, Austria

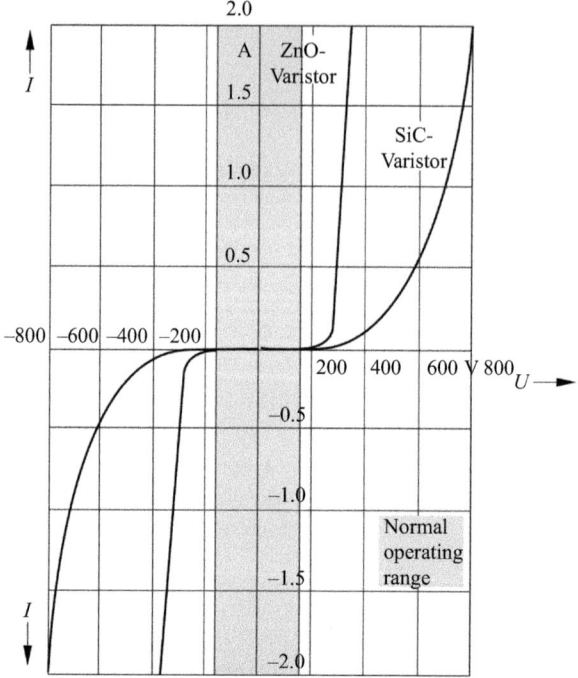

Figure 6.1   Components with continuous U/I-characteristics*

6.1   Technologies

6.1.1   History

The use of low-voltage arresters in overhead lines started between 1920 and 1950. These arresters used a series connection of a silicon carbide varistor and a spark gap. In the later nineteenth century the application of such arresters also expanded to cable distribution cabinets, and with the increasing number of electronic equipment, manufacturers also started providing constructions for indoor application and DIN-rail mounting. These constructions had a quite bad limiting characteristic compared to today's SPDs, because the U/I-characteristic of a silicon carbide varistor shows a quite smooth bend – see Figures 6.1 and 6.2 – and to reach an acceptable protection level, a series spark gap was needed to extinguish the follow current after conduction due to a surge event. Around 1980, the development of metal oxide varistor (MOV)-based low-voltage arresters as well as the development of self-extinguishing pure spark-gap arresters for low voltage started, and in December 1989 the new term surge-protective device (SPD) was agreed within the first meeting of IEC SC37A. Since that the variety of SPD designs and their

*VDE-Verlag GmbH, ISBN 3-8007-2052-3, 1999, modified.

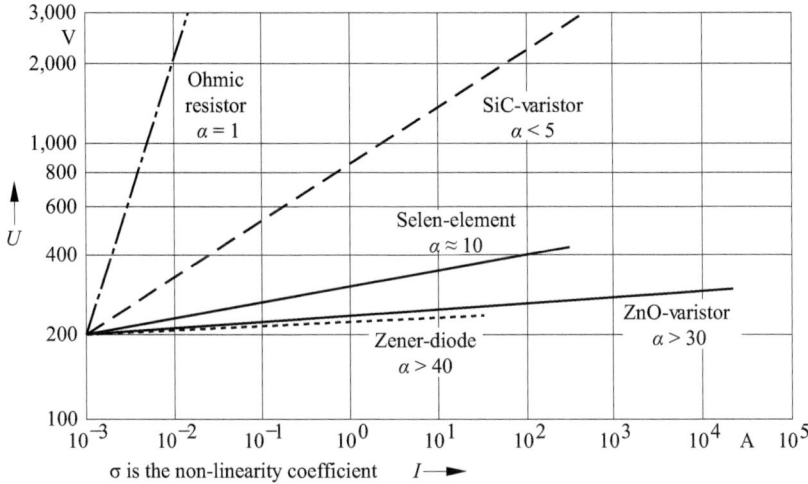

Figure 6.2   Normalised U/I-characteristics of voltage-limiting SPCs.[†]

technology have developed continuously, leading to increased discharge capabilities as well as improved overvoltage-limiting behaviour together with an increased resistibility against stresses due to power frequency overvoltages.

6.1.2   Surge-protective devices (SPDs) versus isolating spark gaps (ISGs)

The major difference between an SPD and an isolating spark gap (ISG) is the ability of an SPD to avoid or extinguish any follow current from the power system by itself during and after a surge event that caused a conduction of the SPD's surge-protective component(s) (SPCs) when a voltage up to the maximum operating voltage is present at its terminals. This section addresses SPDs according to the IEC/EN 61643 series of standards but does not address ISGs according to IEC/EN 62561-3 (former EN 50164-3). Surge arresters and low-voltage limiters for DC traction systems according to the EN 50526 series of standards (former EN 50123-5) are also not addressed here.

6.1.3   Classifications

There are a number of classifications already given by the product standard series IEC/EN 61643 as different application purposes, targets of protection, product topologies and designs partly require the application of different test set-ups or test procedures. Some other or additional classifications have been introduced for application-related parameters and for marketing purposes. The following aims to provide an overview and to describe the interrelation of parameters.

[†]VDE-Verlag GmbH, ISBN 3-8007-2052-3, 1999, modified.

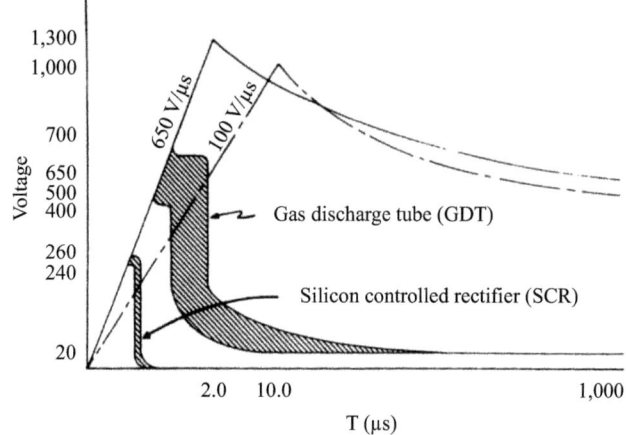

Figure 6.3   U/t-characteristics of voltage-switching SPCs[‡]

6.1.3.1   Overvoltage response characteristics

Voltage limiting (clamping)
When an SPD contains only SPCs with a continuous voltage–current characteristic –
see Figures 6.1 and 6.2 – such an SPD is classified voltage limiting. Examples of
such 'voltage-limiting' components are as follows:

- voltage-dependent resistors
  - selenium elements
  - silicon carbide varistors (SiC)
  - MOVs, e.g. ZnO

- avalanche breakdown diodes, Zener diodes (Z-diodes), transient voltage suppressor (TVS)-diodes (TransZorb-diodes)
  - single, unidirectional
  - antiserial, bidirectional

Voltage switching (crowbar)
When an SPD contains only SPCs with a discontinuous voltage–current characteristic – see Figure 6.3 – such an SPD is classified voltage switching. Examples
of such 'voltage-switching' components are as follows:

- air gaps (spark gaps)
  - simple (single) air gaps
  - multiple air gaps
  - triggered air gaps

  Δ open air gaps (blowing out of ionised gases during the arc extinction process)
  Δ encapsulated air gaps (no blowing out of ionised gases)

[‡]HARRIS semiconductor, Transient Voltage Suppression Devices, 1990, modified.

- zero-point extinguishing (only for AC applications)
- (follow) current limiting (for AC or DC applications)

- hard gas spark gaps
  - simple
  - triggered

- gas discharge tubes (GDTs)
  - with two electrodes
  - with three electrodes (triggering possible)
  - with multiple electrodes (triggering possible)

- thyristors (silicon controlled rectifiers)
  - thyristor with fixed trigger voltage
  - thyristor, gate controlled
  - triacs (bidirectional thyristors) with fixed trigger voltage
  - triacs (bidirectional thyristors), gate controlled

Combination of switching and limiting
An SPD containing any combination of 'voltage-limiting' and 'voltage-switching' components, no matter if these are connected in series or in parallel, is classified a combination SPD. Such SPDs generally show a more complex behaviour as compared to voltage-limiting or voltage-switching SPDs.

6.1.3.2   Occurrence of follow currents

An ideal SPD would divert surge currents without any negative impact on the power supply and the installation and equipment to be protected. In real world, each of the SPCs described in Section 6.1.3.1 has its advantages and disadvantages, and the better the overvoltage protection function is with regard to remaining surge energy (I^2t) 'let through' that stresses the downstream installation and equipment, the bigger is normally the impact on the power supply.

One of these possible impacts is the occurrence of follow currents drawn from the power supply as a result of the overvoltage-induced impedance change of the SPD, specifically when the relevant components either need some time to recover or need a voltage zero-crossing to reach a high-impedance state again.

Therefore, SPDs are classified into the following:

- devices that may cause a follow current and
- devices not causing any follow current ('follow current free').

The amplitude of a follow current may depend on various factors as:

- the remaining internal impedance of the SPD;
- the prospective short circuit current of the supply at the point of installation;
- the trigger point of the transient on the sine wave (phase angle);
- the amplitude of the surge diverted by the SPD.

6.1.3.3   Leakage current

The term leakage current is used synonymously here for continuous (operating) current and for residual current as well as for PE conductor current.

For some applications, it is important that SPDs have very high impedance when no overvoltage is present and either cause only a very low leakage current or almost no leakage current:

- either for functional reasons, e.g. in telecommunication or signalling circuits or
- for safety reasons to avoid any hazards due to potential touch voltages or excessive protective conductor currents or
- to enable insulation resistance measurements without the need to disconnect SPDs.

Therefore, manufacturers are required to provide data and some designations like

- leakage current free or
- no leakage current

are used, when the effective leakage current is in the single µA or even sub-µA range, which can normally only be reached by the use of spark gaps or GDTs.

6.1.3.4   End-of-life behaviour (failure mode)

There is an ongoing discussion and change process with regard to terminology and details in requirements for the upcoming edition of the product standards; therefore, the following description focuses on technical aspects only.

An SPD may reach its end of life for various reasons, e.g.:

- It was exposed to voltages above its allowed continuous operating voltage for a time period exceeding its corresponding temporary voltage withstand.
- The number of surges passing through has exceeded its endurance capability.
- A surge event has exceeded its discharge capability with respect to amplitude or energy (I^2t).
- An SPC has degraded, leading to excessive leakage current – see Section 6.1.3.3 – and/or to an increase in power dissipation and temperature.

In such case the 'SPD assembly' is required to behave safely and may either end up in an 'open circuit condition' or in a 'short circuit condition', whereby the open circuit end-of-life condition is the more usual and the normally expected one by most users, at least in power applications. For a clarification of the terms 'SPD assembly' versus just 'SPD', please refer to Section 6.3.5.

The open circuit condition can be reached by opening of an internal or external 'SPD disconnector'.

A short circuit condition could in principle also be reached by internal or external means that not only provides an appropriate ampacity to carry the currents, which are to be expected from the system as a consequence of the short circuit, but also any further surge that may occur before replacement of the SPD. As an external short-circuiting means (SC-means) would constitute an additional device connected in parallel to the SPD, this is not covered by the SPD standard.

Short-circuiting SPDs

When an SPD integrates SC-means then it is classified a short-circuiting SPD and is required to perform a number of additional tests to check for the previously described current carrying capabilities over time.

Such kind of SPDs can be of specific interest for systems with limited short circuit currents, e.g. photovoltaic installations, telecom or signalling circuits, as the short circuit condition still protects from overvoltage and the faulty SPD can easily be detected, even if no specific external indication exists.

Nevertheless there are short circuiting SPDs on the market, where the manufacturer requires an external SPD disconnector, mostly a fuse, and where that external SPD disconnector will operate as soon as the SPD internal SC-means gets activated. In such case the complete 'SPD assembly' finally reaches an open circuit condition.

Overstressed fault mode for telecom and signalling SPDs

Although the principle is the same, because of the fact that the majority of telecom and signalling SPDs are two-port SPDs (please refer to Section 6.3.5. for clarification of the terms 'one-port SPD' versus 'two-port SPD'), the standard for telecom and signalling SPDs describes three different end-of-life conditions, which are as follows:

- Mode 1: this corresponds to an SPD disconnector function and the already-described open circuit end-of-life condition, but it is further clarified that the voltage-limiting function is no longer present afterwards while the protected line remains operable. Therefore, the SPD disconnector must be located in the shunt path of the relevant protective components as shown in Figure 6.4, unlike the described mode 3.

- Mode 2: this corresponds to a short-circuiting SPD as shown in Figure 6.5, and it is further clarified that, as a result of this behaviour, the protected line is inoperable afterwards, but the equipment is still protected by the short circuit.

- Mode 3: this corresponds again to an SPD disconnector function with open circuit end-of-life condition, but this time the SPD disconnector is located in the incoming line of a two-port SPD as shown in Figure 6.6, resulting in the line being inoperable afterwards, but the equipment is still protected by an open line as long as any incoming overvoltage does not exceed the flashover voltage of the operated SPD disconnector.

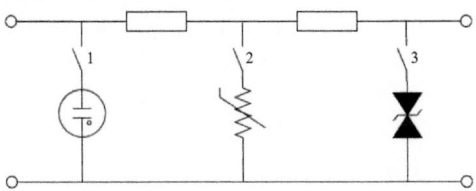

© Hubert Bachl-Hesse

1, 2, 3 ...   SPD disconnectors (open position – activated)

Figure 6.4   SPD according to IEC 61643-21 with overstressed fault mode 1.

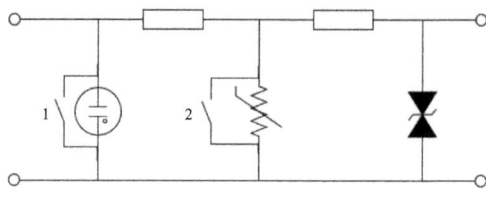

© Hubert Bachl-Hesse

1, 2 ...   short circuiting means (open position – not activated)

Figure 6.5   SPD according to IEC 61643-21 with overstressed fault mode 2.

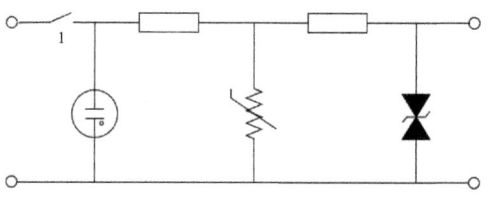

© Hubert Bachl-Hesse

1 ...   SPD disconnector (open position – activated)

Figure 6.6   SPD according to IEC 61643-21 with overstressed fault mode 3.

6.1.3.5   SPD disconnectors, protection functions and status indicators

As already described in Section 6.1.3.4, the main purpose of SPD disconnectors is to protect from any hazards due to the SPD or part of it having reached its end of life.

Attention is drawn to the fact that 'SPD disconnectors' need not provide isolation in the open position, although the term 'disconnector', when used in other IEC standards and specifically in the installation rules and switchgear documents, is defined as a mechanical switching device providing isolating distance in the open position.

Further attention is drawn to the possibility that an SPD disconnector may deactivate and cut off the whole SPD from the system where it is connected but may also just deactivate and cut off a part or section of an SPD which has reached its end of life.

Any SPD disconnector operation must be indicated somehow to enable a timely replacement. Details of indication are mostly up to the manufacturer, whereby local indication is required and remote indication is an option. The details about disconnector function and operation and corresponding indication are required to be given in the installation instructions.

Thermal protection – thermal disconnector

This kind of an SPD disconnector is widely used for voltage-switching and combination SPDs containing MOVs as protective components, to protect from thermal runaway of the varistor and from overheating of the device due to a degradation of the MOV, leading to an increased leakage current and power dissipation.

Leakage current protection

Leakage current detection/measurement primarily serves the same purpose like the thermal protection, just by utilising another parameter as a measure for the thermal behaviour of an SPD.

Additionally, this can be used directly to protect from excessive SPD-caused PE conductor currents in an installation.

This protection approach was used in some countries decades ago already by installing a special kind of voltage-independent residual current circuit breaker – called an arrester disconnector switch – in series with an SPD. At that time the biggest challenge was to find a compromise between the sensitivity of the leakage current detection and the resistivity against nuisance tripping, because of the high surge currents passing through.

Today, this approach potentially recurs, specifically with the increase in 'smart SPDs' entering the market, as these are able to monitor currents through the SPD for various purposes.

Overcurrent protection

Conventional overcurrent protection means, mainly fuses, but also miniature CBs and CBs, are used in conjunction with SPDs since the beginning of overvoltage protection in low-voltage installations to protect from hazardous situations in the case of low-impedance failures of SPDs.

While this overcurrent protection was historically built into the SPD, today only some products still follow that route, and many SPD manufacturers require it to be added externally and specify a maximum overcurrent protection.

It is important to understand that, when it is possible to apply this specified maximum fuse or breaker rating, the full discharge capability of the SPD will be available within the application, without any interruption of supply or of the overvoltage protection to be expected. But often the installation rules and the given wiring cross sections do not allow to use this specified maximum overcurrent protection for other protection reasons.

In the past the focus for SPD branch circuits was on short circuit protection, because in an installation provided with a utility supply and a stable powerful system behind, a low-impedance failure of an SPD was expected to cause significant short circuit currents to flow. Nowadays we are aware that

- in many countries the utility power supply is less stable;
- more and more power electronics like converters and inverters are used in AC systems;
- in DC systems the situation anyway varies depending on the sources used.

This has led to the assumption that more attention must be given to overload protection and to provide a gapless protection for potential SPDs end-of-life conditions and situations in future, ranging from leakage currents to high short circuit currents. For that reason, a separate standard for SPD fusing disconnectors was developed in Japan, and in China a separate standard for SPD-specific disconnectors was established for a special kind of external SPD overcurrent disconnectors.

6.1.3.6   Means for short circuiting an SPD, protection functions and status indicators

The basic purpose of such SC-means is the same like of SPD disconnectors, namely to protect from any hazards due to the SPD or part of it having reached its end of life, but the difference is of course the final condition of the SPD with regard to the system to be protected, namely the permanent short circuit.

Also, any SPD's SC-means activation must be indicated similar to a disconnector operation and the same principles apply with regard to potential protective functions.

6.1.3.7   Number of ports (terminals) and 'feed through' capability

There are two basic SPD connection topologies:

• Shunt devices, connected in/as a branch circuit in parallel to the power source.
• Serial connection devices, inserted into the power supply line and having separate input and output terminals. Such devices may or may not contain intended series impedance between separate input and output connections.

A shunt device is by definition always classified to be a one-port SPD. But a one-port SPD may either be equipped with single terminals or with double/multi-terminal blocks, allowing to pass the load current of the connected line through the SPD's terminals ('feed through' capability), to avoid connection lead length and any corresponding voltage drop leading to an increase in protection level. Such one-port SPDs with double/multi-terminal blocks are treated like two-port SPDs, as it is also necessary for these devices to perform load current and overcurrent testing.

A serial connection device may be a one-port SPD when there is no intended series impedance between the separate input and output connections.

But as soon as a serial connection device contains any intended series impedance between the input and output connections, it is by definition classified to be a two-port SPD.

6.1.3.8   Specific classifications for SPDs used in low-voltage power systems

Discharge capabilities – surge protection versus lightning protection and SPD location

There are basically two different scenarios and protection purposes to the selection and application of SPDs:

• Protection against transient overvoltages coming in through the external power supply line or any other outgoing or incoming electrical service due to indirect lightning effects, when no protection measures against direct lightning are taken.
• Protection against the above and against partial lightning currents due to direct lightning effects.

In addition, the location of the SPD plays an important role on the surges and their energy content to be expected and there is a need to distinguish at least two objectives:

- Primary protection as close as possible to the point where the electrical service enters the structure to be protected.
- Secondary protection within the structure to be protected.

Therefore IEC 61643-11, the standard for SPDs used in low-voltage AC power systems, and IEC 61643-31, the standard for SPDs used in low-voltage photovoltaic installations, distinguish between three classes of tests (types). While UL 1449 also offers the choice of three different types of SPDs, there are different specifications behind and these are not directly comparable to the IEC classification.

Class I tests – Type 1 (T1) SPDs – SPDs marked 'T1': This test class applies when the SPD is designed to divert partial lightning currents as well as other transient overvoltages and requires the highest energy handling capability. Such SPDs are for installation as close as possible to the point where the electrical service enters the structure to be protected.

Testing is performed with impressed current impulses of the 8/20- and 10/350-μs waveshape, and, in the case of voltage-switching or combination SPDs, additionally with 1.2/50-μs voltage impulses.

The most relevant test parameter is the impulse discharge current I_{imp}. During a complete operating duty test procedure, 17 impulses with an 8/20-μs waveshape and a crest value according to the I_{imp} rating, and in addition five impulses with an 10/350-μs waveshape – one impulse each at crest values of 0.1/0.25/0.5/0.75 and 1.0 times the rated I_{imp} – are applied.

A comparable test procedure does not exist in UL 1449.

Class II tests – Type 2 (T2) SPDs – SPDs marked 'T3': This test class applies to SPDs:

- for installation as close as possible to the point where the electrical service enters the structure when only indirect lightning effects are expected and
- for secondary protection.

Testing is performed with impressed 8/20-μs current impulses and, in the case of voltage-switching or combination SPDs, additionally with 1.2/50-μs voltage impulses.

The most relevant test parameter is the nominal discharge current I_n (8/20 μs), which is applied 17 times during the complete operating duty test procedure.

Comparable and similar tests are specified for Type 1 and Type 2 SPDs according UL 1449.

Class III tests – Type 3 (T3) SPDs – SPDs marked 'T3': This test class is primarily intended for, e.g.

- socket-outlet/SPD combinations and SPDs incorporated in socket-outlets,
- SPDs for specific equipment protection, either for the fixed installation or for incorporation into the equipment,
- portable SPDs and SPDs incorporated in extension leads with multiple socket outlets,

but sometimes also SPDs for primary protection are specified according to test class III, because this is the only test class that can be applied correctly to SPDs incorporating any kind of intended series impedance (two-port SPDs).

Testing is performed with a combination wave generator (CWG) that provides at the output a 1.2/50-μs voltage impulse into an open circuit (high impedance) and an 8/20-μs current impulse into a short circuit. The output impedance is specified to be 2 Ω. Test results are not directly comparable with test class II, because the real test current waveform depends on the relationship of the generator output impedance to the SPD internal dynamic impedance under the specific test surge conditions.

The relevant test parameter is the open circuit output voltage of the CWG U_{OC}. In total, 17 impulses at the rated U_{OC} are subjected during a complete operating duty test procedure.

The application of class III tests is limited to SPDs with a maximum open circuit voltage U_{OC} of 20 kV and the corresponding short circuit current of 10 kA.

Comparable and similar tests are specified for Type 3 SPDs according UL 1449.

SPDs complying with more than one test class/type designation: It is allowed to specify an SPD to more than one test class (type). Many manufacturers do already make use of this possibility, but there is of course some cost implication, as in such case all relevant tests for all declared test classes must be performed.

SPDs only for connection between neutral and earth (N-PE SPDs)

When SPDs are applied in an AC power system with a solid grounded neutral conductor (TN-S system, TT-system with a multi-grounded neutral conductor which cannot reach a dangerous potential even under fault conditions), such SPD connected between neutral und PE will never see the full system voltage and can therefore also never be stressed by a high follow current, if applicable, or by a high short circuit current, even under fault conditions. For that reason, these SPDs are exempted from short circuit tests and even when they may cause follow currents, the prospective follow current of the test set-up is limited to 100 A.

Technically the same applies to SPDs in a DC system with a solid grounded line conductor or midpoint conductor.

6.1.3.9   Specific classifications for SPDs used in telecom and signalling systems

Discharge capabilities – surge protection versus lightning protection and SPD location

Similar to the SPDs for power applications, SPDs for telecom and signalling systems follow the principal distinction of the two scenarios and the location-related differentiation described in the 'Discharge capabilities – surge protection versus lightning protection and SPD location' section.

But IEC 61643-21, the standard laying down the performance requirements for SPDs connected to telecommunications and signalling networks, specifies ten different SPD categories as shown in Table 6.1. These SPD categories vary primarily in the impulse waveshape to be applied. In many cases and for most applications,

Table 6.1 Classification of SPD categories for telecom and signalling systems[§]

Category	Type of test	Open circuit voltage	Short circuit current	Minimum number of applications
A1	Very slow rate of rise	≥ 1 kV Rate of rise from 0.1 to 100 kV/s	10 A $\geq 1{,}000$ μs (duration)	
A2	AC			
B1	Slow rate of rise	1 kV 10/1,000 μs	100 A 10/1,000 μs	300
B2		1–4 kV 10/700 μs	25 or 100 A 5/300 μs	300
B3		≥ 1 kV 100 V/μs	10, 25, or 100 A 10/1,000 μs	300
C1	Fast rate of rise	0.5–2 kV 1.2/50 μs	0.25–1 kA 8/20 μs	300
C2		2–10 kV 1.2/50 μs	1–5 kA 8/20 μs	10
C3		≥ 1 kV 1 kV/μs	10–100 A 10/1,000 μs	300
D1	High energy	≥ 1 kV	0.5–2.5 kA 10/350 μs	2
D2		≥ 1 kV	1–2 kA 10/250 μs	5

only the SPD categories C1, C2 and D1 are of real interest. These three SPD categories can be somewhat compared to the test classes for SPDs in power applications and are thus explained in more detail and in a descending discharge-capacity-related order first, followed by a description of SPD category C3, which also has quite some relevance. The other categories listed in Table 6.1 are used for SPDs in dedicated applications and sometimes for historic reasons, e.g. for SPDs in telecommunication networks where category B2 is relevant, and are not further explained here.

SPD category D1 – SPDs marked 'D1': These are SPDs intended for instal-lation at a location where telecommunications and signalling networks enter a building. These SPDs are capable of diverting even partial lightning currents with very high energy content and are tested with impulse currents of the waveshape

[§]IEC 61643-21:2000 +A1:2008 +A2:2012, modified. Copyright © 2019 IEC Geneva, Switzerland. www.iec.ch.

10/350 µs. Therefore, this SPD category directly corresponds to class I tested SPDs for power applications.

The amplitudes of surges occurring in telecommunications and signalling networks are usually quite low in comparison to surges in power systems. Therefore, when applying the SPD category D1, the given discharge capacity in the range between 500 and 2,500 A (10/350 µs) is sufficient.

However, compared to overvoltages due to switching operations, direct lightning and resulting partial lightning currents do not occur as frequently. Therefore, a category D1 SPD is only tested with two impulses of the specified 10/350-µs impulse current.

SPD category C2 – SPDs marked 'C2': These SPDs are intended for installation at a location where telecommunications and signalling networks enter a building when only indirect lightning effects are expected or at a downstream location for secondary protection. They are capable of diverting even high line conducted and induced surge currents and they are tested with impulse currents from a hybrid generator delivering a 1.2/50-µs open circuit voltage and an 8/20-µs short circuit current. This SPD category is comparable to class II tested SPDs for power applications.

Category C2 SPDs typically have a medium-to-high discharge capacity represented by the corresponding impulse current range between 1,000 and 5,000 A, and they are tested with ten impulses of the specified hybrid generator.

SPD category C1 – SPDs marked 'C1': SPDs according to category C1 are primarily intended for secondary protection and typically used in systems with short low-impedance lines. They are still capable of diverting significant line conducted and induced surge currents and they are also tested with impulse currents from a hybrid generator delivering a 1.2/50-µs open circuit voltage and an 8/20-µs short circuit current. This SPD category directly corresponds to class III tested SPDs for power applications.

Overvoltages with low amplitudes occur much more frequently in extended electrical systems and therefore category C1 SPDs have to stand 300 impulses from the specified CWG. As compared to category C2, the category C1 has lower discharge current capacity in the range between 250 and 1,000 A.

SPD category C3 – SPDs marked 'C3': The intended applications of SPD category C3 are in principle similar to category C1, but their capability focuses on systems with long cables, where the high-impedance lines are expected to be exposed to surge currents with longer duration, but lower amplitudes. Such SPDs are tested with a kind of CWG delivering an open circuit voltage with defined 1-kV/µs rise and 10/1,000-µs short circuit current.

They are tested with 300 impulses like category C1 and shall provide a discharge current capacity in the range between 10 and 100 A.

Current-limiting function

An SPD may be provided with one or more nonlinear current-limiting components in the line/signal path, aiming to restrict any currents that exceed a value predetermined by the manufacturer. Such components may be resettable, self-resetting or non-resettable and may therefore just limit the current, e.g. by an increase in impedance like it is provided by a PTC-thermistor, or interrupt the current, like e.g. a fuse or a thermal link.

6.1.4   Specific topologies and designs of SPDs and their characteristics

Although many details can be described via the topology, the design, the classifications, the technical parameters and a description of the characteristics of an SPD, an oscillogram tells more and often provides details about an SPD's behaviour that are not visible from the specs. This is the main reason why reliable and robust information on good SPD coordination can only be expected from SPD manufacturers, as they are normally the only ones having all this information and knowledge about product details.

While the basic behaviour of pure voltage-limiting SPDs is relatively simple and very similar to the principle protective characteristics of the components used, care must always be taken in the design of the internal circuitry, as any loops in the internal current path represent additional internal inductance, and with increasing surge currents, the SPD's internal voltage drop as well as the dynamic forces stressing the construction and housing increase. The bigger this internal inductance is, the more the protective characteristic of the SPD deviates from the protective characteristic of the component used – see Figure 6.7.

Even simple voltage-switching SPDs containing only a single GDT are already significantly more complex in their behaviour than a voltage-limiting SPD. This is due to the significant dependency of their spark-over voltage on the steepness of the surge voltage to be limited. Additional factors for air gaps are the atmospheric pressure and the relative humidity.

Here the most significant advantages of the basic component characteristics become clear already. While the continuous characteristic of a voltage-limiting

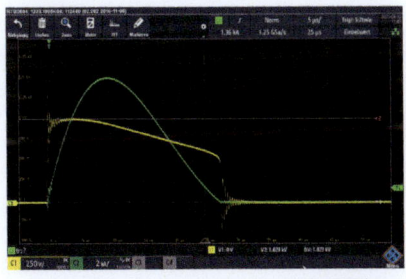

© CTI-Vienna

Residual voltage of an MOV based voltage limiting SPD with a single mode of protection (1,000 V)

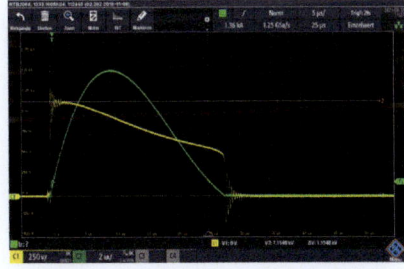

© CTI-Vienna

Residual voltage of an identical MOV based voltage limiting mode of protection of a multimode SPD with 100 mm of additional (internal) wiring (1,150 V)

Figure 6.7   Dependency of the voltage-limiting characteristic on the SPD's internal inductance

Source: DEHN SE + Co KG

*Figure 6.8   Example of a 'historic' combination SPD based on an air gap and a
SiC-varistor*

component avoids any voltage spikes due to spark-over or trigger, the voltage-
switching component due to its discontinuous response curve and low arc voltage
can significantly better limit the energy content of the remaining surge fraction that
passes on to any coordinated downstream SPD or the equipment to be protected.

6.1.4.1   Series combination of switching and limiting components

Some decades ago when no powerful MOVs were available for low-voltage
applications and series combinations of silicon carbide varistors and either air gaps
or GDTs represented the state of the art in SPD design – see Figure 6.8, the following disadvantages of each of the components used really limited the SPDs
protection potential:

- The spark-over voltage of the gap or GDT, which had to be chosen high enough
  to sufficiently quench the follow current to extinguish at the next zero crossing
  on one hand, and on the other hand, low enough to still enable a useful voltage
  protection level considering the relatively high spark-over voltage spike.
- The poor nonlinearity coefficient α of the silicon carbide varistor, which required
  a series gap or GDT for follow current extinction to allow for a useful voltage
  protection with regard to the residual voltage of the varistor during surge passage.
- The quite limited discharge capability of a silicon carbide varistor compared to
  its volume.

- The heavy and frequent stress on the varistor and the gap or GDT due to significant follow currents even after low surges, which limited the endurance of the SPD and required relatively low internal or external overcurrent protection to avoid any uncontrolled failure.

Furthermore, that SPD design could not be used for DC applications.

Nowadays, the use of MOVs enables the design of voltage-limiting SPDs without gaps or GDTs for AC and DC, nevertheless for many applications combination SPDs, mostly comprising an MOV and a GDT in series, are still preferred or even required, e.g.

- to provide galvanic separation under normal system conditions for functional reasons;
- to avoid any noteworthy leakage current, specifically at higher frequencies;
- to enable insulation resistance measurements without having the need to previously disconnect the SPDs.

And by designing combination SPDs with a series connection of an MOV and a triggered air gap or a GDT, most of the disadvantages of the earlier described 'historic' construction can be avoided:

- The spark-over voltage of the GDT has become almost irrelevant, as the MOV characteristic already prevents any significant follow current and it can therefore be chosen sufficiently low, just to avoid conduction under normal system conditions – see Figure 6.10.
- The MOV with its good nonlinearity coefficient α ensures similar or even the same voltage protection level like a pure MOV-based voltage-limiting SPD – see Figures 6.9 and 6.10.
- The discharge capability is the same like for a pure MOV-based voltage-limiting SPD.
- There are no follow currents anymore and any internal or external overcurrent protection can be the same like for a pure MOV-based voltage-limiting SPD.
- These SPDs can be used for AC and DC applications.

Latest developments even realise an MOV in series with a specially designed arrangement of a triggered air gap with an arc chamber and a trigger circuit disconnector, which gets activated when reaching end of life, whereby that specially designed arrangement serves as a voltage-switching component under normal service conditions and as an overcurrent disconnector under end-of-life conditions.

6.1.4.2   Parallel combination of switching and limiting components

This kind of topology is used since long time to improve the voltage protection level of an SPD by combining the benefits of different surge protection components, e.g.

- gaps or GDTs with their high surge current and energy-handling capability and
- MOVs or Zener diodes with their 'sharp' bend in the voltage–current characteristic.

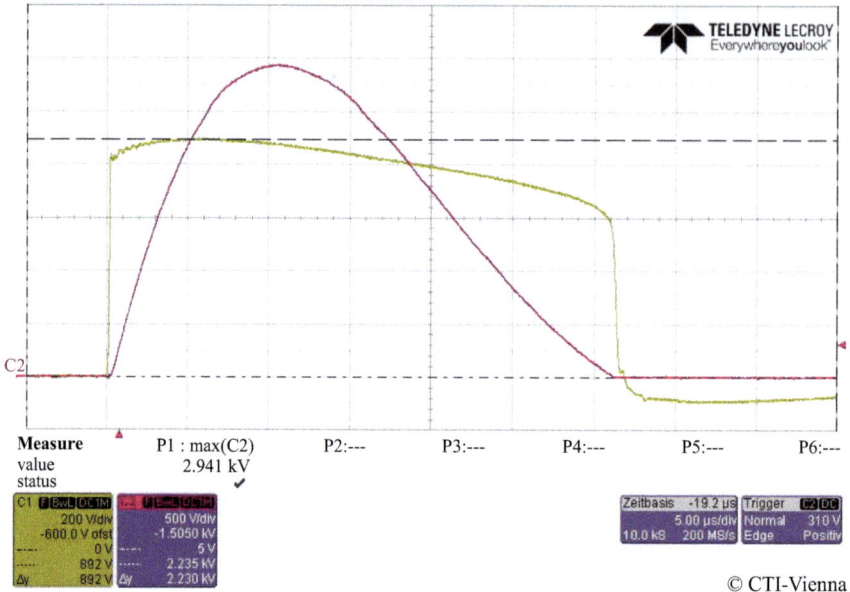

© CTI-Vienna

Figure 6.9   Voltage over time oscillogram (green trace) of a state-of-the-art pure MOV-based voltage limiting SPD during a 6-kV (1.2/50)/3-kA (8/20) hybrid surge (red trace).

Most designs use either resistors or inductances for decoupling of the different surge protection components, to coordinate their characteristic advantages and to avoid overstress to the more sensitive downstream semiconductor-based components.

Often such SPDs are combined with capacitors or even filters at the protected/output port, to reduce the steepness in the voltage rise *du/dt* of the remaining impulse or to provide additional filtering for EMC purposes.

6.1.4.3   Modes of protection and connection types versus configuration

There are some terms from the SPD product standards and from the installation rules which are closely related to each other.

The mode of protection is a current path with protective components between two terminals of an SPD, as declared by the manufacturer. This is somewhat comparable to a pole of a switch, but this does not contain any information about what kind of protective components is/are used and how they are interconnected within a multimode (multipole) SPD.

The installation rules standard IEC 60364-5-53 specifies connection types (configurations) for SPDs, which describe how single mode SPDs shall be connected or the modes of protection of a multimode SPD shall be interconnected, depending on their application in a specific power supply system configuration,

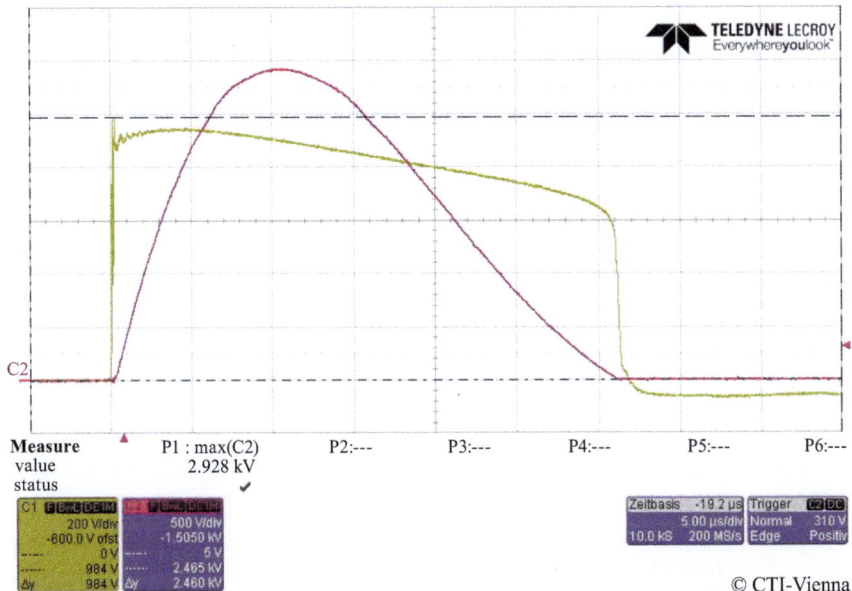

Figure 6.10   *Voltage over time oscillogram (green trace) of a state-of-the-art combination SPD during a 6-kV (1.2/50)/3-kA (8/20) hybrid surge (red trace); the spark-over voltage spike is only visible when using a high-resolution scope.* © *CTI-Vienna*

e.g. single-phase/3-phase, TN-/TT-/IT-system. It needs to be noted that these connection types were developed focusing on single- and three-phase AC systems and they apply to SPDs for primary protection in fixed installations.

Additional or other configurations are used and described, e.g., for secondary protection like socket outlet SPDs, SPDs for equipment, portable SPDs, but also for different kinds of installations like the DC-side of photovoltaic installations.

This leads us to the internal interconnection of an SPD, which can be described as follows:

- related to the internal circuitry of a single mode SPD or of a single mode of protection within a multimode SPD, as described, e.g. in Sections 6.1.4.1 and 6.1.4.2; or
- related to the modes of protection of a multimode (multipole) SPD, e.g.
  - referring to the connection types (configurations) in IEC 60364-5-53 – see Figures 6.11 and 6.12 – or
  - referring to the configurations described in EN 50539-11 (predecessor standard of IEC/EN 61643-31) – see Figure 6.13 – or
  - referring to the configurations used and established in practice – see Figure 6.14.

When combining single-mode SPDs for protection of a dedicated (power) system, for most of the above-described configurations, each SPD/mode of

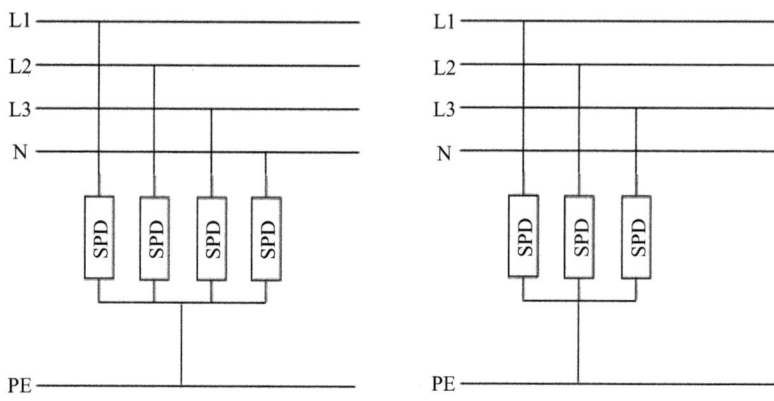

© Hubert Bachl-Hesse   © Hubert Bachl-Hesse

4+0 configuration
for a 4 wire three phase system

4 obvious modes of protection
(L1-PE, L2-PE, L3-PE, N-PE)
optional (if declared) 6 additional
modes of protection
(L1-L2, L2-L3, L3-L1, L1-N, L2-N, L3-N)

3+0 configuration
for a 3 wire three phase system

3 obvious modes of protection
(L1-PE, L2-PE, L3-PE)
optional (if declared) 3 additional
modes of protection
(L1-L2, L2-L3, L3-L1)

Figure 6.11   Connection type 1 (CT1) according to IEC 60364-5-53

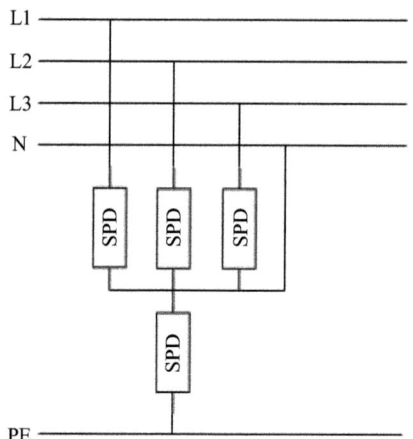

© Hubert Bachl-Hesse

3+1 configuration for a 4 wire three phase system

4 obvious modes of protection (L1-N, L2-N, L3-N, N-PE)
optional (if declared) 6 additional modes of protection (L1-L2, L2-L3, L3-L1, L1-PE, L2-PE, L3-PE)

Figure 6.12   Connection type 2 (CT2) according to IEC 60364-5-53

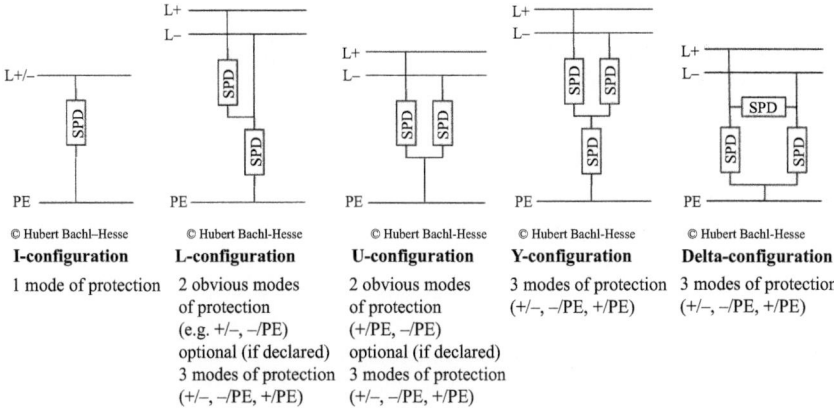

© Hubert Bachl–Hesse | © Hubert Bachl-Hesse | © Hubert Bachl-Hesse | © Hubert Bachl-Hesse | © Hubert Bachl-Hesse

I-configuration | **L-configuration** | **U-configuration** | **Y-configuration** | **Delta-configuration**

1 mode of protection | 2 obvious modes of protection (e.g. +/−, −/PE) optional (if declared) 3 modes of protection (+/−, −/PE, +/PE) | 2 obvious modes of protection (+/PE, −/PE) optional (if declared) 3 modes of protection (+/−, −/PE, +/PE) | 3 modes of protection (+/−, −/PE, +/PE) | 3 modes of protection (+/−, −/PE, +/PE)

Figure 6.13   SPD configurations for photovoltaic installations according to EN 50539

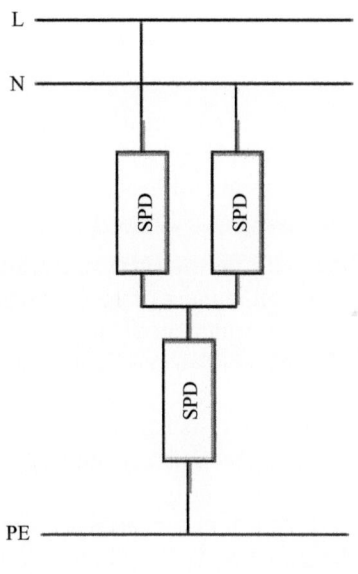

© Hubert Bachl-Hesse

Figure 6.14   Typical Y-configuration for socket outlet and portable SPDs

protection is designed to be directly connected to two conductors of that system, e.g. L–N, N–PE, +/−, −/PE.

But for a Y-configuration, this is different. For single-mode SPDs combined to a Y-configuration, in fact always two SPDs/modes of protection are connected in series, and the interconnection node of these three single-mode SPDs may not be connected to any system conductor. This requires particular attention for testing, as this detail is currently not sufficiently addressed in the IEC/EN 61643 series of standards – see Section 6.3.5.

6.1.4.4   Portable SPDs

Specifically in Europa, some additional conditions and construction requirements must be fulfilled for portable SPDs which are classified as so-called pluggable equipment type A according to IEC 62368-1. This is equipment dedicated to household and similar uses, where the supply connection to the mains, and thus also for the protective conductor, is only done via a non-industrial plug and a socket-outlet or via a non-industrial appliance coupler.

For such SPDs, all the following conditions must be fulfilled:

- All type tests must be passed without any external SPD disconnector.
- Every mode of protection must contain appropriate SPD disconnectors, which may not be resettable or replaceable.
- Line and neutral terminals must be interchangeable and are both tested as line terminals, consequently there cannot be a dedicated N-PE mode of protection.
- The declared short circuit current rating ISCCR (see Section 6.2.2.1) shall not be less than 1,500 A.

Full information and all related constructional requirements, e.g. with regard to component selection, can be found in EN 61643-11, Amendment 11.

The reasons for these stringent and harsh additional requirements are multi-layered, the most important ones are as follows:

- In a number of historically grown cities, there are still old buildings with electrical installations not providing a protective conductor at all; therefore, the probability of an earth fault within an SPD must be kept to a minimum.
- The safe end-of-life behaviour of such SPDs, which are bought and connected by ordinary, unskilled and uninstructed persons, shall not depend on any protective device in the upstream installation.
- Many European plug and socket outlet systems for household and similar uses are non-polarised.
- Investigations have shown that at the installation points of socket outlets for household and similar uses, the available short circuit current may reach up to 1,500 A, but this value is very rarely exceeded.

For sure, similar conditions can also be found in other countries outside Europe and should be generally considered.

6.1.4.5   Topologies and designs only used for SPDs for power systems

The vast majority of SPDs for power applications are one-port devices. More and more of them provide 'feed through' capability and are either equipped with separate input and output terminals or with terminals big enough to clamp two conductors at a time.

Multiple gap constructions

This is used for air gaps as well as for GDTs to increase the follow current extinction capability in combination with a controlled spark-over behaviour, e.g. by capacitive control – see Figure 6.15.

Triggered gap constructions

Triggering allows to partly separate the following technical parameters of an air gap, a GDT or a hard gas spark gap:

- the continuous AC or DC voltage withstand of the main gap;
- the trigger level;
- the dependency of the trigger level from environmental conditions like air pressure, humidity and temperature and from production tolerances;
- the self-extinction capability;
- the amplitude and duration of follow current.

An example schematic is given in Figure 6.16.

Due to the potential high discharge capability of gaps as compared to semiconductor-based overvoltage protective components, this technology, irrespective from its higher costs, is mostly used for SPDs required to handle partial lightning currents (IEC 61643-11, class I tested).

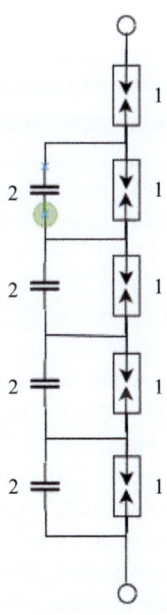

© Hubert Bachl-Hesse

1 ... spark gaps
2 ... capacitors

Figure 6.15   Multiple spark gaps with capacitive trigger control

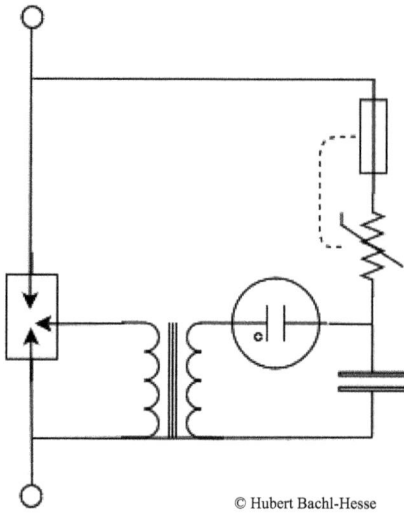

© Hubert Bachl-Hesse

Figure 6.16   Schematic of a spark gap with three electrodes and a trigger circuit with a firing transformer

6.1.4.6   Topologies and designs only used for SPDs for telecom and signalling systems

Plenty of different applications and signal forms exist in telecommunication, instrumentation (signalling) and IT (data) technology. For this reason, an uncounted number of different SPD topologies and designs, many of them specifically tailored to the respective application, are requested by the market, the relevant users and applicants.

Typical components for these protective devices include GDTs and TVS diodes. MOVs are seldom used due to their limited suitability for high-frequency applications, because of their 'ageing behaviour' (increase of leakage current due to degradation) and their larger dimensions.

GDTs usually have a high surge current discharge capacity for impulses up to more than 10,000 A with short duration, e.g. 8/20 μs. Right before the discharge starts, a high voltage peak appears, which can reach several hundred volts depending on the steepness of the voltage impulse. Therefore, the voltage protection level is relatively high.

The main advantage of TVS diodes is their reaction speed and their exceptional voltage-limiting capability as can be seen in Figures 6.2 and 6.3. The surge current discharge capacity of TVS diodes is significantly lower than that of GDTs.

However, modern protective devices are mostly two-port SPDs and use GDTs and TVS-diodes tailored to one another to make a best use of their respective benefits. This way, the GDT offers a high discharge capacity and the TVS diode provides a low limiting voltage and fast response behaviour. To achieve that, a good coordination of the GDT, the TVS-diode and the respective decoupling elements is required. The circuitry of such a two-stage protective device is shown in Figure 6.17.

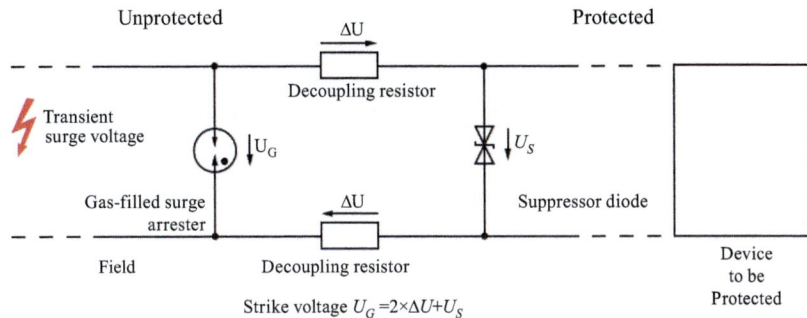

Figure 6.17   Two-stage protective device[∥]

If transient overvoltage occurs between the signal wires, the TVS-diode becomes low-ohmic, after a very short response time. This results in a flow of current through the diode and the decoupling elements included in the signal path. These decoupling elements, mostly resistors, help to share the current between the TVS diode and the GDT and ensure that the TVS-diode does not get overloaded. As soon as the voltage at the GDT ($U_G = 2 \times \Delta U + U_S$) reaches the GDTs spark-over voltage, the GDT becomes conductive and takes over the remaining surge current. During the conduction phase of the GDT, the arc-burning voltage of the GDT is typically in the range between 10 and 30 V (depending on the type of GDT). Because of this low arc-burning voltage, the GDT takes over the main share of the surge current. This principle of sharing the current between different SPCs without overloading any of these components is called 'energetic coordination'. Attention is drawn to the operational currents, which cause permanent power dissipation and heat up the decoupling elements.

6.1.5   Environmental service conditions, mounting and accessibility

As these conditions and requirements for SPDs are mostly consistent with the conditions defined in other standards for installation material and equipment, they are only commented here if different.

- Altitude and atmospheric pressure
  - Normal range is up to 2,000-m above sea level
  - Application at higher altitudes requires particular consideration

- Installation location
  - Indoor is defined as inside closed shelters but does not presume a weather-protected location for the shelter
  - Outdoor is defined as outside closed shelters with full exposure to, e.g., rain, snow, icing, UV-radiation

[∥]R. Hausmann/Phoenix Contact.

- Temperature
  - Normal range is −5°C to +40°C (weather protected, no specific control of temperature)
  - Extended range is −40°C to +70°C (outdoor, non weather protected)

- Humidity
  - Normal range is 5% RH to 95% RH (weather protected, no specific control of humidity)
  - Extended range is 5% RH to 100% RH (outdoor, non weather protected)

- Mounting method
  - Fixed installation, e.g. in a distribution board, with a socket outlet or within the terminal compartment of equipment
  - Portable, e.g. as a plug-in unit, as an adaptor or incorporated in a multiple socket outlet or extension lead
  - Modular unit, e.g. for printed circuit board mounting. There is increasing demand for such devices, although they are currently not explicitly addressed by the SPD standards.

- Accessibility
  - Accessible means partly or fully accessible to ordinary/unskilled persons without the use of a tool or key
  - Inaccessible means
  - Either covered by an enclosure that can only be opened by use of a tool or key or
  - Mounted out of reach, e.g. on an overhead line

- Degree of protection for touchable surfaces (in normal service)
  - Minimum requirement is IP20 according to the SPD standard
  - Minimum requirement is IP20C according to the standard for distribution boards to be operated by ordinary/unskilled persons

6.2   Main parameters for SPDs

6.2.1   *General parameters for all SPDs*

6.2.1.1   AC and frequency or DC

At first glance the answer to this question seems simple, as there is always a clear system designation and the operating currents of a system can normally clearly be assigned to be AC with a declared frequency or DC, besides the fact that DC may be smooth or contain a certain percentage of ripple.

Looking into the details the issue becomes a bit more complex, because there are an increasing number of installations, which are connected to an earthed AC grid via converters or inverters, but which are not galvanically separated from that AC grid. Thus leading to a situation, where, e.g., the voltage between the system conductors is DC, but the voltage between system conductors and earth is a complex AC waveform with a wide frequency spectrum.

Table 6.2   Minimum U_C required according to IEC 60364-5-53, depending in the supply system configuration[¶]

SPO connected between (as applicable)	System configuration of distribution network		
	TN system	**TT system**	**IT system**
Line conductor and neutral conductor	$\frac{1.1U}{\sqrt{3}}$ or $(0.64U)$	$\frac{1.1U}{\sqrt{3}}$ or $(0.64U)$	$\frac{1.1U}{\sqrt{3}}$ or $(0.64\ U)$
Line conductor and PE conductor	$\frac{1.1U}{\sqrt{3}}$ or $(0.64U)$	$\frac{1.1U}{\sqrt{3}}$ or $(0.64U)$	$1.1U$
Line conductor and PEN conductor	$\frac{1.1U}{\sqrt{3}}$ or $(0.64U)$	N/A	N/A
Neutral conductor and PE conductor	$\frac{U}{\sqrt{3}}$[a]	$\frac{U}{\sqrt{3}}$[a]	$\frac{1.1U}{\sqrt{3}}$ or $(0.64U)$
Line conductors	$1.1U$	$1.1U$	$1.1U$

Note: N/A: not applicable. *U* is the line-to-line voltage of the-low voltage system.
[a]These values are related to worst case fault conditions; therefore the typical utility supply voltage tolerance of 10% is not taken into account.

This information is vital for choosing an appropriate SPD and to avoid excessive leakage currents and potential durability issues, or even functional problems when using polarity sensitive SPDs.

6.2.1.2   Continuous operating voltage

Because an SPD is a voltage-sensitive protective device, the service endurance depends significantly on the correct component selection with regard to the continuous operating voltage to avoid premature ageing or degradation. This must consider the following:

- The (nominal) system voltage(s), where the SPD shall be applicable.
- The maximum long-term system voltage variations to be expected.
  While in some parts of our world the utility provides a quite stable supply with typical voltage variations in the range of ±10% of the nominal system voltage, in other parts this variation may exceed 25% or even 30%.
- The temporary overvoltages (TOVs) and their frequency of occurrences to be expected – see Section 6.2.1.4.

An analysis of these system voltage variations and the TOVs and their occurrences to be expected helps deciding, which TOVs should be withstood by the SPD to prevent unwanted frequent failures, and which TOVs may lead to an acceptable end of life of the SPD.

Such analysis was performed for typical three-phase systems complying with IEC 60364 installation rules and resulting requirements for the maximum continuous operating voltage are given in Table 6.2.

[¶]IEC 60364-5-53 ed.4.0. Copyright © 2019 IEC Geneva, Switzerland. www.iec.ch.

In fact the maximum continuous operating voltage for SPDs for power applications has lost importance, because the SPD manufacturers are also required to provide the following data to ensure correct application:

- The nominal system voltage(s) and its/their maximum variation (e.g. ±10%), to which the SPD may be connected.
- Type of system earthing of the power system (e.g. TN-, TT-, IT-), for which the SPD is designed.
- In which connection configuration(s) and between which conductors (e.g. L to N, N to PE, + to −, − to PE) the SPD may be applied.

Therefore, the manufacturer is responsible to perform the above-mentioned analysis already in the product design phase for all common applications.

But this may become quite tricky as soon as unusual voltage variations or TOVs need to be considered.

6.2.1.3   Voltage protection level, limiting voltage, residual voltage

The voltage protection level is 'the' parameter describing the protective function of an SPD, but unfortunately, when tested according to the IEC 61643 standards, this can only be directly compared between SPDs having the same impulse discharge current or nominal discharge current rating, because the value is linked to these discharge current ratings.

In UL 1449, this is called voltage protection rating and the value is determined equally for all SPDs by applying a combination wave impulse at 6 kV – see the 'Combination wave generators' section, but in the end, is again only partly comparable, because based on the measured numbers, the next higher standard value must be taken from a table and be declared.

Therefore, some manufacturers do provide a limiting voltage over discharge current diagram or a set of values for different discharge currents.

The term measured limiting voltage is used for the highest measured number out of the following test results:

- The front-of-wave spark-over voltage determined with a 1.2/50-μs voltage impulse with a crest value of 6 kV or in some cases 10 kV. This applies only for class I tested and class II tested SPDs incorporating voltage switching protective components.
- The highest residual voltage measured during 8/20 impulse current applications with crest values up to the impulse discharge current for class I tested SPDs or the nominal discharge current for class II tested SPDs, respectively.
- The highest voltage recorded during application of combination wave impulses up to U_{OC}, for class III tested SPDs.

As some datasheets for SPDs refer to residual voltage values only, it is important to carefully distinguish here between the voltage protection level, which is a rounded up value of the above-described measured limiting voltage, and the residual voltage, which is the highest voltage measured during passage of the impulse current applied and which does not include any spark-over voltage peaks that may occur before the current flow starts. Just for voltage-limiting SPDs, the measured limiting voltage and the residual voltage will always equal.

6.2.1.4   Temporary overvoltages – power frequency overvoltages

In low-voltage AC power systems TOVs are caused by short circuits or earth faults in the low-voltage system and installations, but the quite severe ones may also be caused by faults in the upstream medium voltage system.

In low-voltage DC power systems, they depend again on the kind of faults to be expected, on the power sources used and their load characteristic and dynamic behaviour, and if there is galvanic coupling to the AC grid or not.

In signalling and telecommunication systems again, they depend on any faults to be expected.

Sometimes, the limits of these events are described by a voltage–time diagram that mostly looks similar to a horizontal funnel, as the potential voltage variations in the transient time domain are the highest and then the amplitudes of the over- and under-voltages to be expected decrease with increasing duration until the two curves end up at the upper and lower limit of the long-term system voltage variations.

For AC power systems complying with the IEC 60364 standard series, the TOVs to be expected are well defined and covered by standard test procedures. But for other systems, this may only partly be the case or may completely depend on the system design, the system characteristics and the protective measures applied.

Therefore, the SPD manufacturer may declare other or additional TOV ratings to enable an installation design engineer, who performs his own analysis for such non-IEC 60364 compliant systems to select an appropriate SPD.

6.2.1.5   Rated load current – rated current

The term and parameter of rated current is commonly used for many devices and equipment, and in most cases, describes which effective value of current can be permanently conducted by a device or be consumed by current using equipment.

For SPDs for power applications that term was modified a little and named 'rated load current' but in fact defines exactly what is described earlier. This parameter is of vital importance for all two-port SPDs, but also for one-port SPDs with either separate input and output terminals or with terminals declared to be suitable for connection of more than one conductor only. For more details on this, please refer to Section 6.1.3.7.

On the contrary, for SPDs for telecom and signalling applications, the rated current is defined with a totally different meaning, as it is only linked to SPDs with current-limiting function as described in the 'Current-limiting function' section and describes the effective current that may flow continuously without causing any change in the impedance of the current-limiting components. In addition, it is stated that this can also be applied to SPDs with linear series components, which then comes closer to the common use of that term.

6.2.1.6   Total discharge current I_{total}

As overvoltages of atmospheric origin are assumed to primarily cause so-called common mode overvoltages, these overvoltages and the surge currents and discharge currents through SPDs which are caused by them will occur simultaneously on all incoming conductors of an electric service entering a structure.

As impulse testing of SPDs is generally performed on a mode-by-mode basis, such common mode impulse stress as assumed for overvoltages of atmospheric origin required additional consideration, and therefore the total discharge current I_{total} was defined to cover the related electrical and dynamic mechanical strain.

Many multimode SPDs or assemblies of single-mode SPDs for power applications provide modes of protection between the lines and the neutral- or a midpoint-conductor, and then one mode of protection between that neutral- or midpoint-conductor and PE, which, in such case, has to discharge the sum of all the common mode impulse currents diverted by the single modes of protection to PE.

This is quite similar for multimode SPDs used in telecom and signalling applications, where often the modes of protection are first referred to internal reference ground and from there then via another commonly used mode of protection to the grounding system of the structure.

The total discharge current I_{total} can be as high as the sum of the impulse ratings of all the line connected modes of protection but may also be less and is not generally required, but optional. This may sound technically illogical but is due to some marketing-driven influence and the never-ending horsepower race in all kinds of impulse ratings.

6.2.2   Specific parameters for SPDs used in low-voltage power systems

In addition to the before-mentioned general parameters, there are a number of additional parameters that are only used for SPDs for power applications.

6.2.2.1   Short circuit current rating I_{SCCR} and follow current interrupt rating I_{fi}

The relevance of this parameter today is twofold, because the historic distinction in ratings between the former follow-current self-extinction capability of voltage switching and combined SPDs and the ability of a failed SPD to withstand the maximum short circuit current to be expected is of no more relevance. Today, the installation rules require that the follow-current self-extinction capability of SPDs and its ability to withstand the maximum short circuit current in failed condition must both be equal or greater than the maximum short circuit current to be expected at the SPDs point of installation.

Exemptions exist for SPDs for overhead lines, because the risk of severe damages due to SPD failures are considered to be reasonably low there, and for SPDs for connection between neutral and PE only, because there no noteworthy short circuit current is to be expected (see the 'SPDs only for connection between neutral and earth (N-PE SPDs)' section).

6.2.2.2 Maximum discharge current I_{max}

As the denomination already expresses, this should provide information about the maximum 8/20-µs impulse current capability of an SPD that the device can divert at least one time without significant degradation or change in behaviour, irrespective of the test class applied. Thus, the ratio of maximum discharge current to the nominal discharge current provides an indication of the 'reserves' built into an SPD, but declaration of this parameter is also just optional and not required, like for the total discharge current.

6.2.2.3 Voltage drop for two-port SPDs

Power consumption and power dissipation are since years becoming more and more an issue. This optional declaration together with the rated (load) current may be used as an indication for SPDs contribution to the total power dissipation of, e.g., a distribution board, but this is to be taken with care, as long as there is no information of what the contributing resistive and the inductive components are.

6.2.2.4 Load-side surge withstand capability for two-port SPDs

In principle, surges may not only come from the supply side but may also be created by equipment inside an electrical installation. This parameter is a measure how sensitive a two-port SPD is for such impulse stress, but it must be mentioned that this optional declaration value, although in the standard since decades, is very seldom used.

6.2.2.5 Voltage rate of rise *du/dt*

Specifically for electronic equipment and power semiconductors, the maximum steepness of any surge voltage to be expected is an absolutely crucial parameter to avoid avalanche breakdown effects in the off-state. All the classic SPCs listed in Section 6.1.3.1 and described in Chapter 5 limit surge voltages but do in fact not significantly change the *du/dt* of the incoming surges. Such reduction in steepness can only be provided by sufficient inductance in the circuit, or in some cases, by significant capacitance in relation to the charge content of the limited surge, that remains from the SPDs protective action. Two-port SPDs often contain such additional components in the current path supplying the load circuits.

6.2.2.6 Transition surge current rating for short-circuiting type SPD I_{trans}

The principles related to short-circuiting type SPDs are already described in the 'Short-circuiting SPDs' section. For such SPDs, the manufacturer must declare

- an impulse discharge current (10/350 µs waveshape) for class I tested SPDs,
- an 8/20-µs discharge current for class two tested SPDs,

that is sufficient to ensure the SPDs internal transition into a dedicated short circuit condition, thus simulating a severe impulse overstress to the SPD.

After having reached that short circuit state the SPD must pass a number of consecutive tests to check for sufficient robustness of the internal current path effectuating the short circuit.

6.2.3   Specific parameters for SPDs used in telecom and signalling systems

While not all of the general parameters given in Section 6.2.1 may always be required, some or even all of the following additional parameters may be necessary for the selection of the appropriate SPD, depending on the intended applications.

6.2.3.1   Impulse reset time

This parameter is a measure for the time interval that a voltage-switching SPD needs to extinguish any follow current and to return to its high-impedance state again after conduction. It always needs to be related to a source with a defined voltage/current characteristic, as the value may differ for different sources.

6.2.3.2   Transmission characteristics

This includes the SPD's capacitance, the insertion loss, return loss, longitudinal balance, bit error ratio (BER) and near-end crosstalk (NEXT). All these parameters are a measure for the SPD's adverse impact on the signal quality and the signal attenuation it causes, e.g. with regard to the systems frequency–response characteristics.

Insertion loss is related to the impact of the SPD's on the transmission system where it gets installed and is given as the ratio of the power delivered with the SPD inserted to the power delivered without the SPD being present. In most cases, this ratio is expressed in decibels (dB).

The return loss is a measure for the signal fraction that gets reflected back to the signal source due to the insertion of the SPD into a matched transmission line. It is normally given for a dedicated frequency range and also expressed in decibels (dB).

Longitudinal balance or longitudinal conversion loss is a measure for the symmetry of a multi-terminal SPD and may be described as the maximum difference in series resistance of the compared lines or as the difference in signal attenuation. It can be expressed as a resistance difference in ohms or as a percentage, or in decibels (dB), respectively, and is also given for a dedicated frequency range.

The BER provides a number for the bit errors in a digital transmission system that are caused by the insertion of the SPD into this system, if any.

NEXT shall describe the amount of signal that gets coupled from one line to another by the SPD and refers to a given frequency range.

Further details about transmission characteristics can be found in Sections 7.8.5 and 7.8.6.

6.2.3.3   Insulation resistance

While the insulation resistance, specifically the insulation resistance to earth, can also be an important parameter for SPDs for power applications in view of the required measurements to be performed during initial and follow-up checks of electrical installations, it almost always is a crucial parameter in telecom and signalling applications to keep signal losses low. In many cases, isolation-like behaviour is required in normal system operating condition and leakage currents are allowed in the µA or even sub-µA range only.

6.2.3.4   Current response time and current reset time

For SPDs with current-limiting function as described in the 'Current-limiting function' section making use of self-resetting components, e.g. PTC-thermistors or self-resetting fuses, there are two important parameters for describing their response characteristics: one for the protection of downstream equipment and the other for the functionality of the system.

The current response time defines how long it takes for the current-limiting component to operate under given current and ambient temperature conditions when starting from cold unloaded condition. Operation is thereby defined as either an increase of the current-limiting component's impedance causing the current to drop to less than 10% of the rated current, or opening of the current-limiting component.

The current reset defines how long it takes after the operation of the current-limiting component as mentioned previously and once the circuit conditions and impedances are back to where rated current would flow under normal system and SPD conditions, until the current limiting component's impedance has either decreased again to allow at least 90% of the rated current to flow or the current-limiting component has closed again.

6.3   Test procedures

The content of this section focuses on requirements established by the IEC 61643 series of standards for low-voltage surge protective devices and only refers to UL 1449 where significant differences are to be mentioned.

It is important to note that for feasibility reasons most electrical testing, including operating duty, dielectric, end-of-life and SPD disconnector testing, is allowed to be performed under usual laboratory conditions and does not include the service temperature and humidity range of the SPD as declared by the manufacturer. Therefore, such general environment-related data is under the sole responsibility of the SPD manufacturer and only for SPDs classified for outdoor application – see Section 6.1.5 – some additional tests addressing UV-radiation, dielectric strength under wet conditions, temperature cycling and corrosion resistance are required. But these are performed separately and without any interconnection to the electric testing mentioned earlier.

Furthermore, the attention is drawn to the fact that impulse voltage and current testing, especially when exceeding the normal EMC surge testing capability with 6 kV and 3 kA, requires real good testing techniques, dedicated measuring equipment and exigent grounding techniques to reach meaningful, robust and proper measurement results. Even when the measurements themselves are flawless, e.g. the correct interpretation of oscillograms sometimes requires long-term experience to end up with the right conclusions and with correct values.

6.3.1   Test impulses and impulse generators

Laboratory impulse testing tries to simulate the surge events occurring in real installations. Therefore, a number of impulse waveshapes that are assumed to adequately represent these events have been standardised. To reach these required waveshapes in testing, a variety of impulse test generators is available on the market, but not all of them fulfil the specific test requirements established within the IEC 61643 series of standards for SPDs, e.g. with respect to waveform tolerances or their ability for AC power supply synchronised impulse triggering. This section aims to provide an overview on the most important aspects and equipment properties.

It must be noted that waveshape parameters and definitions as well as the associated tolerances for testing may differ in UL 1449 and IEEE C62.45 standards but are generally less stringent. In addition, these standards primarily define the waveform parameters to be delivered by the generator but do not describe at which points of a specific SPD test set-up these must be checked.

6.3.1.1   Impulse waveshapes

Voltage impulse of 1.2/50 µs

This voltage impulse is also referred to as lightning-impulse voltage as it shall represent a typical lightning-induced overvoltage.

Although this test impulse is in use since quite long time and defined in IEC 60060-1 as shown in Figure 6.18, some adaptation was necessary to satisfy the testing purposes of power SPDs. This is because the goal was to avoid defining a new kind of test generator and in principle to allow the use of test generators that already exist in many laboratories and that are available on the market already.

As this waveform is used for the spark-over voltage determination on the rising edge of the impulse and because this spark-over voltage significantly varies with the steepness of the impulse, the above-mentioned adaptation and refinement affects the tolerance of the front time and provides an additional descriptive requirement on the rising edge of the impulse as follows:

- The tolerance for the front time was constricted to $\pm10\%$ (from originally $\pm30\%$).
- The rising edge must be smooth and without distortions or superposed oscillations to avoid any discontinuity in the du/dt rise gradient.

These constraints in conjunction with oscillographic documentation are absolutely necessary to reach acceptably consistent, comparable and repeatable test results, but they are only explicitly required by the test standards for SPDs for power applications.

As the crest value of the impulse is not so important, the tolerance was relaxed to $\pm5\%$ (from originally $\pm3\%$) in the standards for power SPDs, and even to $\pm10\%$ in the standard for telecom and signalling SPDs. The time to half value has not been modified and is tolerated with $\pm20\%$.

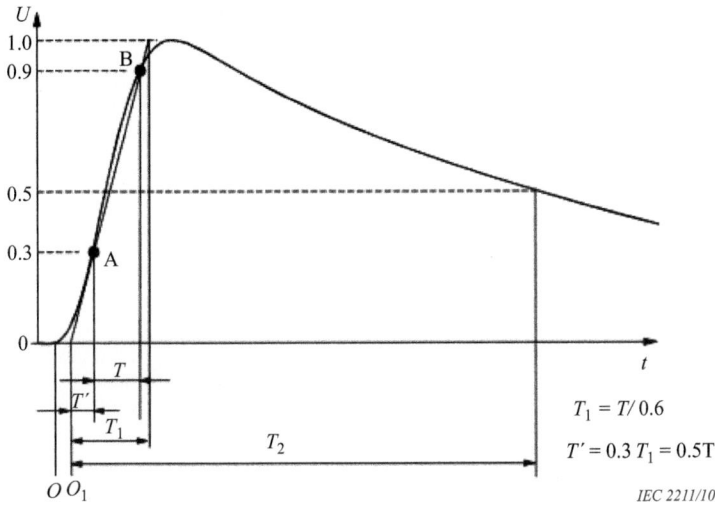

O ...   origin of the impulse

O_1 ...   virtual origin of the impulse

T_1 ...   front time

T_2 ...   time to half value

*Figure 6.18   Voltage impulse of 1.2/50 µs, including voltage time parameters.***

Current impulse of 8/20 µs

This test current impulse shall represent induced currents due to atmospheric overvoltage and also as a consequence of any resulting insulation breakdown or SPD action. It is also defined since long time and today provided in IEC 62475 as shown in Figure 6.19.

Also here some restrictive modification to the tolerances was necessary to satisfy the needs for power SPD testing, because many SPCs, specifically the semiconductor-based ones, are quite sensitive regarding the maximum energy (I^2t) that may pass through without damage. Therefore, the following constrictions are required:

- The tolerance for the front time was constricted to ±10% (from originally ±20%).
- The tolerance for the time to half value was constricted to ±10% (from originally ±20%).

Again these constraints in conjunction with oscillographic documentation are necessary to reach consistent, comparable and repeatable test results, but they are only required by the test standards for SPDs for power applications.

The crest value's tolerance with ±10% is unchanged, and it depends on the generator's switching technology and on the SPD under test, if there is any polarity

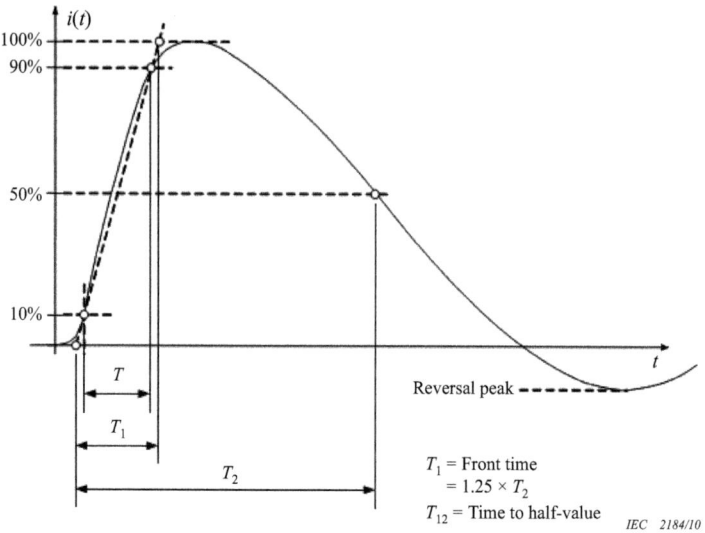

Figure 6.19   *Current impulse of 8/20 μs, including voltage time parameters.*[††]

reversal after the discharge current has decayed to zero. In any case, a polarity reversal must be limited to 30% of the mains impulse crest value.

Lightning impulse current of 10/350 μs (10/250 μs)
This test current impulse was described to simulate partial lightning currents and its ideal waveform also fits with the waveshape as defined in Figure 6.18. Although IEC 61643 only defines a test impulse via three parameters:

- the crest value for I_{imp},
- the charge Q in As with $Q = I_{imp} \times 5 \times 10^{-4}$ and
- the specific energy W/R in kJ/Ω with $W/R = I_{imp}^{2} \times 2.5 \times 10^{-4}$,

these parameters perfectly fit to a 10/350-μs impulse waveshape. But in practice, lightning impulse test currents above 10 kA are mostly and above 50 kA are in fact only produced with so-called crowbar generators for feasibility reasons. That type of generator can fulfil the test parameters given before but produces more a triangular waveshape than a double exponential waveshape in practice. The test tolerances are required to be ±10% for the crest value, +20/−10% for the charge and $^{/+45}/_{-10}$% for the specific energy in the standards for power SPDs. This is necessary because of the unavoidable differences in the test generators used. For more details, see the 'Pure current impulse generators' section.

The standard for telecom and signalling SPD allows to optionally use a 10/250-μs current impulse. There the tolerances for both waveforms are required to be 10% for the crest value, 30% for the front time and 20% for the time to half value.

Combination wave 1.2/50 μs for voltage and 8/20 μs for current
A CWG provides a defined voltage waveshape into an open circuit and a defined current waveshape into a short circuit at its output. The quotient of the open circuit voltage crest value U_{OC} to the short circuit current crest value I_{SC} gives the fictive output impedance of the generator.

For testing SPDs, a CWG with the 1.2/50-μs voltage waveshape, the 8/20-μs current waveshape and a nominal fictive output impedance of 2 Ω is required. Therefore, the crest value of the short circuit current to be adjusted in amperes equals U_{OC} (V)/2 (Ω).

While the principle waveshape descriptions and requirements of the 'Voltage impulse of 1.2/50 μs' and 'Current impulse of 8/20 μs' sections apply, the tolerances are summarised in Table 6.3. These tolerances must be kept for adjustment of the SPDs declared open circuit voltage U_{OC} and the corresponding short circuit current I_{SC} at the connection points of the SPD. For more details, see Section 6.3.2.1. To avoid unnecessary efforts for generator readjustments, it is not necessary to check the waveform parameters again when surges with only fractions of the declared U_{OC} are applied according to the test procedures.

Although not explicitly required by any of the SPD standards, it needs to be reminded here, that similar to the 1.2/50-μs voltage impulse, the smoothness of the rising edge of the output voltage without distortions or superposed oscillations is absolutely crucial for consistent, comparable and repeatable test results, no matter if testing is performed on SPDs for power, telecom or signalling applications. Unfortunately, also the tolerances for the voltage impulse front time have not been restricted for the combination wave.

Other waveforms used for SPDs for telecom and signalling systems only
The different line impedances of many telecom and signalling cables as compared to power lines have led to some different and additional waveform specifications in this area. The tolerance requirements from IEC 61643-21 are shown in Table 6.4.

Table 6.3   Combination wave tolerances of 1.2/50 μs/8/20 μs

Impulse test parameter	Tolerance for testing SPDs according to IEC 61643-11		Tolerance for testing SPDs according to IEC 61643-21	
	1.2/50-μs voltage (%)	8/20-μs current (%)	1.2/50-μs voltage (%)	8/20-μs current (%)
Crest value	±5	±10	±10	±10
Front time	±30	±10	±30	±20
Time to half value	±20	±10	±20	±20

Table 6.4   Impulse tolerances according to IEC 61643-21

Impulse test parameter	Open circuit voltage for 1.2/50- and 10/700-µs waveform (%)	Short circuit current for 8/20- and 5/320-µs waveform (%)	All other voltage/current waveforms (%)
Crest value	±10	±10	±10
Front time	±30	±20	±30
Time to half value	±20	±20	±20

Combination wave 10/1,000 µs for voltage and current: The principle of a CWG providing a 10/1,000-µs waveform is the same like the one described in the 'Combination wave 1.2/50 µs for voltage and 8/20 µs for current' section, but in this case the open circuit voltage and the short circuit current have the same double exponential waveshape. There are actually two generators standardised, in IEEE and the other, most common one, in Telcordia GR-1089-CORE.

This waveform represents surges of atmospheric origin on long tele-communication lines with high impedance.

Combination wave 10/700 µs for voltage and 5/320 µs for current: Again the principle of a CWG providing a 10/700-µs open circuit voltage and a 5/320-µs short circuit current is the same like the one described in the 'Combination wave 1.2/50 µs for voltage and 8/20 µs for current' section, but here for voltage and current, a double exponential waveform is defined. A schematic for this kind of generator is given in ITU-T Recommendation K.44, but it is also referenced in a number of IEC standards.

This waveform also represents surges of atmospheric origin on long tele-communication lines with high impedance.

Combination wave 100-V/µs linear rise for voltage and 10/1,000 µs for current: While the basic current waveform requirements are the same like for the combination wave 10/1,000 µs, here for the open circuit voltage, there is only a defined linear rate of rise requirement, but no waveshape definition. Details are given in Telcordia GR-974-CORE.

This waveform represents surges with a slow rate of rise in telecommunication applications.

Combination wave 1-kV/µs linear rise for voltage and 10/1,000 µs for current: Here again the basic current waveform requirements are the same like for the combination wave 10/1,000 µs, but for the open circuit voltage a steeper linear rate of rise is required. More details can also be found in Telcordia GR-974-CORE.

This waveform represents surges with a fast rate of rise in telecommunication applications.

6.3.1.2   Generator topologies and their implication on testing

The discharge capability of SPDs varies over a wide range, from some 100 to some 100,000 A, and it is impossible to cover that just with one set of generators. The typical advantages and disadvantages of specific generator topologies as well as non-technical aspects shall be mentioned here.

Pure voltage impulse generators

This is the generator type with the lowest impulse energy and price range as compared to the other generator types mentioned later in this section. But even here, careful selection is required to ensure fulfilling the needs for testing SPDs. This specifically refers to the following:

- The switching technology used to apply the voltage impulse to the test sample, as this is critical to reach the required smoothness of the rising edge of the output voltage without distortions or superposed oscillations.
- The circuit design and components used in the circuitry forming the wave, as these are critical to avoid superposed oscillations again.
- The tightened tolerance requirement for the front time in comparison, e.g. to a general use 1.2/50-μs impulse generator for dielectric testing or a CWG designed primarily for EMC testing.

The output impedance of such voltage impulse generators can typically be as high as 500 Ω, and as soon as the load (test sample) connected constitutes an internal impedance of less than ten times the generator output impedance, the output voltage waveform as well as the output current waveform and amplitude are no longer defined and will be arbitrary.

Therefore, this type of generator is only used for determining an SPD's front-of-wave spark-over voltage.

Pure current impulse generators

There is an incredible variety of products and designs on the market, and considering the SPD product range, topologies and discharge capabilities that should be covered in testing, a focus on the possibilities to adjust the impulse parameters and the potential range for adjustments is recommended. This includes considering

- powered and unpowered SPD testing respectively,
- testing with or without coupling elements for the SPD sample,
- testing with or without decoupling elements, networks or filters to protect the energising power source.

The output impedance of such generators can be in the range of milliohms up to some thousand ohms and depends primarily on the quotient of the total charging voltage to the intended amplitude of the current impulse. The lower that quotient is, the less are the options for impulse adjustments.

The open circuit output voltage of such generators is typically as high as the total charging voltage of the generator and the steepness of the voltage rise du/dt right after impulse initiation is only limited by the generators internal stray capacitances, if no particular measures are taken. This must be taken into account for any attached probes and measurement circuitry.

Direct discharge type: For pure voltage impulse generators and for generators producing short current waveforms, like the 8/20-μs impulse, or longer impulse current waveforms with limited surge amplitude, generally a design is used that directly discharges a pre-charged capacitor mostly into a passive series output circuit consisting of resistors and/or inductances.

For generators with very low charging voltages in relation to the intended current impulse amplitudes, sometimes even the resistance or inductance of the internal interconnections is already sufficient to form the wave, but impulse adjustments are more or less impossible then.

In addition, shunt components like resistors or capacitors may be used in the generator circuitry to define the open circuit voltage waveform or to limit the output voltage rise du/dt.

Crowbar type: Such generator type is used to realise high-energy impulses like medium and high current amplitudes with longer waveforms as well as very high current amplitudes with shorter waveforms, because this technology requires far lower capacitance and volume in design as compared to direct discharge generators.

A crowbar generator as shown in Figure 6.20 typically comprises two switching elements (S1, S2), mostly realised with triggered spark gaps. The first one (S1) is for initiating the impulse within a direct discharge output circuit that delivers a heavily oscillating current wave (I_1), which determines the front time and the crest value of the impulse current. The second one (S2) called crowbar switch is located in a shunt branch circuit of the generator, which consists of a well-balanced R/L series combination, and needs to be synchronised and be triggered at the first peak of the oscillating current wave (I_1), causing a superimposed highly damped current (I_2) because of the inductance in the circuit. The total current appearing at the generator output then can be close to a double exponential waveform or be almost triangular, depending on the load (test sample) impedance and characteristic in relation to the generators effective impedance – see Figure 6.21.

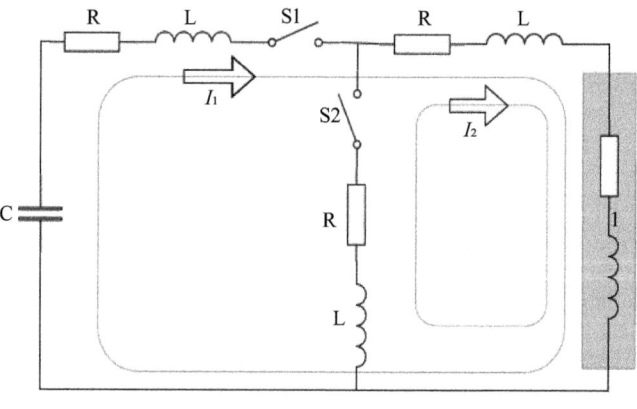

© Hubert Bachl-Hesse

1 ...   SPD under test (after conduction)

Figure 6.20   Schematic of a crowbar type generator. 1: SPD under test (after conduction).

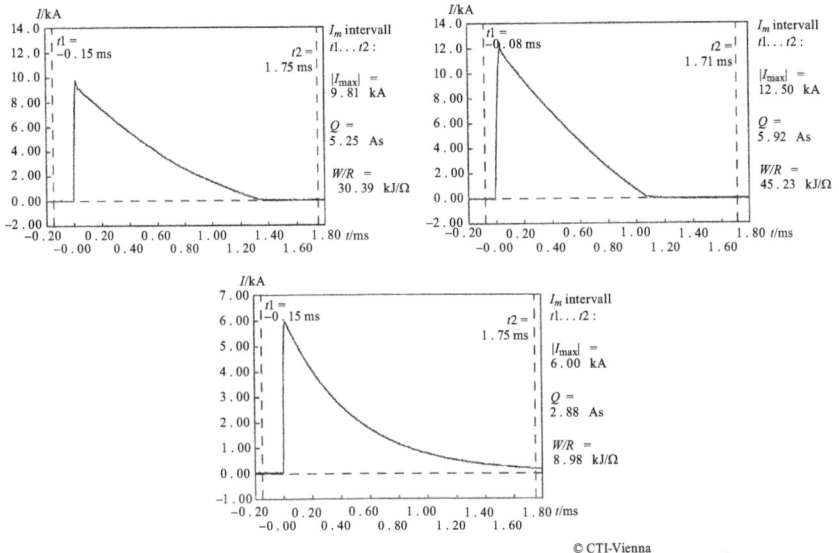

Figure 6.21   *Output waveforms of a 10/350-μs crowbar generator with different test samples.*

It becomes clear that the control and parameter adjustment of such a crowbar generator is far more complex than for a direct discharge generator, and it is much more sensitive to any change in load (test sample) characteristics.

Especially when testing voltage limiting SPDs care must be taken to avoid any current fraction from the generator exceeding 5 ms, but preferably even in excess of 2 ms, as such currents easily overstress, degrade or even destroy the voltage limiting components, leading to an unjustified negative test result.

Combination wave generators
The behaviour of a CWG is already described in the 'Combination wave 1.2/50 μs for voltage and 8/20 μs for current' section and the basic topology is a direct discharge generator as described in the 'Direct discharge type' section with appropriate shunt components like resistors or capacitors in the output circuitry, to define the waveform of the open circuit output voltage.

6.3.2   Coupling of test impulses and decoupling of power supply

6.3.2.1   Coupling of the test impulses to the test sample(s)
For energised impulse testing, many generators require a kind of coupling element between the generator output and the test sample, respectively, the power source connected to the test sample. This is necessary to avoid excessive power frequency or DC current flow from the power source into the generator, as such current may

damage the generator or prohibit the impulse-initiating switch inside the generator from opening again after the impulse has decayed.

The preferred components used as coupling elements are capacitors or MOVs, depending on the impulse energy to be conducted. The advantage of the capacitor is that there is no voltage dependency in its characteristic, but its application is limited to low impulse energy applications and therefore primarily to pure voltage impulse generators and CWGs. Powerful MOVs and parallel arrays of them can be used even for very high current impulses and provide very good decoupling as soon as the generators driving voltage drops below the MOVs conduction area, but their impact on the impulse waveform is generally bigger due to the nonlinear voltage–current characteristic. In any case such coupling components must be carefully considered as in many cases, they require a readjustment of the generator to keep the required wave shape tolerances at the SPD's connection point.

6.3.2.2 Decoupling of the test impulses from the power supply (back filters)

A major concern in test laboratories is the protection of the power supply for energised impulse testing from the surges applied to the test samples, although the preferred and more or less universal test set-up, irrespective of the SPD's internal circuitry, is without a decoupling network.

In principle, there are three scenarios:

- For SPDs not containing any reactive components, there is no difference, if testing is performed with or without a decoupling network.
- For two port voltage limiting SPDs containing one or more reactive component (s), no matter if the SPD is for AC or DC applications, a DC supply via a blocking diode according to Figure 6.22 may be used, whereby the test impulse shall be triggered 100 ms after an application of the DC voltage and the DC voltage should be switched off 10 ms after impulse application, if critical for the test sample.
- For two-port voltage switching and combination SPDs containing one or more reactive component(s), no decoupling network shall be used.

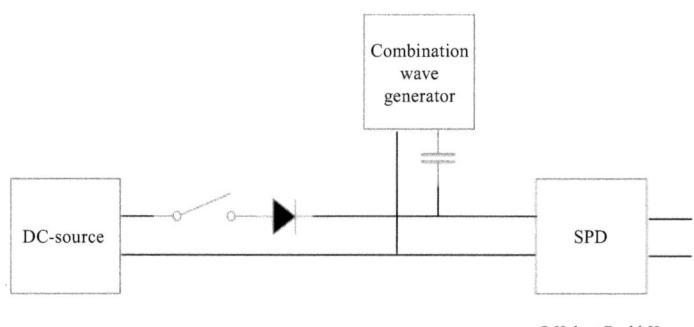

© Hubert Bachl-Hesse

Figure 6.22   Decoupling for two-port voltage limiting SPDs

This preference to use no decoupling network or even prohibition to use one in some cases in fact requires the use of robust power supplies that are designed to withstand the recurring surge stress.

Careful consideration must be given to the impulse current fraction flowing towards the power source, specifically when testing voltage switching or combination SPDs with follow currents and with high short circuit current ratings.

6.3.3 Impulse testing and tolerances

The impulse tolerances mentioned in Section 6.3.1.1 must be met at the connection points of the SPD, no matter if testing is performed energised or un-energised, and irrespective if impulse coupling elements or decoupling elements/filters are used. This may require different generator adjustments for energised and un-energised testing.

The only exemption from that is for the combination wave, where the tolerances must only be kept when adjusting the open circuit voltage U_{OC} for which the SPD is rated, to avoid unnecessary efforts for generator adjustments. No further check is then required when testing in the same set-up, but only with fractions of that previously adjusted U_{OC} rating.

The connection points of the SPD are defined to be either the SPDs field wiring terminals themselves or, in the case of pigtail connections (SPDs with integrated connection leads instead of field wiring terminals), a lead length of 150 mm.

6.3.4 Measurement accuracy and measurement uncertainty

Let us first have a look at the definitions of these terms to estimate their importance for testing. While measurement accuracy is a quality criterion that describes the ability of a measuring instrument to provide a reading close to the true value of a measurand, measurement uncertainty describes the spread of values that could reasonably be assigned to the measurand. In other words, accuracy is related to the measuring instrument and uncertainty considers all potential influencing factors to the measurement.

From that it can be summarised that the bigger the potential operator's and external influencing factors are, the more must be focused on measurement uncertainty and not only on the instruments accuracy. For impulse measurements, it can generally be assumed that there are quite a number of serious 'environmental' influencing factors in addition to the usual ones like temperature, humidity, etc.:

- Bandwidth and frequency response (over-/undershoot, resonance phenomena).
- All devices included in the measurement chain, their interaction, mutual influence and adjustment, as the measuring instrument normally requires external probes/ sensors and sometimes converters for wireless signal transmission are used.
- The location, placement and attachment of the probes/sensors.
- The routing and shielding of the measurement connections.
- (Reference) earthing and bonding concept for the test set-up – impulse generator, power source(s) for testing and supply of measurement equipment, measurement connections.

The standard requires a bandwidth of at least 25 MHz and an overshoot of less than 3% for the measuring instrument used for impulse testing and in addition refers to IEC 60060-1 for high-voltage test techniques. This is imprecise and the future edition will most probably differentiate to require

- an overall measuring system bandwidth of 10 MHz only for residual voltage measurements, which is sufficient and
- an overall measuring system bandwidth of 25 MHz for all other impulse tests,

and will in addition refer to IEC 61180 for high-voltage test techniques for low-voltage equipment and to IEC 62475 for high-current test techniques.

Considering that typical high voltage probes used for peak voltage measurements above 6 kV have a -3-dB bandwidth declared to be 25 MHz and a tolerance of $\pm 3\%$ for the nominal attenuation ratio, this already completely uses up the previous accuracy requirements and to avoid a further degradation by the oscilloscope connected to the probe, this would require the scope to be at least ten times better than the probe with regard to bandwidth and vertical resolution, resulting in an oscilloscope with a bandwidth of ≥ 250 MHz and a sampling rate ≥ 1 GS for limiting voltage measurements with the combination wave at 6-kV measurement range. And this example up to here only considers the measurement accuracy of the measuring equipment used but does not respect any of the above-mentioned factors impacting measurement uncertainty.

In conjunction with this, it may be of interest to know that there is also IEC 60060-2 for high-voltage test techniques and related measuring systems for deeper consideration.

6.3.4.1   Voltage probes and their arrangement and attachment

The choice of the preferred voltage probe together with the reachable performance parameters is influenced by one major selection criteria related to the peak voltage it shall be rated for, which is the following:

- Either the open circuit voltage of the generator or of a protective element placed in parallel to the test sample, in case the probe shall survive a non-conducting sample or a sample failure.
- The maximum limiting voltage of the test sample to be expected, in case the probe may be destroyed in the case of a non-conducting sample or a sample failure.

This selection criterion also determines the physical dimensions of the probe and as a result of that the frequency behaviour and the minimum loop area of its connections, which is feasible when attaching the probe to the test sample. As the area of that loop, due to the voltage induced there, falsifies the voltage measurement, it is in many cases the crucial factor for the measurement uncertainty.

A number of measurements were executed by different laboratories to find a solution as to minimise the influence of that. The results have finally led to a recommended probe arrangement and attachment according to Figures 6.23 and 6.24:

- with the measurement connection wires starting to be twisted as close as possible to the test sample according to Figure 6.24,

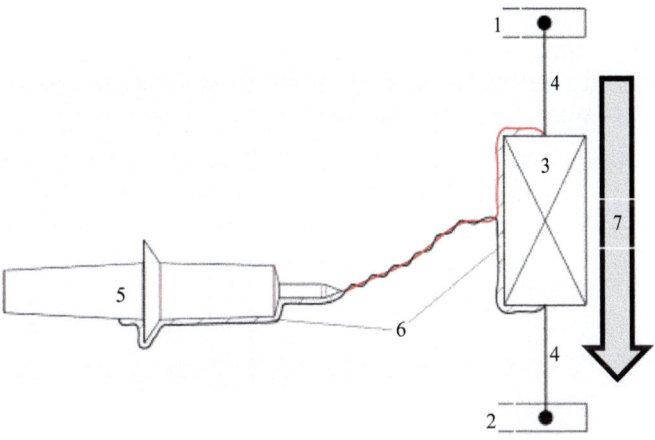

1 ...   HV output connection of impulse generator
2 ...   Reference ground plane of impulse generator
3 ...   Test sample (SPD)
4 ...   Test sample connections to the impulse generator
5 ...   Voltage probe
6 ...   Loop area created by the measurement connections (shaded area)
7 ...   discharge current path

Figure 6.23   Preferred probe arrangement and attachment.[‡‡]

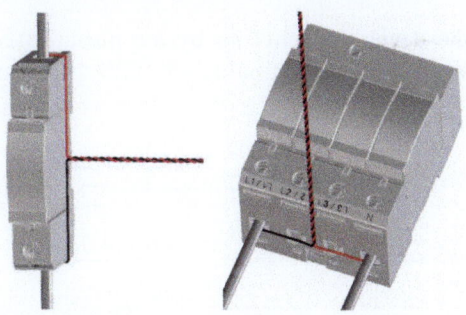

Figure 6.24   Examples of preferred arrangement of measurement connection wires[§§]

- the voltage probe being placed at a perpendicular distance between 200 and 500 mm from the discharge current path, and
- the loop area being formed by the probe inherent connections still being kept to a minimum.

Other voltage probe arrangements and specifically the use of two single ended voltage probes in a differential mode arrangement cannot be recommended and may lead to significant measurement errors.

6.3.4.2 Current probes

For impulse current measurement, primarily appropriate wide-band current transformers or coaxial shunts are used, whereby the current transformers have the advantage of delivering a measurement signal which is galvanically separated from the generator circuitry. For the impulse current measurement chain, it is much easier to comply with the accuracy requirements than for impulse voltage measurements.

For energised impulse testing, the impulse current probe must be located in the SPD branch circuit of the test set-up according to Figure 6.25 to ensure that only the impulse current through the test sample is monitored and any fraction of the impulse current driven towards the power supply does not falsify the measurement. This needs to be carefully checked when the impulse generator includes the current probe.

6.3.5 *SPD versus SPD-assembly and mode of protection versus current path*

During the revision and improvement process of the actual IEC 61643 series of standards it became evident that there is a need to distinguish between four principal objects for test application:

1. The SPD is the device as delivered by the manufacturer, which may be a single-mode or a multimode (multipole) SPD, without any external SPD disconnectors.

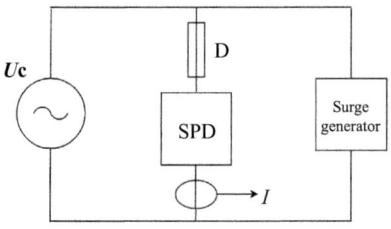

U_C ...	power source for energised testing
D ...	external SPD disconnectors, if specified by the manufacturer
SPD ...	Test sample
I ...	Current probe

Figure 6.25 Typical operating duty test set-up for energised impulse testing.

This is addressed mainly for mechanical tests, but also for some electrical tests, e.g. insulation resistance and dielectric withstand.

2. An 'SPD assembly' is either a multimode device or a complete set of (single mode) devices, intended to provide overvoltage protection for a specific power system configuration, e.g. a three-phase 5 wire TN-system, including the external SPD disconnectors as required by the SPD manufacturer for that specific power system configuration.

    This is required, e.g., for the measurement of the protective conductor current.

3. A 'mode of protection', as already addressed in Section 6.1.4.3, addresses an SPD's internal protective circuitry between two connections/terminals and must be declared as such by the manufacturer for testing, see Figures 6.11 to 6.13.

    Many of the electrical tests are applied separately to each mode of protection, e.g. the measured limiting voltage tests and the operating duty test. Special attention is required when testing single-mode SPDs that shall be combined to a Y-configuration, because in such case whenever testing on the mode of protection is required, in fact any required sample preparation must be done and the test be applied to the relevant series combination of two single-mode SPDs, that is finally intended to be connected between two conductors of the system it is intended for.

4. A 'current path' is sometimes addressed for sample preparation or testing, when the specific test cannot or is not useful to be applied to a complete mode of protection, because the internal protective circuitry of that mode of protection splits up into two or more protective components being connected in parallel.

For a mode of protection with, e.g., only one protective component, there is only one current path, which is then identical with the mode of protection. But for a three-stage two-port SPD according to Figure 6.26 with a GDT shunt branch (1), an MOV shunt branch (2) and a Z-diode shunt branch (3) connected in parallel, there are three potential current paths to be tested separately.

This is addressed, e.g., in the thermal disconnection test and in the short circuit tests.

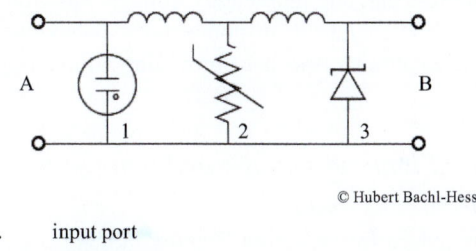

© Hubert Bachl-Hesse

A ...   input port
B ...   output port
1, 2, 3 ... current paths

Figure 6.26   Example of a three-stage two-port SPD containing three current paths

6.3.6   Power supplies

Depending on the kind of test, there are different test voltages, but also different current capabilities required by the SPD standards.

While for powered impulse testing, mainly the maximum continuous operating voltage (U_C) is applied to check for the limit capabilities of the SPD with respect to surge withstand and the consecutive reaching of a 'reliable' high-impedance state again, for most other tests the reference test voltage (U_{REF}) is used to simulate maximum supply voltage conditions of a power supply system, where the SPD may be connected.

It is important to note here that these maximum supply voltage conditions are primarily based on systems, which comply with installation rules of the IEC 60364 series of standards, with the National Electrical Code in North America, or with the Japanese requirements for power distribution systems. In most cases a maximum tolerance for voltage regulation is therefore considered with plus 10% of the nominal system voltage. The other regions or countries did not provide specific information on the operating conditions and the expectable voltage regulation so no detailed information is available so far, but it is known that especially in rural areas, some countries face voltage variations up to plus 30% or even more. This is not covered by the tabulated reference test voltages given in the SPD standards but must be considered for testing, as far as it is known, that such applications should be covered by the SPD concerned.

The source current capabilities required for the various tests relate, as far as possible, to real world and in some cases even worst case conditions from typical supply systems. Where this is not possible, e.g. because no such test facilities are available or it does not seem feasible to require for building up such facilities just for SPD testing, a kind of synthetic test set-up is used to simulate specific conditions as close as economically possible. Good examples for that are the additional duty test for class I tested SPDs, where decoupling of a high-power source and the impulse generator providing that simulated long partial lightning current impulse is no longer really possible, or the failure mode or end-of-life behaviour tests.

Having this said, also the ripple specification for DC test voltages and currents still causes repeated discussions, as an acceptable compromise needs to be found between the awareness that any momentary voltage variation may have significant influence on the SPD's and its disconnector's behaviour with regard to arc extinction, and the economic aspects of providing almost ripple free and powerful DC supplies.

6.3.7   Measured limiting voltage and voltage for clearance determination (U_{max})

Details on the procedure for measured limiting voltage determination can already be found in Section 6.2.1.3; therefore, the focus here is on the background for the additional determination of the 'voltage for clearance determination (U_{max})'.

An SPD does not only protect the downstream installation and equipment from transient overvoltages, but of course also itself, as long as it is in working order.

Following this, the necessary clearances to be respected in SPD design depend primarily on the maximum peak voltages that occur during the most unfavourable impulse stresses, but as some of the highest impulse stresses that an SPD is required to withstand depend on optional parameters like the maximum discharge current (I_{max}) or the total impulse discharge current (I_{total}), the determination of that maximum peak voltage for clearance determination (U_{max}), became a bit complex. U_{max} is no declared parameter and is only needed for decision on the minimum required clearances during the type testing process. It is the highest peak voltage measured during the following tests:

- measured limiting voltage – as applicable
  - residual voltage
  - front-of-wave spark-over voltage
  - limiting voltage with combination wave

- maximum discharge current
- total discharge current (currently not considered in the standard for U_{max} determination).

Acknowledgement

The author thanks Mr. Ralf Hausmann (Phoenix Contact) for kindly providing additional information for the sections addressing telecom and signalling SPDs. The author thanks CTI-Vienna, Gesellschaft zur Prüfung elektrotechnischer Industrieprodukte GmbH, for kindly providing the oscillograms reproduced in this chapter.

The author thanks VDE Verlag GmbH for providing the permission to use and the assistance in modifying figures from the book "Schutz in elektrischen Anlagen – Band 5: Schutzeinrichtungen, Biegelmeier, G., Kiefer, G., Krefter, K.-H – VDE-Schriftenreihe 84. 1999"

The author thanks the International Electrotechnical Commission (IEC) for permission to reproduce Information from its International Standards. All such extracts are copyright of IEC, Geneva, Switzerland. All rights reserved. Further information on the IEC is available from www.iec.ch. IEC has no responsibility for the placement and context in which the extracts and contents are reproduced by the author, nor is IEC in any way responsible for the other content or accuracy therein.

Bibliography

General:

IEC 60364 series *Low-voltage electrical installations.*
IEC 62368-1:2018 Audio/video, information and communication technology equipment – Part 1: Safety requirements.

NFPA 70 National Electrical Code (NEC).

UL 1310:2018 Standard for Safety for Class II Power Units.

Telcordia GR-1089-CORE, Issue number 07, 2017 Electromagnetic Compatibility (EMC) and Electrical Safety – Generic Criteria for Network Telecommunications Equipment.

Surge protective components, surge protective devices:

Schutz in elektrischen Anlagen – Band 5: Schutzeinrichtungen, Biegelmeier, G., Kiefer, G., Krefter, K.-H – VDE-Schriftenreihe 84. Berlin, Offenbach: VDE VERLAG, 1999.

EN 50539-11:2013 Low-voltage surge protective devices – Surge protective devices for specific application including d.c. – Part 11: Requirements and tests for SPDs in photovoltaic applications.

IEC 61643-11:2011 Low-voltage surge protective devices – Part 11: Surge protective devices connected to low-voltage power systems – Requirements and test methods.

EN 61643-11:2012 + A11:2018 Low-voltage surge protective devices – Part 11: Surge protective devices connected to low-voltage power systems – Requirements and test methods.

IEC 61643-12:2020 Low-voltage surge protective devices – Part 12: Surge protective devices connected to low-voltage power systems – Selection and application principles.

IEC 61643-21:2000+AMD1:2008+AMD2:2012 Low voltage surge protective devices – Part 21: Surge protective devices connected to telecommunications and signalling networks – Performance requirements and testing methods.

IEC 61643-22:2015 Low-voltage surge protective devices – Part 22: Surge protective devices connected to telecommunications and signalling networks – Selection and application principles.

IEC 61643-31:2018 Low-voltage surge protective devices – Part 31: Requirements and test methods for SPDs for photovoltaic installations.

IEC 61643-32:2017 Low-voltage surge protective devices – Part 32: Surge protective devices connected to the d.c. side of photovoltaic installations – Selection and application principles.

IEC 61643-311:2013 Components for low-voltage surge protective devices – Part 311: Performance requirements and test circuits *for gas discharge tubes* (GDT).

IEC 61643-312:2013 Components for low-voltage surge protective devices – Part 312: Selection and application principles for gas discharge tubes.

IEC 61643-321:2001 Components for low-voltage surge protective devices – Part 321: Specifications for avalanche breakdown *diode* (ABD).

IEC 61643-331:2020 Components for low-voltage surge protection – Part 331: Performance requirements and test methods for metal oxide varistors (MOV).

IEC 61643-341:2020 Components for low-voltage surge protection – Part 341: Performance requirements and test circuits for thyristor surge suppressors (TSS).

IEC 61643-351:2016 Components for low-voltage surge protective devices – Part 351: Performance requirements and test methods for telecommunications and signalling network surge isolation transformers (SIT).

IEC 61643-352:2018 Components for low-voltage surge protection – Part 352: Selection and application principles for telecommunications and signalling network surge isolation transformers (SITs).

Telcordia GR-974-CORE, Issue number 04, 2010 Generic Requirements for Telecommunications Line Protector Units (TLPUs).

UL 1449, Edition 4 Standard for Safety for Surge Protective Devices.

Testing techniques and test equipment:

Report on Critical investigation on the impulse current specification for test class I in IEC 61643-1 and type 1 in EN 61643-11 when testing Metal Oxide Varistors, *Hubert Bachl/Thomas Dötzl – Cooperative Testing Institute Vienna (CTI-Vienna)*, 2007.

Testing of Low Voltage Surge Protective Devices today and in the near future, *Hubert Bachl – Cooperative Testing Institute Vienna (CTI-Vienna), Highvolt colloquium, Dresden* 2011.

Impulse generators used for testing low-voltage equipment, *Michael J Maytum, SPDC Tutorial,* 2012.

Ju-Hong Eom, Sung-Chul Cho, and Tae-Hyung Lee. Parameters Optimization of Impulse Generator Circuit for Generating First Short Stroke Lightning Current Waveform, *J Electr Eng Technol* Vol. 9, No. 1: 286–292, 2014.

IEC 60060-1:2010, Edition 3.0 (2010-09-29) High-voltage test techniques – Part 1: General definitions and test requirements.

IEC 60060-2:2010, Edition 3.0 (2010-11-29) High-voltage test techniques – Part 2: Measuring systems.

IEC 60060-3:2006, Edition 1.0 (2006-02-07) High-voltage test techniques – Part 3: Definitions and requirements for on-site testing.

IEC 61000-4-5:2014 +A1:2017 Electromagnetic compatibility (EMC) – Part 4-5: Testing and measurement techniques – Surge immunity test.

IEC 61083-1:2001, Edition 2.0 (2001-06-11) Instruments and software used for measurement in high-voltage impulse tests – Part 1: Requirements for instruments.

IEC 61083-2:2013, Edition 2.0 (2013-03-20) Instruments and software used for measurement in high-voltage and high-current tests – Part 2: Requirements for software for tests with impulse voltages and currents.

IEC 61083-3 (currently 42/380/FDIS) Instruments and software used for measurement in high-voltage and high-current tests – Part 3: Requirements for hardware for tests with alternating and direct voltages and currents.

IEC 61180:2016 High-voltage test techniques for low-voltage equipment – Definitions, test and procedure requirements, test equipment.

IEC 62475:2010, Edition 1.0 (2010-09-29) High-current test techniques – Definitions and requirements for test currents and measuring systems.

IEEE C62.45:2002 Guide on surge testing for equipment connected to low-voltage AC power circuits.

ITU-T K.44 (10/2019) Resistibility tests for telecommunication equipment exposed to overvoltages and overcurrents – Basic Recommendation.

ITU-T Guide on the use of the overvoltage resistibility Recommendations, 2012.

Chapter 7

Application rules

Alain Rousseau[1] and Ralf Hausmann[2]

Selection of surge-protective devices (SPDs) and their installation rules are generally not complex. For most of the cases, the path to follow is clear and straightforward. In a few circumstances, there are complexities either for selection of the appropriate SPDs (such as coping with superimposed high-frequency impulses, switching surges, high temporary overvoltages (TOVs)) or more generally during the installation process (lack of space in a panel board, need to install SPDs in another place instead of originally scheduled, difficulties to keep lead length short, etc.). There is then a need to accommodate these complexities by selecting other SPDs (such as with a better voltage protection level) or adding more SPDs to further reduce the overvoltages in front of equipment to be protected. Except in these cases, the rules are simple (it does not mean that the application of these rules is always simple). However, these rules are based on the assumption that protection is efficient. Even if true in a large majority of cases, the application of standard rules may not be sufficient in a few cases. Sometimes (see Sections 8.7 and 8.8), it is necessary to demonstrate protection efficiency. Then tests or calculations are needed and those make the selection process more complex.

Note: Sur-imposed surges may degrade the protection level. For example, IGBTs used for power conversions will, by design, generate harmonics at different frequencies and voltage surges which will also apply to the SPD. To take care of such high values, it is often necessary to increase the SPD continuous permanent voltage U_c and/or its TOV withstand, that will, as a consequence, increase the protective level that may then exceed the requested voltage protection level. This solution can only be used after a careful study of the surge withstand of equipment to be protected.

The most advanced document for AC SPD application is IEC 61643-12. This chapter is primarily based on rules presented in that standard. For other applications, such as DC or PV, there are either other existing standards or specific rules established by the industry. These are detailed in Chapter 8. IEC standards are consistent and SPD application rules are based on other standards describing the type of systems covered by IEC and their application rules. However, other systems exist in the world and the specific parameters of a few of them are also presented in Chapter 8.

[1]SEFTIM, Vincennes, France
[2]Product Marketing Surge Protection TRABTECH, Blomberg, Germany

This chapter will describe as follows:

- Selection of SPD based on main parameters: in many cases, the selection of the SPD at the entrance of installation following these rules will be enough. These rules include installation rules and if these installation rules cannot be fully fulfilled, additional or different SPDs can be necessary.
- Current rating: there are general values for SPD rating but a calculation can be performed especially for Type 1 SPDs as a part of a lightning-protection system (LPS). For Type 2 SPDs, the risk calculation can help in determining the best nominal discharge current I_n.
- Protective distance: unfortunately, except in a few cases, a single SPD at the line entrance will not be sufficient and equipment located at a large distance from the entrance SPD may not be efficiently protected. Other SPDs, nearer to equipment to be protected, are needed.
- Coordination between SPDs: when there is more than one SPD on a line, they need to be coordinated, first to ensure safety and then to ensure protection.
- Coordination with equipment to be protected: an SPD is not a black hole. The surge that enters the SPD will not disappear downstream of the SPD. Not only an overvoltage is generated at the SPD output terminals but also a surge current will continue its way towards equipment to be protected that will depend on the impedance of equipment, SPD's characteristics and impedance between SPD and equipment. This residual surge should be considered for providing efficient equipment protection.
- Selection of SPD disconnector: this is the trickiest part of SPD selection. However, the SPD disconnector selection should come last. An SPD will be operational 24 h a day for many years and the parameters selected earlier will help to provide the best protection. One day, the SPD may fail due to too many surges, a surge too large or other causes such an unexpected switching surge or TOV or for any other causes. When it fails, the SPD should not create a hazard to the installation. This requirement to stay safe at the SPD's end of life is, of course, important but should not challenge the protection efficiency of the SPD along all its operational life.
- Surge-protection philosophy: a summary is presented based on six main points that should be taken into account for easy SPD selection.
- Specific rules for data and signal SPDs: these are based on IEC 61643-22 standard.

7.1   Selection based on main parameters

In this clause, main points will be marked by an arrow ($\rightarrow$). When needed, additional explanations will follow and indented like the arrow.

The selection of SPDs starts with a selection of the SPD located at the entrance of the installation.

What is the purpose of that SPD? When an LPS is installed, all incoming lines should be equipotentially bonded with the LPS earthing system. Type 1 SPDs are used to provide this equipotential bonding function. When the lightning current (assuming a 10/350 waveshape) flows through the earthing system, the earthing voltage will increase. As a consequence, the current will also flow through incoming lines that are referenced to remote earthing systems (these remote earthing systems are at 0 V because they are located at some distance from the structure protected by the LPS and not influenced by the local earthing system voltage rise). For example, in a TT system, the transformer is connected to the utility earthing, which is separate from the structure earthing. Equipment connected both to the incoming line and to the structure earthing will experience a large overvoltage that will create a sparkover in the absence of a Type 1 SPD. Thanks to this sparkover, a part of the lightning current will flow through the incoming line to an earth point where the voltage is still at 0 V. In the absence of the SPD, this partial lightning current would create a fire hazard and mechanical damages. In terms of protection zones (see Section 4.2) these SPDs will determine lightning-protection zone (LPZ) 1. This is the main use of Type 1 SPDs for incoming lines.

For surges that are coming from the incoming lines, SPDs at the entrance are generally of Type 2 and they will ensure that most part of the surges will be directed to the earthing system avoiding large EMC disturbances within the installation.

Note: SPDs Type 1 may be also useful for surges that are coming from the incoming lines, especially in rural areas with overhead lines (see next).

Note: SPDs are able to protect against surges and can tolerate certain power frequency overvoltages (temporary overvoltages). They are generally not able to protect against the power frequency overvoltages. A few devices are able to provide this protection. They are called 'power frequency overvoltage protective devices' (POP) but they are not SPDs and not discussed in the book.

→ SPD at installation entrance should be Type 1 in the case of LPS (even if natural LPS) to protect the structure at the location of the electrical installation and Type 2 when only induced surges are considered.

It is interesting to note that Type 1 SPDs may be used in other cases. For example, when an overhead LV line directly supplies a structure, there is a possibility that a direct strike occurs on the line. Generally, LV standards consider that the last span is more critical but experience with 20 kV lines has shown that up to three spans it may be considered where a significant current still propagate to the structure. Another case that is probably more common is the interconnection of structures from the same supply line. When lightning strikes a structure protected by an LPS associated with Type 1 SPDs, a 10/350-μs impulse current will propagate towards the power utility network. The presence of a Type 1 SPD in front of the LV side of the utility HV/LV transformer may then be useful. Other structures connected to the same power distribution will also experience a partial lightning current stress that can create damage to the electrical installation in these structures. The current will depend on the line length and the various earthing resistances (LPS, transformer, other structures, etc.). Structures located closest to the structure protected by LPS will generally experience a greater partial current. When protecting a facility where there are many structures interconnected, Type 1 SPDs are then necessary at line

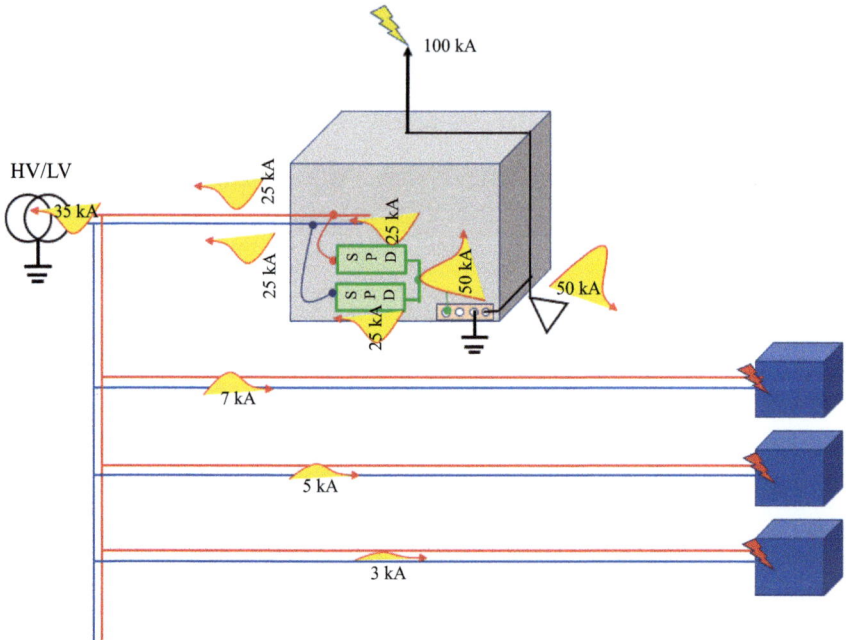

Figure 7.1   Sharing of lightning current between interconnected structures

entrance on structures that are not protected by an LPS, if one of the structures is protected by an LPS in the vicinity (Figure 7.1).

Note: The phenomenon is the same if a structure not protected by an LPS is impacted by a lightning strike, but in that case the affected structure will be damaged.

Case study

An interesting case has been investigated a few years ago that needs to be explained to show that a Type 1 SPD may be needed even if there is no LPS on the structure.

This case is presented in Figure 7.2.

A private house located on a hill top was supplied by a transformer 20 kV/400 V through an underground cable. The transformer was supplied by a 20-kV overhead line on a wooden pole connected down the hill. Wooden poles are known to be an aggravating factor because there is less sparkover at the poles (a surge can then propagate over a larger distance than with a concrete or metal pole). The transformer was protected by a 20-kV surge arrester rated 65 kA 8/20 μs (maximum discharge current). Due to bad soil conditions, a 15-Ω resistance was created by the utility at the transformer

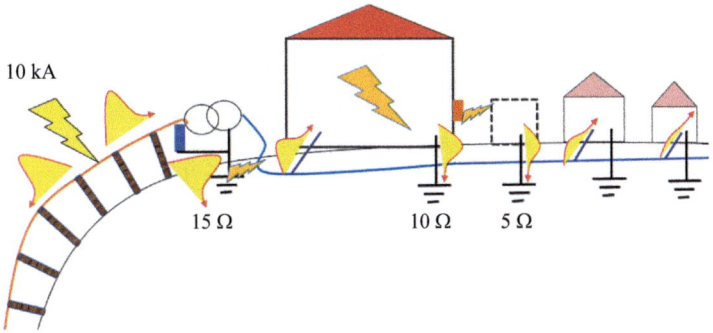

Figure 7.2   Investigated case

location. The private house had a better earth resistance (10 Ω) and a small metal shelter located 2 m from the wall had a 5-Ω resistance earthing system. No LPS was present on the roof and no SPD was installed.

The initial lightning current was investigated to be 10 kA 10/350 μs. It shared between the transformer circuit, the supply line and the sparkover at each pole. This current flowed through the 20-kV surge arrester. Due to relatively bad earthing resistance at the transformer, the generated voltage was large enough to sparkover to the LV underground cable. Thus, a partial lightning current flowed through this LV cable and reaches the first connected house. At this level the consumer panel was severely destroyed. The surge continues its way inside the installation, leading to cables to be pulled down from the ceiling where cables were running in plastic tubing. Sparkover occurred in many power sockets creating a fire in the kitchen. Thankfully the house owner was at home and could stop the fire quickly. All electronics equipment (including telecom and TV) were destroyed (Figures 7.3–7.5).

There was a waterproof external power socket outlet on the wall facing the metal shelter (the shelter was not supplied by power) and the partial lightning current made an unexpected jump in air (2 m) to reach the shelter and escape by the shelter earthing. The outdoor wall socket was destroyed during that jump.

Other houses connected to that line experience damages but to a lesser extent (only a few electronics equipment failed): either due to a longer line length, higher earth resistance or both.

A single Type 1 SPD at installation entrance would have avoided most of these damages except electronics equipment that would have needed additional Type 2 SPDs.

It must be noted that the utility circuit breaker (CB) was an RCD (residual current device) that would have failed in any case due to the fact that Type 1 SPD could only be installed only downstream of the RCD because the utility rules do not allow the installation of SPDs in front of the RCD. The

Figure 7.3   Consumer panel

Figure 7.4   Kitchen power socket

Figure 7.5   Cabling pulled down from the ceiling tubing

house would have then lost its power supply (and possibly its fridge content) but most of the significant damage would have been avoided. If the owner of the house was not present during this event, the fire could have spread and the house burned down.

→ SPD at installation entrance should be installed as near as possible to the installation entrance to avoid the detrimental effect of large surge currents flowing inside installation.

Ideally, they should be connected at the interface between outside and inside. This exists for coaxial SPDs (panel feedthrough) but not for power lines. Thus, SPDs should either be outside or when inside, downstream the entrance point. Located outside, for example on a power distribution pole, is a good option if SPDs remain near installation entrance (the line between the SPD and installation entrance should remain protected by this SPD, so distance much smaller than 10 m). For overhead lines, this means either an outdoor SPD which limits the choice or installing a box for accommodating a din rail mounted SPD. For underground cables, SPD has no other choice but to be located near or even at the main panel board. Generally, the structure's installation starts inside the structure after the meter and the main overcurrent protective device (CB). The power utility rarely accepts to have SPDs installed before the private installation starts. Then, in most of the cases, SPDs will be installed downstream of the overcurrent protection device (OCPD) and in the main panel board (MPB) wherever it is installed. This MPB may be far away from the point where the cable enters the structure for practical reasons. An extreme case encountered was an MPB located on the third floor of structure. This requires that a surge coming from an external line will flow over a large distance before being treated by the first SPDs. If current is high, a large magnetic disturbance will be created along the path and in all circuits in vicinity. Primarily, the circuits routed in the same conduit will be disturbed. This should be taken into consideration in the EMC plan.

Note: An overcurrent protection device (OCPD) is used in power installations to cover the risk related to overload, short circuit or ground fault.

SPDs at line entrance are either Type 1 or Type 2. They are determined mainly by their surge capability. For Type 1, it is given by impulse discharge current I_{imp} when for Type 2 it is given by nominal discharge current I_n. The usual value for I_n is 5 kA per pole whereas the usual value for I_{imp} is 12.5 kA per pole. Other values may be used or need to be used (see Section 7.2). The current magnitudes given earlier only apply to SPD connected between earth and active conductors (equi-potential bonding function).

→ Type 1 SPDs at line entrance are generally rated $I_{imp} = 12.5$ kA per pole or more when Type 2 SPDs are generally rated $I_n = 5$ kA per pole or more.

Note: Surge current values given per pole are valid for SPDs connected between earth and active conductors. As it will be indicated later, SPDs can also be connected between phase and neutral and between neutral and earth. In that case, the value given previously remains true for SPDs connected between phase and neutral but for SPDs connected between neutral and earth, the rating will be multiplied by the number of phases (current that is coming from lines or escaping from the earth can flow through each phase but also through the neutral thus the current is summed) (Figure 7.6).

Note: In general, each line connected to another structure need to be protected by a Type 1 SPD. However, when there is a multiplicity of external lines connected

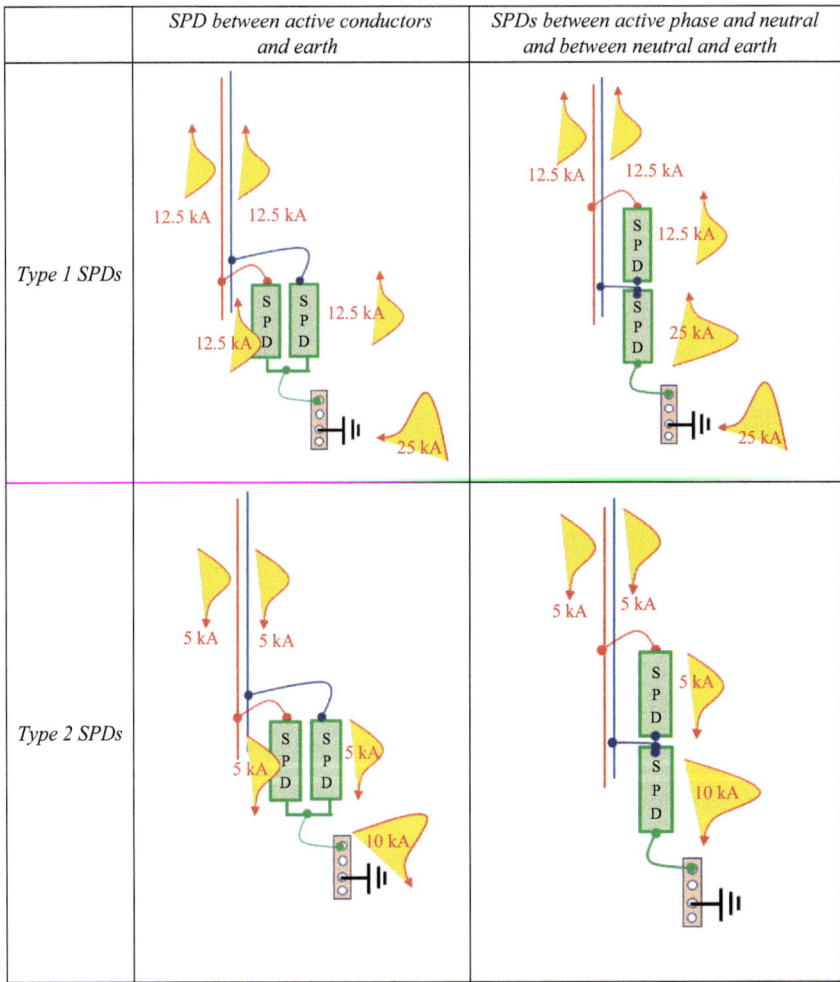

Figure 7.6   Sharing of surge current as a function of SPD connection type – example of a single-phase system

to the same distribution panel board, this solution may be difficult to apply due to lack of space or other causes. It could be tempting to protect only the panel board with a single Type 1 SPD. This can provide basic safety but will not necessarily provide the expected protection. For example, when each line is supplied by a circuit breaker, an SPD in front of each CB will protect the CB as well as the panel. Installing a single SPD on the main distribution bus bar will protect the panel but not protect each CB. In the case of surge on one line, the CB supplying this line may be damaged and the power supply for this external line lost. In addition, for large panel boards connecting various external equipment, a surge coming from a line will propagate towards the bus bar SPD but can also share between the various external lines, depending on the earth resistance of each equipment. A single SPD connected to the main bus bar may not be able to protect all the external lines.

Another surge current was popular for Type 2 SPDs: I_{max}. This current is the maximum discharge current that the SPD was able to handle. It is now an optional value. It may be declared or not. When it is not declared, it should be assumed that I_n is the maximum capability of the SPD. Often, I_{max} is equal to two times I_n. The residual voltage at I_{max} is generally not declared.

Note: I_{max} has been removed from mandatory parameters in 61643-11 standard because it was leading to a horsepower race where the biggest was assumed to be the best. However, a large surge current is often connected to poor protection levels. Current as high as 100 kA 8/20 μs or even more could be found in the market and users were misled by the assumption that using the biggest SPD they would protect well their installation.

Of course, SPDs located at the entrance should survive the permanent voltage and also most of the TOVs, especially the TOVs due to fault within the LV system. For the permanent voltage this will fix the maximum continuous operating voltage U_c that is in general 10% above the line to neutral voltage (the line to neutral voltage is usually the line-to-line voltage of the low-voltage system divided by $\sqrt{3}$).

For TOVs, we need to consider TOV due to fault within the LV system and TOVs due to fault within the HV system. According to IEC 61643-11, SPDs have to withstand TOVs due to fault in the consumer LV installation. However, it is possible to select SPDs that will not withstand the stress but will fail safe (non-violent destruction) in the case of a fault on the distribution system, loss of neutral or fault in the HV installation.

SPDs need to be adapted to the system configuration (TN, TT or IT) and to the TOVs (mainly those of moderate amplitudes generated by the LV system itself). For that purpose, U_c and U_T need to be selected in accordance with Table 7.1.

→ In practice, the indication of the system configuration on which the SPD can be used (TT, TN or IT) appears clearly on data sheets and the user no longer needs to refer to the U_c voltage nor to U_T.

Note: A few countries impose in their national code that SPD should withstand all LV faults. This may seem sensible, because poor handling of temporary over-voltages is one of the main causes of SPD failure. In addition, temporary over-voltages are often more frequent than lightning surges.

Table 7.1   Typical U_c and TOV values for SPD on 230/400 V system with 10% voltage tolerance[*]

Type of system	TNC	TNS	TT	IT
U_c				
Phase-PEN	255	–	–	–
Phase-neutral	–	255	255	255
Neutral-earth	–	230	230	255
Phase-earth	–	255	255	440
Phase–phase	440	440	440	440
TOV (U_T 5 s)				
Phase-PEN	335	–	–	–
Phase-neutral	–	335	335	335
Neutral-earth	–	–	–	–
Phase-earth	–	335	440	–
Phase–phase	–	–	–	–
Maximum stress (U_c, U_T 5 s)				
Phase-PEN	335	–	–	–
Phase-neutral	–	335	335	335
Neutral-earth	–	230	230	255
Phase-earth	–	335	440	440
Phase–phase	440	440	440	440
TOV (LV distribution fault and loss of neutral 120 min/HV fault 200 ms)				
Phase-PEN	440/–	–	–	–
Phase-neutral	–	440/–	440/–	440/–
Neutral-earth	–	–	–/1,200	–/1,455
Phase-earth	–	440/–	335/1,455	–/1,640
Phase–phase	–	–	–	–

Note: TOV may have duration up to 200 ms for fault in HV system, 5 s for fault in consumer installation and 120 min for fault in distribution system or in the case of loss of neutral. TOVs that have a duration exceeding 5 s represent a stress that is considered as a permanent voltage for SPDs (mainly for SPD based on MOV, this does not apply to gapped SPDs).

Note: Selecting an SPD with a $U_c = U_T$ is a temptation to cover all the stresses that may occur. This is possible when considering faults in the consumer system but in the case of faults in the distribution system or loss of neutral and TOV dues to fault in HV system, this may lead to a bad level of protection (U_p). It is then necessary to check that the voltage protection level is still compatible with the surge withstand of installation. Alternatively selecting an SPD that will fail safe in a few fault case may only be considered when application does not require continuity of protection. The most stringent case is generally coming from TOVs having a duration up to 120 min because it is much longer than lightning surges. If a

[*]IEC 61643-12 ed.3.0. Copyright © 2020 IEC Geneva, Switzerland. www.iec.ch.

significant current flows through the SPD in that particular case then it is unlikely that the SPD will withstand.

The next and fundamental question is related to the protection provided by this SPD (even though the SPD at the entrance has a primary function to divert the primary stress and provide equipotential bonding, they of course also provide surge protection). This is characterized by the voltage protection level U_p.

This voltage may appear sometimes confusing. It is the voltage that is maintained between SPD terminals during fix surges conditions. It is based on sparkover voltage (for gaps) and also on residual current (for varistors). However, this value will be degraded by the installation rules. Unless the SPD is connected to the MPB with lead conductors of negligible impedance, these conductors will add a residual voltage that can add to the residual voltage of the SPD. In addition, for Type 2 SPDs the voltage protection level is defined at I_n when a higher surge current can occur that will not degrade the SPD (e.g. I_{max}). At a higher current, the residual voltage may be higher than the voltage protection level.

The first criterion for U_p at the line entrance is to be lower than the rated impulse voltage according to overvoltage category II (i.e. $U_w = 2.5$ kV for 230/400 V system). In the case of category I equipment (reduced impulse withstand), a lower U_p could be selected. However, we will see that very often, one SPD is not enough and protection for such equipment can be provided by a downstream SPD.

The second criterion is to keep as much as possible, a 20% safety margin between U_p and U_w. This safety margin is necessary due to

- current greater than I_n (but lower than I_{max} when declared),
- voltage drop along the lead conductors depending on the lead length: 50-cm rule (see next),
- voltage drop due to the SPD disconnector when it is in the SPD branch of the electrical installation.

A better parameter than U_p is $U_{p/f}$ (effective voltage protection level) that will include the voltage drop along the lead conductors and, when necessary, the voltage drop due to the SPD disconnector. U_p is based on the SPD manufacturer declaration (can be found in the data sheet and on the SPD nameplate). $U_{p/f}$ is determined by the electrical contractor based on U_p. However, voltage drop due to the SPD disconnector and the residual voltage at I_{max} is rarely known and a margin between U_p and U_w is then always recommended (Figure 7.7).

Note: Generally, an inductance of 1 μH/m is assumed for conductors. In the case of a surge, the current through the conductor (inductance L) will generate a voltage equal to L × di/dt, where di/dt is the front of the wave. Assuming a current of 10 kA and a waveshape 10/350 μs, a simplified calculation leads to a voltage drop along 1 m of conductor = 1 μH/m × 1 m × 10 kA/10 μs = 1 kV. In general, a voltage drop of 1 kV/m is assumed. If current is greater or steeper, the voltage drop can be larger. However, due to approximations made, this value can be considered general enough for most applications. It is furthermore generally accepted that for MOV-based SPDs this voltage drop due to lead conductors will add to the voltage protection level (even if the voltage drop and residual voltage generally do not

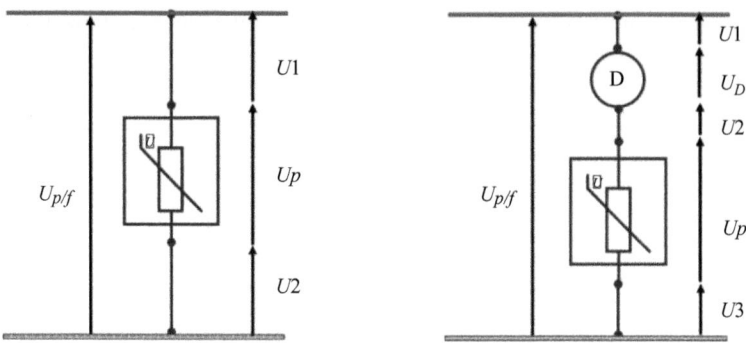

Figure 7.7   *How installation will degrade the protective effect of SPD (D is the SPD disconnector)*

reach their peak value at the same time). For gapped type SPDs, the situation is a bit different. Until the gap operates, there is no current flow and therefore no voltage drop along the conductor. As soon as the gap reaches its sparkover level, the current can flow but the voltage at the SPD terminals decreases to the arc voltage plus the wiring effect inside the SPD. The total voltage is generally much lower than U_p plus the voltage drop along the lead conductors. For combination SPDs (SPDs combining many surge-protection components as, for example, a GDT in series with an MOV), the effect may be more complex to evaluate and a good practice is to sum U_p with the voltage drop due to lead conductors.

Note: Disconnectors may be included inside the SPD, or external to the SPD. In that case, they are specified by the manufacturer based on type tests it has performed. This external disconnector may be in the SPD branch or in series with the line. The SPD branch is in parallel to the line. When a disconnector is located in the branch, it is expected that this disconnector will operate first and leave the line alive even if unprotected. When the disconnector is located in series with the line, the operation of that disconnector will disconnect both the SPD and the pro-tected loads. In the past it was called priority to surge protection vs. priority to power supply. As it will be seen later on (see Section 7.6), this concept is rarely valid in practice and a compromise needs to be reached.

Of course, such a margin is not needed for SPDs:

- gapped type, for which there is no residual voltage, just an arc voltage that is much lower than U_p;
- having the SPD disconnector built-in allows this voltage drop to be included in the U_p;
- connected to MPB with conductors of negligible impedance.

Note: Surges lower than the sparkover voltage of a spark gap-based SPD will stress the insulation of the device to be protected. It is then important to reduce the voltage protection level of gapped SPD as much as possible.

When a margin is needed and cannot be maintained, protection will be degraded and alternative solutions are needed:

- use an SPD with a smaller U_p,
- reduce lead length (see next, 50-cm rule),
- use an SPD with a built-in disconnector or install the SPD disconnector in the main installation circuit and not in the SPD branch,
- install an SPD downstream to further reduce the protective level.

→ Install an SPD with U_p between active conductors (phases and neutral) and earth lower than the rated impulse withstand of category II: 2.5 kV for a 230/400-V system.

→ Keep a margin of 20% between U_p and U_w ($U_p \leq 2$ kV for a 230/400-V system). When not possible, another SPD will be needed downstream. In a few cases, this margin is not needed but, unless one is sure to be in one of these exceptions, it is wiser to keep a margin (use an SPD with a smaller U_p) or install another SPD downstream to be in the safe side.

It is recommended to keep the total lead length to a maximum of 50 cm to meet the 20% margin rule. It is often assumed that this rule known as the 50-cm rule will cover all needs and cases.

If an SPD has a voltage protection level of 2 kV and the lead length is 50 cm (thus 500 V assuming 1 kV/m) then $U_{p/f} = 2.5$ kV $= U_w$.

But when providing protection for sensitive equipment, $U_w = 1.5$ kV with 20% margin would result in U_p not exceeding 1.2 kV. If the total lead length is 50 cm (500 V when assuming 1 kV/m), then $U_{p/f} = 1.7$ kV $> U_w$. To keep a 20% margin, it is necessary to limit the lead length to only 30 cm. Limiting the total lead length to 50 cm is already a challenge especially in large panel boards, so it was decided to generalize the '50 cm rule'. However, for sensitive equipment, it its frequent that additional SPDs will be needed, installed downstream of entrance SPD.

Note: It is difficult for IT system configuration to find SPDs with a voltage protection level equal to 1.5 kV or less because of the high value needed for U_c and the severe TOVs. The range of U_p values is usually between 1.8 and 2.2 kV and can be as high as 2.5 or 2.8 kV. In that case, it is needed to either install a second SPD downstream that will reduce U_p in front of sensitive equipment thanks to protection level coordination or to keep the residual voltage at 5 kA lower than 1.5 kV even if U_p is higher (in most of the cases where current is lower than 5 kA, protection will be efficient) or a combination of both solutions.

→ Follow the 50-cm rule by using lead length as short as possible. If not possible, use alternative methods to maintain $U_{p/f}$ below the maximum acceptable value (U_w).

Generally, SPDs are located at the top of the panels where the line enters and is distributed. It is often possible to keep the conductor length between active conductors and SPD disconnector short. It is called $L1$. A 10-cm length restriction is reasonable to keep conductors short and as straight as possible, avoiding the SPD conductors

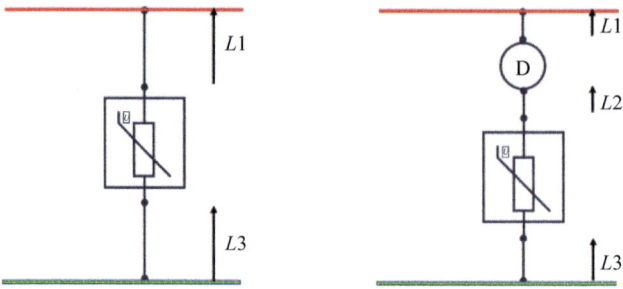

Figure 7.8   Connecting lead lengths L1, L2 and L3 with or without external SPD disconnector in the SPD branch (D is the SPD disconnector)

following the other conductors along the din rail path (efficient SPDs wiring is rarely nice looking because conductors should go straight and then visible). It is recommended to install the SPD right beneath the SPD disconnector and not on the same rail as the disconnector because it will avoid loops and will reduce lead length. A 10-cm length is also possible for that (called $L2$). If there is no SPD disconnector in the SPD branch (either no disconnector is needed, there is one built-in or there is an SPD disconnector in series with the line), the only length to consider is between active conductors and SPD terminals ($L2 = 0$). This allows the remaining 30–40 cm to reach the earth bar ($L3$) that is generally located at the bottom of the panel. In general, it is impossible to keep 50 cm for the total length due to the large dimensions of an MPB where the earthing bar is at more than 1 m from the SPD location (Figures 7.8 and 7.9).

Many possibilities are offered to keep this distance short or to use alternatives (they are not sorted by importance or easiness and are all valid provided they meet the goal to reduce $U_{p/f}$):

- Alternative 1: select an SPD with a lower U_p: if $U_w = 2.5$ kV and SPD $U_p = 1.5$ kV, there is a margin of 1 kV. Based on 1 kV/m, this can be transferred into a total 1-m lead length. This is easier to achieve than 50 cm, even if not always enough.
- Alternative 2: use a V-connection on active conductors and/or on earth (PE, protective earth) (Figure 7.10). When a disconnector is needed it is more difficult to do it for active conductors but when no disconnector is used in the SPD branch, it is possible to connect the line to the SPD terminal in and out. Many SPDs provide this facility with two terminals interconnected inside the SPD, especially at earth terminals. In that case $L1 = 0$ and $L3 = 0$. Of course, this is only possible for small ratings load currents because SPD can only accommodate limited conductor sizes. For an MPB using bars, it is difficult to apply this solution.
- Alternative 3: install another SPD downstream of the entrance SPD, to use the combination effect of two SPDs to reduce the voltage at equipment terminals. Of course, this solution is only valid when there is no sensitive equipment in the MPB.
- Alternative 4: use an intermediate earth bar in the MPB to reduce $L3$ (Figure 7.11).

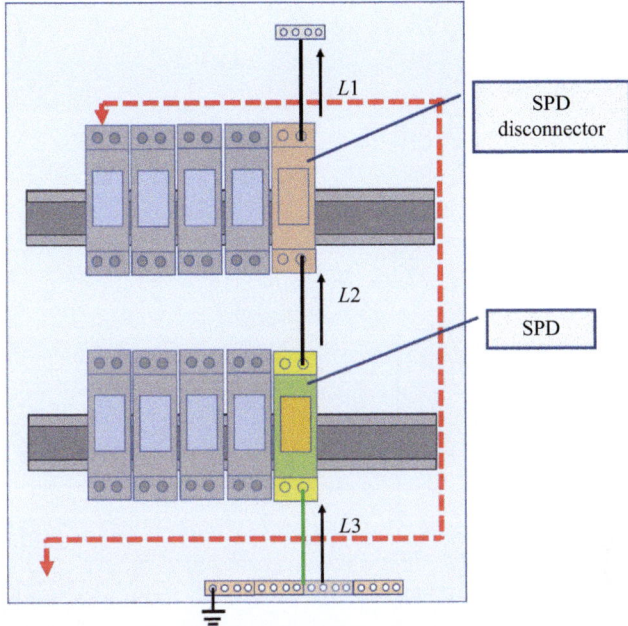

Figure 7.9   View of a panel with an SPD disconnector located on a din rail above the SPD. Lead conductors are straights and do not follow path of other cables (red dotted arrow)

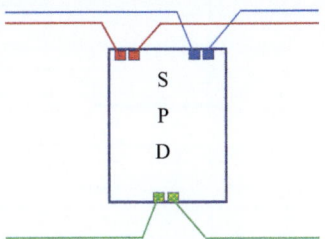

Figure 7.10   'V'-connection both side (active conductors and PE conductor)

This solution is often misunderstood and should be explained in detail for an appropriate use. The basis is simple: everywhere a surge current flows in a conductor (high frequency leads to voltage drop in inductances) in a conductor, there is a voltage drop (typically 1 kV/m). If no current flows, there is no voltage drop. Let us apply this basic principle. Figure 7.12 shows where the current flows and where it does not flow (indicated by 0 even if it is not 0 but a limited current).

Equipment connected to the MPB main earth bar using a PE (green/yellow) conductor is referenced to this main earth bar but the earth bar does not experience

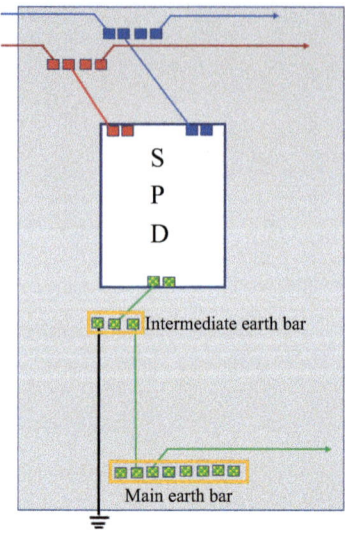

Figure 7.11   Intermediate earth bar principle

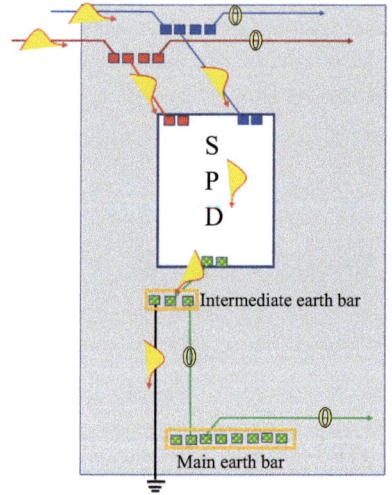

Figure 7.12   Flow of surge current in the case of an intermediate earth bar

any surge current. Thus, the reference voltage of the equipment is not only the main earth bar but also the intermediate earth bar. The voltage the equipment will experience will be between the phase (or neutral) distribution point and the intermediate earth bar. Then $L3$ is measured between points C and D and $L1$ between points A and B. It is likely that the earth bars will be exposed to a high voltage during the surge flow compared to that on the earthing system of the structure, but

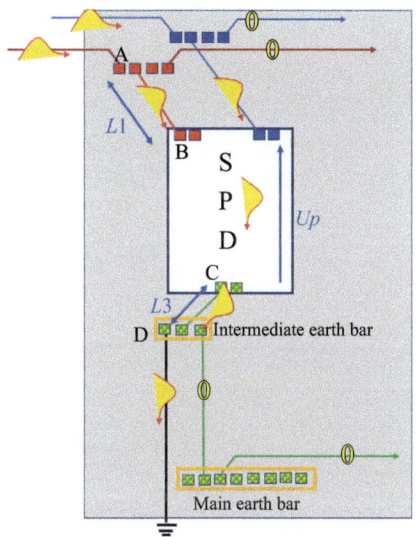

Figure 7.13   Determination of L3 in the case of an intermediate earth bar

it is not important because the SPD and equipment are referenced to the same equipotential location (Figure 7.13).

Note: This rule may be partially true when some equipment is connected to the PE conductor on one side and to the grounded metal frame of the structure on the other side. In such a case, the voltage drop between point D and E is not equal to zero and should be considered. In fact, the reference of the equipment is point F and the voltage drop between E and F should also be considered when investigating the protection at equipment terminals. Hopefully, in most of the cases, the inductance along D–E–F-earth of the structure is large (long conductors) and the current flowing through it will be quite reduced. The voltage drop is no longer 1 kV/m but much less (most of the current will flow through the shorter path between point D and earth of structure) and is generally negligible (Figure 7.14).

- Alternative 5: use the low-inductance effect of the metal frame of the MPB to reduce $L3$: in this case the surge current is separated in two parts. The front of the impulse current where frequencies are high, leading to high $L \times di/dt$ and the tail of the waveform where the frequencies are low but associated with large energies. The efficiency of the SPD is then split into two: first the front of the wave where $U_{p/f}$ is the main parameter and after the tail of the wave where the surge withstand (e.g. I_{imp}) is crucial. A metal plate (frame of the panel board) has a low inductance so the voltage drop will be negligible. As far as energy is concerned, it is wiser to provide a direct path thanks using the PE conductor between the SPD earthing terminal and the main earth bar of the panel (this will also allow 50-Hz fault current, in the case of SPD failure, to flow directly to earth). Provided the metal plate is connected to PE and meets the requirements

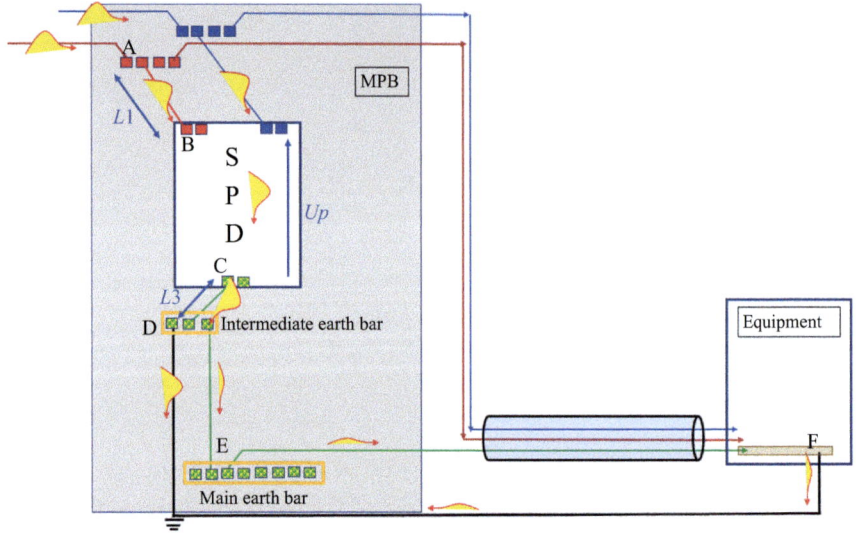

Figure 7.14   Influence of direct grounding of equipment enclosure on the voltage drop along L3

for a protective conductor, connecting the SPD earthing terminal to the main earth bar of the panel and to the metal frame will help fulfil the 50 cm rule. For the tail of the waveform (and fault current) the current will mainly flow through the copper conductor between the SPD earthing terminal and the main earth bar of the panel, but this has no effect on $U_{p/f}$ (Figure 7.15).

Note: Earthing and grounding are similar terms that describe the system of conductor in direct contact with soil to provide a reference voltage for equipment and for the structure. Depending on the country one term or the other is preferred. A grounded equipment is connected to a reference point that is itself connected to the earthing system. The green/yellow conductor (PE conductor) is used to provide this electrical connection. Equipment can also have its metal enclosure connected to the earthing system through a bare copper cable.

Note: The link to the metal plate can be direct, provided the conductor and the terminal have at least a cross-section area of 16 mm^2 for Type 1 and 6 mm^2 for Type 2 and provided a PE conductor between the SPD earthing terminal and the main earth bar of the panel has been installed. It is also possible to connect the SPD earthing terminal to the din rail thanks to a specific terminal, provided the din rail is connected to the metal frame through to a mechanical fixing of the appropriate size. In that case, the SPD should be located as close as possible to this connection point between din rail and metal frame because tests have shown that the inductance of the din rail is not as low as the inductance of the metal plate.

- Alternative 6: use low-inductance connecting cables to reduce *L3*: cables have been specifically designed (e.g. by the French Alternative Energies and

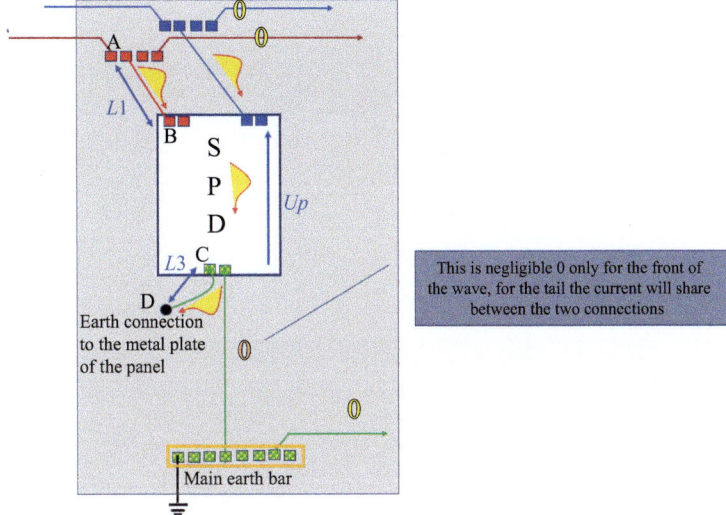

Figure 7.15   *Determination of L3 when the metal plate of the panel is used as an additional PE conductor*

Atomic Energy Commission (CEA)) to reduce the inductive effect of a surge current, allowing much longer cabling length (the less inductance, the lower the voltage drop; much smaller than 1 kV/m).

   Note: Multiple conductors have been used in the past to reduce L3. If one conductor of 1 m is equivalent to 1 µH, we can expect to reach an inductance of 0.5 µH by using two conductors in parallel. But this approach is only valid if the coupling between these conductors and the metal frame of the panel can be reduced by keeping them far away from each other. In practice, the benefit of such a solution has never been demonstrated for panel boards.

- Alternative 7: use a bar to connect the SPDs instead of a conductor (in fact use a bar for the longest path and, of course, a small conductor, to connect the bar to the SPD terminal). Experience has shown that a bar, if large enough, has a much lower inductance than a conductor (much less than 1 µH/m). Minimum width for the bar should be 5 cm to allow inductance reduction. Wider bar will lead to lower inductance.

When there is no space in a panel for installing an SPD, it is possible to install it in another cabinet located upstream to the panel to be protected but at a small distance from the panel (less than 10 m) according to Figure 7.16. Of course, this scheme is easier to use when the cross-section area of panel board conductors is small.

When conductor cross-section area is wider, it is possible to use the following scheme but the protection provided is rarely adequate due to too long connecting conductors (*L*1 and *L*3 in Figure 7.17).

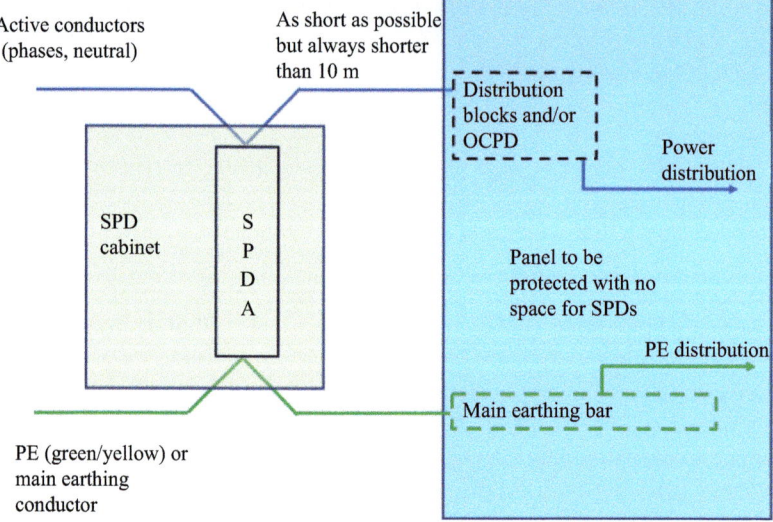

SPDA = SPD with its external SPD disconnector (when required)
OCPD = overcurrent protection device

Figure 7.16   External SPD cabinet for small panel board conductors

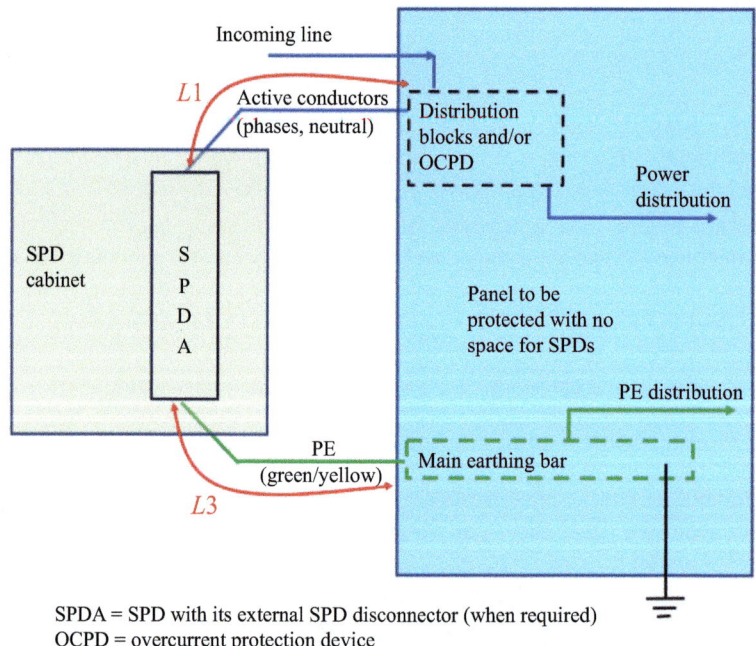

SPDA = SPD with its external SPD disconnector (when required)
OCPD = overcurrent protection device

Figure 7.17   External SPD cabinet for bigger panel board conductors

Note: When the panel board and SPD cabinet are both metallic, provided the metal plates are connected to the PE in a manner that meets the requirements for a protective conductor, and the panel and cabinet structure are continuous, it is possible to reduce L3 using the SPD cabinet as an additional conductor as presented earlier.

The protection efficiency is related to the connection type. As already said, what is mandatory for safety is to protect between active conductors and earth. However, this is not where equipment to be protected is the most sensitive. For example, Class II equipment has no earth conductor and therefore the impulse withstand voltage between active conductors and earth is generally large. Between phase and neutral, internal circuitry of the equipment will have generally a weaker impulse withstand voltage. To mix these two needs, protect for safety between active conductors and earth and protect for operational purpose between phase and neutral; a connection type for SPDs has been developed and is nowadays quite popular: it is called CT2 as opposed to the older scheme that is called CT1 (Figure 7.18).

Note: Protection between phase and neutral is often called differential mode protection where protection between active conductors (phase and neutral) and earth is generally called common mode. CT2 is also colloquially called 3+1 due to the shape of protection and by analogy CT1 is then called 4+0.

CT1 offers only safety protection (insulation breakdown) while CT2 combines both. To achieve the same protection with a CT1 connection type, three more SPDs are needed between each phase and neutral (in the case of a three-phase plus neutral system). However, CT2 may also present a drawback. U_p between phase and earth is not always given as it is a combination of two SPDs. In that case, if U_p between phase and earth is not declared by the manufacturer, it is needed to sum the U_p between phase and neutral and U_p between neutral and earth (both of them being declared). In such a case, if each U_p is 1.5 kV for a 230/400-V system (a rather good value because 2.5 kV maximum is required), then U_p between phase and earth is 3 kV, which is not acceptable (>2.5 kV). A more in depth reading of the nameplate or technical brochure is sometimes needed.

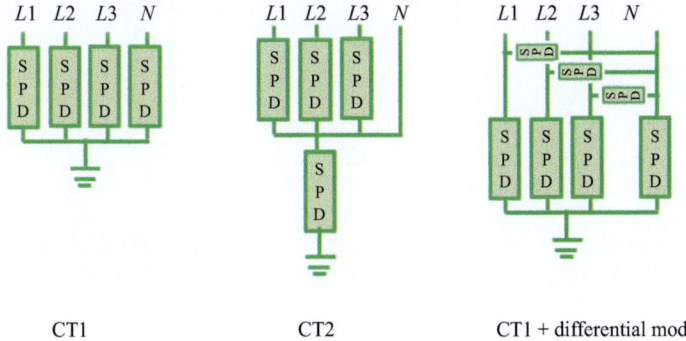

Figure 7.18   Connection types

Note: SPDs are defined by their modes of protection, i.e. where protection is provided and declared by the manufacturer. A manufacturer may declare only phase–neutral and neutral–earth as modes of protection and thus in that case, the phase–earth voltage protection level is not declared.

→ Generally, a CT2 connection scheme provides a more efficient global protection. However, all the parameters and especially the voltage protection level between phase and earth should be considered.

Surge protection can be based on single-pole SPDs that the electrical contractor will combine to provide the requested protection using a CT1 or CT2 scheme. Surge protection may also be provided using a multipole SPD where elementary surge-protection components are directly combined (CT1 or CT2) by the SPD manufacturer. An SPD, either unipolar or multipolar contains, in addition to protective components, other components such as wiring, internal disconnectors, fault indicators, contact for remote information, etc. When an SPD disconnector is needed in addition to the ones already included, they are specified by the SPD manufacturer (at least the maximum rating allowed) but need to be provided and installed by the contractor. The group of an SPD and its disconnector, when located in the SPD branch, is called an SPD assembly. When the SPD disconnector is not located in the SPD branch but in series with the line, it is generally an OCPD already installed and then not called an SPD assembly. Unipolar SPDs or multipolar SPDs may be combined with their needed SPD disconnector in an enclosure to facilitate wiring and keep the lead lengths short (Figure 7.19).

Finally, the SPD end of life should be considered. The short-circuit current rating I_{SCCR} of the SPD shall be greater or equal than the maximum prospective short-circuit current at the MPB. This can be obtained by the SPD itself, by an SPD

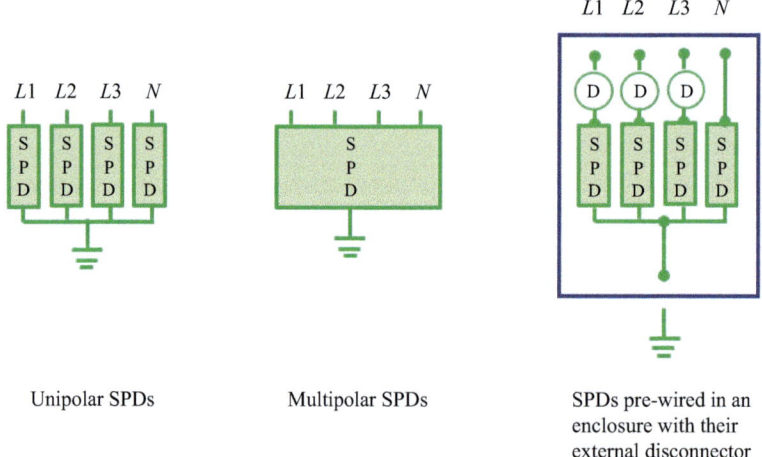

Unipolar SPDs   Multipolar SPDs   SPDs pre-wired in an enclosure with their external disconnector

Figure 7.19   *Various types of SPDs used to provide protection in a three-phase plus neutral TT system*

disconnector located inside the SPD or outside of the SPD but in the SPD branch or upstream in series with the line.

For SPDs that have a power follow current (current supplied by the power system flowing through the SPD after a surge current, e.g. for a gapped type SPD), the capacity to cut this power follow current (called I_{fi}) should also be greater than or equal to the maximum prospective short-circuit current at the MPB.

Note: These rules regarding I_{SCCR}, I_{fi} and prospective short-circuit current do not apply to SPDs connected between neutral conductor and earth in TN or TT systems. For SPDs connected between the neutral conductor and earth in IT systems, I_{SCCR} and I_{fi} shall not be lower than the maximum prospective short-circuit current in the case of a double earth fault.

→ The short-circuit current rating I_{SCCR} of the SPD and when appropriate the power follow current interrupting rating I_{fi} shall be greater than or equal to the maximum prospective short-circuit current at the MPB.

When considering direct lightning impact on a structure and thus Type 1 SPDs, a differentiation must be made between different cases:

1. Line entering the structure connected to another remote structure or a power utility.
2. Additional line entering the structure connected to equipment directly located on the structure (roof or façade) that cannot be impacted by direct lightning because protected by the structure's LPS.
3. Additional line entering the structure connected to an equipment directly located on the structure (roof or façade) that can be impacted by direct lightning or that is too close (less than separation distance) to the LPS (for a structure equipped with an LPS) (Figure 7.20).
4. Additional line entering the structure connected to equipment located nearby at ground level and equipotentially bonded with the structure that cannot be impacted by a direct strike (protected by the structure or its LPS) (Figure 7.20).
5. Additional line entering the structure connected to equipment located nearby at ground level that can be impacted by a direct strike (Figure 7.21).

In case 1, the situation has been described earlier. The lightning current will share between the lightning earthing system and the power line. SPDs at line entrance need to be Type 1.

$I_{imp} = 12.5$ kA per pole, in general. A more precise calculation often needs to be performed. U_p is generally ≤ 2.5 kV for Type 1 and ≤ 1.5 kV for Type 1+2 (230/400 V system).

In case 2, another line supplies equipment on the roof or the façade. This equipment and its connecting line cannot be impacted directly by lightning. There is no need to install Type 1 SPDs at the line entrance. However, equipment and its cabling are subjected to a large magnetic stress (LPZ 0_B) that will induce surges on the external part of the circuit. Type 2 SPDs are needed at line entrances in the structure and at the equipment entrance (if equipment needs to be protected), unless

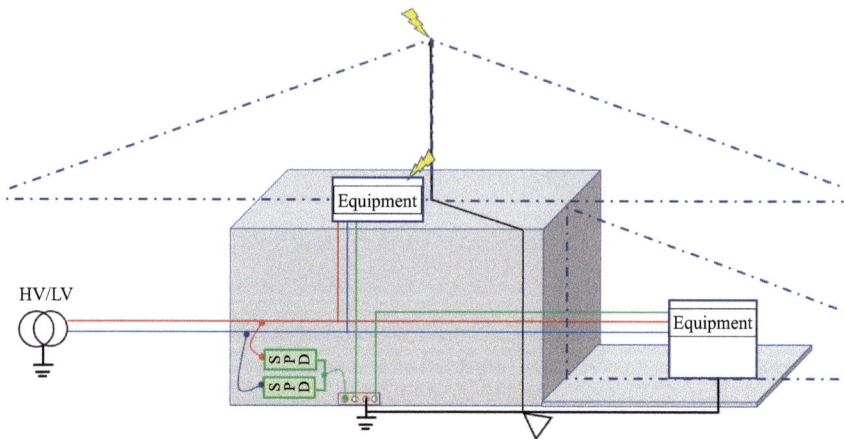

Figure 7.20   Case of a structure protected by an LPS and an equipment on the roof too close to the LPS circuit (less than separation distance) and nearby equipment protected by the structure

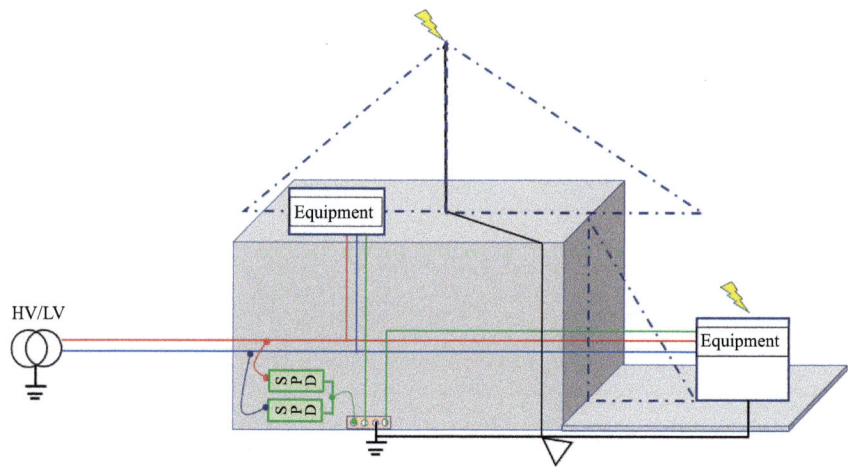

Figure 7.21   Case of a structure protected by an LPS and an equipment on the roof at a distance from the LPS circuit greater than separation distance and nearby equipment not protected by the structure

line length on the roof is short (much less than 10 m). This SPD is not necessary if the cabling on the roof is shielded and grounded on both sides (at structure entrance and at equipment) or if the line runs in a metal conduit grounded at both ends. Directly induced surges on equipment also need to be considered (e.g. if equipment is not in a metal grounded enclosure or if a part of the internal cabling is running outside the metal enclosure). When the cable running outside of the structure is short (typically less than 1 m), Type 2 SPDs can be omitted only if the cabling does not form loops.

$I_n = 5$ kA per pole or higher is needed, U_p generally ≤ 1.5 kV (230/400-V system).

In case 3, equipment on the roof or the façade is supplied by the structure. Equipment or its connecting line can be impacted directly by lightning. If they cannot be impacted but are too close to the LPS circuit (less than separation distance), a sparkover may occur between this equipment and the LPS, resulting in a partial lightning current flowing in the connecting line. In both cases, a Type 1 SPD is needed at the line entrance and Type 1+2 at the equipment entrance (if equipment needs to be protected), unless the line length on the roof is short (much less than 10 m). In the latter case, a single Type 1+2 may fit. This SPD can be avoided if the cabling on the roof is shielded and grounded both sides (at the structure entrance and at the equipment) or if the line runs in a metal conduit grounded both ends. The shield or metal conduit needs to be able to withstand the partial lightning current (or full current in the case where the equipment is not protected against direct lightning strike, but it is very rare to have equipment on the roof not protected when an LPS is present for the structure).

$I_{imp} = 12.5$ kA per pole, in general. A more precise calculation often needs to be performed. U_p is generally ≤ 2.5 kV for Type 1 and ≤ 1.5 kV for Type 1+2 (230/400-V system).

Note: IEC 62305-3 provides a calculation method for determining the voltage drop along the shield resistance and acceptable current limits.

In case 4, equipment near the structure is located on a platform that is equipotentially bonded to the structure. This equipment is connected to the structure by a power line (a pump, an air conditioning device or an external power generator). The equipment and its connecting line cannot be impacted directly by lightning. They are at the same potential as the structure (equipotentiality) and thus it is not an incoming line (to be considered an incoming line, the source of the line needs to be from another structure with a different earthing system. A line for which the source is from another part of the same structure is not considered an incoming line because the two ends are at the same potential). Type 1 SPDs are not needed at line entrance.

However, the equipment and its cabling are subjected to a large magnetic stress (LPZ 0_B) that will induce surges on the external part of the circuit. Type 2 SPDs are needed at line entrance of the structure and at the equipment entrance (if equipment needs to be protected), unless its line length is short (much less than 10 m). This SPD is not required if the line is shielded and grounded on both sides (at structure entrance and at equipment) or if the line runs in a metal conduit grounded on both ends. Directly induced surges on equipment also need to be considered (e.g. if equipment is not in a metal grounded enclosure or if a part of its internal cabling is running outside the metal enclosure).

$I_n = 5$ kA per pole or higher is needed, U_p generally ≤ 1.5 kV (230/400-V system).

Note: It is considered that the earthing system is densely meshed and acts as an equipotential plane. For poorly meshed earthing systems or earthing system not meshed at all – earth loops or local earthing pits – or for large structures (e.g. wider than 40 m), this may not be always true. The voltage drop along the earthing system will depend on the lightning injection points and need to be analysed with simulations tools. In that case, a line connecting two parts of the same building or an equipment on an equipotential platform may need to be protected by Type 1 SPDs but with a much lower rating than incoming lines.

In case 5, equipment near the structure is located on a platform that is equipotentially bonded to the structure. This equipment is connected to the structure by a power line (a pump, an air conditioning device, or a back-up energy source such as an external power generator). The equipment or its connecting line can be impacted directly by lightning. Type 1 SPDs are needed at line entrance and Type 1+2 at equipment entrance (if equipment needs to be protected), except is line length supplying that equipment is short (much less than 10 m). In that later case a single Type 1+2 may fit. This SPD is not required if the cabling is shielded and grounded on both sides (at structure entrance and at equipment) or if the line runs in a metal conduit grounded on both ends. The shield or metal conduit needs to be able to withstand the partial lightning current. The lightning current will share between the Type 1 SPDs or the shield and the bonding conductor to structure's earthing system.

I_{imp} = 12.5 kA per pole, in general. A more precise calculation often needs to be performed. U_p is generally ≤ 2.5 kV for Type 1 and ≤ 1.5 kV for Type 1+2 (230/400-V system).

It is often difficult to install these additional SPDs (Type 1, Type 1+2 or Type 2) because there is no space to install them inside equipment and generally no panel available outside (the line comes directly from the MPB or secondary panel located inside the structure). Installing these SPDs only in a secondary panel or in the MPB located inside the structure may not avoid sparkover at equipment (SPDs are too far from equipment) but will merely avoid damages in these panels. A large current will propagate inside the structure and thus creating disturbances inside what was supposed to be a clean area. Coupling between these lines connected to external equipment with other lines and especially lines supplying sensitive equipment needs to be addressed with care.

Note: The case of an HV/LV transformer inside a structure protected by an LPS need to be addressed. In general, when the transformer neutral earthing is bonded to the LPS earthing system, there is no possible path for partial lightning current to flow. The flow of current through the windings of the transformer (for separate windings transformers) can only be done by capacitive or inductive coupling and only high frequencies will be transmitted (20%–30% coupling is generally assumed). The only problem that can occur is the sparkover of the LV side of the transformer (the transformer being connected to the earth of the structure a high voltage will be applied to the windings). Thus Type 2 SPDs are needed at the transformer LV side for avoiding this sparkover and also to limit the surges transmitted from the HV system. However, as soon as an LV line from the transformer supplies an external equipment bonded to another earthing system, this line needs to be protected by a Type 1 SPD (incoming line). If there is no Type 1 SPD in a panel supplying that equipment in the protected structure, then the SPD at the LV side of the transformer needs to be Type 1. In the same way, in the rare case where the transformer neutral would not be bonded to the LPS earthing system, Type 1 SPD should be needed on the LV side of the transformer.

Case study

A structure protected by an LPS includes various SPDs (Type 1, Type 2 and so on). The power-line circuit and the data-line circuit are nearby and form a loop. A study of the influence of a direct strike to the LPS and the current injected in the power line is needed. The loop area is determined precisely on site (2.4 m^2). The structure is protected by an LPS level of protection I. There are two LPZ: the structure itself and an inner room coated with metal sheets. The magnetic field induced inside the structure can be evaluated using formulas given in IEC 62305-4 for the cases of a first positive stroke, first negative stroke and subsequent strokes. It appeared that due to the tight mesh forming the skeleton of the structure, the magnetic field inside the structure remains below 7 A/m in LPZ 1, which is negligible for the equipment. In LPZ 2, it is smaller than 1×10^{-3} A/m which is negligible for that zone. A loop of a conductor inside LPZ 2 was evaluated, using the criteria of IEC 62305-4 and a maximum induced voltage generated in the loop was found to be 15.5 V that is negligible for the 48-V data system and for the power supply port in the zone.

To evaluate the disturbance caused by surge flowing through the power line to the data-line circuit, the current injected downstream of the T2 SPD installed at LPZ 2 entrance must be evaluated. This current $I_{residual}$ depends on the stress injected on the line that itself depends on the LPS level of protection and the number of connected lines. It also depends on the coordination between the T1+2 SPD located at the entrance and the Type 2 SPD. For that purpose, electrical model software was used, based on data provided by the SPD manufacturer (what was existing in the SPD data sheet was not detailed enough to allow an accurate calculation). The mutual inductance M between the power and data circuit has been evaluated. Knowing M and $I_{residual}$ it was possible to calculate the voltage generated in the loop. This voltage was found to be 44.1 V much larger than the induced voltage generated by direct lightning current in the LPS. However, this voltage remains below the surge withstand of a 48-V system.

This case demonstrates the importance of evaluating the disturbances generated by 10/350 injected current, coming from a line connected to equipment located on the roof. There is then a significant risk of disturbance of the most sensitive equipment.

Two examples are presented in the following to demonstrate cases where SPDs are needed (see Figures 7.22 and 7.23).

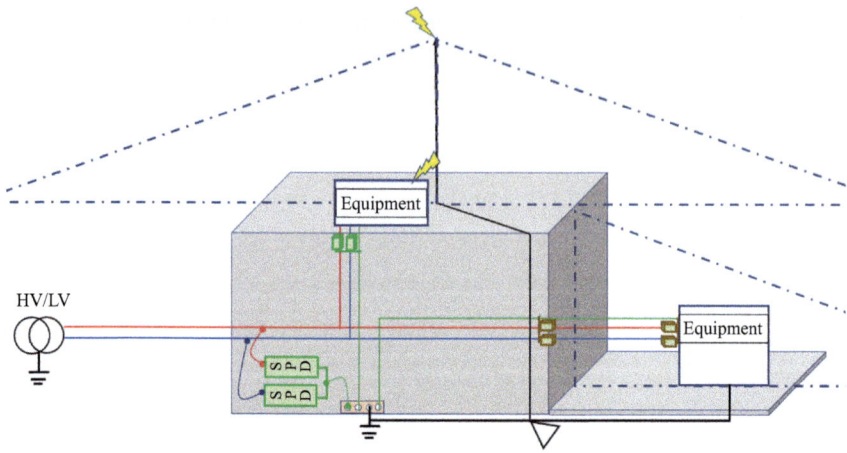

Figure 7.22   *(See Figure 7.20.) Additional Type 1 or Type 1+2 SPDs are represented by green boxes and Type 2 by brown boxes. It is assumed equipment on roof is connected with a short cable when equipment on the platform is located at more than 10 m of structure and need to be protected due to its important role*

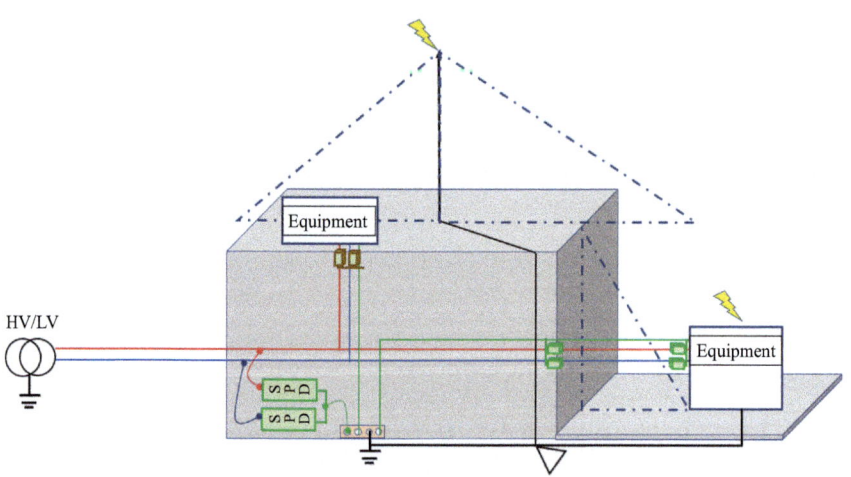

Figure 7.23   *(See Figure 7.21.) Additional Type 1 or Type 1+2 SPDs are represented by green boxes and Type 2 by brown boxes. It is assumed that equipment on roof is connected with a short cable when equipment on the platform is located at more than 10 m of structure and needs to be protected due to its important role*

→ Additional SPDs are needed for all lines entering the structure even if located on the roof, the façade or near the structure and bonded to it. The type of SPD Type 1, Type 1+2 or Type 2 will depend on the case (LPS or not, separation distance maintained or not, equipment protected against direct lightning or not) and importance of equipment. Shielded lines or line routed in metal conduit can be used in place of SPDs if specific requirements are met.

→ These additional SPDs should at least be located near line entrance in the structure. Where these SPDs are not located at the line entrance but instead in a panel inside the structure (especially if it is the main panel), specific care should be taken for potential coupling between this affected line and sensitive circuits inside the structure.

Protection at the line entrance (in the MPB) is then provided. As indicated earlier it is likely that this SPD will not be sufficient to protect the complete installation:

- $U_{p/f}$ may be greater than requested for equipment category II.
- Equipment category I may exist in the installation that requires better protection.
- Higher surges (e.g. I_{max}) could degrade the protection level.
- Equipment to be protected is too far away to be effectively protected by the entrance SPD (more than 10 m of cable away from front SPD, called the 10 m rule).
- Entrance SPDs providing only the mandatory protection between active conductors and earth.

Note: When the distance between the SPD and the equipment to be protected is greater than 10 m, the voltage at the equipment terminals may be as large as twice $U_{p/f}$.

There is generally no need to protect each piece of equipment within the installation. Basically, what need to be protected is preferentially determined by a risk assessment process. If not, sensitive equipment, expensive equipment and equipment having a safety function should be protected with first priority. Protection can be obtained by other SPDs located in secondary panel boards and if necessary, in front of the equipment. One SPD in a subsidiary panel can protect multiple pieces of equipment. It is good practice to protect a UPS because circuits supplied by a UPS are by nature important and protecting one UPS generally helps protect many pieces of equipment.

The SPDs used to protect equipment are Type 2 or Type 3 SPDs. How these SPDs are determined?

There is one simple rule to apply for determining the characteristics of these SPDs: they need to be coordinated with the SPD at the service entrance.

Note: Coordination means two things – from one side, energy coordination (the oldest concept) and from the other, the protection level coordination. It is necessary to pay attention, because energy coordination is related to safety and as such considered mandatory while protection level coordination is generally seen as an option. The fact that protection level coordination is a new concept helps propagate confusion on the need of such coordination. Basically, it is enough to understand the importance of protection level coordination thanks to two aspects:

- *Generally, two SPDs that are coordinated with protection level are also coordinated with energy sharing. Protection level coordination is a more stringent concept.*
- *Energy coordination is for protecting secondary SPDs when protection level coordination is for protecting equipment.*

Energy coordination means that the stress between two SPDs is shared in such a way that none of them are destroyed by an excess of energy (the energy injected in each SPD remains below their energy withstand).

Protection level coordination means that the voltage protection level of the secondary SPD is not exceeded when the entrance SPD is submitted to its maximum acceptable stress. In addition, this coordination can lead to a reduction of the residual voltage of the secondary SPD thanks to a current lower than its nominal discharge current flowing through the secondary SPD (the voltage protection level is partly determined by the residual current at the nominal discharge current, if the current flowing through the SPD is lower than I_n, the residual voltage is lower than U_p, in the same way that when the current is equal to $I_{max}>I_n$, the residual voltage is greater than U_p).

By using coordination rules, the user simplifies its specification for down-stream SPDs and needs only to specify the SPD at entrance and what is the target in terms of voltage protection for the other SPDs. These SPDs should additionally be specified by other system parameters (single phase, three phases with or without neutral distributed, system configuration: TN TT or IT, prospective short circuit current, etc.). Other parameters such as I_n and I_{max} are not needed for downstream SPDs. Their voltage protection level also needs to be fixed but what is more important is the voltage needed to protect the subsidiary panel or equipment. If equipment has a rated impulse voltage U_w equal to 1.5 kV, an SPD with a voltage protection level $U_p = 1.5$ kV may not fully protect due to phenomena presented earlier: lead length influence, distance between SPD an equipment, disconnector voltage drop, etc. To keep a 20% margin, the U_p of this SPD should be equal to 1.2 kV that is sometimes difficult to find in the market especially for IT systems. If, thanks to protection level coordination, it can be demonstrated that the voltage at the SPD location is 1.2 kV (lower than what is written on the nameplate), this SPD can be accepted thanks to coordination rules. Another example; a Type 2 SPD has a nominal discharge current $I_n = 20$ kA. Its voltage protection level is 1.8 kV. It is based on MOV and due to the non-linear curve of this SPD, residual voltage at 5 kA is 1.2 kV. Once again, this SPD that apparently did not fit, will do the job if coordination rules show that the current through the secondary SPD is lower than 5 kA. Of course, coordination rules need to consider the worst case conditions for the entrance SPD and be based on the cabling distance between MPB and the secondary panel, entrance and secondary SPD's lead length. If coordination rules provided by the SPD manufacturer do not allow such a demonstration, there is no other choice but to request at 5-kA nominal discharge current I_n and $U_p = 1.2$ kV for secondary SPDs ($U_p = 1.5$ kV may be acceptable if the cabling length is short, $\ll 50$ cm) and the disconnector voltage drop negligible.

Note: In general, the voltage drop due to disconnectors (and lead length to some extent) is mainly to be considered for SPDs located at entrance where the front of the wave is steeper. When the surge propagates downstream the installation, the front becomes slower and the voltage drop has less influence.

→ Additional SPDs are generally needed because SPDs at the entrance cannot protect a complete installation. Rules for these SPD selections are similar but secondary SPDs should be coordinated with the entrance SPD (protection level coordination) and will be Type 2 or Type 3.

It may appear to be a good idea to install an SPD with a lower voltage protection level U_p at the installation entrance to provide protection for downstream equipment and avoid downstream SPDs. In the case of Type 1 SPDs, such an SPD is generally called Type 1+2 in the market or Type 1+Type 2: a single SPD tested according to Type 1 tests and Type 2 tests. More generally, it will also describe a Type 1 SPD having a voltage protection level compatible with sensitive equipment (e.g. 1.5 kV or less for 230/400 V systems). However, the protective distance is generally short (the 10-m rule indicates that after 10 m, protection is no longer ensured) and such an SPD will not be able to efficiently protect all sensitive equipment in medium-to-large installations. For small installation, such as a telecom shelter for mobile telecommunication, such SPDs are very efficient. Even for large installations, they will reduce the surge-propagated downstream of the front SPD and is also quite interesting. It is cautioned that, in general, to install only one SPD, although a Type 1+2 SPD, will not be enough:

- when the installation cables are longer than 10 m;
- when surges can occur from inside the installation (switching surges or equipment connected to an external equipment on the roof or façade);
- when induced surges can be generated directly inside the installation.
  - → Using a Type 1+2 at installation entrance instead of a Type 1 is generally a good idea but will rarely prevent the use of additional SPDs downstream in the installation.

Note: Type 1 SPDs tend to be replaced in the market by Type 1+2.

The mandatory protection is for safety and thus between active conductors and earth. Why differential mode protection needs also to be considered (between phase and neutral)?

In many cases, the surges due to lightning are common mode surges. In the case of a direct strike to an overhead line, phases and neutral will experience almost the same surge current. In the case of an induced surge, the loop formed by phase conductors and ground is almost the same as the loop between the neutral conductor and the generated voltage equivalent. A differential mode may appear if the neutral is grounded. A differential mode may also appear when the SPDs at the entrance of the installation on phases and on neutral are different for a CT1 connection type or because a scheme CT2 is used. In a TT system configuration, it is also possible to generate a differential mode surge (between a phase and neutral), even when a common mode protection is present (between phase or neutral and earth). The HV/LV transformer's neutral is connected to a low earthing resistance (typically between 5 and 15 Ω). If the installation's earthing resistance is a much higher value (such as 30 to hundreds Ω), the surge current will preferentially flow from a phase to earth terminal through the phase SPD and from there back through the neutral SPD. In the worst case scenario, the voltage between phase and neutral is approximately the sum of the level of protection of each of the common mode SPDs and in general, twice the level of protection. This justifies the need for the installation of SPDs between phase and neutral for the TT systems. Similar phenomena can also occur in a TNS system configuration, when the distance between the common PE-neutral point and the SPD exceeds a certain length. It is considered today that when this distance is greater than 50 cm, an SPD between neutral and the PE is necessary. This means that in general an SPD between neutral and PE is needed for TNS systems (Figure 7.24).

Generally, as soon as the neutral conductor is distributed by the utility and grounded, there is a differential mode whether for direct strikes or for an induced surge (because most of the surge current will flow through the neutral). Furthermore, SPDs are neither exactly the same between neutral-PE and between phase-PE, either by design or construction (manufacturing tolerances) leading to a differential mode surge. Providing surge protection between phase and neutral (differential mode DM) is then generally needed. In considering a Type 1 SPD at the entrance, the absence of such an SPD would lead to a 10/350 impulse between phase and neutral. If a Type 1 SPD is not installed, the stress will appear on equipment terminals. For Type 1 SPDs, it is then recommended to also install a Type 1 SPD between phase and neutral at line entrance in addition to the mandatory active conductors to earth SPDs.

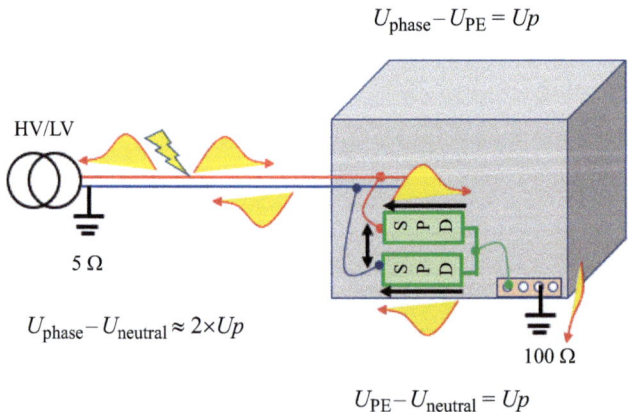

$$U_{phase} - U_{PE} = Up$$

HV/LV

$5\,\Omega$

$$U_{phase} - U_{neutral} \approx 2 \times Up$$

$100\,\Omega$

$$U_{PE} - U_{neutral} = Up$$

Figure 7.24   Need of a differential mode protection in a TT system

Note: Pure symmetrical system (such as three phases without neutral) protected by a Type 1 SPD with tight tolerances (based on gaps for which the arc voltage will be similar) may not need a Type 1 SPD between phase and neutral.

To provide this protection between phase and neutral (where equipment is more sensitive to surges), it is possible to apply many protection schemes:

- Differential mode protection at the MPB (using CT2 connection type or differential mode+common mode protection: CT1+DM) for entrance SPD and differential mode protection at secondary panel. This applies for T1 and T2 SPDs (Figure 7.25).
- Differential mode protection at the MPB (using CT2 connection type or differential mode+common mode protection: CT1+DM) for entrance SPD, if the distance between the MPB and equipment is short (less than 10 m). This applies for T1 and T2 SPDs.
- Common mode protection at the installation entrance (CT1) and differential mode protection (DM) only at equipment level. This applies mainly for Type 2 SPDs, because the differential mode SPD may not be able to withstand the differential mode stress when entrance SPD is Type 1 (Figure 7.26).
- Common mode protection at the installation entrance (CT1) and differential mode+common mode protection (CT1+DM) at the secondary panel. This applies mainly for Type 2 SPDs, because the differential mode SPD may not

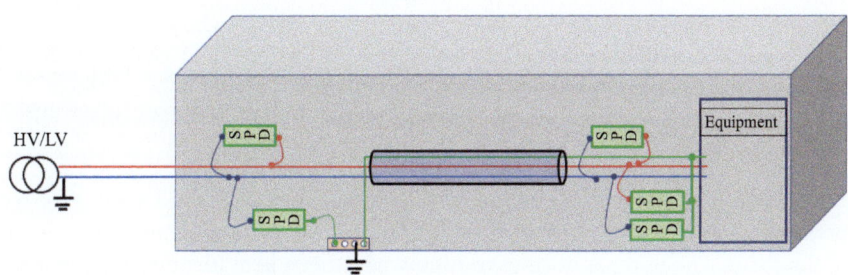

Figure 7.25   Case where differential mode protection is provided at entrance (scheme CT2) and also in front of equipment (CM+DM)

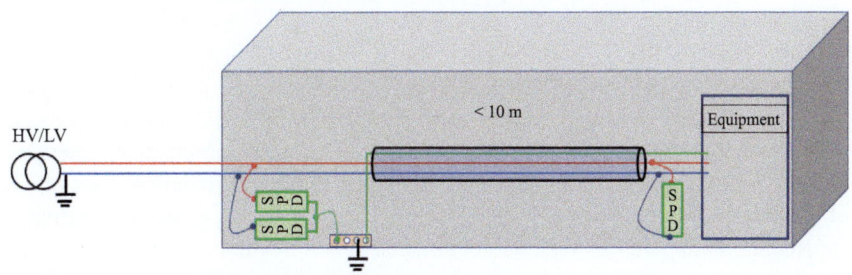

Figure 7.26   Case where CT1 scheme is used for entrance SPD and differential mode protection is provided only in front of equipment (T2 SPDs)

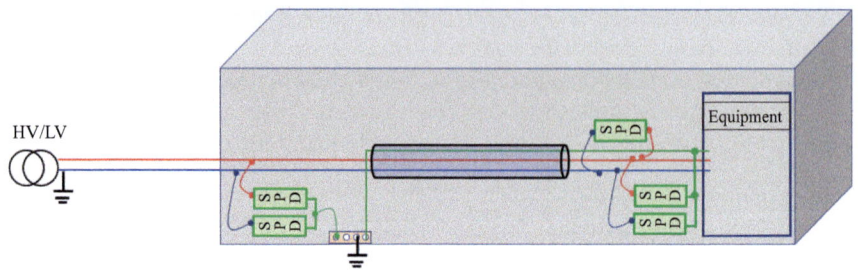

*Figure 7.27   Case where CT1 scheme is used for entrance SPD and
common + differential mode protections are provided in a secondary
panel (T2 SPDs)*

be able to withstand the differential mode stress when entrance SPD is Type 1
(Figure 7.27).

→ When SPDs are Type 1 at the installation entrance, it is recommended to
install also Type 1 SPD in differential mode or use a CT2 connection type.
When only Type 2 SPDs are used at installation entrance, differential mode
protection (DM) may be provided only at secondary panels but remains
necessary to protected sensitive equipment.

→ All additional SPDs (other than at the MBP) should also meet the
general rules: be adapted to the system voltage and configuration, be
adapted to the number of active conductors and be able to resist to the
prospective short circuit current at the location where they will be
installed.

In summary, it may appear complex because many cases (and many pages) are
presented but following the arrows (→) as a guideline will be enough to solve most
of the cases. The number of rules to follow have been kept to a minimum but there
are many possible cases. Most of the practical cases have been presented and the
user should be able to identify their own case amongst them and if necessary,
expand the rules to their own cases.

Application of these rules will generally lead to a satisfactory result in terms of
surge protection.

SPD end of life is a complex matter but as previously said, even if it is
important that in 20 years or more the SPD fails safe, it is much more important that
the SPD efficiently protects during these 20 years. End of life generally relies on
SPD disconnector selection but this does not fit in this sub-chapter, where only
general and simple rules are presented. Selection of SPD disconnector became one
of the most complex parts of SPD selection, mainly due to misunderstanding and
conflicting statements in various documents. When needed, the user has to select
the disconnector recommended by the manufacturer or follow the rules that are
proposed in the manufacturer application guides. For more experienced users,
Section 7.6 provides additional details on this subject. It is enough at this stage to

understand that using an inappropriate SPD disconnector may degrade the surge-protection plan and in the worst case the end-of-life mode.

Note: At the end additional rules need also to be applied:

- *SPD should be designed to work in the installation conditions (IP code, pollution index, temperature range, mounting method, etc.).*
- *Lead conductor acceptable gauge depends on the manufacturer declared minimum and maximum values. In general, 16 mm² for Type 1 SPDs, 6 mm² for Type 2 SPDs and 1 mm² for Type 3 SPDs are the minimum requested values for the SPD to earth lead conductors.*
- *Specific rules exist for two ports SPDs (refer to IEC 61643-11 and IEC 61643-12).*
- *The total permanent current flowing to earth should be limited to avoid unnecessary operation of overcurrent devices or other protective devices such as RCDs (in general the leakage current for gapped SPD is almost nil when for MOV-based SPDs, it remains very low. However, the cumulative current from many SPDs, added to potential other products in installation such as filters, may lead to RCD tripping and should be considered. The permanent current to earth should never exceed the RCD IΔn/3 taking into account the SPD active components and any potential additional circuits such as fault monitoring inside the SPD).*

7.2 Current rating

7.2.1 Type 2 SPDs

The current rating for Type 2 SPDs is based on the nominal discharge current I_n and on a 5-kA minimum value. The maximum discharge current a Type 2 SPD can withstand is not a required parameter. When it is declared, it is often between one to two times I_n.

For MOV-based SPDs the residual current at I_n determines the voltage protection level of the SPD. By increasing I_n, the same SPD will provide a larger U_p, degrading the level of protection than that provided for a smaller I_n. Conversely, for an SPD that has a U_p value equal to 2.2 kV and a nominal discharge current of 20 kA, the residual voltage at 5 kA would be lower (e.g. 1.5 kV). This is a primary reason the two values U_p and U_{res} (5 kA) are given in manufacturer's catalogues. This means that the SPD is robust (I_n greater than 5 kA) but at the same time, the protection offered at 5 kA would be much better than the published U_p (Figure 7.28).

Note: For Type 2 SPDs based on MOV technology, there is only one U_p value that is defined at the nominal discharge current (other SPDs using gaps, for example, have U_p defined by the front-of-wave sparkover voltage and combination SPDs with gap and MOV are tested with both residual voltage and front-of-wave sparkover voltage, see Chapter 6). Often, SPD catalogues give various values for U_p at different values of current (e.g. 1.0, 1.10, 1.30 and 1.45 kV at, respectively, 1, 5, 10 and 20 kA). There are residual voltages at various values of current. The

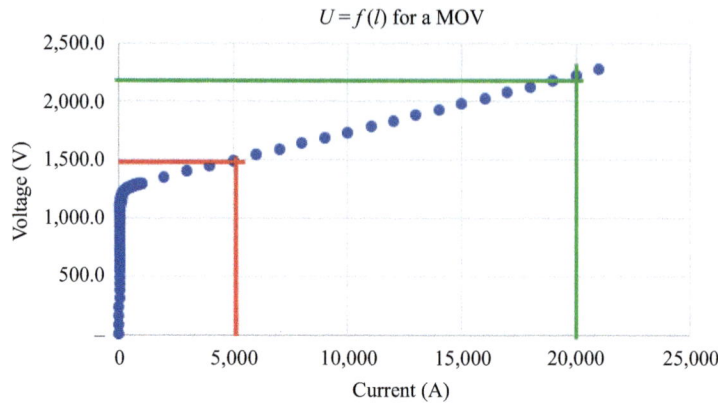

Figure 7.28   Influence on the current on the residual voltage and U_p for an MOV-based SPD

published U_p is defined at I_n and in this example would be 1.45 kV. The value at 5 kA allows easy comparison with other SPDs. The values given for the residual voltage allow plotting a curve that will be helpful, especially for coordination between SPDs and between SPD and equipment to be protected.

At the line entrance, the minimum nominal discharge current is 5 kA. For other places inside the installation, the nominal discharge current will be determined by energy coordination or protection level coordination rules. In general, it is good practice to compare voltage protection levels of various Type 2 SPDs at the same nominal discharge current I_n (typically at 5 kA).

It is more difficult to compare with Type 3 SPDs. As previously indicated, the current flowing through the SPD during the test with a combination wave generator depends on the impedance of the SPD under test. If it was a short circuit, the current would be equal to $U_{oc}/2$ Ω and the waveshape 8/20 μs. But with the impedance of the SPD (MOV based) that is low but not equal to zero, the current flowing through the SPD will generally be lower. The residual voltage will also be lower along with the declared U_p. For a Type 2 SPD, the current I_n is injected through the SPD during the test and thus, the maximum residual voltage at I_n (called U_p) and nominal discharge current are known and published in the SPD data sheet. For a Type 3 SPD, what is generally published is the current I_{cw} and not the current flowing through the SPD during the tests. Comparison becomes more difficult.

What is the purpose in using a larger I_n than 5 kA when only the minimum value of 5 kA is requested? A nominal discharge current greater than 5 kA will give an indication of the energy withstand and consequently gives an indication of life expectancy of the SPD. However, there is no easy way to determine by calculation what the most appropriate value of I_n should be at a certain location. The expected lifetime of an SPD depends on the probability of occurrence of surges exceeding the maximum discharge capability of the SPD. As I_{max} is related to I_n, a larger I_n will, in general, lead to a larger I_{max}. In practice, the SPD lifetime will depend on many other parameters such as switching surges or TOVs. It only relates to a prospective lifetime.

Note: Field experience has shown that a nominal discharge current $I_n = 5$ kA would cover most of the needs for an LV system. For example, the French utility EDF published a paper at CIRED in 1999 regarding their field experience 20 kV surge arrester with $I_n = 5$ kA and I_{max} 40 kA 8/20 µs (65 kA 4/10 µs): About 800,000 polymeric surge arresters were installed on EDF distribution networks. The electrical failures (energy overstress of the SA) represented less than 0.005% a year for 7 years of experience. This is negligible. These SAs have been installed on overhead lines and much more subjected to surges than LV SPDs in buildings. In addition, most of the SA failures are not due to lightning but to other causes (TOVs, environment, mechanical damage, etc.): five failures due to other causes for one failure due to lightning. France has an average lightning activity typical of Western Europe. Of course, more severe areas could need bigger SPDs. In addition, the situation is different between a power utility that has to use hundreds of thousands of SA and single structure owner that will use tens of SPDs.

There is no published rule for helping select a different value for I_n. We propose Table 7.2 for use either with the simplified risk method (see Chapter 3) or the complete method (IEC 62305-2, when there is no need to install Type 1 SPDs at the line entrance). For the simplified method, I_n depends on the calculated risk level (CRL) value and for the complete method I_n depends on P_{SPD} needed to reduce the risk.

To show how to use this table, let us consider one of the examples calculated previously (see Chapter 3):

This example involves a farm in a place where $N_g = 1$, supplied by an LV underground cable 400-m long to the limit of the property and from that point an LV overhead line 600-m long is connected to an LV/HV transformer protected by an SA.

CRL was calculated and is equal to 60.7.

According to the table below, a 15-kA I_n would be recommended.

For the same example, when N_g is equal to 4, CRL would become 15.2 and a 30-kA I_n would be recommended.

Table 7.2 Estimation of I_n based on CRL and P_{SPD}

CRL	P_{SPD}	I_n (kA 8/20 µs)
$1{,}000 > CRL \geq 400$	0.05	5
$400 > CRL \geq 200$	0.02	7.5[*]
$200 > CRL \geq 100$	0.01	10
$100 > CRL \geq 40$	0.005	15
$40 > CRL \geq 20$	0.002	20
$CRL < 20$	0.001	30

*This value does not exist in the market so far and thus should be replaced by 10 kA.

7.2.2   Type 1 SPDs for incoming lines connected to another structure

The rating for a Type 1 SPD is given by the I_{imp} value. For incoming lines connected to another structure (see case 1 of Section 7.1), it can easily be determined by a simple calculation based on a few assumptions. When the subject building is struck by lightning, its local earth will rise in potential. For example, an earthing system with an earth resistance as low as 1 Ω will rise in potential to 100 kV when is subjected to a 100-kA lightning current. This is well in excess of the LV voltage or sometimes even of the HV voltage installations. The result is a risk of flashover to an incoming line. A partial lightning current, assumed to be as a first approximation, 50% of the lightning current that struck the building, will then flow to the adjacent structures connected by the incoming lines. All of the connected structures (power utility, telecom provider, etc.) have their own earthing system that is not yet impacted by lightning and therefore is still at 0 V. The current will flow like in a normal electrical circuit from the source at a high voltage (earthing system of the struck building) to the zero volt of the circuit (all the other earthing systems).

It may be considered a first approximation that the current shares equally between all these paths (including the metal pipes). Inside each path the current will also share equally between all the conductors (e.g. the current of the path will be divided by 5 in a three-phase+neutral+PE cable). More precise calculations for the sharing of current between paths and conductors can be performed using simulation tools but as a first approximation, this approach gives a reliable value. Very often, many paths and conductors will limit the current to a low value. In general, the minimum value found in the market for SPDs is $I_{imp} = 12.5$ kA and the calculated value may be much less than that (Figure 7.29).

Note: It happens frequently that this value is smaller than the minimum level found in the market (12.5 kA). However, even in that case, a calculation is suggested to define the SPD disconnector (see Section 7.6).

This simple calculation is based on assumptions:

- Only 50% of the lightning current will flow to the local earth: in practice, a higher resistance earthing path will result in a lower current value in comparison to a lower resistance earthing, such as moist soil near the sea will lead to a higher share of the current.
- The remaining part of the lightning current will share equally between all connected paths: in practice, the current will depend on the type of path (overheard or underground) and the resistance at the other end of the path. For an overhead path, the resistance at the extremity is crucial for current dissipation and reduction of possible sparking over on pole along the path needs to be considered. For underground conductive path, the current can dissipate all along the path.
- The current in one path will share equally between conductors: in practice, more current will flow though the grounded neutral than through the phase. For a three-phase+neutral system, the neutral conductor can carry 50% of the total

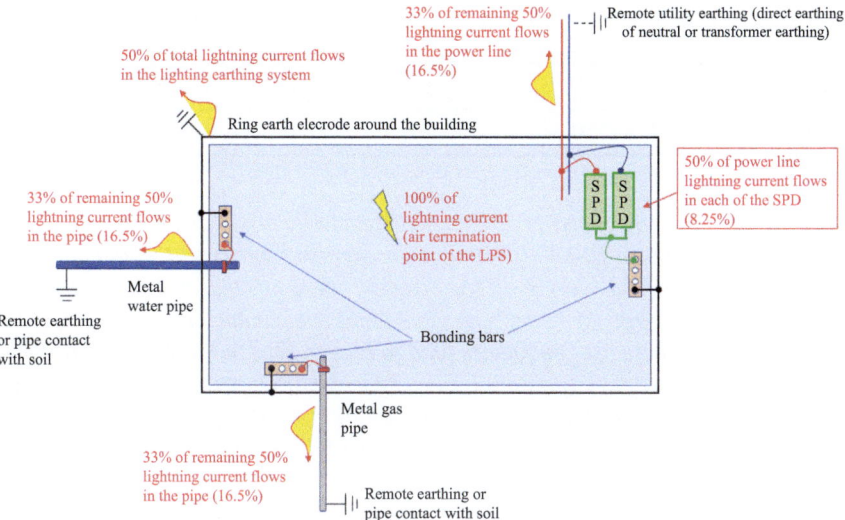

Figure 7.29   Example of simple current sharing between many paths (two pipes and one hypothetical singe phase power line) to determine current I_{imp} through single pole T1 SPDs

current injected in the service when each of the phase conductors will carry only 17% of the total current.

- The waveshape of the partial lightning current injected in each incoming line remains the same as the initial lightning current (10/350 µs) and only the peak current changes. However, this current waveshape may be modified by the various impedances, including the impedance of the SPDs (mainly for limiting and combination type SPD. Gapped SPDs may weakly change the tail part of the waveshape due to a low arc voltage). However, the waveshape tail remains long and a 10/350-µs waveshape is more appropriate than an 8/20-µs one. Assuming a constant waveshape and current magnitude divided by the number of paths may sometimes overestimate the stress to the SPDs.

Note: A good example of waveshape change is described in IEC 61643-32 (application standard for PV SPDs). For PV panels mounted on a roof protected by an LPS when the separation cannot be maintained (thus the panel metal frame is bonded to the LPS circuit), a partial lightning current will flow through PV SPDs located at roof level and then through PV SPDs located on the front of the inverter at the bottom of the building. PV SPDs in that example are all MOV-based SPDs. There are many impedances in parallel, including the LPS down-conductors and the circuit made of PV SPD on roof – PV conductor – PV SPD in front of the inverter. The impedance of these SPDs in series will change the current waveshape flowing through this circuit compared to a 10/350-µs waveshape. When the LPS lightning protection level is III and there are three down-conductors, simulations have shown that the peak current of the impulse (used to determine the residual

Table 7.3   Maximum current for each lightning protection level[†]

LPL	I	II	III	IV
Maximum current (kA 10/350 µs)	200	150	100	100

voltage at the SPD terminals) is 17 kA when the energy dissipated through the SPD is well represented by a 10-kA 10/350 impulse current. The current magnitude of 17 kA can be found by considering all the circuits: there are three down-conductors, two PV conductors (one for + and one for −) and one conductor for grounding the PV metal frame. These six conductors lead to a current equal to 100/6 = 17 kA in each conductor. There are two possibilities to select SPDs, either use Type 1 SPDs with a peak current of 17 kA or use a Type 1+2 SPD with $I_{imp} = 10$ kA and $I_n = 17$ kA. By selecting a 17-kA 10/350 Type 1 SPD, the SPD exceeds the required rating and there is no need to perform any additional calculations.

It may then be needed to analyse in a more precise way the sharing of current. IEC 62305-1[†] discusses the calculation of a coefficient k for each path, being a percentage of the total current given by the lightning protection level (Table 7.3).

where k depends on the following:

- the number of parallel paths n ($n = n1 + n2$, with $n1$ the number of underground paths and $n2$ the number of overhead paths),
- the conventional earthing impedance for underground paths $Z1$,
- the earth resistance for overhead paths $Z2$,
- the conventional earthing impedance Z of the earth-termination system.

For underground paths,

$$k_{underground} = Z \times Z2/(Z2 \times Z1 + Z \times (n2 \times Z1 + n1 \times Z2)) \qquad (7.1)$$

For overhead paths,

$$k_{overhead} = Z \times Z1/(Z2 \times Z1 + Z \times (n2 \times Z1 + n1 \times Z2)) \qquad (7.2)$$

This formula assumed that each underground path has the same resistance and each overhead path has also the same resistance. The formula can be expanded to cover the case of various earthing resistances for various paths.

Normally Z is measured on site (in spite of being called impedance, it is the usual earthing resistance measured with a low-frequency earth meter. Most of the energy will be dissipated during the tail that has mainly a low-frequency content. The earth surge impedance is important to define the overvoltage generated at the front of the waveshape and the earth resistance is important for energy sharing).

Table 7.4 in the standard gives values for $Z1$ as a function of the soil resistivity that needs to be measured.

[†]IEC 62305-1 ed.2.0. Copyright © 2010 IEC Geneva, Switzerland. www.iec.ch.

Table 7.4   Z1 function of soil resistivity[‡]

Soil resistivity (Ω m)	Conventional earthing impedance Z1 (Ω) for underground paths longer or shorter than 100 m	
	≤ 100 m	>100 m
100	8	8
200	11	11
500	16	32
1,000	22	44
2,000	28	56
3,000	35	70

The resistance of the overhead line must be measured or, as a first approximation, may be assumed to be given by the same table as $Z1$.

Let us take three examples to compare these formulas with simplified formula based on the assumptions that the current always share equally (in earthing system and paths, between paths and between conductors).

First example,

The earthing resistance of the LPS is 10 Ω. The structure is connected to an overhead line connected to a 15-Ω resistance and to a metal pipe 100-m long in a soil resistivity 500 Ω m.

Applying the k factors, $k_{underground} = 16\%$, $k_{overhead} = 34\%$ and current through LPS earthing system = 50% and comparing to the simplified formulas: $k_{underground} = k_{overhead} = 25\%$ and current through LPS earthing system = 50%. In this case, the total current compares well between both methods but the current through each underground path is 9% smaller than that calculated with simplified formulas. On the other hand, the current through the overhead line is 9% larger than that calculated with simplified formulas. The SPD calculated based on the simplified formula will be underrated by around 10%.

Note: Generally, a low earth resistance, and if possible lower than 10 Ω, is requested for LPS. There is generally no request for low resistance for SPDs in the other cases. The building earth resistance value is generally influenced by electrical safety rules. A higher resistance will limit the current that can flow to the local earth in the case of a surge on the incoming lines and force the current to flow elsewhere. Provided all elements on an installation are equipotentially bonded, this high resistance is not a problem. However, a low resistance is always a good option, especially if the equipotentiality is not perfect.

Second example,

The earthing resistance of the LPS is 10 Ω. The structure is connected to two overhead lines each one connected to a 15-Ω resistance and to a metal pipe 200-m long in a soil resistivity of 1,000 Ω m. Applying the k factors, $k_{underground} = 9\%$, $k_{overhead} = 26\%$ and current through LPS earthing system = 39% can be compared to the simplified formulas: $k_{underground} = k_{overhead} = 17\%$ and current through LPS

[‡]IEC 62305-1 ed.2.0. Copyright © 2010 IEC Geneva, Switzerland. www.iec.ch.

earthing system = 50%. In this case, the current flowing through the LPS earthing system is much smaller than that expected. The current through each underground path is 8% smaller than calculated with simplified formulas and conversely the current through the overhead line is 9% larger than that calculated with simplified formulas. The SPD calculated based on the simplified formula will be underrated by around 10%.

Third example,

The earthing resistance of the LPS is 10 Ω. The structure is connected to an overhead line connected to a 15-Ω resistance and to two metal pipes 200-m long in a soil resistivity of 100 Ω m. Applying the k factors, $k_{\text{underground}} = 30\%$, $k_{\text{overhead}} = 16\%$ and current through LPS earthing system = 24% can be compared to the simplified formulas: $k_{\text{underground}} = k_{\text{overhead}} = 17\%$ and current through LPS earthing system = 50%. In this case, the current flowing through the LPS earthing system is much smaller than that expected. The current through each underground path is 13% larger than calculated with simplified formulas and on the other hand, the current through the overhead line is 1% smaller than that calculated with simplified formulas. The SPD calculated based on the simplified formula will be well rated.

These examples show that the simplified formula assuming equal current sharing may be too simplified for a few cases, leading to Type 1 SPDs underrated in terms of impulse current I_{imp}. It is noted that the hypothesis that 50% of the current flows in local earth may not be true when there is competition between this local earth and other earthing systems that are the reference for incoming services into the structure. In the third example the current in the local earth was around 25% when the general simplified hypothesis implies 50% current in local earth.

In most of the cases, the simplified formula will be appropriate but, in a few cases, a more specific calculation will be needed using the previous formulas or more precisely computer simulations. The main benefit of the formulas presented before (see the previous text) is that they show the influence of the parameters (the number of services connected, underground or overhead services, resistance of the service and of the soil for buried service, etc.).

There are many parameters that influence the sharing of the lightning current in incoming lines:

- The line length: this will influence the front of the wave mainly due to its impedance and the sharing of energy due to its resistance that is proportional to the line length.
- The transformer earthing resistance: if the resistance is low compared to the earth resistance of the structure, more current will be injected in the line.
- Number of connected lines: this will also increase the partial lightning current flowing into the lines.
- The electrical circuits inside the structure and type of SPDs: this may affect the equivalent impedance of the building and the current injected in each conductor within an incoming line.

Computer simulations are needed if one wants to determine more precisely the stress associated with these equipotential bonding SPDs and as a consequence, the correct rating of the Type 1 SPDs. These simulations need to take care of all parameters and are mainly justified when the waveshape through the SPDs needs to be considered. For example, to define the protection efficiency of an SPD, it is not possible to simply consider the peak current the SPD will have to conduct under surge conditions but also the front of the waveshape for residual voltage and the tail of the waveshape for the energy handling.

In most of the cases, when the simplified rules are not enough, a simple method based on resistance sharing is sufficient to determine the energy stress for Type 1 SPDs. Such a method is presented in IEC 61643-12. This method is similar to the one presented before (see the previous text) but concentrates only on the power line. Both methods may be combined to take care of metallic ducts connected to the building and multiple incoming lines. This method is illustrated in Figure 7.30.

Note: A pure current sharing method based on earth resistance may over-estimate the required SPD rating. The waveshape in all the circuits may be different from the assumed 10/350 waveshape (see the PV example earlier).

R_B is the building earth (LPS earthing system), R_N is the resistance of neutral earthing of the power utility and R_{En} is the earthing resistance of building n. R_{eq} is the equivalent resistance of all the connected structures, including the neutral earthing.

Assuming a current $I_{lighting} = 100$ kA (LPL IV), $R_N = 15$ Ω, a building earth $R_B = 10$ Ω (LPS earthing) and ten connected structures with a resistance of 30 Ω each, the result is $I_{power} = 80$ kA and $I_{building\ earth} = 20$ kA only. With two SPDs, as shown in Figure 7.30, each SPD would be rated $I_{imp} = 40$ kA 10/350 μs instead of 25 kA given by the simplified formulas, a significant 60% increase.

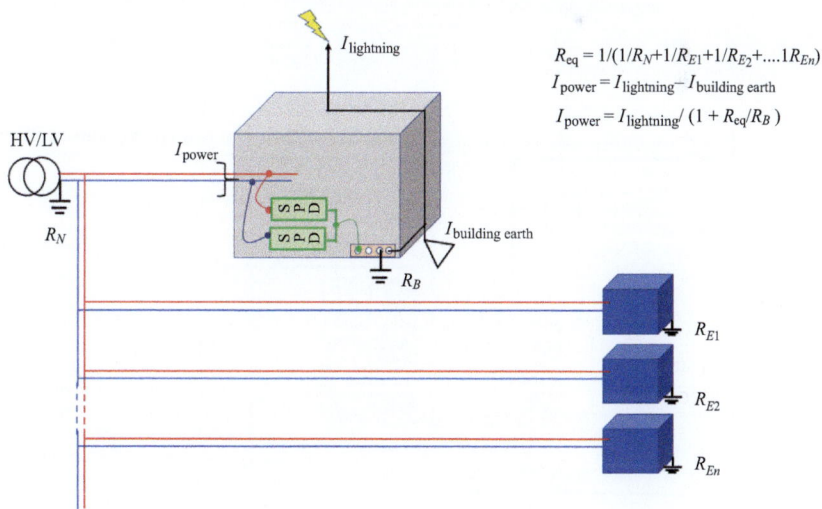

$$R_{eq} = 1/(1/R_N + 1/R_{E1} + 1/R_{E2} + \ldots 1R_{En})$$
$$I_{power} = I_{lightning} - I_{building\ earth}$$
$$I_{power} = I_{lightning}/(1 + R_{eq}/R_B)$$

Figure 7.30   Simple calculation of current flowing through Type 1 SPDs based on resistance sharing

Case study

A structure not protected by an LPS is supplying external lighting luminaires on metal poles in the company extended parking lot. The risk study has indicated, a Type 2 SPD rated $I_n = 30$ kA should be installed in MBP, and another SPD on the line supplying the external lighting ($I_n = 20$ kA). The installation is presented schematically in Figure 7.31. The system configuration is IT with neutral distributed. SPD disconnectors are made of fuses. All SPDs are MOV based.

A strike occurred on one of the lighting poles, leading to SPD damages. All SPDs were found to be damaged to some extent. A few SPDs had only minor characteristics degraded, but one T2 SPD on a phase that was supplying the lighting circuit was completely destroyed (shown in black in Figure 7.31) and led to damage to neighbouring SPDs (grey in Figure 7.31). One the same phase, the entrance SPD was also found to be damaged (also black in Figure 7.31). The fuse melted at the T2 SPD that was protecting the destroyed SPD.

This example shows that Type 1 SPDs may be needed even in the absence of an LPS on the structure and that extended lighting circuits may be collecting more lighting flashes than the structure itself if they are high enough. These Type 1 SPDs should be on the line supplying the external lighting, between the lighting circuit and the CB protecting this circuit, to

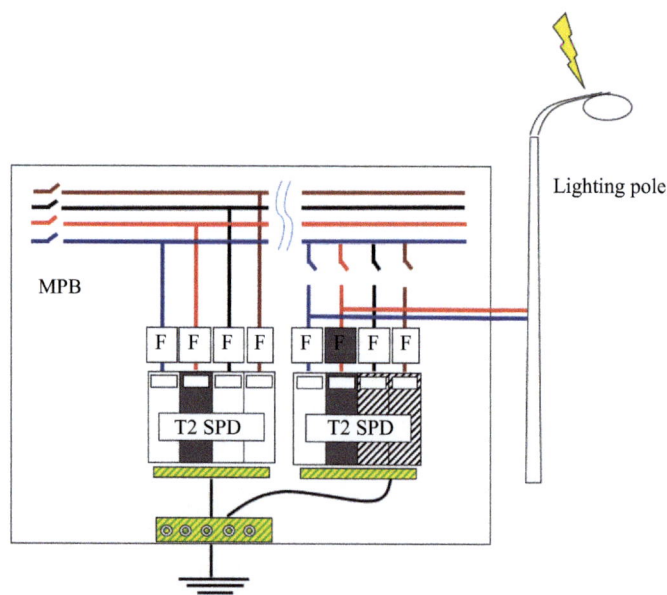

Figure 7.31   Investigated case

also protect this CB. In that case, the T2 SPDs at the MPB entrance should be coordinated with the T1 SPDs or, if necessary (e.g. because the earthing for the structure has a high ohmic value, leading to a high earthing voltage injecting a 10/350-μs current inside the incoming line), all these SPDs can be changed to use only T1+2 SPDs on both circuits (line entrance in the MBP and line supplying the lighting circuit).

7.2.3   Type 1 SPDs for other cases

For Type 1 associated with other cases (other than incoming lines discussed earlier), current I_{imp} could also be calculated. For example, in the case of a line connected to an equipment located on the roof that can be impacted by lightning in the absence of any LPS on the structure (see case 3 in Section 7.1), the I_{imp} current value can be determined as follows: the rolling sphere (based on electrogeometric model see IEC 62305 series) may help to determine the maximum value of current that can impact directly the equipment. There are generally many items on the roof, such as a chimney, that can reduce the probability of a direct strike to the equipment and as a consequence the maximum current to consider for I_{imp} calculation. To avoid the burden of performing the rolling sphere calculations, it is possible to consider the maximum current for the weakest level of protection (level IV associated with a maximum current of 100 kA).

Note: It is possible to ignore direct impact and Type 1 SPDs when no LPS is provided on the structure (including natural components LPS).

When an LPS is provided, it should normally protect the structure and equipment on the roof from a direct attachment. In such cases, it must be ensured that the separation distance between the LPS and the equipment and associated wiring is met (then no need of Type 1 SPD) or potential equalization must be implemented between the systems (equipotential bonding associated with partial lightning current, see case 3 in Section 7.1).

Note: If the LPL for the structure's LPS is not critical (such as LPL IV) but the location of the equipment on the roof is protected against a direct strike associated with a better LPL (such as LPL II), the value of maximum current associated with that LPL should be used for the calculation.

Note: Sometimes the LPS protects only the structure roof and only a few equipment on the roof (e.g. when the protection to the structure's roof is provided by natural components). In that case, the line associated with the non-protected equipment needs to be protected by Type 1 SPD in order to protect the structure.

The current I_{imp} in the Type 1 SPDs is determined by a current $I_{10/350}$ kA determined by Table 7.5, divided by the number of conductors of the supply line (including PE conductor and possibly a conductor connecting equipment metal frame with earthing bar).

Table 7.5 Determination of current trough SPD

Case	LPL of the LPS	Determination of current $I_{10/350}$
		Case 1: fixed value
Impact on equipment for a structure not protected by an LPS	None	$I_{10/350} = 100$ kA (associated with weakest LPL = IV)
		Case 2: calculated value based on rolling sphere model considering also other items located on roof
Impact on equipment for a structure protected by an LPS but not protecting equipment on the roof	IV to I	$I_{10/350} = 200$ kA for LPL I, 150 kA for LPL II and 100 kA for LPL III and IV
		Make a geometric calculation to determine the maximum sphere that can impact equipment and use the electro-geometric model (see IEC 62305-1) to determine the maximum current $I_{10/350}$ associated with that sphere
		Make a geometric calculation to determine the maximum sphere that can impact equipment and use the electro-geometric model (see IEC 62305-1) to determine the maximum current $I_{10/350}$ associated with that sphere
		Case 3: separation distance is maintained with equipment and its circuit
		Case 4: separation distance is not maintained with equipment or its circuit
Impact on equipment for a structure protected by an LPS protecting also equipment on the roof	IV to I	No need of Type 1 SPD
		$I_{10/350} = k_c \times (200, 150$ or 100 kA) with a value of current depending on LPL (see Table 7.3) and k_c is the part of lightning current in % flowing along the relevant part of the LPS circuit (see IEC 62305-3)

Let us take an example. An LPS level IV (current 100 kA) is installed on a roof. There are only two down-conductors and in such a case an easy value is $k_c = 50\%$. Equipment on the roof is too close to one of the LPS down-conductors. Equipotential bonding is necessary and a Type 1 SPD is provided on the power supply of that equipment. This equipment is supplied by a phase+neutral+PE cable. The current where the equipment is located is $I_{10/350} = 50\%$ of 100 kA = 50 kA. This current will share equally between the down-conductor and the phase+neutral+PE conductors. I_{imp} for each pole will be equal to 12.5 kA.

Note: The partial current flowing through the cable will protect the insulation but the current will flow through the cable to reach the earthing system of the structure. Thus, similar SPDs are needed at the subsidiary board that supplies this equipment.

When equipment bonded to the earthing system for the structure may be impacted by direct lightning (see case 5 in Section 7.1), the I_{imp} current needs to be evaluated. In that case, the current that can impact the equipment (determined as earlier by the lightning protection level (LPL) of the LPS or the rolling sphere) will share between the local earthing (assuming in general 50% through the earthing system), and the supply cable (including PE and bonding conductor, if any) and pipes.

An example is given for equipment supplied by a phase+neutral+PE cable connected to a platform directly connected to the earthing system for the structure. A metal water pipe also connects this equipment to the structure. There is no protection provided by the structure or any object in vicinity that will prevent a lightning flash to occur on the equipment. The lightning current to consider the equipment could experience is given by the LPS for that structure. In this case, an LPL level II (150 kA) is selected. Current will share 50% through the earthing system and the remaining 50% equally through the phase+neutral+PE conductors and water pipe. I_{imp} for each SPD is then equal to 18.75 kA (150 kA/2/4).

Note: It may be necessary to use simulation tools to better define the current I_{imp}.

In the absence of calculations, it is possible to use the values proposed by IEC 62305-1 in its Annex E. This standard provides generic values for both power and telecommunication lines for the various LPLs associated with direct flash to a line, induced overvoltage on a line, induced surge inside a structure due to direct strike to the structure or to nearby strike. As discussed previously, the stress inside the structure due to nearby direct lightning is negligible. According to the annexe, direct strike to a power line (i.e. for 1–3 spans of overhead line) does not lead to a surge current greater than 10 kA 10/350 μs and induced surges on the line do not exceed 5 kA 8/20 μs. Induced surges due to a direct strike to the building (for a circuit not concerned by Type 1 SPDs) can be up to 10 kA 8/20 μs. Values for telecom lines are generally lower (e.g. 2 kA for direct strike to telecom lines). The use of these generic values does not prevent more precise calculations to be performed because many assumptions are made to obtain these values.

7.3   Protective distance

It is well known for HV long lines that an overvoltage propagating along a long open line (end of the line neither connected to ground nor to a load) will double the voltage at the end of the line. This is due to the propagation of the voltage impulse reflected at the open end and doubling (the incoming wave $+U$ ◣ added to the reflecting wave of the same magnitude but the opposite direction ◢ $+ U$).

Note: The situation is similar for the impulse current. The incoming wave $+I$ ◣ is reflected but with the opposite sign $-I$ ◥, leading to a current equal to 0 at the open end. In the same way, on a short-circuited line, the voltage becomes equal to 0 and the current impulse is doubled.

The overvoltage generated at the terminals of an HV surge arrester follows the same rule. As a consequence, surge arresters are able to provide protection of a limited area. This is true whatever the system voltage may be (see next for LV systems). The protective distance, which is the distance d from the surge arrester where the overvoltages are still sufficiently reduced, may be quite short. Sometimes it is only metres. In this case, the surge arrester must be located on the equipment to be protected or even inside this equipment (such as a surge arrester immersed in a transformer). Not only does the protective distance, d, need to be small but the total distance, including the connecting conductors, also need to be small. The total length, d + connecting conductors (between phase and SA and between SA and earthing system) + the SA height defines a loop into which the surge current will generate an additional voltage that will add to the protection level U_p of the surge arrester.

Similar studies have been carried out by the French utility EDF for various system voltages. It provided the following result to be used as an example. As can be observed, the distance is generally in the range of a few tens of metres but when insulation is reduced the distance becomes very short (Table 7.6).

The phenomenon is the same for LV systems. The protective distance is in general in the order of 10 m (generally called the 10 m rule). It is assumed that at 10 m from the SPD equipment may not be protected and another SPD will be necessary to protect the equipment satisfactorily.

An incoming surge will generate step voltage in the case of MOV-based varistor (usually in the range 10–20 kV/μs). The step voltage will lead to oscillations on the circuit made of the line (resistance and mainly inductance) and the load

Table 7.6   Example of protective distance for HV surge arresters

Nominal voltage of HV system (kV)	63	90	225 with reduced insulation	225
Insulation level (BIL) of transformer (kV)	325	450	650	900
U_p of surge arrester (kV)	180	235	475	475
Protective distance: maximum distance between surge arrester and transformer (m)	30	30	Directly on the transformer	40

(capacitance at high frequency) representation of such a system. The voltage experienced at the load will depend on the front of the wave of the surge, the length of the conductors and the type of load. Depending on the value of the resistance, oscillations between inductance and capacitance will increase the voltage generated at the SPD terminals by a factor that can be as high as 2. These oscillations are damped by the line and load resistance. On the reverse, high capacitance or disconnected loads will lead to shorter protective distance.

In general, oscillations may be disregarded for distances less than 10 m but this value is not a fixed one. It has been found as reasonable but if the voltage is doubled at 10 m, it will not be too much different at 9 m. This is illustrated by Figure 7.32 where the amplification factor between the peak voltage U_p at equipment terminals and the voltage protection level U_p has been calculated as a function of the protective distance d (load capacitance and cable resistance have been selected purposely to reach an amplification around 2 for 10 m).

The purpose of the 10-m rule is to draw the attention of the surge-protection planner that SPDs need to be near equipment that they are supposed to protect.

Note: When equipment is internally protected by built-in surge-protective components integrated will significantly reduce oscillations even at longer distances.

It is important to understand that the voltage to be considered is not the voltage at the SPD terminals but the voltage that also includes the lead conductors. If an SPD with a voltage protective level $U_p = 2$ kV is connected with a total length of 50 cm to phase and earth, the voltage that will propagate towards the equipment will be $U_{p/f} \approx 2.5$ kV. At more than 10 m, a 5-kV voltage may be expected in the worst case at the equipment level. Even, if the voltage is not doubled, it clearly

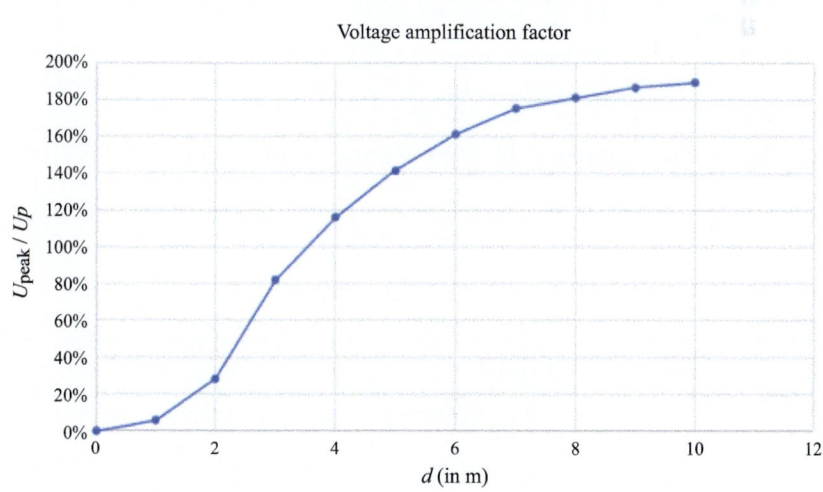

Figure 7.32   Amplification factor between U_{peak} at equipment and U_p

appears that as soon as the distance between SPD and equipment exceeds a few metres, equipment may not be fully protected. This is very similar to what has been presented earlier for HV surge arresters.

An experiment was made in the laboratory to inject a 10-kA 8/20-μs impulse from a current generator to a circuit made of an MOV-based SPD with a $U_p = 1.2$ kV and $I_n = 10$ kA. A 10-m straight conductor and a fridge were located at the other end of the line, as shown in Figure 7.33.

Note: The surge current (in blue on the figure) propagating to the fridge in the experiment was lower than the one flowing through the SPD but the voltage at the fridge was potentially damaging.

Voltage measured at the fridge terminals had a peak voltage of 2.2 kV (1.8 times the voltage protection level U_p). As can be seen in Figure 7.34, an oscillation occurred that was responsible of this peak voltage. In that figure, the bold line is the voltage at the SPD terminals (with short lead conductors) and the oscillating wave is the voltage at the fridge terminals. The fridge was an old one with a 2.5-kV impulse withstand voltage and thus was not destroyed. A more recent fridge connected to the Internet with a lower impulse withstand of 1.5 kV (considered sensitive equipment) would have been destroyed.

If the front of the wave is not very steep, the oscillation level will be lower. For example, for a 15-kV/μs steep front when it is possible to obtain an amplification factor around 2 at 10 m, the amplification factor will drop to around 1.3 for a 1.5-kV/μs steepness and almost negligible for a steepness equal to 0.15 kV/μs.

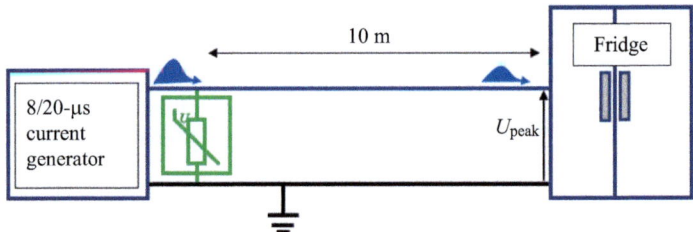

Figure 7.33   Circuit tested in laboratory

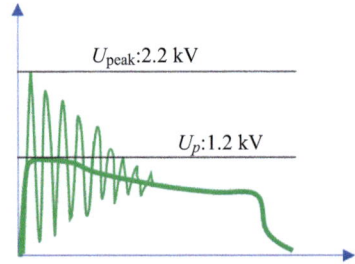

Figure 7.34   Voltage measured during test in laboratory

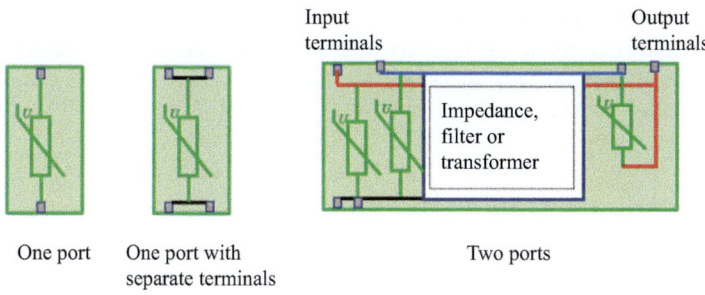

Input terminals   Output terminals

Impedance, filter or transformer

One port   One port with separate terminals   Two ports

Figure 7.35   Internal design of one- and two-ports SPDs

Generally, the steepness at the output of two ports SPDs is much smaller than for a one-port SPD (Figure 7.35).

A two-port SPD is an SPD that has a series impedance connected between input and output. It may also include a filter or even a transformer. IEC 61643-11 defines an optional parameter for two-port SPDs: the voltage rate-of-rise of a two-port SPD that is defined as the rate of change of voltage with time measured at the output terminals. At the output of two-port SPDs, the generated voltage generally has a slow front of wave (du/dt) that will generate little oscillations and thus will extend the protective level to much more than 10 m. The limitation of such a two-port SPD in terms of protective distance is primarily the disturbances that are generated downstream of the SPD due to switching surges, disturbances coming from another circuit (e.g. when running near a circuit where a partial lightning current is flowing) or induced surges in the loop between SPD and equipment to be protected. In general, it is admitted that a much longer protective distance (50 m or more) is achievable for a two-port SPD with a voltage rate-of-rise of 150 V/μs.

Note: A recent concept of two ports SPDs, even if existing since many years in a few places, is the surge isolation transformer (SIT, see Chapter 5). This is typically an insulation transformer (a two windings transformer with a grounded screen between them) protected by an SPD at the input side to ensure there is no flashover of the insulation that would bypass the decoupling between input and output. Such a transformer is limiting the surges transferred (a simple two-winding transformer will leave 20%–30% of the initial surge-propagated downstream) to the output. In the past, such products with an output SPD or in a few cases, a capacitor or diode, have been used satisfactorily and provided a very low protective level at the output.

Note: Most of the two-port SPDs present a lot of advantages in terms of reduced voltage protection level at the output an extended protective distance. In fact, it really depends on the internal circuit and the decoupling between input and output. The output du/dt is an important parameter to extend the protective distance. It may be declared or not. The U_p at the output of the SPD may be directly compared with U_p provided by comparable one-port SPDs. However, such two-port SPDs also present a drawback: the load current flows through the SPD circuit.

If the load is 10 A, the SPD will fit on a din rail. If the load is a few hundred amps, the SPD becomes bulky (especially if it includes a decoupling inductance and furthermore a transformer) and heavy and will only fit in a cabinet or a dedicated panel.

7.4   Coordination between SPDs

There are generally more than one SPD in a circuit. These SPDs need to be coordinated together. The main reason for such coordination is to avoid most of the surges to propagate downstream and to possibly damage SPDs located downstream near equipment.

Generally, for a line connected to an external structure, the surge is supposed to come from that line, enters the structure and be mainly diverted to ground thanks to the SPD located at the entrance. This SPD is then rated to withstand a high surge current (either 10/350 or 8/20) and may have a voltage protection level U_p equal to 2.5 kV (for a 230/400-V system). There is a relationship between the surge withstand and the voltage protection level and, in general, SPDs that can withstand large surge current have generally a higher U_p than SPDs that can withstand only a limited surge current. As a consequence, the trend is to have large surge current withstand at line entrance and accept potentially a high value for U_p and to have lower surge current withstand downstream in order to obtain a better U_p value near equipment.

The risk is that the front SPD and the downstream SPD badly share the surge current leading to a weak stress on the front SPD and a too high stress on the second SPDs. Such a stress may not be handled by the second SPD that would fail.

To avoid this, the primer goal of coordination is to share energy between SPD in such a way that each SPD is not stressed more than what it is able to handle. The process to assess this is called 'energy coordination'.

A simple example will help understanding the problem. A Type 1 SPD with a surge withstand $I_{imp} = 12.5$ kA per pole is installed at the entrance of the installation. It has a voltage protection level $U_p = 2.5$ kV that is not able to protect sensitive equipment in the MPB. Another SPD, Type 2, needs to be installed in the MPB. This SPD is just able to withstand 8/20 surges ($I_n = 10$ kA) but has a $U_p = 1.5$ kV. The front SPD is based on a gap that will sparkover at 2.5 kV when the second SPD is based on MOVs. Let us assume that the Type 2 SPD is able to withstand limited 10/350-μs impulses, for example 0.5 kA 10/350 μs. As soon as the impulse flowing through the T2 SPD exceeds 0.5 kA 10/350, the SPD will fail. If a 0.5-kA 10/350 surge occurs, it will not be diverted by the front SPD until the voltage reaches 2.5 kV. The surge will then flow first through the second SPD that will try maintaining the voltage between its terminals to 1.5 kV. The voltage at the entrance SPD depends on the voltage across the Type 2 SPD (assumed to be constant and equal to U_p, i.e. 1.5 kV) and the voltage drop between the two SPDs. However, in an MBP the conductors are generally bars that have a low inductance. Then the voltage drop will be probably too low and the voltage will never reach

2.5 kV at the front SPD terminal. The total surge will then flow through the second SPD only that will be destroyed.

If there is a conductor between the two SPDs assuming the usual value of 1 μH/m, then the maximum voltage drop along the conductor will be $U = L \times 1$ μH/m $\times 0.5$ kA/10 μs $= L \times 0.05$ kV/m.

To obtain a voltage drop high enough to reach 2.5 kV at the front SPD, it is then necessary to have $U = U2 - U1$ and then a conductor length, $L = (2.5$ kV $- 1.5$ kV)/0.05 kV/m $= 20$ m. When the voltage reaches 2.5 kV at the front SPD, it will spark over and divert most of the surge current, protecting the second SPD.

If the injected current was higher, it would be easier to obtain good energy coordination. If we assume that this Type 2 SPD is able to handle the front part of a 10-kA 10/350 μs surge (most of the energy of the 10/350-μs waveshape is coming from the tail and not from the front), then for this 10 kA surge the length of conductor becomes $L = (2.5$ kV $- 1.5$ kV)/1 kV/m $= 1$ m. It is easier in that case to obtain a good coordination with a higher current than with a lower current. With higher current magnitude, only the front part of the tail flows through the second SPD when with lower magnitude, the complete waveshape or a significant part of the waveshape will flow through the second SPD that will be destroyed. This result may seem to be curious because it is expected that the highest magnitude incoming surge will lead to the more severe result. This is known as the blind spot phenomenon. At a lower surge current magnitude, the SPD coordination is more difficult than at high level (Figure 7.36).

Note: Considering 1 μH/m is for simplicity sake. The inductance to consider may be lower and depends on the cable running (all conductors in the same cable or not) as well as the distance between the conductors and the earth plane. The value can then be reduced to 0.8 μH/m or even as low as 0.5 μH/m. Assuming 1 μH/m

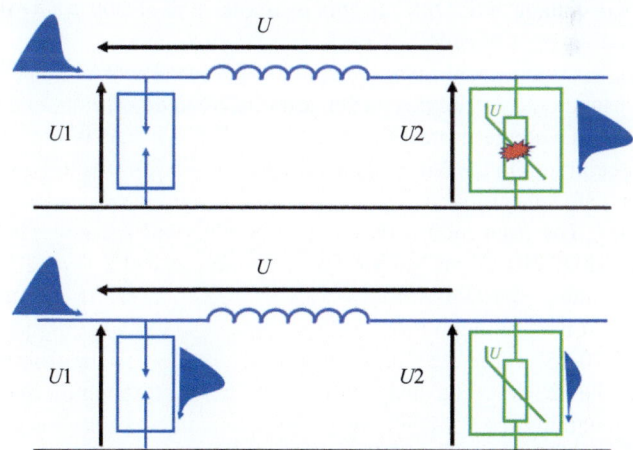

Figure 7.36   Coordination between a spark-gap-based SPD and an MOV-based SPD thanks to a decoupling conductor. Top: bad energy coordination; bottom: successful energy coordination

leads to the worst case conditions for the voltage drop. The length that needs to be taken into account when 1 µH/m is used is the line length (there is no need to add the length of the PE return conductor).

Let us finish to discuss this simple example by addressing the voltage protection provided by the Type 2 SPD. The voltage at the level of the second SPD is not constant but is a curve U vs. I. For low values of current, the voltage at the Type 2 SPD terminals will be lower than U_p. With a distance of 20 m between the two SPDs, a 0.5-kA 10/350 µs will lead in fact to a lower voltage at the Type 2 SPD terminals than 1.5 kV. At this level of current, the voltage at the Type 2 SPD will be lower than U_p, for example 1.1 kV. Then there is no sparkover (the total voltage is only 1.1 due to the second SPD +1 kV due to the voltage drop = 2.1 kV lower than 2.5 kV needed for the operation of the first SPD) and coordination is not guaranteed. Coordination should take care of the real SPD characteristics and not the U_p values that are almost never reached. This is why coordination studies are not simple and generally made by the SPD manufacturers.

Figure 7.37 shows an MOV curve U vs. I, the voltage drop along a 20-m conductor and the total voltage obtained. It can be seen that a current magnitude of around 700 A is needed to reach a total voltage of 2.5 kV. At this value of current, the T1 sparkover will occur at the peak of the waveshape (it is assumed that at this level of current the Type 2 is still able to survive that stress. If it is not the case, a longer length would be necessary but the principles will remain the same). When the injected current magnitude increases, the sparkover of the Type 1 SPD will occur on the front of the wave and not on the peak, and the voltage at Type 2 SPD terminals will remain the same.

The voltage drop along the conductor between the two SPDs increases with the injected current magnitude and as a consequence a lower voltage is requested at the T2 SPD terminal. There is a maximum case, where the current through the SPD leads to a total voltage = 2.5 kV. In this example, it is at 700 A. At this level of

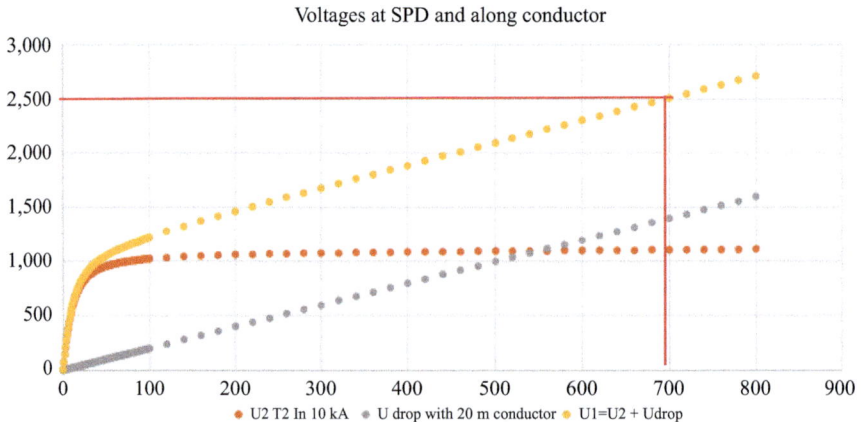

Figure 7.37   Voltage at SPD terminal plus voltage along conductor

current, the voltage of the Type 2 SPD is around 1.1 kV and the voltage drop along the conductor is 1.4 kV. The maximum voltage at the T2 SPD terminals is lower than U_p. Not only energy coordination is provided by furthermore, the global protection efficiency of the two SPDs is better than expected. This is what is called 'protection voltage coordination'.

It must be noticed that the voltage drop along lead conductors will also play an important role in the SPD coordination and especially the voltage protection coordination. In this example, it has not so much impact because the current is very low (700 A). The voltage drop along the SPD lead conductors will not influence so much the voltage for the protected equipment. However, for higher current flowing in the Type 2 SPDs, this may have a detrimental influence on protection provided. In addition, this may change the coordination rules, making it either easier when long lead conductors connect the Type 2 SPD to the line (the voltage increases at the Type 1 SPD level) but can also make it more difficult when long lead conductors connect the Type 1 SPD to the line (the stress injected to the second SPD may be bigger).

Note: The lead length of an SPD has generally more influence on MOV based or combination SPDs than on gapped SPDs. For gapped SPD, there is no voltage drop along conductors until the current flows then until the SPD operates. There are then two situations: before the SPD operates, there is no influence of the lead length. Later, the SPD operates, the voltage decreases quickly to the arc voltage (may be a few hundreds of volts) and the lead length plus the arc voltage remain generally low. For MOV-based SPDs, the voltage drop along lead conductors will add to the residual voltage.

Note: For a usual 10 kA 10/350 µs waveshape, 50-cm wiring on the front SPD will lead to an additional voltage of 500 V which can increase the stress on the second SPD (in terms of energy and protection level). With steeper waveshapes as defined in standard IEC 62305-1 (valid only for protection level coordination because these waveshapes are not associated with high energies), there will be an additional voltage even greater such as 2,500 V for a 5-kA 1/200-µs waveshape (negative first impulse) and this additional voltage is no longer negligible (Figure 7.38). Of course, such steep

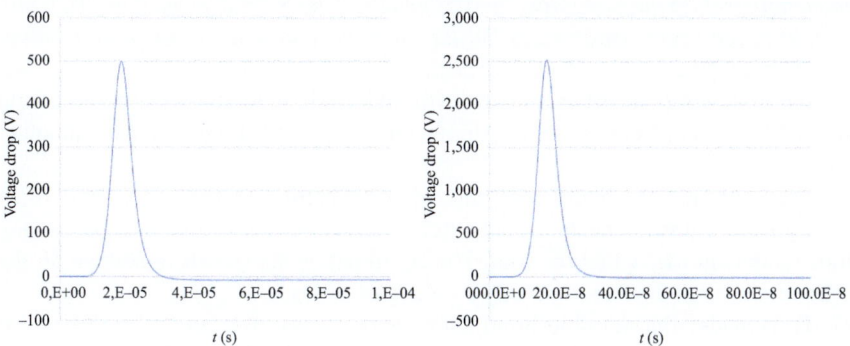

Figure 7.38   Comparison between voltages drops due to a 10/350- and 1/200-µs waveshapes

fronts only exist on current injection point because due to the installation impedances, the current wave will lie down while propagating and become less steep inside the installation. On the reverse, a slow front may delay the sparkover of an entrance gapped SPD and inject a bigger stress to the Type 2 SPD located downstream.

It is then clear that coordination study needs to consider many parameters. As coordination between two SPDs can be complex to study, it is often minimized or ignored in practice. But it is necessary to ensure the protection will work as expected. It can be done through simulations, but this requires knowing system parameters and also the actual characteristics of SPD (real curve and not only a few points indicated in the SPD documentation of the SPD).

Another method is to build a representative installation in a laboratory and to inject the maximum expected current (generally the maximum discharge current of the front SPD) in the circuit and check that the second SPD is not destroyed (energy protection) and what is its level of protection (voltage protection coordination). This phenomenon is described in detail in IEC 61643-12.

In general, to avoid the burden of such demonstration either by simulations or tests, it is recommended to follow the manufacturer's indications of SPDs with regard to coordination. This indicates that minimum distances between SPDs (or equivalent decoupling induction values are based on the commonly accepted assumption that one metre of conductor is equivalent to 1 μH). This obviously requires that only one manufacturer be retained for all the SPDs of a protected circuit.

Note: Coordination rules are valid for stress coming from the line and flowing to earth or for stress coming from the earth (e.g. for current injected in the installation from the LPS earthing system) and flowing in the line.

Note: The coordination information provided by the SPD manufacturer generally relies on lead conductors = 0. This may require adjustment with real conditions observed in the installations especially when MOV-based combination SPDs are used. A few published coordination instructions assume a 50-cm lead length that gives flexibility when installing the SPDs. When nothing is noted regarding coordination rules assumptions, it should be assumed that considered lead length is 0. In the same way, there is generally no SPD disconnector taken into account in the coordination rules but they may have an impact when the residual voltage according to the SPD disconnector is not negligible.

The following part of this clause will address how to study the coordination when SPDs are not from the same manufacturer or provided rules are not enough to cover a specific case.

The coordination should take care of the direction of the surge. It is often coming from a line connected to an external structure, but it may also be coming from equipment located on the roof. The coordination also needs to address all the SPDs on the same line and not only a couple of SPDs.

To consider the possible blind sport, it is necessary to inject in the circuit surge current with a magnitude ranging from the highest possible stress that the front SPD can handle (either I_{imp} for a Type 1 SPD, or I_{max} when declared for a Type 2 SPD).

Table 7.7 Example of coordination between two MOV-based SPDs

Front SPD			Second SPD	
I_n (kA)	U_p (kV)	Minimum distance (m)	I_n (kA)	U_p (kV)
20	2.5	20	5	1.5
20	2.0	10	5	1.5

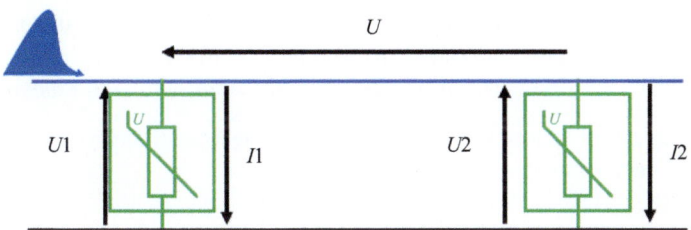

Figure 7.39 Basic scheme for coordination of two MOV-based SPDs

There are general coordination rules:

- The more the distance between the two SPDs increases, the more the current injected downstream decreases.
- The more the impedance between the two SPD increases (i.e. the distance between the two SPD increase), the more the second SPD is decoupled from the front SPD.
- The lower the protection level of the second SPD compared to the first, the more the distance will have to be increased to optimize the sharing of current.
- The weaker the withstand of the second SPD compared to the first, the longer the distance needed to decouple them correctly.

For example, Table 7.7 can be established for energy coordination between two MOV-based SPDs (assuming $I_{max} = 2 \times I_n$) (Figure 7.39).

Case study

A 20-m high concrete tower is equipped with antennas, lights, cameras and aircraft warning lights on the roof (with built-in Type 2 SPD of an unknown brand). An LPS protects the roof and equipment on the roof are equipotentially bonded to the LPS circuit (it was not possible to maintain the separation distance). The suppliers of equipment on the roof provided a power panel protected by Type 1+2 SPDs of brand A. The power supply is coming from a power utility line and due to surge experience on that line, outdoor Type 2 SPDs of brand B are provided by the utility. The main power panel board and an external power generator had built-in Type 2 SPDs from brand C. The electrical contractor for the tower electrical installation provides a front Type 1+2 SPD of brand D due to the presence of the LPS and Type 2 SPD of the same brand in secondary panels. The selection of four brands of SPDs for that tower is not made purposely but is the result of various equipment providers and no requirement at the project level to ensure SPD coordination. As the tower was playing an important safety role, surge protection needed to be demonstrated. Unfortunately, no information is published for the coordination of SPD of two different brands as a fortiori for four brands. The simulation study was too complex due to the four brands involved and too many missing detailed information about the SPD characteristics. The final decision was as follows:

- Replace all accessible SPDs by SPDs for a single brand where coordination can be demonstrated for stress coming either from the line or from the roof. Keeping brand C was the most cost-effective choice but needed to be approved for keeping the guarantee for the various equipment and panel boards.
- Accept that the power utility SPD be destroyed knowing that an external power generator (protected by SPD from brand C) was existing to ensure continuity of power supply.
- Accept that the aircraft warning light be destroyed (unknown coordination) and provide quick repair in the case of damage to avoid losing this equipment for a long period.

There are a few difficulties to consider when performing coordination studies:

The first one is related to the definition of the maximum energy withstand of the Type 2 downstream SPD (for Type 3 it is more complex yet due to the Combination Test Generator used to test the SPD). When I_{max} is declared by the SPD manufacturer, this value is considered the maximum stress the SPD can handle. But this value is optional. When it is not declared, either it is considered that $I_{max} = I_n$ or I_{max} is given by a fixed ratio (e.g. $I_{max} = 2 \times I_n$).

The second difficulty is coming from the SPD real characteristics. What is known is generally a couple of points (U_p, I_n or I_{imp}), a few points ($U_{res}(I)$: residual voltage for different values of the current I) or a curve U vs. I. However, all these characteristics are maximum ones. U_p is a value guaranteed by the manufacturer. To guarantee that this value will never be exceeded in spite of the manufacturing tolerances, the SPD manufacturer should take a margin. One SPD has a voltage U vs. I that remains below the maximum values but that can be much lower in practice. To perform a coordination study, it is necessary to take care of the tolerances and generally the most severe case is when the first SPD has the highest characteristics and the second SPD the lowest characteristics.

The possible methods to demonstrate the coordination are as follows:

- Analytical studies: applicable only for simple cases.
- Graphical method: applicable only for simple cases.

- Let through energy: valid for energy coordination provided data are available. Tolerances should be considered by a margin on the results.
- Tests: valid for all cases but installation may be difficult to reproduce and tolerances should be considered by integrating a margin on the results.
- Simulations: valid for all cases provided data are available. This method is flexible and allows to consider various cases as well as tolerances and to determine influence of parameters.

Analytical method: It consists in making simple calculation: see, for example, the case shown before with a gapped SPD at the entrance and an MOV-based SPD downstream. This method has its limitations because it oversimplifies the problem. For example, the calculations have shown that 20 m were needed but in fact with 20 m, the current through the second SPD was 700 A, greater than the expected surge withstand 500 A 10/350-µs. It is necessary to know precisely the curve U vs. I of the second SPD not mentioning the tolerances as discussed earlier.

Graphical method: in this case, it is necessary to compare the curve U vs. I of two MOV-based SPDs.

An example is given in Figure 7.40 for front SPD (I_{n1} 20 kA, I_{max1} 40 kA, U_{p1} 2.2 kV) and for second SPD (I_{n2} 5 kA, I_{max2} 10 kA, U_{p2} 1.3 kV), when distance is nil. In this case, the voltage at the second SPD terminals when $I_2 = I_{max2}$ is 1,500 V and at this voltage the curve shows that the current through the front SPD is only 5,000 A. The total current is then $I_1 + I_2 = 15$ kA. For a higher current such as I_{max2}, the second SPD will be overstressed.

The situation is more complex than the previous coordination between a gap and an MOV. For the gap, the front of the wave was most important and as soon as the gapped SPD operates, the MOV-based SPD is relieved. When two MOVs are concerned, the question is the sharing of energy and most of the energy is in the tail where the di/dt is negative and thus will decrease the total voltage. It is then difficult to use a graphical method with SPDs except to demonstrate that coordination

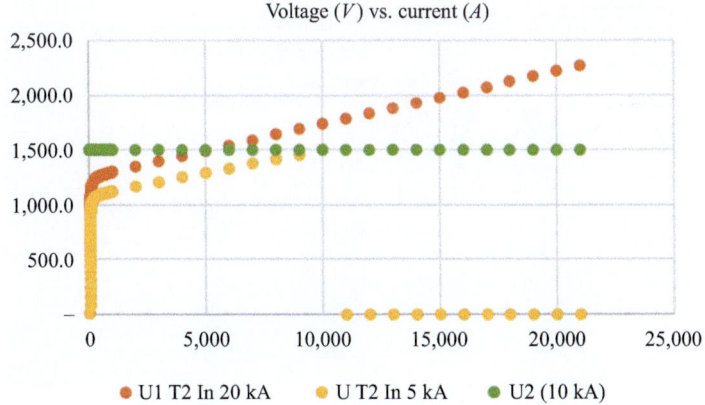

Figure 7.40   Coordination of two MOV-based SPDs when distance = 0 m

is obtained with no decoupling element or at the reverse that a decoupling element is necessary. But to determine what is the necessary decoupling element is quite impossible by this method.

Let through energy: the principle of this method is to transform the output stage of the first SPD into an equivalent combination wave generator defined by an open circuit voltage, a short-circuit voltage and 2-Ω impedance. This equivalent CWG is then applied to the second stage SPD and a comparison is made of these CWG characteristics to the second-stage SPD capability. This method is fine as it assumes that the SPD is a black box and the there is no need to know what are the components inside. However, this implies transformation of each SPD under a certain stress into an equivalent CWG and this means either performing tests or simulations. Withstand capability of the second-stage SPD also needs to be known based on an equivalent CWG. Due to the fact the method is determined only for energy coordination and request in any case performance of tests/simulations to determine the equivalent CWGs, it is probably easier to perform either coordination tests or simulations and get more possibilities such as protection level coordination. The main interest of this method is the systematic approach that allows to address coordination between many SPDs and also to include the equipment to be protected.

Tests: These tests are performed with either a 10/350- or 8/20-μs surge current generator.

The tests are performed with the maximum values declared for the front SPD (I_n, I_{max} if declared or I_{imp}) and also with portion of these values to explore if no blind spots up from low stress to maximum stress are existing. Portions of current are generally 10%, 25%, 50% and 75%. When I_{max} is not declared, I_n should be used or alternatively it may be possible to investigate the behaviour at two times I_n (typical default value for I_{max}).

Note: Figure 7.41 represents an example of coordination test between a Type 1 SPD and a Type 2 SPDs with 10 m distance between them. SPDs are energized

Figure 7.41   Example of a coordination test configuration at SHLPC laboratory

during the tests. These tests have shown that, for this particular case, energy coordination was achieved (no blind spot) but not the voltage protection coordination. The maximum measured voltage at the Type 2 SPD terminals was 2.2 kV for a declared $U_p = 1.8$ kV for the Type 2 SPD. This is quite understandable, because as previously indicated, the voltage at the SPD terminals can exceed U_p when the current exceeds I_n.

During the test, the current and voltages at the second SPD level are measured. The forward and return conductors between the two SPD should be installed following the configuration that will be used on-site or according to manufacturer instructions. Generally, it is recommended that these conductors are not twisted and do not form loops. When the applied voltage may have an influence, SPDs need to be energized at their maximum continuous operating voltage with a short circuit current high enough to allow detecting a failure of one of the SPDs (minimum 5 A). When SPDs have multiple modes of protection, all the modes should be tested.

For energy coordination, the energy through the second SPD is determined based on measurement and compared to the maximum energy that the second SPD can handle. If the energy remains below the maximum energy, the test is satisfactory. This maximum energy is determined either by tests or by simulations. The stress to take into account for determining the second SPD maximum handling energy is I_{max} when declared, I_n or two times I_n as indicated earlier. In addition, and especially if this maximum energy is not known or determined by a preliminary test/simulation, it is necessary to check if the second SPD is degraded or not (either damage such as puncture or flashover or degradation, with a higher leakage current for example). It is possible to perform coordination tests with more than two SPDs and also with Type 3 SPDs or two-port SPDs.

For voltage protection coordination, the voltage at terminals of the second SPD will be compared with the voltage protection level U_p of the second SPD. If the voltage remains below U_p, the test is satisfactory. Alternatively, it is possible to compare the voltage at the second SPD level to the impulse withstand given for the equipment to be protected.

The main drawback of such tests is the difficulty to consider the SPD tolerances.

Note: It is also possible to perform the test with the equipment to be protected connected to the downstream SPD but such a test should be performed with care to avoid damage to this equipment (e.g. starting with low current injection).

Simulations: this is probably the easiest tool to demonstrate coordination but it requires generally to know what is are the main components of the SPD. It is possible to use general computer tools for electrical circuitry simulations or dedicated software. Such dedicated software takes into account SPD based on their internal design (series–parallel association of usual components: spark gaps, spark gaps controlled with electronic circuit or not, MOV, and passive components R, L, C. Therefore, filters sometimes integrated in Type 3 or two-port SPDs can be taken into account). The line length between each SPD and the wiring length of each SPD are taken into account. Usually, the worst case is the entrance SPD with highest tolerances and long wiring and a downstream SPD with lowest tolerances and short wiring. It is necessary to carry out a parametric study to show the influence of the important parameters on the result of the coordination and possibly give wiring restrictions (the coordination as defined in the standard does not impose anything on the wiring lengths of each SPD because it just defines the phenomenon to be taken into account. However, as indicated previously, this wiring length may have influence on the result). Typical current sources should be 10/350, 8/20 µs but also 1/200 or 0.25/100 µs. The input data are generally limited:

- type of SPD with characteristics as given in manufacturer data sheets (if necessary, low current tests should be carried out in addition to refine part of the curve *U* vs. *I*);
- wiring length of SPDs;
- type, routing and length of line between SPDs;
- a maximum magnitude of the injected current and waveform (by default based on the data sheet of the front SPD).

At the output, dedicated software provides current, voltage and energy associated with each SPD in the circuit. When using generic electrical simulation tools, it may be necessary to make a few additional calculations and select the important parameters.

Figure 7.42 shows, as an example, coordination between a Type 2 gapped SPD and an MOV-based SPD with a distance $= 5$ cm between them, for two injected current 4 kA 8/20 µs and 6 kA 8/20 µs. In the first case, there is a blind spot because the gapped SPD does not operate.

Another example is shown in Figure 7.43. At entrance, a gapped powerful Type 1 SPD (calculated $I_{imp} = 57$ kA) is installed followed by 2 Type 2 MOV-based SPDs. The Type 1 SPD is disconnected; thus a 57-kA 10/350-µs propagates towards the two Type 2 SPDs. The first Type 2 SPD (SPD 2) takes most of the injected current. The resulting voltage at the third SPD (SPD 3) is 2 kV and thus the protection level coordination is not achieved.

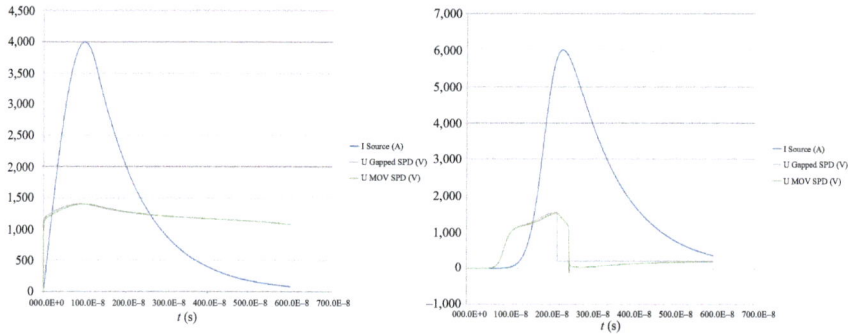

Figure 7.42   Coordination simulation between two SPDs with distance = 5 cm (blue, injected current, green voltage at the MOV-based SPD and grey voltage at the gapped SPD) – left blind spot – right the gap operates

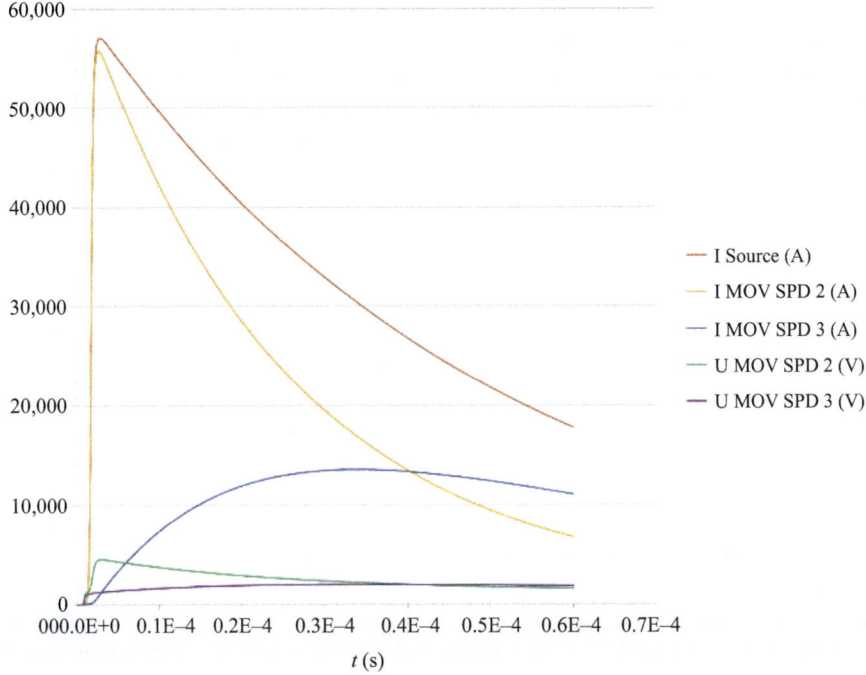

Figure 7.43 Worst case situation where the front Type 1 SPD does not operate leading to overstress Type 2 SPDs

The software gave the following additional information regarding energy: 41,495 J for SPD 2 and 12,144 J for SPD 3. The maximum energy withstand of these Type 2 SPDs is 745 J, so the two Type 2 SPDs 2 will be severely damaged if the Type 1 SPD does not operate. In practice, it would be necessary to do a parametric study to find the maximum blind spot current. In this example, it was assumed that the Type 1 did not operate, for example, because it was disconnected but it could also be due to a failed tripping circuit.

7.5 Coordination with equipment to be protected

Equipment to be protected are characterized from a surge point of view by their rated insulation withstand U_w. But another parameter is important: their immunity to surge. Immunity is the ability of a device, equipment or system to perform without degradation in the presence of an electromagnetic disturbance. This is defined by the IEC 61000 series of standard dealing with ElectroMagnetic Compatibility. The IEC standard 61000-4 series defines the immunity requirements for residential, commercial and industrial apparatuses in relation to electrical fast transients and surges. An important document is IEC 61000-4-5: testing and measurement techniques – surge immunity test. This standard gives immunity requirements and test procedures related to surge voltages and surge currents.

Immunity levels are defined for various test levels that shall be selected according to the installation conditions. Test levels range from 1 to 4 for usual cases with tests voltages ranging from 0.5 to 4 kV. Tests voltages are defined for each of these levels, between line and also between line and ground. Except for special cases, immunity of powered equipment is assumed to be 1 kV line to line and 2 kV line to ground for 230/400 V systems. The generator used by this standard is similar to the CWG used for testing Type 3 SPDs but with different coupling elements and series impedance. Surge immunity of equipment or systems may be achieved by design but more generally thanks to built-in surge-protective components (SPC), see Chapter 5 or external SPDs.

The presence of SPCs inside equipment may cause coordination problems with upstream SPDs, see next.

The U_p value of this upstream SPD should not only be below the rated withstand voltage U_w but also below the voltage immunity U_i. In general, U_i is lower than U_w. U_w relates to safety when U_i relates to operation (normal operation, temporary loss of function or performance or permanent loss of function). When protection between active conductor and earth is mandatory for safety (even if surge protection between phase and neutral remains recommended), the crucial part for immunity is generally between phase and neutral, where the function is powered. Permanent damage or fire hazard is generally non-acceptable for the user. This is why insulation withstand is so crucial. However, loss of function or degradation of equipment being part of critical systems is generally non-acceptable as well and immunity should be considered.

Note: As a generic rule, it may be assumed, in the absence of any other information, that an equipment installed on 230/400-V system and without any component such as filter or SPD between active conductors and earth has a rated impulse withstand U_w equal to 2.5 kV between active conductors and earth and 1.2 kV between phase and neutral and an immunity level U_i equal to 1 kV between phase and neutral. When a circuit exists between active conductors and earth, immunity should be assumed to be equal to 2 kV at this level. It may be considered that sensitive equipment has a lower rated impulse withstand (1.5 kV) between active conductors and earth and possibly lower immunity levels.

Installing an SPD in front of equipment to provide its protection is important for insulation coordination. However, it is generally considered that insulation between active conductors and earth is passive (distance on a printed circuit board (PCB) or in air, insulating material, insulating liquid or even insulating gas and others). In that case, until the rated impulse withstand is exceeded, the insulation is safe. It should be noted that insulation withstand is a time function: a lower voltage is needed to break the insulation with a long duration overvoltage compared to a short lightning surge. The SPD parameter to consider is U_p associated with the impulse withstand of equipment with the usual 1.2/50-μs surge (U_w).

When there are active components such as SPC or filters, between the active conductors and earth, a part of the surge current can flow through these components in the case of a surge between active conductors and earth (common mode surge). Then there is a possibility that the surge withstand of these internal components is exceeded in spite of being protected by the SPD at the appropriate level that would be enough for pure insulation. There is then a need to perform a coordination study between the SPD in front of equipment and these internal components. This is, of course, also valid for SPDs between phase and neutral (or between phases), where the main part of active circuits is located. Very often, protection efficiency is expected and no coordination study is made to prove it due to difficulty to perform such a coordination study. The cases where coordination studies are quite simple are when there is an internal SPC at the entrance of equipment. In that case, coordination is mainly reduced to coordination between an SPD and an SPC and coordination rules and methods are similar to coordination between two SPDs.

As indicated previously there are different ways to achieve coordination studies.

- let-through energy method
- coordination tests
- simulations

The let-through energy method is well appropriate to address this coordination issue between an SPD and internal SPCs or other components. The last stage SPD in front of equipment is represented thanks to that method by an equivalent combination wave generator. As the equipment is also tested for immunity with a similar generator, it is easy to determine if the output level of the last SPD is compatible with the input circuit of the equipment. See Section 7.4.

Coordination tests are also adapted to equipment with built-in SPCs: sharing of energy between internal varistors and external SPD can be checked by tests but it would be very costly to test coordination between this SPD and all the possible equipment that an SPD may protect in practice.

When the same configuration is used in many places (same SPD, same equipment to be protected) the user or manufacturer of equipment may decide to perform tests in real configuration.

It is also possible to perform generic coordination tests. In most of the cases, the entrance impedance of equipment can fit in one of the following categories:

- power supply through a transformer,
- power supply through a rectifier with in-rush current limitation,
- power supply with power factor correction,
- power supply enabling power line communication.

To perform these generic coordination tests, it is necessary to build, based on experience, typical equivalent input impedances for equipment and test an SPD for coordination with these impedances. Less than ten typical impedances are necessary with present equipment. With a limited number of tests, it would then be possible to give guidance to the user and to all parties about the real protective capability of the SPD when used in field.

Coordination simulation: it is possible to simulate the input circuit of equipment when it is known or, as described earlier, defined generic models (based on typical equivalent input impedances). In most of the cases, the main component to consider for the input circuit is the SPC when it exists.

Note: To obtain coordination with equipment internal components, a basic rule is to use an SPD with U_p as low as possible taking care of voltage limitations due to the power system (especially TOVs). Two-port SPDs and especially those with diodes or capacitors at the output port or Type 3 SPD with low U_p are generally well adapted.

Case study

A building that hosts sensitive and expensive equipment is located on a hill with high resistivity soil conditions. The building is supplied by an HV/LV transformer thanks to an underground long cable (more than 1 km). A telecom line of the same length is also connected to the building. The LV line and telecom line both enter the building in technical room where is located the MPB, a few subsidiary panels and the telecom main distribution frame. The MBP is protected by a Type 1 SPD ($U_p = 1.8$ kV and $I_{imp} = 12.5$ kA). The SPD disconnector is a CB C40A. The telecom line is protected by a Type 2 SPD ($I_n = 5$ kA). Following a lightning event, sensitive pieces of equipment were found destroyed even if they are supplied by UPS. Damages were limited (failed component on PCBs) but inhibited the operation of the equipment until the PCB was repaired. UPS were not damaged. Telecom SPDs were found disconnected. The grounding wire of technical room goes to a high resistance earthing system outside and run in an open cable tray where is also located the cabling from an air conditioning unit supplied by a panel located in the room where the sensitive pieces of equipment are located (this room was shielded). The following scenario has been established. An induced surge occurs both on the telecom line and power line at a level below the power SPD withstand but above the telecom SPD withstand. The telecom SPD fails. The current flows to ground. Due to high earth resistance, a part of the surge current flows up to the panel supplying the air conditioning but also other circuits such as lighting and power sockets in the sensitive equipment room. Coupling in the room between normal power circuit and UPS power circuit leads to damage to a few components on PCB inside sensitive equipment.

The installation presented a few problems: Type 1 SPD instead of Type 1+2 (in any case, the SPD disconnector was not compatible with the Type of SPD). No secondary SPDs in various panel boards and in front as well as at the output side of UPS. Grounding cable running inside a cable tray where other circuits are located. No separation between UPS and normal power circuits in sensitive equipment room. No SPD on the air-conditioning entering line. A probably weak induced surge caused an expensive damaged in a remote room, in spite of SPD at the entrance of the installation. Damages occurred at the back side of the UPS. An SPD in front of the UPS or even on the UPS circuit would not have been sufficient to avoid the problem, because the surge occurred between the UPS output and the equipment. It was necessary to avoid the penetration of a surge inside this room that was coming from an auxiliary circuit. A risk analysis would have been able to identify the sensitive equipment and the possible threat sources. A systematic approach of shielding and installation of SPDs on each incoming line would have been the solution, with U_p value compatible with the circuit immunity.

Sometimes, it is possible to use shielded cable instead of protecting a circuit with an SPD. Such a solution is presented in Section 7.1.

Ideally a shielded cable should be able to transmit energy or signals without any disturbing signal being coupled to the useful signal. However, the effect of the shielding is limited, even if it is homogeneous. For homogeneous shielding, immunity is only due to penetration of the electric field which is limited by absorption. Absorption is related to the thickness of the shield and the nature of the material constituting it. Absorption is related to the conductivity and magnetic permeability of the material. Indeed, for a thickness of a shield made of a material having a given electrical conductivity and magnetic permeability, the absorption depth is related to the frequency. In addition to the fact that it is advantageous to use materials with high conductivity and magnetic permeability, it is observed that the effectiveness of the shielding increases with the frequency, i.e. the absorption thickness decreases. It is therefore not useful at high frequencies to have thick shields.

If I_s is the current flowing through the shield, the electric field E_i along the surface of the shield is produced by an attenuated current density. The attenuation is determined approximately by the depth of penetration into the material forming the shield. The transfer impedance characterizes the effectiveness of the shielding with respect to the interference current I_s.

If V_i is defined as in Figure 7.44 (voltage along an elementary line length dx), the transfer impedance Z_t in Ω/m is given by the following formula:

$$Z_t = \left(\frac{1}{I_s}\right) \times \left(\frac{dV_i}{dx}\right) \tag{7.3}$$

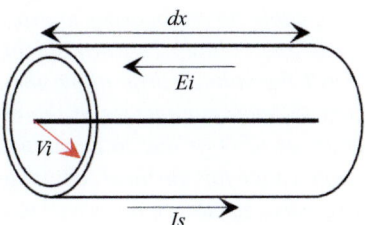

Figure 7.44   Transfer impedance

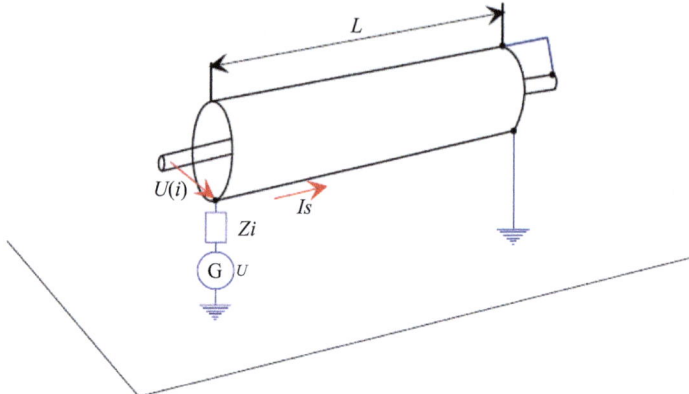

Figure 7.45   Laboratory measurement of transfer impedance $Z_t = U(i)/(L \times I_s)$

Except for homogeneous shields, the performance of the shields is not only due to the absorption thickness, but also to the geometry of the shield itself when the shield consists of discrete conductors at the periphery of the cable. These conductors are themselves of varied shapes but mainly in the form of flat ribbon wound on the periphery (ribbon or strip wound in a single or double helix) or cylindrical (many wires forming a braid with overlapping of the turns according to a more or less tight pitch). As it may be difficult to calculate the transfer impedance, tests can be performed at the frequency of lightning surges in the following way (Figure 7.45).

Tests in laboratory on various shielded cables used in the industry have shown that the transfer impedance can be as low as 2 m Ω/m for best cables up to 75 m Ω/m for a few of them. Assuming a 4-kV withstand between the shield and the inner conductors, a sparkover inside the cable will occur when a 10-kA 10/350 flows through the shield for a maximum distance of 200–5 m, respectively.

Type 2 SPDs may then be necessary to avoid this insulation breakdown in spite of the shield being used to carry partial lightning current, avoiding Type 1 SPDs on inner conductors.

Note: This transfer impedance gives a different result than the voltage drop calculated based on shield resistance as described in IEC 62305-3 'Minimum cross-section of the entering cable screen in order to avoid dangerous sparking'. In that clause of the standard, the voltage between shield and inner cable at one extremity of the cable is just the voltage drop along the screen that is being supposed to be short-circuited to inner conductors at the extremity (worst case condition) or via a low impedance SPD or due to capacitive coupling if the cable is very long. It should be noted that for an ideal shielding (homogeneous, without discontinuity), the surge transfer impedance measured must tend towards the ohmic resistance of the shield. For perfect shields, the difference is low (a few %) but for more usual shielded cables, it can be around two or even seven times bigger.

Case study

A shelter is protected by an LPS made of catenary wires supported by four metallic pylons. An earthing system is connected to each of the four pylons. The shelter houses sensitive equipment connected to both a power line and a telecom line, both lines being underground. Due to the presence of the LPS, the two incoming lines are protected by Type 1+2 and the smaller size of the shelter allows to preclude other SPDs inside the structure. The two lines do not enter from the same side. When the power line SPD is bonded to the LPS earthing system, the telecom SPD is bonded to another earthing system for communication reasons. The two earthing systems are bonded together via a 10-m conductor. A lightning event occurred nearby that transmitted a partial lightning current to the power line. In spite of the two SPDs protecting well, sensitive equipment was destroyed. This led to a lot of investigations because the small structure was so well protected, there should be no damage due to a nearby flash. It appears that there was a voltage drop along the bonding conductor between the two earthing systems that was sufficient to create a damage to sensitive equipment between the two ports (Figure 7.46).

Simulations have shown that, depending on the following (Figure 7.47):

- the earthing impedance of the two systems,
- the bonding conductor length,
- the location of strike (power or telecom line),
- the configuration of the power system and the neutral connection to ground.

A 10-kA 8/20-µs surge could create a voltage drop between the two ports ranging between 8 and 35 kV.

To avoid this phenomenon, there are two main possibilities. Either improve the bonding between the two earthing systems or install a multi-service SPD downstream of the entrance SPD. A multiservice SPD provides protection between the two services at the same time as the protection of each of the services (usually a Type 2 or 3 SPD) and possibly a bonding (direct or via an isolating spark gap) between the two earthing systems.

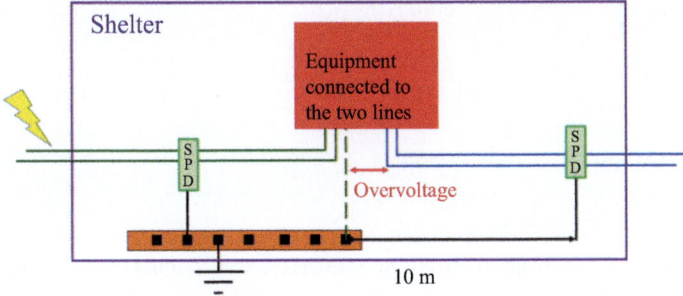

Figure 7.46   Investigated shelter case

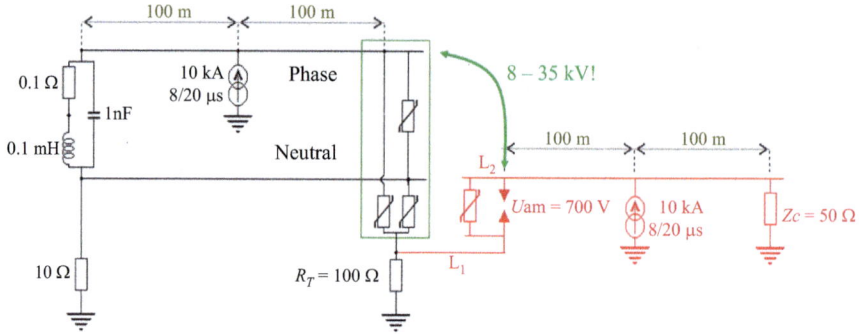

Figure 7.47   Simulation model

7.6   Selection of an SPD disconnector

When first SPDs were present on LV lines, they were derived from HV surge arresters and mainly dedicated to overhead lines. At this time SPD and SA were made of SiC blocks (silicon carbide) and due to low resistance of these blocks, a series gap was added. These preliminary SPDs were having a disconnector once again because HV SAs were having one. The disconnectors were either a metal part that was melting when the temperature increased too much and pushed back thanks to a spring or a black powder cartridge that was ignited by high temperature. The main aim was to avoid putting fire to the environment if the temperature was getting too hot. Evolution of basic protection components for HV SAs from SiC blocks to ZnO ceramics (zinc oxide, also known as metal oxide varistor) led to the same trend for LV SPDs. Due to this technology evolution, the SPDs moved from outside to panel boards.

The main differences between SAs and SPDs are as follows:

- Cooling thanks to the porcelain housing for SAs compared to heating due to panel boards temperature, coating of varistors and resin used in SPDs.
- Highest voltage for HV compared to LV and this means that small disturbances may have an impact on LV SPDs compared to HV SAs.
- Different compositions of the ceramics for LV compared to HV due to different specifications that make them more sensitive to repetitive small surges.
- Smaller volume for LV ceramics compared to HV ceramics and this means that the temperature increase of an LV ZnO ceramic has a different cinnetic than HV ZnO ceramics.

For all these reasons, a problem occurred on LV ZnO SPDs known as the thermal runaway of the MOVs (this phenomenon can also occur with HV SAs but rarely and in this case the end of life is a short circuit that is in most of the cases addressed by HV CBs). The first step of this process is coming by the fact a few junctions (the same concept as for electronic components) inside the varistor are destroyed due to too many small surges or ageing. With fewer junctions, the resistance of the varistors decreases slightly. With the operating voltage remaining the same, the current through the varistor also increases slightly. Due to higher current and thus higher power losses, the temperature increases locally. MOVs have specificity: the resistance of the MOV decreases with temperature and then the current will further increase leading to additional power losses and so on. As soon as this phenomenon starts, it will never stop. It can take weeks or more depending on ambient temperature and cooling possibilities of the SPD but will inexorably lead to high temperature. When the temperature reaches 120°C, the SPD is burning and this can cause panel board fire and possibly an extended fire in the structure.

Note: At the earlier days, the SPD state could be checked my measuring the temperature on the front of the SPD. This clearly shows that the degradation process for the first step of the thermal runaway was pretty slow.

As soon as this phenomenon appeared, the standard organizations started working to find a solution and this leads very quickly to the mandatory use of thermal disconnector for MOV SPDs. This requirement is of course always valid even if the progress in varistors has reduced the occurrence of such a problem.

This thermal disconnector (thermal fuse or equivalent solution) generally works once temperature exceeds 80°C and as soon as the SPD is disconnected, it cannot be used again and should be replaced. A local fault indicator is then mandatory to indicate that the SPD does not protect anymore.

Note: From that point started a frequent misunderstanding. This internal disconnector does not cover all the possible failure modes of an SPD but only the thermal runaway that is due to ageing or cumulative small stresses. Other modes such as short circuit of the varistor due to a too high surge are not covered. A few SPDs have been developed with two parallel MOVs each of them protected by a thermal disconnector. Due to uneven current distribution between the two MOVs, one MOV was supposed to be destroyed first and then disconnected. The second MOV was still present and provided a reduced but still existing protection. If the principle is fine, it just covers the thermal problems and not the short circuit problems. When there is a short circuit, there is no more protection.

The short-circuit failure mode was first covered by an external disconnector that would disconnect the SPD in the case of a short circuit inside the SPD. To check the efficiency of this disconnector, the SPD was prepared by replacing active part by copper blocks and this means that only the highest short-circuit current was critical. Due to the fact, the prospective short-circuit current can be low in the electrical installation and a damaged MOV block may have a resistance that is not nil, new tests are nowadays performed to cover the wide range of short-circuit currents (Section 6).

The external disconnector should work in conjunction with the SPD (this is what is called SPD assembly): the surge withstand of the disconnector should be at least equal to the surge withstand of the SPD. For that reason, at the origin, this SPD disconnector was expected to be a specific device specifically developed for SPDs. At the end of the day, it appears that everywhere in the world (except in Japan first and then in China) usual current protection components such as fuses or CBs have been used. This is not a problem for the tested disconnectors selected by the SPD manufacturer and used during the SPD test, but in practice only a few disconnectors are used during the tests (generally one fuse and/or one CB with a maximum current rating). There is a general rule: a high surge current rating for the SPD requests to use a high current rating of the disconnector. For example, when a 160-A gG fuse can withstand 13 kA 10/350 μs, a 16-A gG fuse will withstand only 0.7 kA 10/350. The SPD data sheet indicates the maximum rating allowed for the external disconnector. For higher rating, the SPD may not pass the short-circuit tests. The problem comes from the fact that very often, this maximum rating external disconnector cannot be used in the electrical installation because the rating of the OCPD is lower and it is needed that selectivity rules apply between the OCPD and the SPD disconnector (the SPD disconnector should operate first when the SPD fails). If the installation is protected by a 25-A fuse, there is no benefit to install downstream a 160-A fuse to protect the SPD. In practice in many cases, the maximum external disconnector declared by the SPD manufacturer cannot be used and a lower rating is necessary.

The question is then: is this lower rating external SPD disconnector able to withstand the surge rating of the SPD? And the answer is 'no' by using the general rule presented earlier. The second question is: what is the surge withstand of this lower rating SPD disconnector? And the answer is 'who knows?' because most of the overcurrent protective devices standard do not include a surge test. Using the rule given earlier (lower current rating means lower surge withstand), it is sure that for the highest surges the SPD can handle, the disconnector will operate, and this may cause a problem and consequence should be analysed.

The main problem occurs for Type 1 SPDs because for Type 2 SPDs, it is easy to find an SPD disconnector that will handle a nominal discharge current $I_n = 5$ kA 8/20. For example, a 20-A curve C CB easily withstands such a stress.

Let us take an example: the electrical installation is protected by a 100-A gG fuse OCPD (assumed to have a surge withstand equal to 7 kA 10/350 μs). An $I_{imp} = 12.5$ kA Type 1 SPD is installed downstream of the OCPD. The recommended SPD disconnector is a 160-A gG fuse. To ensure selectivity, the SPD disconnector current rating is selected equal to 63 A gG (assumed to have a surge withstand equal to 4 kA 10/350 μs) (Figure 7.48).

Table 7.8 summarizes the situation as a function of injected surge current into the installation.

Note: Percentage of cases for 10/350 μs surges is based on lightning current distribution (see IEC 62305-1).

Most of the lightning surges are multiples surges (80% of them have two impulses, then a majority of events). As soon as the SPD disconnector opens due to a first impulse, the second impulse is no more diverted to ground and will damage equipment. As a consequence, the supposedly protected equipment is not protected.

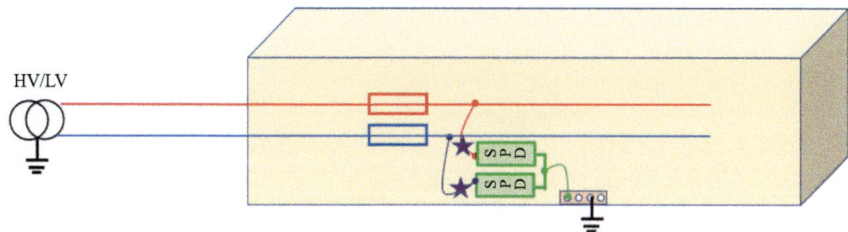

*Figure 7.48 Example with lower rated external SPD disconnector (purple star) –
scheme 1*

Table 7.8 Consequence of a lower rating for the external SPD disconnector

Surge current magnitude (10/350 µs waveshape)	OCPD status (100 A gG fuse)	SPD external disconnector status (63 A gG fuse)	Type 1 I_{imp} = 12.5 kA 10/350 µs SPD status	Global status	Percentage of cases based on LPL IV
$I \leq 4$ kA	Close	Close	Ok	Protected	95
4 kA$<I \leq 7$ kA	Close	Open	Ok	Not protected after the first impulse	2
7 kA$<I \leq 12.5$ kA	Open	Open	Ok	Not protected after the first impulse and no more powered	1
$I>12.5$ kA	Open	Open	Damaged	Not protected	2
Situation in the event of an SPD failure					Probability of occurrence
Failure due to other cause	Close	Open	Damaged	No more protected but powered	Very low

From the example of Table 7.8, we can derive that the installation that was supposed to be protected up to 12.5 kA 10/350 µs is in fact protected only up to 4 kA. A significant change in terms of surge withstand even if it will represent a small percentage of the cases (high current 10/350 surges represent a small amount of the surges).

Figure 7.49 will show an example (derived from a real case and simplified for easier understanding) where such a change in rating appeared to be unacceptable. An industrial building includes a cleaning process of fumes that are released to the atmosphere because the fumes may be toxic. As an additional safety measure, a sensor is located in the chimney to stop the process should toxic fumes be detected. This chimney is considered essential service equipment from safety point of view and as such protected by a lightning rod. A Type 1+2 SPD rated 12.5 kA per pole is

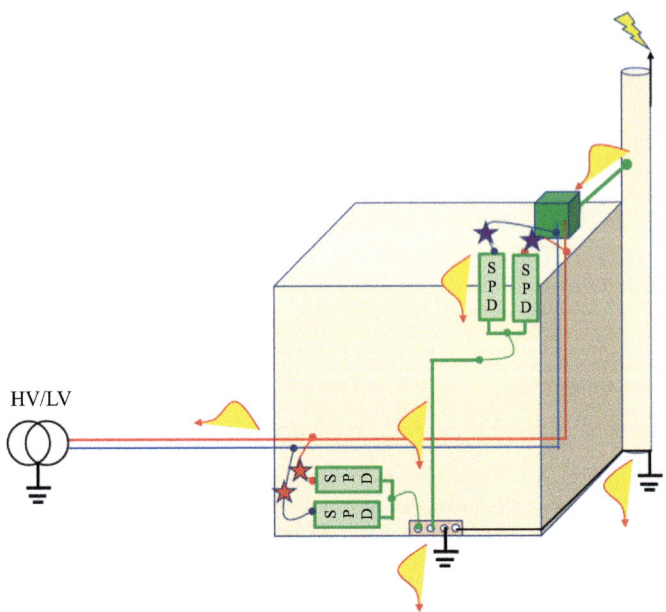

Figure 7.49   Example of a situation where lower rating external disconnector cause safety problems

installed at line entrance and the disconnector (red star in Figure 7.49) is a 160-A gG fuse. To protect the cabinet hosting, the sensor (sensor, power supply and cabinet are also essential services from safety point of view), another Type 1 SPD with the same rating is installed. This circuit is supplied from a protected by a 100-A gG fuse and to allow selectivity the SPD disconnector in the sensor box is rated 63 A gG – purple star in the figure (of course in the real case, there were CBs and not fuses). As soon as the current in the SPD exceeds 4 kA, the sensor will be not be protected. The Type 1+2 SPD was installed to protect the sensor with an efficiency of 98%, but at the end of the day it protects only with an efficiency of 95%. This was not acceptable from the point of view of the safety study.

It may appear that for less sensitive cases, this situation of installation surge withstand degradation is acceptable but this should be explained and recorded in a document and approved by the user.

For that reason, the selection of SPD disconnector has been discussed in length and becomes a confusing issue. Selection of a lower rating of the SPD disconnector compared to what is requested by the SPD manufacturer based on the type test that have been performed has always an influence either on protection or on continuity of power supply.

Let us take the same example that in Figure 7.48, but in that case the SPD and its disconnector is installed upstream of the OCPD (100 A gG fuse assumed to have a surge withstand equal to 7 kA 10/350 μs). An $I_{imp} = 12.5$ kA Type 1 SPD is installed. The recommended SPD disconnector is a 160-A gG fuse (Figure 7.50).

Note: It is assumed that the SPD recommended disconnector is able to meet the short-circuiting conditions at the SPD location and that other safety requirements related to overload current and indirect contact are met. When installation is directly supplied by the utility network, utility rules apply.

Table 7.9 summarizes the situation as a function of injected surge current into the installation.

If we consider Table 7.9, we can see that in that case, the situation is much better because the unprotected situations have been reduced. The limitation is by now coming from the OCPD of the installation.

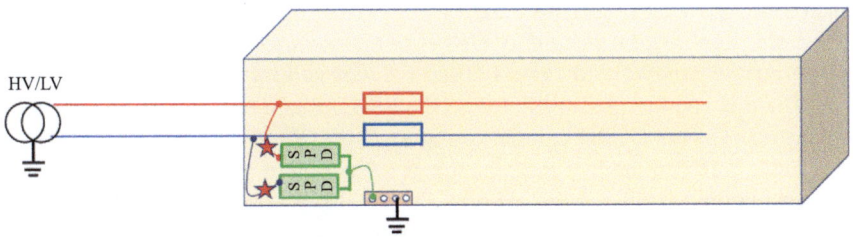

Figure 7.50   Example with normal rating of external SPD disconnector (red star) in front of the OCPD – scheme 2

Table 7.9   Consequence of SPD and its recommended disconnector upstream of the OCPD

Surge current magnitude (10/350-µs wave-shape)	OCPD status (100-A gG fuse)	SPD external disconnector status (160-A gG fuse)	Type 1 $I_{imp} = 12.5$-kA 10/350-µs SPD status	Global status	Percentage of cases based on LPL IV
$I \leq 7$ kA	Close	Close	Ok	Protected and powered	97
7 kA $< I \leq 12.5$ kA	Close	Close	Ok	Protected and powered	1
$I > 12.5$ kA	Close	Open	Damaged	Not protected	2
Situation in the event of an SPD failure					Probability of occurrence
Failure due to other cause	Close	Open	Damaged	No more protected but powered	Very low

Note: In practice the OCPD surge withstand is generally unknown for the reasons given earlier: there is no surge test request in most of the OCPD standards. RCDs time delayed Type S withstand 3 kA 8/20 μs and sometimes 5 kA 8/20. For gG fuses, IEC 61643-12 provides a comprehensive table that allows to determine the surge withstand with 8/20 μs or 10/350-μs waveshapes. Moulded case circuit breakers with large rated current normally have a high 10/350-μs surge withstand when miniature CBs may have a low surge withstand with 10/350-μs waveshapes. CBs have a higher (30%–100% more) surge withstand for 8/20-μs surges than gG fuses having the same rated current (in A). For weaker OPCDs, high 10/350-μs surges can lead to a severe damage of the OCPD when not protected by an upstream Type 1 SPD (Figure 7.51).

Note: The surge withstand of CBs is not correlated with its short-circuit current breaking capability (a 100-kA 50-Hz CB may withstand only 10 kA 10/350 μs).

There is another possibility to use the SPD without any external disconnector downstream of the OCPD (similar to Figure 7.48) (Figure 7.52).

Table 7.10 summarizes the situation as a function of injected surge current into the installation.

If we compare the three tables and concentrate on the lifetime of the SPD and its capability to protect against surges during that period, we can see that the best

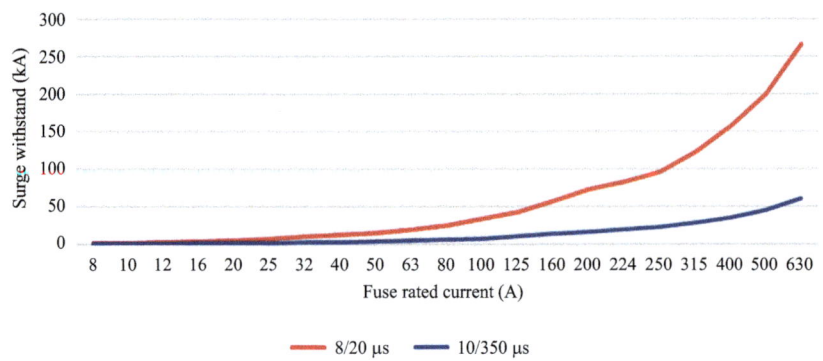

Figure 7.51   gG fuse surge withstand

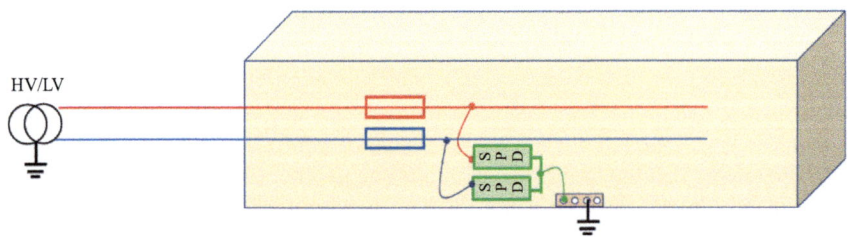

Figure 7.52   Example with no external SPD disconnector – scheme 3

Table 7.10   Consequence of an SPD without any external SPD disconnector except the OCPD

Surge current magnitude (10/350-µs waveshape)	OCPD status (100-A gG fuse)	Type 1 $I_{imp} = 12.5$-kA 10/350-µs SPD status	Global status	Percentage of cases based on LPL IV
$I \leq 7$ kA	Close	Ok	Protected and powered	97
7 kA$<I \leq$ 12.5 kA	Open	Ok	No more powered but protected	1
<12.5 kA Situation in the event of SPD failure	Open	Damaged	Not protected	2 Probability of occurrence
Failure due to other cause	Open	Damaged	No more protected nor powered	Very low

protection scheme is scheme 2, then scheme 3, and scheme 1 provides the weakest protection.

If we concentrate on the end of life of the SPD, we can see that the best schemes are schemes 1 and 2.

So globally scheme 2 is the best for all concerns when Type 1 SPDs are concerned.

Note: Normally the main OCPD defines the origin of the installation. SPDs are supposed to be installed on the downstream side of the OCPD of the circuit they are protecting. It is not always possible to install SPDs in front of an OCPD due to safety or regulation and this may require an application of special rules. IEC 60364-5-53 indicates, for example, when overload and short-circuit protection can be located downstream the origin of a branch circuit.

Note: In TT system, SPD in front of RCDs should use connection type CT2 only. In addition, the SPD between neutral and earth should be an SPD for which the manufacturer specifically declares in its installation instructions that they may be installed in TT systems between neutral and PE upstream the main RCD (purpose is to avoid that a failed SPD between neutral and earth change the system configuration). In TT systems, the RCD is normally installed at the origin of the installation (where the power supply enters the building or the MPB), except if the part of the installation between the origin of installation and the RCD complies with the requirements for class II equipment or equivalent insulation.

Knowing that the failure rate of SPDs due to other causes than excess of energy is generally low (1 per thousand or less) compared to the failure due to high surge current (a few %), the general best scheme, when scheme 2 cannot be used, is scheme 3.

Of course, the scheme to select also depends on the OCPD surge rating. The decision depends on the priority for the installation and user and will be different if the user wants to prioritize:

- power supply (user accept to maintain power supply even if no more protected) or
- surge protection (do not accept to keep power supply if not protected because devices to be protected are important or expensive and powered devices are more sensitive than unpowered ones).

The examples given earlier and their associated tables can be used to extrapolate the consequences to other cases and take decision accordingly. In any case, if for any reason, the recommended external SPD disconnector cannot be used, an analysis of the consequence should be done. In general, the decision on what to do in terms of location and selection of SPD external disconnector is a matter of compromise on which consequences are acceptable or not. Continuity of power supply and continuity of surge protection are generally never completely achievable and usually they are associated with the term 'priority' (priority to power supply or priority to surge protection) to show that the expected consequences cannot always be maintained.

It should however be noted that the selection of an external disconnector is mainly a problem for Type 1 SPDs.

The examples later are showing the influence of location of the SPD and its potential SPD disconnector on the surge withstand of installation for direct strike to the LPS and induced surges on the incoming power line.

The power line is single phase and neutral and the system configuration is TT. The LPS earthing is bonded to the electrical earth. The LPS LPL is IV (100 kA 10/350 μs with 50% in the earthing system and 50% equally shared by phase and neutral, so I_{imp} of the SPD = 25 kA 10/350). For induced surges, the expected level of surges is 10 kA 8/20 μs. A Type 1+2 SPD with I_{imp} = 25 kA and I_n 10 kA is then selected at line entrance. The installation is protected by an RCD Type S with overcurrent protection assumed to withstand 5 kA 8/20 μs and 1 kA 10/350 μs.

Note: Any overcurrent protective device that exists upstream in the installation, for example at the utility side, is supposed to have a higher surge rating and is not represented in the figures.

Three cases are investigated depending on the SPD disconnector rating and location:

- Case A: the SPD equipped with its recommended SPD disconnector (250 A gG fuse) is installed on the downstream side of the RCD. The SPD uses CT1 scheme (Figure 7.53).
- Case B: the SPD equipped with its recommended SPD disconnector (250 A gG fuse) is installed upstream of the RCD. Being in front of the RCD, SPD should use CT2 scheme (Figure 7.54).

    Note: The SPD disconnector is supposed to meet the requirement for short-circuit conditions at the SPD location and safety requirements related to fault

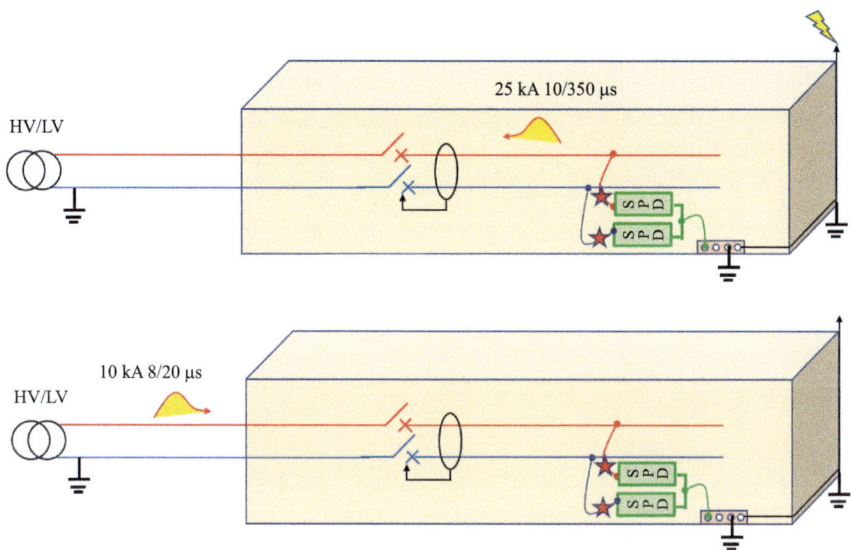

Figure 7.53   Case A – SPD with recommended SPD disconnector (red star) installed downstream of the RCD

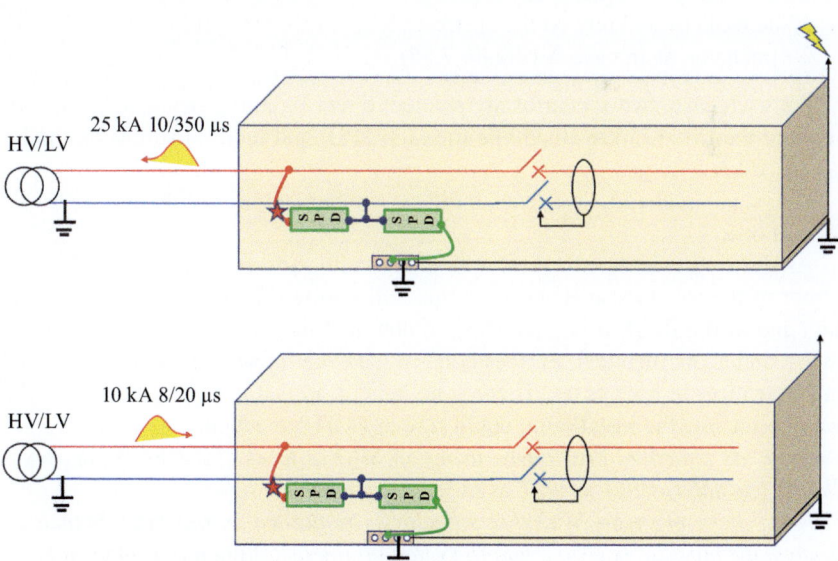

Figure 7.54   Case B – SPD with recommended SPD disconnector (red star) installed upstream of the RCD

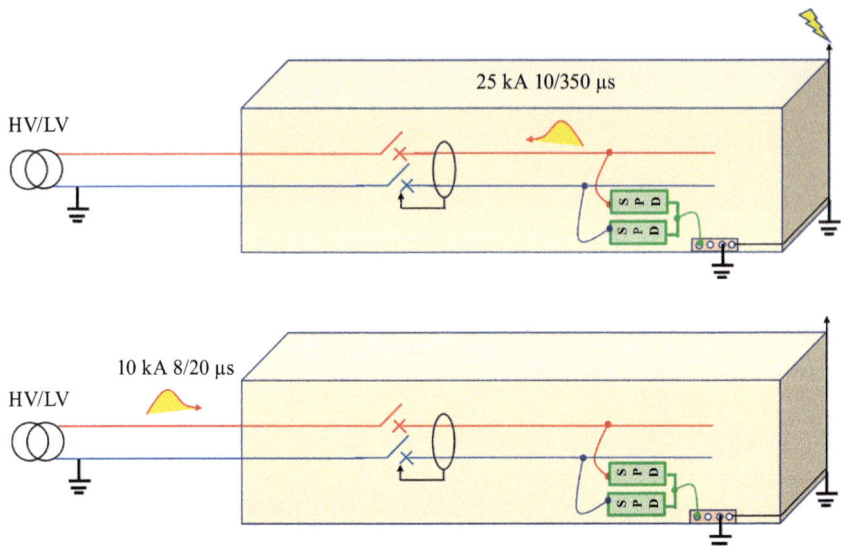

Figure 7.55   Case C – SPD without any SPD disconnector installed downstream of the RCD

conditions (indirect contact and overload) are met. When installation is directly supplied by the utility network, utility rules apply.

- Case C: the SPD without any disconnector is installed on the downstream side of the RCD (it will rely on the upstream RCD as a disconnector). The SPD uses CT1 scheme as in case A (Figure 7.55).

For each case, two scenarios are studied: direct lightning strike at 100 kA 10/350 μs to the LPS (25 kA 10/350 μs for each SPD) and induced surges on the LV lines at 10 kA 8/20 μs.

For each scenario, we can build a similar table (Table 7.11) to the previous one.

As shown previously, the best option (when it is allowed) is to install the SPD in front of the RCD (case B) because this will ensure the protection of the installation and of the RCD. It is interesting to note that the status cells of the table for cases A and C are identical. Furthermore, in the event of an SPD failure, the result is also identical in both cases. There is no benefit, in that example, to use an SPD disconnector and the installation could rely in RCD protection.

Note: As indicated previously, using an SPD in front of the RCD imposes to follow a few safety rules as well as to select appropriate SPDs.

Note: Very often an SPD disconnector is requested in the SPD branch for maintenance purpose (e.g. to avoid to switch off the installation to replace a failed SPD). However, to achieve this, it is possible to use SPD with removable active part (cartridges) or to install a simple disconnector without any overcurrent protection that will be more or less non-affected by surges.

Table 7.11 Consequence of SPD location in the case of direct impact on LPS

	Direct impact on LPS				
Case		**A (SPD with recommended SPD disconnector installed downstream of the RCD)**			
Surge current magnitude (10/350-μs wave-shape)	**RCD status**	**SPD external disconnector status**	**Global status**	**Percentage of cases**	**Situation in the event of an SPD failure (due to another cause than surge)**
$I \leq 1$ kA	Close	Close	Protected and powered	10	No more protected and no more powered because there is no selectivity between RCD and OCPD
1 kA$<I\leq 25$ kA	Open	Close	No more powered but protected	88	
Case		**B (SPD with recommended SPD disconnector installed upstream of the RCD)**			
Surge current magnitude (10/350-μs wave-shape)	**RCD status**	**SPD external disconnector status**	**Global status**	**Percentage of cases**	**Situation in the event of an SPD failure (due to another cause than surge)**
$I \leq 1$ kA	Close	Close	Protected and powered	10	No more protected
1 kA$<I\leq 25$ kA	Close	Close	Protected and powered	88	
Case		**C (SPD without any SPD disconnector installed downstream of the RCD)**			
Surge current magnitude (10/350-μs wave-shape)	**RCD status**	**SPD external disconnector status**	**Global status**	**Percentage of cases**	**Situation in the event of an SPD failure (due to another cause than surge)**
$I \leq 1$ kA	Close	/	Protected and powered	10	No more protected and no more powered because RCD is used as the SPD disconnector
1 kA$<I\leq 25$ kA	Open	/	No more powered but protected	88	

It must be noted that installing an SPD with an I_{imp} given by current sharing rules remains necessary even if, when located on the downstream side of the RCD, 88% of the cases will lead to an installation no more powered. In the absence of such a high rating SPD, not only the installation would be unpowered but probably also destroyed in full or in part. Using an SPD with a low value of I_{imp}, for example 1 kA, is not a valid option.

Note: Percentage of cases for 8/20 surges is derived from Table 7.2 when percentage of cases for 10/350-μs surges is based on lightning current distribution (see IEC 62305-1).

The situation regarding induced surges is presented in Table 7.12. The conclusions remain the same as for the direct lightning scenario. Case B provides the best protection options and cases A and C are also identical. There is no benefit in that example to install an SPD disconnector when the SPD needs to be located on the downstream side of the RCD, and it is possible to rely on the RCD.

Such an analysis should be, of course, adapted to each case taking care of

- type of stress the installation is submitted (direct lightning from LPS, direct lightning from line, induced surge on line and so on);
- type of system (TN, TT, IT);
- SPD connection type (CT1, CT2);
- safety rules related to overcurrent protection (short-circuit, overload and earth fault) and indirect contact;
- local regulation.

The main benefit of these examples is to show that the location of the SPD and the selection of its disconnector will influence the efficiency of the surge protection as well as the continuity of power supply.

Note: Selection of the SPD disconnectors depends on the surge current that is supposed to flow through the SPD (e.g. determined by current sharing rules or simulations) and not to the I_{imp} value written on the nameplate. For example, if the calculated surge current (sometimes called $I_{imp\ calc}$) is 4 kA and the selected SPD has a surge withstand $I_{imp} = 12.5$ kA, the gG fuse needed to both protect the SPD and withstand $I_{imp\ calc}$ is not rated 160 A (that would be needed for 12.5 kA) but 100 A only.

To avoid these limitations and the burden to study the consequences, an ideal external SPD disconnector should

- withstand high surge current (at least as high as the SPD surge withstand);
- operate before the upstream OCPD (based on comparison between the time vs. current operating curves);
- have a low residual voltage (not add to the SPD U_p when located in the SPD branch).

Note: The SPD disconnector is not required to have isolating capability.

Note: This search for an ideal SPD disconnector presently leads to the development of specific product specifications (see Section 7.6).

Table 7.12 Consequence of SPD location in the case of induced surge

Case	RCD status	Induced surge on the incoming line		
		A (SPD with recommended SPD disconnector installed downstream of the RCD)		
Surge current magnitude (8/20-μs waveshape)		**SPD external disconnector status**	**Global status**	**Percentage of cases**	**Situation in the event of an SPD failure (due to another cause than surge)**
$I \leq 5$ kA	Close	Close	Protected and powered	95	No more protected nor powered because there is no selectivity between RCD and OCPD
5 kA$<I \leq$ 10 kA	Open	Close	No more powered but protected	1	
Case	RCD status	**B (SPD with recommended SPD disconnector installed upstream of the RCD)**		
Surge current magnitude (8/20-μs waveshape)		**SPD external disconnector status**	**Global status**	**Percentage of cases**	**Situation in the event of an SPD failure (due to another cause than surge)**
$I \leq 5$ kA	Close	Close	Protected and powered	95	No more protected
5 kA$<I \leq$ 10 kA	Close	Close	Protected and powered	1	
Case	RCD status	**C (SPD without any SPD disconnector installed down stream of the RCD)**		
Surge current magnitude (8/20-μs waveshape)		**SPD external disconnector status**	**Global status**	**Percentage of cases**	**Situation in the event of an SPD failure (due to another cause than surge)**
$I \leq 5$ kA	Close	/	Protected and powered	95	No more protected nor powered because RCD is used as the SPD disconnector
5 kA$<I \leq$ 10 kA	Open	/	No more powered but protected	1	

Note: This will, of course, not solve the problems related to the unknown behaviour of a few OCPD under surge conditions that play an important role in the surge behaviour of the installation.

To solve the problem caused by external SPD disconnector selection, a few solutions are presently offered:

- SPD external disconnector being specific fuses with larger surge current withstand than given in IEC 61643-12: these may be used only with an approval of the SPD manufacturer because in general these surges have a slower time of response compared to gG fuses and this may overload the SPD in the case of its end of life.
- SPD disconnectors directly built-in the SPD (both thermal and overcurrent SPD disconnector are included inside the SPD enclosure): generally, these disconnectors are either CBs or fuses. There are then two cases:
  - Either the current rating and curve time vs. current are published or can be obtained from the SPD manufacturer: in that case selectivity between the SPD internal overcurrent disconnector and upstream OCPD can be demonstrated.
  - Or this information is missing: in that case, selectivity can neither be demonstrated or ensured and this should be considered in the end-of-life behaviour analysis.

Of course, an SPD disconnector, either external or built-in, should provide a status indicator to show that it is connected or disconnected. In this later case, the SPD is not protecting anymore and needs to be replaced and the user needs then to be warned. In any case, this status indicator needs to be local but in addition, it may also be remote for facilitating the maintenance.

Note: A few SPDs provide an intermediate status indicator. This is generally the case when two varistors are in parallel, each of them being associated with a thermal disconnector to avoid thermal runaway of MOVs. In that case, when a varistor fails first due to ageing, the second varistor may provide a reduced protection for a certain time (if the first varistor failed, it is likely that the second varistor will also fail and the SPD needs in any case to be replaced). Of course, this is not valid for all types of damages and is well adapted to ageing of MOVs. For a very high surge the two varistors may fail at the same time and the intermediate status indicator will not operate. This type of intermediate status indicator is rarely associated with the short-circuit disconnector. For sites where maintenance is difficult (e.g. remote sites) this intermediate status indicator may allow to delay the SPD replacement. For sites where SPD are not provided with remote fault indicator, a long time may occur before the next inspection cycle (generally yearly) allows to identify the failed SPD. This intermediate status indicator protection scheme may then increase the probability that the installation is still protected. For sites where maintenance is easier and an SPD remote fault indicator is provided, it is safer to replace the SPD as soon as it has been damaged, whatever is the cause of damage.

Note: When SPDs are located inside panels and not visible without opening the panel, or when non accessible or easily accessible (e.g. when it is needed to switch off power supply to access the SPD), it is recommended to use a remote contact to provide information on the SPD status on the front door. Additionally, for panels where SPDs cannot be seen from outside, a label indicating that there is an SPD inside is always a good idea to facilitate maintenance.

7.7   Surge-protection philosophy: six main points to take into account for easy SPD selection

There is an easy six-step selection process to select an SPD (when, for a few steps of the process, there are two paths to follow regarding SPD at entrance of installation, one for Type 1 and the other for Type 2, the Type 1 related items are written in blue whereas the Type 2 related items are written in green. In that case, the text in black is valid for all entrance SPDs).

* The first step is related to the presence of an LPS to protect the structure or not (an LPS may be a natural LPS with protection provided by the structure, provided by a risk assessment showing the need for an LPS and the natural component meets the criteria of an LPS). It may appear that an LPS only protects equipment and not a complete structure. In that case, SPDs should consider this local LPS as well.
  – If the structure is protected by an LPS then Type 1 SPDs are needed at each line entrance in the structure.
     The use of Type 1 SPDs can be extended to cases where direct lightning impact on the line near the structure is considered or when outdoor tall objects (such as lighting poles) are connected to the structure and direct lightning expected.
  – If the structure is not protected by an LPS (or equivalent cases), SPDs at line entrance are still needed but a Type 2 SPD at the line entrance will be usually enough.

* The second step relates to the rating of the entrance point SPDs.
  – Type 1 SPDs may be rated based on the usual values given in Section 7.1 (typically 12.5 kA per pole) but a more detailed calculation is generally necessary requested and the simplified method to calculate I_{imp} is presented in Section 7.2. When needed, a more accurate calculation can be done using explanations given in Section 7.2 depending on the need of accuracy and available data to make more precise calculations.
  – The usual rating for Type 2 SPD at the entrance is 5 kA per pole but a higher rating may be used (see Section 7.2).

* The third step is related to the selection of parameters relative to the system:
  – System configuration: TNC, TNS, TT, IT, other?
  – Nominal voltage of the system?

> If there is a doubt in selecting the appropriate SPD based on system configuration and nominal voltage only, refer to maximum continuous operating voltage U_c and TOV rating of the SPD in Table 7.1.

- Short-circuit current rating higher than the prospective short-circuit current at the location of the SPD.
- Miscellaneous: if appropriate (e.g. indoor use, temperature range, IP code and pollution index) when the installation location request a special care

- The fourth step is related to the protection efficiency of entrance SPDs. There are usually many possible SPDs in SPD catalogues meeting the parameters relating to steps 1–3. Data sheets in the catalogues provide voltage protection level (U_p) values. It is necessary to use the rules presented in Section 7.1 regarding the necessary margin between U_p and U_w to ensure the proper protection efficiency.

  The question is then: is the selected entrance SPD alone able to protect the installation or not?
  - If only safety is required (equipotential bonding), a single SPD in main panel board may be enough.
  - If the installation is small and U_p is adequate to protect equipment, a single SPD in the main panel board may be enough (generally, this SPD is then Type 1+2 instead of Type 1).
  - If there is no sensitive equipment or equipment that needs to be protected as requested by a risk assessment or safety equipment, at more than 10 m and the U_p is adequate to protect equipment, a single SPD in main panel board may be enough (generally, this SPD is then Type 1+2 instead of Type 1).

    For each of these cases, use rules presented in Section 7.3 to confirm that the selected single SPD is really protecting well when far away from equipment requiring protection.
  - If none of these cases applies, more than one SPD on the same line will be needed, especially closer to equipment to be protected or in secondary panels. These additional SPDs are Type 2 or Type 3.

- The fifth step relates to the selection of these other SPDs.
  - These secondary SPDs should also be selected using the third step (system voltage and configuration, short-circuit current rating, etc.).
  - They should be energy coordinated with the entrance SPD on the same line, following rules provided by the SPD manufacturer or following the rules presented in Section 7.4 (there is no need to define a specific value for I_n).
  - Select U_p of these secondary SPDs:
    o either U_p of the secondary SPDs is selected keeping a margin between U_p and U_w as indicated in Section 7.1,
    o or SPDs are voltage protection coordinated (preferred method) following rules presented in Section 7.1. In this case, it is likely that the

voltage at equipment terminal be lower than U_p written on the SPD nameplate.

- The sixth and the last step relates to the SPD disconnector selection (when requested by the manufacturer), see Section 7.6:

   Reminder:

   The disconnector tested during SPD type tests has the same surge withstand as the SPD itself. It may be a CB or a fuse or both. Sometimes a few possible SPD disconnectors are given in the SPD data sheet. A disconnector for a Type 2 SPD has generally a low rating but for a Type 1 SPD, the rating is generally higher to be able to withstand the surges at I_{imp}.

   The SPD data sheet gives the maximum rating that has been used during the tests. Lower ratings may fit because they will disconnect the SPD earlier than the maximum rating. The maximum rating cannot be exceeded because in such a case the SPD may not be disconnected safely. This maximum rating is sometimes too high, especially for Type 1 SPDs to be used in the installation, because the upstream OCPD has a lower rating.

   As the possible difficulties in selecting the SPD disconnector mainly occur with Type 1 SPDs, the following text use Type 1 as an example but this may be expanded to Type 2 as well.

   - If the installation OCPD upstream of the SPD has a lower rating than the one given in the SPD data sheet, in general the SPD disconnector may be omitted. Alternatively, a lower rating SPD disconnector may be used.

      o If the SPD disconnector is omitted (relying on the upstream OCPD to disconnect the SPD from the mains in the case of SPD end of life), the surge behaviour of the installation will be degraded compared to what was expected: at a lower current than I_{imp}. The OCPD may trip and the installation is no more powered. In addition, when the SPD fails, the OCPD will open and the installation will no longer be powered. In worst case the OCPD may be damaged by the I_{imp} surge and may need to be replaced before restoring the power.

         ▪ If this is acceptable, then no problem but the consequences in terms of continuity of power supply should be analysed.

         ▪ If not, another protection plan has to be established, such as locating the SPD in front of the OCPD when it is permitted (best option because it protects the OCPD) or using an SPD disconnector with a lower rating allowing selectivity with the OCPD (when a lower rating SPD disconnector is used, the consequences in terms of surge protection should be analysed, see next).

      o If the SPD disconnector rating needs to be lower than the maximum proposed in the data sheet, to ensure selectivity between SPD disconnector and the OCPD, the SPD may be disconnected by a surge current lower than I_{imp} (before it fails). The surge-protection behaviour will be degraded compared to what was expected: at a lower current than I_{imp}, the SPD may be disconnected and another surge occurring a few ms later (multiple surge) equipment may be destroyed at a level less than I_{imp} (for which the surge protection was designed).

- If this is acceptable, then no problem but the consequences in terms of surge protection should be analysed.
- If not, another protection plan has to be established, for example locating the SPD in front of the OCPD when it is permitted (best option because it protects the OCPD).

–   If the OCPD rating allows selectivity with the maximum rating of the SPD disconnector, the installation is fine, no further analysis is needed.

The systematic approach of the six steps allows to guarantee that all significant parameters for SPD selection are taken into account and that the expected protection is provided.

Now the selected SPDs need to be installed according to the rules presented in Section 7.1 (especially the 50 cm rule). When installation rules cannot be totally fulfilled, another selection of the SPDs may be needed, based on principles presented in Section 7.1, following the six-step process.

7.8   Specific rules for data and signal SPDs

7.8.1   General

Reliable transmission of signals and data plays a central role in the field of instrumentation, control and IT. A smooth operation of building services management, manufacturing or process technology requires a high level of quality and availability of the signals and data transmitted. However, these technologies are being exposed to an increasingly active electrical environment. This is especially true for the rather weak signals emitted by sensors and for high-speed data transmission. Small voltages or electric currents that must be securely transmitted, carefully conditioned or evaluated are increasingly being subjected to electromagnetic and radio frequency (RF) interference. Reasons for this are as follows:

- An increasing number of electrically operated components in all performance classes, especially motors operated via frequency inverters and other actuators.
- The increasing miniaturization and packing density of device components.
- A growing volume of wireless communication and control equipment.
- Digital systems that work with ever higher transmission frequencies and transmission rates.

Insufficient consideration of these sources of disturbances, inadequate adjustments to remedy faults or other planning deficiencies can all affect the reliable transmission of measurement values, signals and data.

Surge voltages, such as those caused by the effects of lightning, can also have a negative impact on the function and reliability of electronic modules in instrumentation or IT technology. Interference and damage caused by surge voltages in instrumentation or IT systems can, however, be effectively prevented by using tailor-made SPDs.

Depending on the potential for risk and the requirements of the desired level of protection, single-stage SPDs or multi-stage SPDs are used. These are installed directly upstream of the signal inputs to be protected. The circuits of the SPDs to be used are adapted to the various signal types.

7.8.2 Selection of SPDs in general

To be able to achieve an optimized protection effect against surges, SPDs are always designed application specific. This is especially true for SPDs for instrumentation, control and IT applications – because of the many different signals used in this kind of applications. The function of a typical protective circuit is described in Section 6.1.4.5.

To select the appropriate SPD following steps are necessary. First, a distinction is made between floating signals without common reference potential and signals with a common reference potential or a shared return conductor. Floating signals are insulated from the ground potential for interference immunity. A frequently encountered application of this type is the well-known 4–20-mA current loop for the *analogue* transmission of measured values and for the *digital* transmission of data – i.e. via the so-called HART protocol. The SPDs are designed to ensure continued isolation in the application (Figure 7.56). Gas discharge tubes guarantee isolation between the signal wires and the ground potential during operation. In the event of surge voltage, the GDT effectively discharges the transients to ground and limits the voltage differences – so that the dielectric strength of the end device is not exceeded. Typical dielectric strength of end devices is 1.5 kV (1.2/50 µs) between active lines and ground. In addition to protecting the dielectric strength between active lines and ground, protection between individual active lines is especially important for instrumentation applications in order to prevent exceeding the dielectric strength. The end devices are often much more sensitive to voltage differences of this nature, as sensitive semiconductor components in the end device are directly affected. Often, the corresponding dielectric strength of the

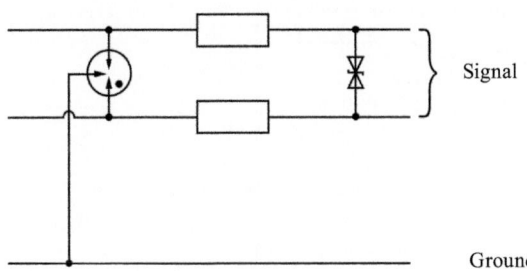

Figure 7.56 Basic circuit for floating signals – where decoupling resistors are permissible

devices is – between active lines – below 100 V. To achieve a good protection effect, in SPDs for the protection of 'sensitive' applications, frequently fast response transient voltage suppressor diodes (TVSDs) are installed between active lines. By using multi-stage protection schemes, with decoupling resistors between the individual stages, it is possible to achieve a high discharge capacity and to optimize the protection effect. For certain applications, it is not desirable to use decoupling resistors in SPDs. For this kind of applications, multi-stage SPDs without decoupling resistors are commercially available (see Figure 7.57). This can be the case with Pt-100 two-conductor temperature measuring circuits in which the decoupling resistors of SPDs can have an impact on the measurement accuracy. For actuator circuits with higher nominal currents, SPDs without decoupling resistors are also used. It has to be considered that multi-stage SPDs without decoupling resistors have a lower maximum discharge capacity between active lines, in comparison to multi-stage SPDs with decoupling resistors.

Applications with a common reference potential require a specially designed protective circuit, as the sensitive semiconductor components in the end devices can also be damaged by transient overvoltages between active lines and the reference potential. For this reason, in such cases the TVSDs are located between each active line and the reference potential. In cases where the reference potential is grounded, the SPD can be used, as shown in Figure 7.58. In most cases, a direct

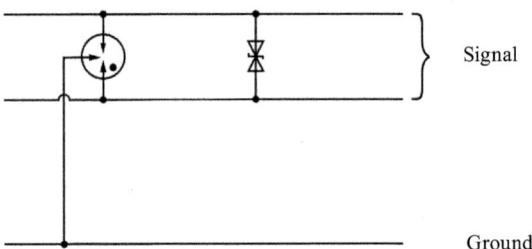

Figure 7.57   Basic circuit for floating signals – where decoupling resistors are not permissible

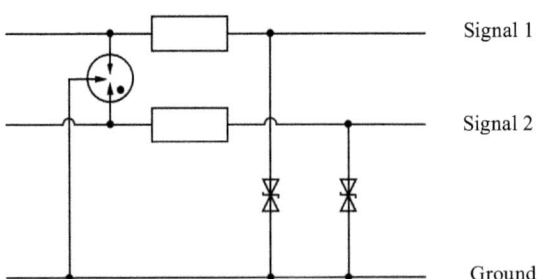

Figure 7.58   Basic circuit for applications with common reference potential – where the reference potential is directly grounded

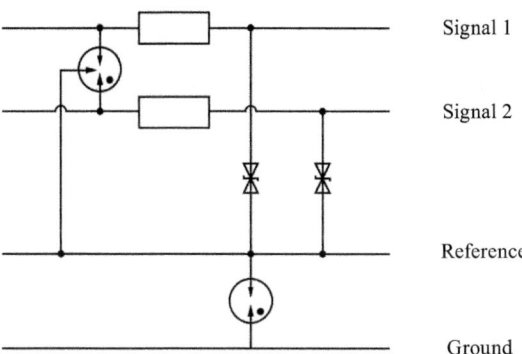

Figure 7.59   Basic circuit for applications with common reference potential –
where the reference potential is indirectly grounded

connection between the common reference potential (e.g. signal ground) and the ground potential is not permitted or desired. Circuit versions with an additional GDT between the reference potential and the ground are used for this kind of applications (see Figure 7.59). This is referred to as indirect grounding.

7.8.2.1   Easy selection of suitable SPDs

Due to the wide range of signals in real-world applications, it can be challenging to select the right SPD. Some companies, in the surge-protection industry, offer selection guides for SPDs. An example of such a guide provides comprehensive recommendations for the proper selection of SPDs and covers more than 150 types of applications in the field of instrumentation, control, telecom, data communications and antenna systems. For the selection of SPDs for complex applications, manufacturers of SPDs usually provide detailed application diagrams.

7.8.3   Selection of SPDs based on lightning protection zones

The necessity for implementing surge protection is determined based on a risk analysis. SPDs are then selected in accordance with the applicable IEC test categories. Usually, an SPD is installed at the boundary of two different LPZs (see Figure 7.60). To achieve an optimal protection effect, all active and passive lines need equipotential bonding at the boundary of two different LPZs. During the discharge of surge voltage impulses and surge current impulses, SPDs become low-ohmic. Only for this very short moment of time, when there is a surge, SPDs make sure that there is sufficient equipotential bonding between all active and passive lines the respective SPD is connected to.

In particular, LPZs are defined in facilities which are equipped with an external LPS. For example, the first stage of protection (D1) primarily offers protection against destruction for installation directly at the entrance to the building. The SPDs used are to be rated according to the expected level of threat. Subsequent SPDs (C2, C1) then only need to reduce the remaining surge voltages and redirect

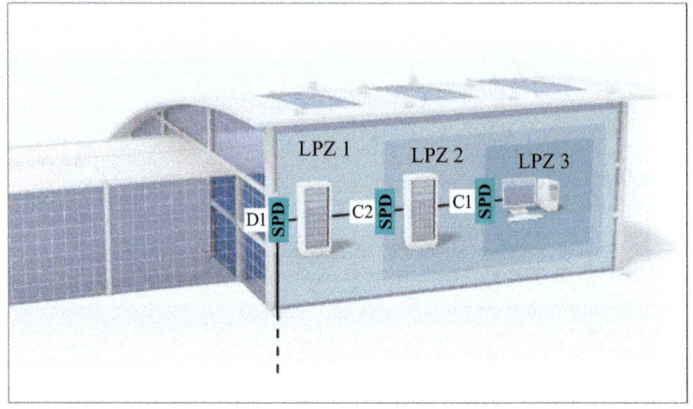

Figure 7.60   Lightning protection zones in accordance with IEC 61643-22

surge currents to a value acceptable for the end device. SPDs for the protection of power systems get installed at each location where an electric line goes from one LPZ to another LPZ (see IEC 61643-12). SPDs for instrumentation, control and IT applications usually also get installed at locations where lines go from one LPZ to another LPZ (see IEC 61643-22). In comparison to electric lines for power systems, it is not always mandatory to install SPDs for instrumentation, control and IT applications at each location where lines go from one LPZ to another LPZ.

Multiple stages of protection can be combined in one SPD for signalling, control and IT applications. Therefore, in practice, the choice is often made not to install SPDs at each location where lines go from one LPZ to another LPZ. This keeps the cost of installation low. As a practical solution, this SPD can be installed upstream of the device to be protected, such as the inputs of a controller. SPDs for the protection of power systems are classified as Class I (T1), Class II (T2) or Class III (T3) SPDs (see IEC 61643-11). For SPDs for the instrumentation, control and IT applications in total ten different categories designations are used (see IEC 11643-21). The major categories or SPDs are as follows:

- D1 SPDs – at the boundary between LPZ 0_A and LPZ 1; can carry lightning currents (10/350 μs)
- C2 SPDs – at the boundary between LPZ 1 and LPZ 2
- C1 SPDs – at the boundary between LPZ 2 and LPZ 3

The selection list (Table 7.13) from IEC 61643-22 provides information about the location at which each SPD class must be installed.

7.8.4   Selection of SPDs in instrumentation, control and IT applications

There is a great variety of SPDs in the field of instrumentation, control and IT applications. The selection criteria range from the obvious installation

Table 7.13 Lightning protection zone transitions and corresponding SPD types

Zone transition	$0_A \rightarrow 1$	$1 \rightarrow 2$	$2 \rightarrow 3$
SPD type corresponding to IEC-61643-21	D1	C2	C1
SPD type corresponding to IEC-61643-11	1	2	3

characteristics of the SPD and advantageous product features to the technical parameters of the application.

Installation characteristics:

- Mounting type

  SPDs are installed on the DIN rail as standard. For installing SPDs on field devices, it is sometimes easier to screw the SPD directly onto the sensor head.
- Connection technology

  Many SPDs feature the familiar screw connection. As many wires are connected in instrumentation, control and IT applications, the quicker, tool-free push-in connection is preferred.
- Overall width

  The number of signals to be protected in an instrumentation, control or IT application is often very large. A narrower SPD can therefore contribute considerably to allowing the entire control cabinet to be dimensioned in a smaller form.

Product features:

- Local status indication and remote indication

  Overloaded SPDs no longer offer protection and must be replaced. A failed device can be detected with the aid of a status indicator on the SPD. Optional remote indication makes it possible to transmit the status of an SPD to the control room and replace the SPDs quickly. Remote indications increase the quality of protection for the overall surge-protection system.
- Pluggability

  Pluggable SPDs can be replaced easily without the use of tools. It is recommended to use such kind of pluggable SPDs where signals are neither interrupted nor affected during plugging and disconnecting.
- Knife disconnection

  SPDs with knife disconnection offer the possibility of opening the signal path on the SPD. This way, SPD wiring to the field can be disconnected from wiring to the electronics. Maintenance work also becomes exceptionally simple, such as conducting insulation measurements to identify a fault in the field cabling, for instance.

Application parameters:

- Type of signal

  A distinction is generally made between signals with and without reference potentials. Signals with reference potential require coarse and fine protection elements between the active lines and the reference potential. Floating signals (without reference conductors), e.g. 4–20 mA current loops, require a fine protection element between both active lines, as this is where the sensitive electronics are installed, and coarse protection to the ground potential. The protective circuits are to be selected for the SPDs accordingly.

- Nominal voltage

  The nominal voltage of the application has a significant influence on the ability of sensitive end devices to withstand surge voltage impulses. As a rule, the lower the nominal voltage of the application is, the lower the voltage protection level of the SPD should be. The maximum voltage of the application may not exceed the highest continuous voltage U_c of the SPD, however, as it can otherwise lead to an overload.

- Rated current

  The rated current of SPDs for instrumentation, control and IT application is limited by the type of protective circuit. Since the nominal current in instrumentation, control and IT applications is generally low, an SPD with a low rated current is sufficient in many cases. The protective circuit must be varied for applications with higher nominal currents. As a rule, the nominal current of the application is not allowed to exceed the rated current of the SPD.

- Number of signal wires

  A separate SPD can generally be used for any pair of signal wires. In order to increase the packing density, SPDs that protect multiple signal wires are useful, e.g. two digital inputs with one common reference conductor.

- RF application or data interface (>1 Mbps).

  Many surge-protection circuits have a low-pass characteristic. Therefore, for RF applications, protective circuits are required where the signal attenuation is barely noticeable.

- Resistance-dependent measurement circuits

  In multi-stage protective circuits commonly used for instrumentation applications, decoupling resistors are used in the signal path that serve to coordinate between fine and coarse protection elements. Circuits without decoupling resistors are available for resistance-dependent measurements circuits which do not influence the impedance of the signal path.

  The following is a description of SPDs for the most important interface types.

7.8.4.1   Surge protection for current loops

Measured values are usually transmitted using standardized signals. The 4–20 mA signal is used especially often for applications where longer cables are in use. The measured value at the sensor is converted into a current signal. The ohmic resistance of the cable has no influence here on the current of the measured value transmission. For current loops, two signal wires are often used which do not necessarily require an additional reference potential and are routed isolated from the ground potential. In order to protect an application of this kind from transients, an SPD is needed at both end points. The respective SPD is equipped with a multi-stage protective circuit. Transient normal-mode voltages between signal wires and common-mode voltages to ground are effectively limited at both end points as a result (see Figure 7.61).

7.8.4.2 Surge protection for binary signals

In control technology, modules are often used that feature a higher number of signal inputs and outputs (digital in/digital out). Furthermore, there is a common reference potential that is often simultaneously used as a common return conductor from the field. The protective circuit suitable for this type of application is designed with two stages of protection. Between two 'neighbouring' signal wires, there is always protection through series connection of two suppressor diodes. Moreover, there is protection to the ground via a GDT so that, together, all conceivable transients are limited (see Figure 7.62).

7.8.4.3 Surge protection for temperature measurements

If a temperature measurement is taken using a temperature-dependent resistor, like Pt-100, the ohmic portion of the additional cables as well as the decoupling

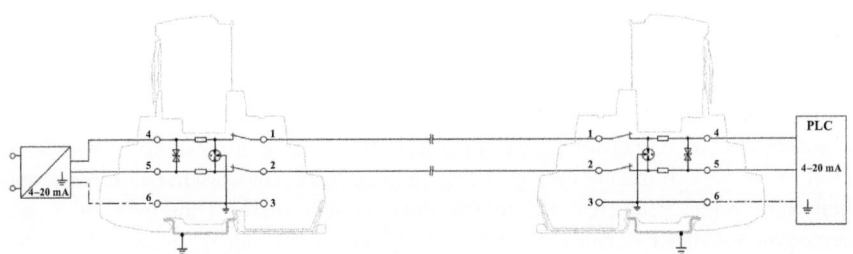

Figure 7.61   Example of a measuring signal transmission (4–20 mA) with surge protection at both ends of the line

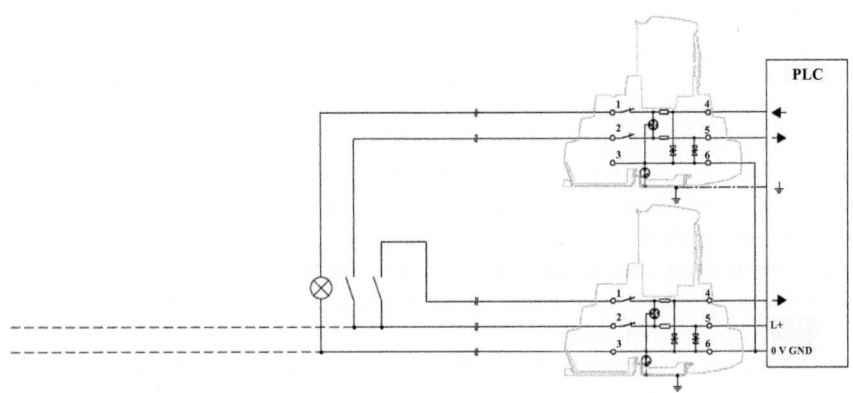

Figure 7.62   Example of protected binary inputs and outputs of a controller

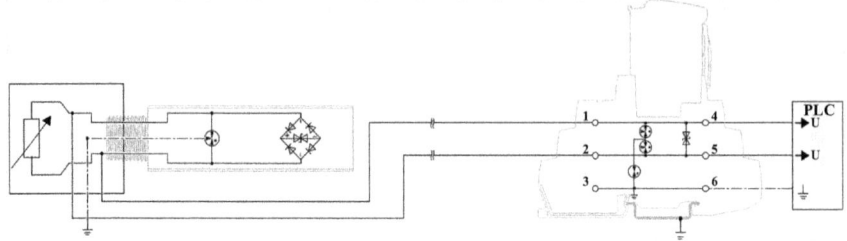

Figure 7.63   Example of protected two-wire temperature measurement (Pt-100)

resistors of SPDs specifically need to be taken into account. In the case of two-wire measurement, the resistance value of an SPD with decoupling resistors can cause a measurement inaccuracy. If the sum of the decoupling resistances in the measured circuit is, for example, 4 Ω, there is then a measuring error of 4% for a measurement of 0°C, as instead of 100 Ω, 104 Ω is detected. For this reason, the two-stage protective circuits are available as a version without decoupling resistors in order to minimize the influence of the SPD in this application (see Figure 7.63).

7.8.4.4   Surge protection in explosion-protected areas

Explosive atmospheres can frequently occur in the chemical and petrochemical industries due to industrial processes. They are caused, for example, by gases, fumes or vapours. Explosive atmospheres are also likely to occur in mills, silos, and sugar and fodder factories due to the dust present there. Therefore, electrical devices in potentially explosive areas are subject to special directives. This also applies to SPDs that are used in these types of applications.

Potentially explosive areas are divided into standardized zones. Classification for explosive dust and gas zones is found in the standard IEC/EN 60079-11. Zones are classified based on the probability that an explosive atmosphere will arise.

The Ex (i) intrinsic safety type of protection is used often in the field of instrumentation and control. Intrinsic safety, as opposed to other types of protection (such as increased safety), refers not only to individual items of equipment but to the entire circuit. A circuit is described as intrinsically safe if the current, voltage, the capacitance and inductance are limited to such an extent that no spark or thermal effect can cause an explosive atmosphere to ignite. The voltage is limited in order to keep the energy of the spark below the ignition energy of the surrounding gas (or dust). A thermal effect, such as a surface that is too hot, is prevented by limiting the current. Energy may also be stored in the form of capacitances or inductances within the intrinsically safe circuit. This must also be taken into consideration when examining the intrinsically safe circuit. Safety level ia, ib or ic defines whether protection is maintained with two faults or one fault in the protective circuit, or whether no protection is provided in the event of a fault. Intrinsic safety is based on the limitation of the amount of electrical energy in a circuit and the limitation of the generation of heat – during normal operation and under fault conditions. SPDs for intrinsic safe applications [Ex (i)] are designed in a way that the energy and the heat generation is limited properly. For all SPDs, used in intrinsic safe applications, type certification for hazardous environments is mandatory. All SPDs for intrinsic safe applications are marked accordingly. More detailed information about explosion protection measures, and about intrinsically safe circuits, can be found in the IEC/EN 60079-11.

7.8.4.5   Zone 0

Area in which a hazardous explosive gas atmosphere is present continuously or for long periods or frequently. These conditions are usually found inside containers, pipelines, equipment and tanks.

7.8.4.6   Zone 1

Area in which a hazardous explosive gas atmosphere is likely to occur periodically or occasionally in normal operation is to be expected only occasionally during normal operation. This includes the immediate area surrounding zone 0, as well as areas close to filling and emptying equipment (Figure 7.64).

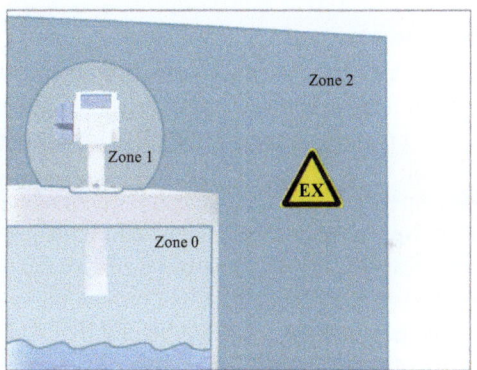

Figure 7.64   Zone classification at explosive gas atmospheres

7.8.4.7   Zone 2

The area in which a hazardous explosive gas atmosphere is not likely to occur in normal operation but, if it does occur, it will exist for a short period only. Zone 2 includes areas that are used exclusively for storage, and areas around pipe connections that can be disconnected and generally the intermediate area surrounding zone 1.

7.8.5   Selection of SPDs for the protection of local area networks

For SPDs in local area networks (LAN), all selection rules as mentioned in Section 7.8.4 are also valid. However, for LAN application, additional/special rules are necessary because of high data transmission rates. Communication via data networks is a part of daily life in all areas of business.

LANs operate with low signal levels at high frequencies. This makes them particularly sensitive to surge voltages and can lead to the destruction of electronic components in IT systems. In addition to protection that is tailored to these systems, SPDs must also exhibit high-quality signal transmission behaviour, as otherwise malfunctions are to be expected in the data transmission. This aspect is becoming increasingly important in the face of constantly increasing data transmission rates. To this end, when developing new SPDs for IT systems, the focus is on implementing high-quality signal transmission behaviour. It is evaluated based on the ISO/IEC 11801 or EN 50173 standards.

Furthermore, a wide range of connection technology is seen in this area of application. For this reason, the protective devices must correspond to the electrical specifications and must also be adapted to the networks to be protected. The SPD versions often differ only in their design and connection technology.

The protective circuits usually combine fast responding, low-capacitive suppressor diodes with powerful gas discharge tubes. Wherever required by the circuit technology, ohmic decoupling resistors are used between individual stages of protection.

7.8.5.1 Ethernet and token ring

Ethernet and token ring have been used for years. Ethernet systems have prevailed, however, due to their transmission speed and compact connectors. The transmission behaviour of the Ethernet system is defined in standard IEEE 802.3. The transmission speed is up to 10 Gbps.

The transmission speed is defined (Table 7.14) depending on the performance categories (cats. 5–7). Newer systems with high data transmission rates function in accordance with cats. 6 and 7, and eventually cat. 8.1 or 8.2.

Protective devices with Rj45 connection, where all eight signal paths are protected, are universally suited to the Ethernet, PROFINET and token ring (see Figure 7.65).

Table 7.14   Data transmission rates vs. performance categories

	Area of application	Category (EN 50173)	Mbps	Cable	Connection
100 Base TX (Fast Ethernet)	LAN, structured building cable	5	100	2–4-pair twisted pair	Rj45, pairs: 1–2, 3–6, or 4–5, 7–8
1000 Base T (Gigabit Ethernet)	LAN, structured building cable	5e, 6	1,000	4-pair twisted pair	Rj45, pairs: 1–2, 3–6, + 4–5, 7–8
10 GBase T (Gigabit Ethernet)	LAN, structured building cable	6a	10,000	4-pair twisted pair	Rj45, pairs: 1–2, 3–6, + 4–5, 7–8
10 GBase T (Gigabit Ethernet)	LAN, structured building cable	7	10,000	4-pair twisted pair	Rj45, pairs: 1–2, 3–6, + 4–5, 7–8

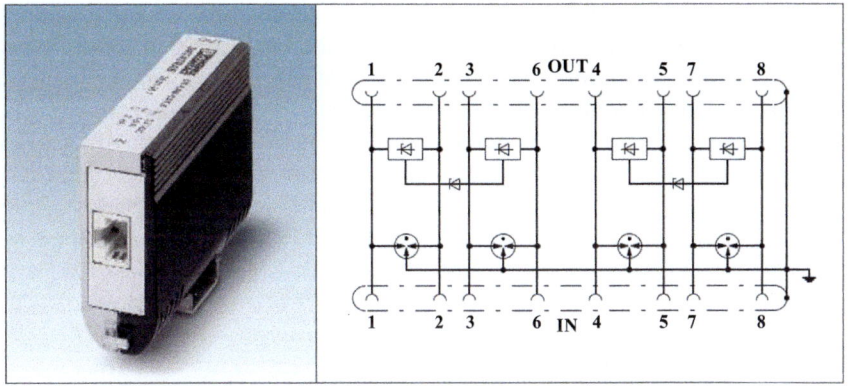

Figure 7.65   SPD for information technology

7.8.5.2   Power over Ethernet

Power over Ethernet (PoE) is an operation mode in which the auxiliary power for the connected devices is also transmitted via the Ethernet data cable.

The auxiliary power is applied either to the unused wire pairs or fed as phantom power between the signal wire pairs. In-line with IEEE 802.3af, a maximum power of 13.5 W can be transmitted using this method. The subsequent IEEE 802.3at standard now allows 25.5 W and with PoE+ and with PoE++ (IEEE 802. bt) even higher powers up to 100 W can be transmitted. Figure 7.65 shows a protective circuit which protects the data lines and also PoE.

7.8.6   Selection of SPDs in telecom application

For SPDs in telecom application, all selection rules as mentioned in Sections 7.8.4 and 7.8.5 are valid. The transmission systems are mainly called DSL (digital subscriber line). DSL interfaces provide Internet connections with speeds of 1 Mbps (ADSL) to 400 Mbps (VDSL). The transmission frequency is between 2.2 and 35 MHz. The nominal voltage for the protective circuit on suitable protective devices depends on whether a DC supply is also transmitted.

If telecom overhead lines together with power overhead lines are used, it is recommended to use SPDs with an additional protection against voltages from the power supply network (so-called power cross). The gas discharge tubes are equipped with additional thermal protection.

7.8.7   Selection of SPDs in antenna application

The selection rules of SPDs in antenna application are similar as mentioned in Section 7.8.5. However, the transmission speed and the thread are often higher. Antenna cables which extend beyond the building and are generally particularly long, as well as the antennas themselves, are directly exposed to atmospheric discharges. For this reason, cables with a coaxial design and associated favourable

EMC properties are used. The shield of the antenna cable can be either grounded or floating, depending on the system conditions. However, the risk of surge voltage coupling in antenna cables is not completely eliminated. Surge voltages can even reach the sensitive interfaces of transceiver systems via this cable path.

The high frequencies of wireless transmission require the use of protective devices with low stray-capacitance or low insertion loss with good impedance matching. Nevertheless, a good level of protection is required together with high discharge capacity. For this reason, most protective devices are equipped with powerful gas-discharge tubes or with the Lambda/4 (also called quarter wave) technology.

7.8.7.1   LAMBDA/4 stubs

SPDs with Lambda/4 stub use a shorting stub between the inner conductor and the shield. The length of the shorting stub matches a quarter of the wavelength of signals that are allowed to pass through without attenuation. A great advantage in this Lambda/4 SPDs is achieving a very good (low) voltage protection level, as the protective device functions as a (nearly perfect) short circuit in the typical frequency range of surge voltage impulses. However, it must be considered that the coaxial cable that is connected to the Lambda/4 SPDs cannot be used for the transmission of auxiliary power. Relatively wide bandwidth signals (e.g. 2.2–5.9 GHz) can be transmitted by means of RF-optimized Lambda/4 protective devices. Figure 7.66 shows a typical design of a protective device with a Lambda/4 shorting stub.

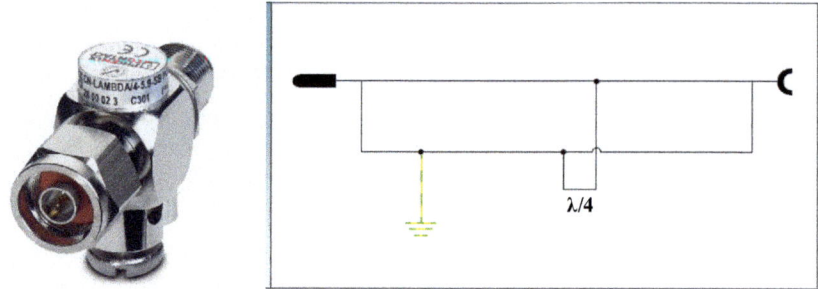

Figure 7.66   CN-LAMBDA/4 – protective device with Lambda/4 shorting stub

7.8.8   Function monitored protection

Monitoring protection elements combined with a remote indication function is especially useful for protective devices in difficult to reach locations. A status-oriented, even pre-emptive maintenance strategy is possible, thanks to the continuous status acquisition and evaluation of ageing indicators. Relating to the observed components of the TVSD (transient voltage suppressor diode) and GDT (gas discharge tube) of the protective circuit, the methods described later, based on physical and statistical principles as well as the links between the two, can be used to detect the 'health state' of the component and the evaluation of ageing processes.

- A 'direct' evaluation of the component state, i.e. a direct physical evaluation, is possible if there is a direct relationship between the measured variables and the ageing state to be detected. This kind of physical relationship exists, for example, between the leakage current generated by a TVSD and the extent of its damage.
- Statistical evaluation procedures are used if there is established knowledge on the load-dependent ageing and failure behaviour for the observed component. In this case, statistical statements can be made about the state by detecting the load (surges) and comparing it with load limits described in IEC 61643-21. Here, for example, the visual detection of gas discharge inside a GDT, associated with a flow of current through the GDT, is useful.

The visual detection of the flow of current through the GDT and detection of leakage current through the TVSD are used for the technical implementation of state acquisition and evaluation for a two-stage protective circuit. By collecting these measured variables, it is possible to make statements regarding prior component loads and physical component parameter changes continuously using suitable algorithms. This information is displayed via a status message. In order to provide the option of retrieving this information in a control room, for example, it is advantageous if the protection state can be indicated remotely. For this purpose, a floating contact – that can be evaluated by a PLC – is often used on the protective

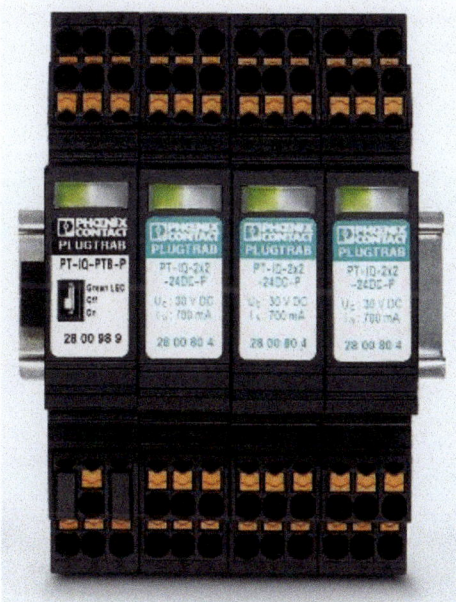

Figure 7.67   SPD with local and remote fault indication

device. The result can then be forwarded to the control room using various trans-mission media (bus or wireless systems). An example of SPDs with local status detection and remote indication contacts is shown in Figure 7.67.

7.8.9   Protection with increased security functionality

Electrical overloading, e.g. through high-energy transients, can lead to a destruction of TVSDs. Exemplary studies carried out to determine the ageing and failure behaviour of such diodes show that an overloading by surge-current impulses leads to an irre-versible degradation, which results in significant decrease of the 1-mA point voltage of the TVSD. Tests performed have confirmed that a correlation exists between the waveshape as well as the amplitude of the surge-current impulse, and the decrease of the 1 mA point voltage of the TVSD (impedance drop).

When using TVSD in powerful applications, especially with high operating voltages, such degrading leads to increased leakage currents through the diode and, linked to this, to increased temperature. In detail, investigation using pre-damaged TVSDs in a test set-up with adjustable DC operation currents up to 1,000 mA show that a current flow of more than 500 mA through pre-damaged TVSDs leads to sur-face temperatures which exceed the maximum operation temperatures of standard insulation materials and circuit boards. In the case of applying a DC current of 500 mA (mean dissipation power: 1.9 W), a mean surface temperature of 124°C was reached in thermal balance. Thereby, degrading of the insulation material could not be detected. When applying higher currents, the start of degrading of the insulation material up to the evolution of gas or in some cases a fire could be observed.

To avoid this critical situation in powerful applications, SPDs with a thermal disconnection function should be used.

Figure 7.68 shows the circuit diagram and the design drawing of an SPD with an integrated thermal disconnector for a TVSD. In undamaged state the TVSD is electrically connected to the signal path to be protected. The fundamental

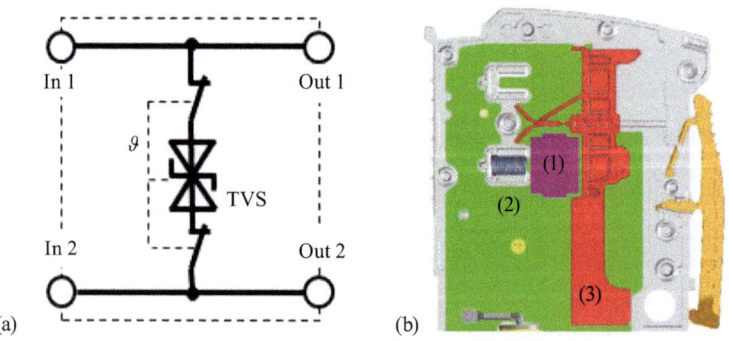

(1) Diode; (2) spring; (3) slider and status-indicator

Figure 7.68   SPDs with thermal disconnectors for a TVSD: (a) circuit diagram,
(b) design drawing

operational principle is to disconnect the TVSD in the case of overload via shifting it away from its position on the PCB. To achieve this, TVSDs in SMD technology are used.

Bibliography

CIGRE 2013 WG C4.408 "Lightning protection of low-voltage networks".

CIGRE-CIRED Guide July 2005 Working Group C4.4.02 "Protection of MV and LV networks against lightning – Part 1: common topics".

A.J. Surtees, F.D. Martzloff, and A. Rousseau "Grounding for surge-protective devices ground resistance versus ground impedance" IEEE Power Engineering Society General Meeting, 2006, pp. 8. doi:10.1109/PES.2006.1709080.

A.J. Surtees, M. Caie, and A. Rousseau "SPD – grenades without pins? coordination of an SPD's short-circuit current rating with the system to which it is connected" 29th ICLP 2008.

G.B. Lo Piparo, R. Pomponi, T. Kisielewicz, C. Mazzetti, A. Rousseau, and V. Crevenat "Performance evaluation of a coordinated surge protective devices system" IEEE International Conference on Environment and Electrical Engineering 2020.

T. Kisielewicz, G.B. Lo Piparo, F. Fiamingo, C. Mazzetti, B. Kuca, and Z. Flisowski "Factors affecting selection, installation and coordination of surge protective devices for low voltage systems" *Electric Power Systems Research*, 2014, vol. 113.

T. Kisielewicz, G.B. Lo Piparo, F. Napolitano, C. Mazzetti, and C.A. Nucci "SPD dimensioning in front of indirect flashes to overhead low voltage power lines" IEEE 15th International Conference on Environment and Electrical Engineering (EEEIC), 2015, pp. 1216–1221. doi: 10.1109/EEEIC.2015.7165342.

G.B. Lo Piparo, R. Pomponi, T. Kisielewicz, C. Mazzetti, and A. Rousseau "Protection against lightning overvoltages: approach and tool for surge protection devices selection" EPSR, Vol. 188, November 2020, 106531.

F. Fiamingo, G.B. Lo Piparo, C. Mazzetti, and A. Rousseau "A method to determine the need of SPD for the protection against lightning overvoltages of electrical installation supplied by power lines" 29th ICLP 2008.

A. Rousseau, G. Rougier, Z. Yang, and Z. Qibin "Coordination tests between SPDs and between SPD and equipment" 31st ICLP 2012.

A. Rousseau "Choice of low voltage surge arresters based on risk analysis" Power Quality, 1995.

A. Rousseau and V. Crevenat "Protective levels at equipment terminals for various SPDs" GROUND, International Conference on Grounding and Earthing & 4th International Conference on Lightning Physics and Effects, 2010.

A. Rousseau and T. Perche "Coordination of surge arresters in the low voltage field" INTELEC, 17th International Telecommunications Energy Conference: Intelec '95, October 29–November 1, 1995, Netherlands Congress Centre, the Hague.

A. Rousseau, P. Auriol, and A. Rakotomalama "Lightning distribution through earthing systems" Hobart Lightning Protection Workshop, 1992.

H. Altmaier, D. Pelz, and K. Scheibe "Computer simulation of surge voltage protection in low voltage systems" 21st ICLP 1992.

M. Clement and J. MIichaud "Overvoltages on the low voltage distribution networks" CIRED, 12th International Conference on Electricity Distribution, 17–21 May 1993, Brighton Centre, Brighton, UK.

J. Schonau, F. Noack, and R. Brocke "Coordination of fuses and overvoltages protection devices in low voltage mains" Fifth International Conference on Electrical Fuses and Their Applications ICEFA, 25th–27th September 1995, Technical University of Ilmenau.

P. Hasse, P. Zahlmann, J. Wiesinger, and W. Zischank "Principle for an advanced coordination of surge protective devices in low voltage systems" 22nd ICLP 1994.

F. Martzloff and J.S. Lai "Coordinating cascaded surge protective devices: High-low versus low-high" IEEE IAS, Annual Meeting, Dearborn, 1991.

J. Huse, O.T. Hostfet, T. Hervland, and B. Nansen "Coordination of surge protective devices in power supply systems: Need for a secondary protection" 21st ICLP 1992.

A. Rousseau and R. Gumley "Surge protection" *Wiley Encyclopedia of Electrical and Electronics Engineering*, vol. 21, 53–165.

H. Li, N. Xu, W. Chen, and R. Brocke "Advanced requirements on SPDs protecting sensitive equipment" 32nd ICLP 2014.

P. Zahlmann and R. Brocke "Selection and installation of SPDs in low voltage systems" VIII International Symposium on Lightning Protection (SIPDA), 2005.

B. Josef and Z. Peter "How to verify lightning protection efficiency for electrical systems? testing procedures and practical applications" IX International Symposium on Lightning Protection (SIPDA), 2007.

J. Birkl and P. Zahlmann "Lightning currents in low-voltage installation" 29th ICLP 2008.

F. Schork, R. Brocke, T. Böhm, and M. Rock "Requirements on surge protective devices in modern DC-grids" 34th ICLP 2018.

G.B. Lo Piparo and G. Carrescia "Protezione contro le sovratensioni" Tutto Normel Nuova edizione aggiornata alla più recente evoluzione normativa edizione, 2020.

IEC TR 62066 2002-06 "Surge overvoltages and surge protection in low-voltage a. c. power systems – General basic information".

IEC 61643-11:2011 "Low-voltage surge protective devices – Part 11: Surge protective devices connected to low-voltage power systems – Requirements and test methods".

IEC 61643-12:2020 "Low-voltage surge protective devices – Part 12: Surge protective devices connected to low-voltage power distribution systems – Selection and application principles".

IEC 61643-21:2012 "Low voltage surge protective devices – Part 21: Surge protective devices connected to telecommunications and signalling networks – Performance requirements and testing methods".

IEC 61643-22:2015 "Low-voltage surge protective devices – Part 22: Surge protective devices connected to telecommunications and signalling networks – Selection and application principles".

IEC 60038:2009 "IEC standard voltages".

IEC 60269 Series "Low-voltage fuses".

IEC 60269-1:2006+AMD1:2009+AMD2:2014 "Low-voltage fuses – Part 1: General requirements and other standards of the series".

IEC 60364-4-42:2010+AMD1:2014 "Low-voltage electrical installations – Part 4-42: Protection for safety – Protection against thermal effects".

IEC 60898-1:2015+AMD1:2019 "Electrical accessories – Circuit-breakers for overcurrent protection for household and similar installations – Part 1: Circuit-breakers for a.c. operation".

IEC 60947-2:2016+AMD1:2019 "Low-voltage switchgear and controlgear – Part 2: Circuit-breakers".

IEC TR 61000-5-6:2002 "Electromagnetic compatibility (EMC) – Part 5-6: Installation and mitigation guidelines – Mitigation of external EM influences".

IEC 61008-1:2010+AMD1:2012+AMD2:2013 "Residual current operated circuit-breakers without integral overcurrent protection for household and similar uses (RCCBs) – Part 1: General rules".

IEC 61009-1:2010+AMD1:2012+AMD2:2013 "Residual current operated circuit-breakers with integral overcurrent protection for household and similar uses (RCBOs) – Part 1: General rules".

IEC 60099-4:2014 "Surge arresters – Part 4: Metal-oxide surge arresters without gaps for a.c. systems".

IEC 60099-5:2018 "Surge arresters – Part 5: Selection and application recommendations".

IEC 61936-1:2021 "Power installations exceeding 1 kV a.c. – Part 1: Common rules".

IEEE 1159-2019 "IEEE recommended practice for monitoring electric power quality".

IEEE C62.41.1-2002 "IEEE guide on the surge environment in low-voltage (1000 V and less) AC power circuits".

IEEE C62.41.2-2002/Cor 1-2012 "IEEE recommended practice on characterization of surges in low voltage (1000 V and less) AC power circuits".

IEEE C62.72-2016 "IEEE guide for the application of surge-protective devices for use on the load side of service equipment in low-voltage (1000 V or less, 50 Hz or 60 Hz) AC power circuits".

Lightning and surge protection basics, PHOENIX CONTACT 2017.

IEC/EN 60079-11:2011 "Explosive atmospheres – Part 11: Equipment protection by intrinsic safety 'i'".

G. Finis, S. Pförtner, and M. Wetter "Safety-related functions and status indication for surge protective devices for the use in MCR applications" APL 2014.

CENELEC EN 50173-1:2018 "Information technology – Generic cabling systems – Part 1: General requirements".

Chapter 8

Specific application rules

Ralph Brocke[1], Nicholas Kokkinos[2], Alain Rousseau[3] and Antony Surtees[4]

8.1 DC network specificities

8.1.1 Introduction

The increasing use of regenerative energy sources together with upcoming new applications like electrical vehicles and storage technologies shift the nowadays dominant AC-supply to DC. This is also justified by the rise of efficiency when using DC-grids (Weiss *et al.*, 2015) and calls for an industry-led transformation of technologies from AC to DC.

Besides batteries storage applications and grids-fed unregulated rectifiers, modern DC-sources are characterized by power electronic converters. In industrial grids (Kaiser *et al.*, 2019) and modern building grids, the DC-link is forming the grid.

The voltage in these grids is an indicator of the grid power comparable to the frequency in AC-grids. Therefore, the power in the grid is regulated via the grid voltage value in a wide voltage band. So, the voltage can, e.g., vary from 180 to 440 V in a system at a nominal voltage of 380 V (Kaiser *et al.*, 2019). These voltage band regulations presuppose that there are no unregulated constant voltage sources in the grid (e.g. batteries). Typically, the regulating power electronic converters limit the short-circuit current (Schork, 2019) to $1.1–1.5 \cdot I_n$ and operate in a constant current source failure mode. The following differentiation could be taken:

1. Linear DC-grids fed by batteries or rectifiers with high short-circuit currents $I_{SC} \gg 1.5 \cdot I_n$.
2. Current controlled DC-grids fed by converters with short-circuit current $I_{SC} < 1.5 \cdot I_n$.

The functionality provided by modern power electronics (e.g. in AC/DC-converters and wind turbine converters) already changed the behaviour of modern AC-grids. With

[1]DEHN SE + Co KG, Neumarkt, Germany
[2]ELEMKO SA, Athens, Greece
[3]SEFTIM, Vincennes, France
[4]RAYCAP Inc., Post Falls, ID, USA

the upcoming energy storage applications and electro-mobility, the supply grid will be transformed from a pure AC-grid to a mixed AC- or DC-grid where both grids are coexistent. In consequence also the requirements for protection devices used in these grids will differ. In general, DC-grids could be distinguished into power grids having an almost constant voltage source characteristic and modern (controlled) DC-grids with limited short-circuit current capability. Furthermore, the used power electronics in DC-grids generate new requirements regarding protection levels, surge capability and frequency of protection. In addition, the expected temporary overvoltages (TOVs) are different to those known in AC-grids. Due to the coexistence of both AC- and DC-grids failure, feedback effects of both grids must be considered as well. Figure 8.1 shows a DC-distribution grid with AC- and DC-loads and sources.

8.1.2   *DC-grid categorization*

Structure of common DC-grids independent of the feeding source are shown in (Schork, 2019), (Dinesh *et al.*, 2017) and (IEC, 2020). Table 8.1 shows the DC-grid topologies divided into bipolar and unipolar grids. The topologies themselves do not differ from AC-grids and IT, TN-C-S and TT topologies are the same. The main differences between AC- and DC-grids are the source characteristic and the grid operation.

Figure 8.2 shows reasonable connection schemes used in the different grid topologies (Schork *et al.*, 2018).

Generally, all grid topologies are threatened by failure events causing overvoltages (Table 8.2) which define the requirements on surge-protective devices (SPDs).

The failure events may occur in both types of DC-grids, in current controlled DC-grids as well as in linear DC-grids. In many applications, the DC-grid is a mixture of both grid types. Figure 8.3 shows a controlled grid where a grid part of it is characterized by a linear source characteristic. Due to the high flexibility of

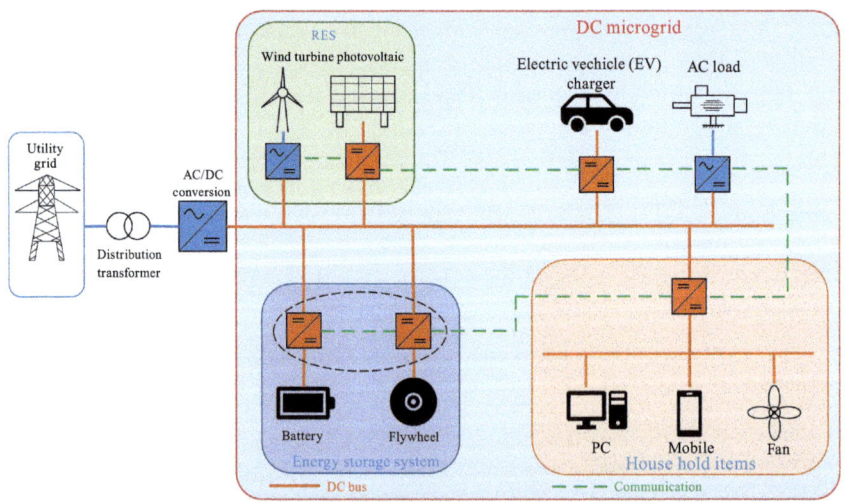

Figure 8.1   DC-distribution grid with AC- and DC-loads (Dinesh et al., 2017)

Table 8.1 Overview of DC-grid topology's (Schork, 2019)

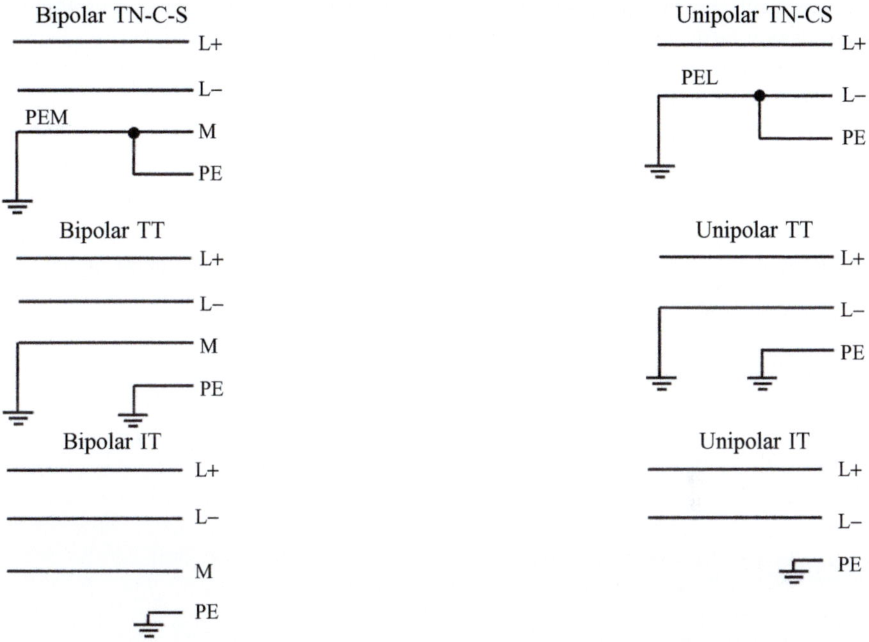

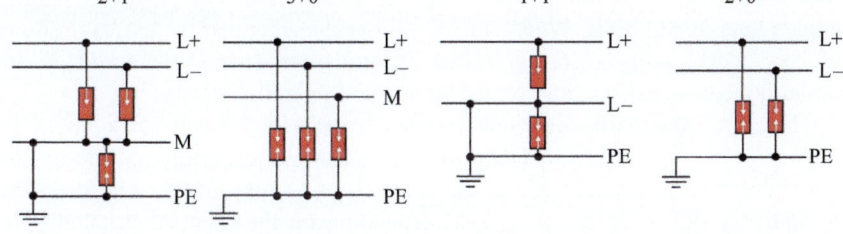

Figure 8.2 Connection schemes of SPDs in bipolar and unipolar DC-grids (Schork et al., 2018)

Table 8.2 Potential failure events causing overvoltages in DC-grids (Schork, 2019)

Transient	Temporary	Secondary currents
• Switching	• Power crossing	• Short-circuit currents
• Induction	• Loss of M	• di/dt
• Direct lighting events	• Earth fault	• High capacitive discharges
• Indirect lightning events	• Overcurrent	• Controlled sources

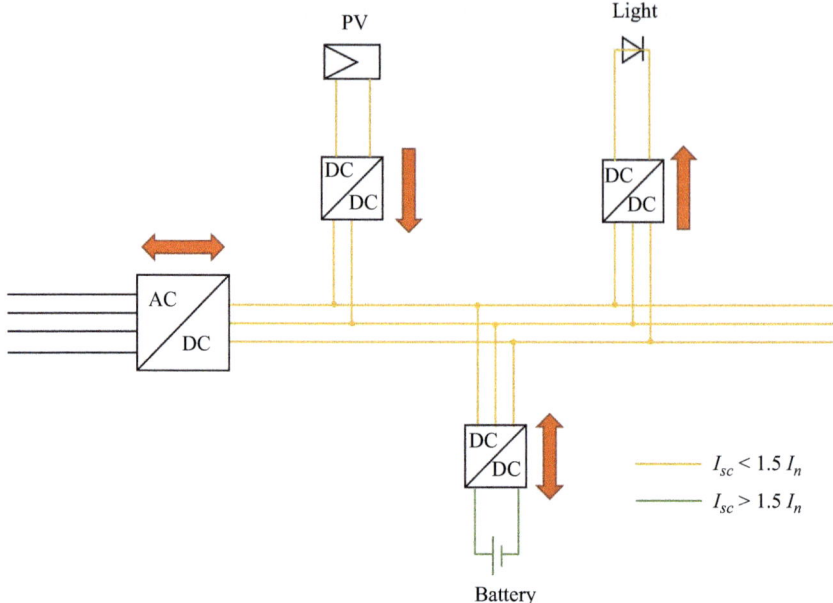

*Figure 8.3   DC-grid with a power part, a current-controlled part and different
controlled loads*

power electronic based converters, a general classification of DC-grids is more
complex than in AC-grids. Where an AC-grid is characterized by a voltage source
with low source impedance, controlled DC-grids can change their voltage and
internal impedances depending on the source and operation mode.

Therefore, a classification only related to a voltage level as common for
AC-grids is not applicable to DC-grids. An additional criterion considering the
short-circuit current capability, e.g. $I_{SC} < 1.5 \cdot I_n$ or $I_{SC} \gg 1.5 \cdot I_n$ is needed. This
criterion may differ for the same grid depending on the point of installation or
grid part.

8.1.3   Short-circuit evaluation

In AC-systems the course and level of the short-circuit current is determined by the
short-circuit power and the impedance of the supplying network and the design of
the transformer. The contribution of the power electronics to the continuous short-
circuit current is rather small compared to the supplying network.

In current-controlled DC-grids, the short-circuit current time behaviour consists of a transient part fed by the capacitive buffer capacities and a continuous part
fed by the power electronics. The transient short-circuit current is determined by
output capacitance, line impedances and the fault impedance, whereas the continuous short-circuit current is determined by the maximum currents of the AC/DC-
and DC/DC-converters (depending on its operating point).

Equation (8.1) describes the transient part and (8.2) represents the constant part.

$$\frac{di^2}{dt^2} + \frac{R}{L}\cdot\frac{di}{dt} + \frac{1}{C\cdot L}\cdot i = 0 \text{ start conditions } i(t=0)=0; \dot{i}\,(t=0)=\frac{U_{DC}}{L}$$

$$(8.1)$$

$$i(t) = \frac{U_{DC} - U_{\text{counter}}}{\sqrt{R^2 - \dfrac{4\cdot L}{C_S}}} \cdot \left(e^{-t/\tau_2} - e^{-t/\tau_1}\right)$$

aperiodic case   (8.1a)

$$\text{with } \tau_{1,2} = \frac{1}{\dfrac{R}{2\cdot L} \pm \sqrt{\left(\dfrac{R}{2\cdot L}\right)^2 - \dfrac{1}{L\cdot C_S}}}$$

$$i_{\text{trans}}(t) = \frac{U_{DC} - U_{\text{counter}}}{L} \cdot e^{-t/\tau}\cdot\ t$$

exponentially damped case   (8.1b)

$$\text{with } \tau = \frac{2\cdot L}{R}$$

$$i_{\text{trans}}(t) = \frac{U_{DC} - U_{\text{counter}}}{\omega\cdot L}\cdot e^{-t/\tau}\cdot\sin\left(\omega\cdot t\right)$$

periodic case   (8.1c)

$$\text{with } \tau = \frac{2\cdot L}{R} \quad \text{and } \omega = \sqrt{\frac{1}{L\cdot C_S} - \frac{1}{\tau^2}}$$

The transient short-circuit current according to (8.1) may cause oscillations having a negative current part (8.1c) resulting in also negative overvoltages at the converter as shown in Figure 8.4.

Depending on R, L and C, one of the three cases (8.1a) to (8.1c) will show up during a short circuit with a transient current amplitude in the range of 1 to 30 kA.

Continuous current $\quad i_{\text{cont}}(t) = 1.1-1.5\cdot I_n = \text{const.}$ (8.2)

The constant-current part of the short circuit is defined by the DC/DC-converter that usually operates under short-circuit conditions in a constant current

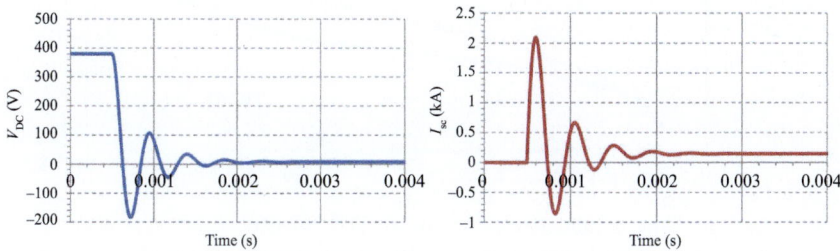

Figure 8.4   Short-circuit current (periodic case) without freewheeling diode

mode for a fixed period of time until it gets shut down. The amplitude of the short-circuit current is in the range of 1.1–1.5 times the nominal current.

In the case of oscillations the negative part is normally absorbed by the converter-internal freewheeling diode. Depending on the oscillation amplitude and energy, the converter may get damaged even if the voltage does not exceed the nominal voltage (Schork, 2019). Figure 8.4 shows an example of a periodic oscillation case with $R = 50$ mΩ, $C = 500$ µF and $L = 10$ µH where the negative overvoltage is about 60% of the nominal voltage.

Converters with bidirectional power flow always have this internal diode for commutation and are more sensitive to these events (Schork, 2019).

8.1.3.1   Linear DC-grids

At a first glance the grid short-circuit behaviour seems to be easier in DC-grids than in AC-grids. The challenge in DC-grids is to characterize the source diversity. This is standardized in IEC (1997) for rectifiers, batteries and capacities. At power grids with low source impedance the short-circuit current can be described as follows:

$$i(t) = \frac{V_{DC}}{R} \cdot \left(1 - e^{-t/\tau}\right) \text{ with } \tau = \frac{L}{R} \tag{8.3}$$

If a counter-voltage $V_{counter}$ occurs, the grid follow current changes to (8.3) as long as $V_{counter} < V_{DC}$:

$$i(t) = \frac{V_{DC} - V_{counter}}{R} \cdot \left(1 - e^{-t/\tau}\right) \tag{8.4}$$

Figure 8.5 shows a comparison of peak currents at comparable AC- and DC-grids during a short-circuit event when taking the impedance values of IEC 60974-1 Table 16 (IEC, 2020) into account. This comparison is permissible because the basic structure is comparable with regard to the cross sections and lengths of cables in AC and DC

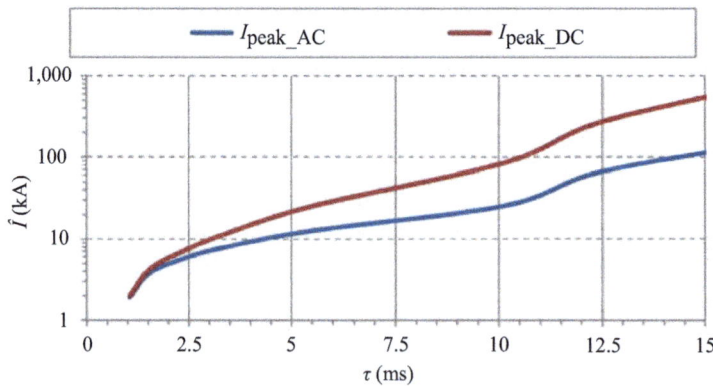

Figure 8.5   Comparison between AC and DC short-circuit currents having equal τ values and grid voltage amplitudes (IEC, 2020)

networks. It is obvious that even if the internal transient impedances are the same, the permanent short-circuit current in the DC-grid is higher due to the missing limiting reactance ($f = 0$ Hz).

8.1.4 Transient overvoltages

Generally, in controlled and in linear DC-grids the following four sources of transient overvoltages should be considered.

8.1.4.1 Overvoltage caused by switching operations

Switching overvoltages occur as a result of a high current rate of rise di/dt together with the presence of an inductive component in the grid. The switching dynamic and the inductance values due to installation length will not differ significantly compared to AC-grids. Therefore, the same switching overvoltages as shown in (Schork, 2019) can be expected in DC-grids as well. However, the values can be reduced for DC-grids with high capacitive buffering as they have a damping effect.

8.1.4.2 Overvoltage caused by lightning

A lightning event does not differ between AC- and DC-grids; therefore, the expected lightning amplitudes are the same and standardized in (IEC, 2010). However, there is a difference in the lighting current sharing within the grid. As an example, Figure 8.6 shows a comparison between a four-wire AC- and three wire-DC-grid with a TN-C topology. As there is one conductor missing generally, the stress per SPD in DC-grids is higher than in AC-grids.

8.1.4.3 Overvoltage caused by induced effects

DC-networks are not fundamentally different from AC-networks in their physical structure. To transfer the same amount of energy into the electrical system, nearly the same installation size is needed. Therefore, induced overvoltages due to indirect lightning strike effects will have similar amplitudes (IEC, 2010).

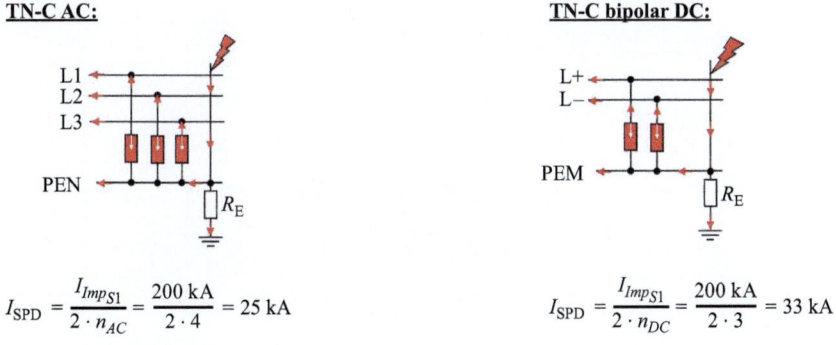

$$I_{SPD} = \frac{I_{Imp_{S1}}}{2 \cdot n_{AC}} = \frac{200 \text{ kA}}{2 \cdot 4} = 25 \text{ kA} \qquad\qquad I_{SPD} = \frac{I_{Imp_{S1}}}{2 \cdot n_{DC}} = \frac{200 \text{ kA}}{2 \cdot 3} = 33 \text{ kA}$$

Figure 8.6   Comparison of the effect of a lightning strike into an AC- and an equal DC-grid

8.1.4.4   Oscillation phenomena

The effect of oscillations differs in DC-grids. DC-grids with high capacitive buffering will oscillate when overvoltage is induced in differential mode. The oscillations may result even at small line length (1 m) in a voltage overshot factor of 2 (Schork, 2019). Basically, this effect can be seen in AC-grids as well, but due to the smaller capacitive values only at longer cable lengths (10 m). This phenomenon results in a reduced protection range of the SPDs used. In DC-grids with a capacitive buffer of more than 1 μF, SPDs show a significantly lower protection range than in networks with a small capacitive buffer.

8.1.4.5   Specific effects in applications with DC/DC-converters

In addition to the above-mentioned general transient overvoltage effects, one additional phenomenon must be considered in controlled DC-grids. Overvoltages generated as differential mode disturbances in applications with DC/DC-converters represent a specific additional load for the power electronic components. This is particularly true for negative surge currents (Figure 8.7). During the discharge of a negative surge current, the freewheeling diode prevents the SPD from being activated and the entire surge current flows through the converter.

Further investigations with different converter concepts are shown in (Schork and Brocke, 2020). Since the converter can feed both linear and current controlled DC-grids, this effect is independent of the source.

8.1.5   Temporary overvoltages

TOV in DC-grids differ from TOVs observed in AC-grids due to the different source and grid characteristic. Investigations on this topic are shown in (Kaiser, Schork, *et al.*, 2017) and (Kaiser, Gosses, *et al.*, 2017). Figure 8.8 shows the possible source for TOVs in DC-grids.

TOV feedback effects from the supplying AC-grid to a DC-grid only occur if a common galvanic coupling between both grids exists, i.e. when using an unregulated bridge rectifier. In this case, the known AC-TOV caused by a failure in the

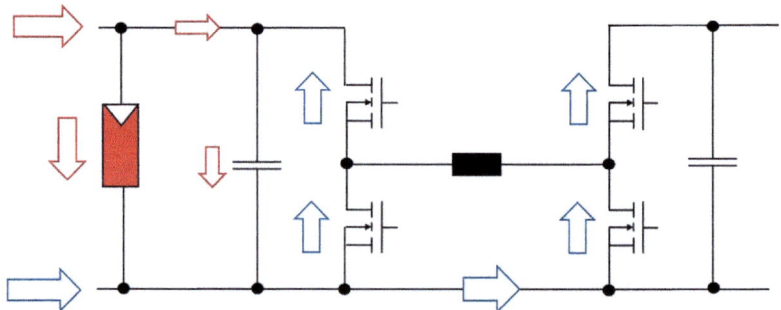

Figure 8.7   Effect of negative differential mode disturbances when using an SPD with unipolar protection level U_p to protect DC/DC-converters

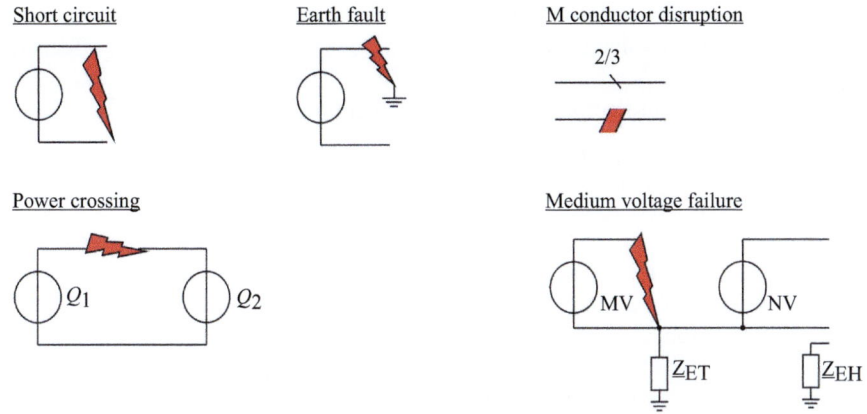

Figure 8.8   TOV failure situation according to Schork (2019)

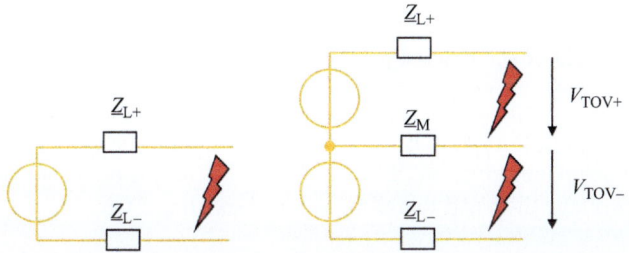

Figure 8.9   TOV at short circuits in unipolar DC-grids (left) and bipolar DC-grids (right) (Schork, 2019)

medium voltage can also affect the voltage between L+, L− and M to PE. Furthermore, single-phase rectifiers will forward AC-TOVs between L and N to the DC-grids.

When isolated DC/DC-converters are used the feedback effect of AC-TOVs will be handled by the converter isolation. This behaviour is independent of the DC-grid topologies (TN, TT or IT).

The power crossing (PC) effects from DC++ (high DC potential) to DC+ (lower DC-potential) or a direct connection from the AC-grid to the DC-grid due to isolation breakdown failures may also occur.

8.1.5.1   TOV caused by short circuits

During grid short circuits, the situation shown in Figure 8.9 occurs. In general, the TOV during a short circuit does not depend on the grid form, as short circuits show the same effects in grounded (TN, TT) or ungrounded (IT) systems.

In unipolar DC-grids, TOVs due to short circuit do not occur. However, in bipolar DC-grids, a voltage collapse results in a short-circuit current that affects the non-faulty pole.

In bipolar DC-grids, three short-circuit situations may occur:

1. short circuit between L+ and L−
2. short circuit between L+ and M
3. short circuit between L− and M

In case 1, no TOV can occur. The situation is equal to a short circuit in a unipolar grid. Cases 2 and 3 will lead to an overvoltage which is defined by the following equation:

$$V_{TOV} = V_{R_L} = V_M \pm V_{DC} = R \cdot i(t) + L \cdot \frac{di}{dt} \pm V_{DC} \tag{8.5}$$

The voltage drop over the M-conductor impedance will be added to the grid voltage and lead to a TOV. Theoretically in the case of asymmetric line impedances ($\underline{Z}_M \gg \underline{Z}_{L+}, \underline{Z}_M \gg \underline{Z}_{L-}$), a TOV of factor 2 is possible. A more realistic scenario is nearly symmetric line impedance ($\underline{Z}_M \approx \underline{Z}_{L+} \approx \underline{Z}_{L-}$) which leads to a shared (50%) voltage drop during short circuit.

$$U_{TOV} = 1.5 \cdot U_{L+} \quad \text{bzw.} \quad U_{TOV} = 1.5 \cdot U_{L-} \tag{8.6}$$

In controlled bipolar DC-grids with limited short-circuit currents, the TOV situation according to (8.5) may be limited even more. The transient short current follows the behaviour given in (8.3), but for TOV, (8.5) is relevant. This leads to a very small voltage drop and therefore to a non-existing TOV.

8.1.5.2   TOV caused by M-conductor disruption

When in linear DC-grids, loss or disruption of the M-conductor occurs, the load voltage shifts proportionally to the load asymmetry (Kaiser, Schork, *et al.*, 2017). The TOV can then be calculated as follows:

$$R_{L+} > R_{L-} \ \rightarrow \ V_{TOV} = \frac{R_{L+}}{R_{L+} + R_{L-}} \cdot V_{L+L-} \tag{8.7}$$

$$R_{L+} < R_{L-} \ \rightarrow \ V_{TOV} = \frac{R_{L-}}{R_{L+} + R_{L-}} \cdot V_{L+L-} \tag{8.8}$$

In controlled DC-grids, the source characteristic of the converter determines the course of the grid voltage during the M-conductor disruption (Kaiser, Schork, *et al.*, 2017). At a maximum asymmetry that means $R_{L+} \gg R_{L-}$ or $R_{L+} \ll R_{L-}$, the TOV could become

$$V_{TOV} = 2 \cdot V_{L+/-M} \tag{8.9}$$

Asymmetric load conditions in controlled DC-grids can be avoided when using specific converters (Han *et al.*, 2016). Consequently, TOVs can be neglected, if these types of converters are used.

8.1.5.3   TOV caused by earthing faults in bipolar DC-grids

TOVs only occur during an earth fault in bipolar DC-grids. The situation does not differ from that in an AC-grid.

In ungrounded bipolar IT systems, the occurring TOV depends on the source characteristic and can reach up to two times the nominal grid voltage at symmetric earthing conditions. As earth fault in an IT-system represents a first failure which does not threaten the load, a permanent TOV between lines of an earth may occur.

In grounded system however, any earth fault is like a short circuit and is described before.

8.1.5.4   TOV caused by power crossing

Figure 8.10 shows the possible PC situations in DC-grids. PC may occur from AC to DC or DC++ to DC. The reason for a PC can be a defect in an AC/DC- or DC/DC-converter as shown in Kaiser, Gosses, *et al.* (2017) or an isolation fault in grounded systems (Makkieh *et al.*, 2018). The occurring TOV in the case of a PC depends on the source characteristic, the load and on the line impedance. Therefore, a PC event in a controlled DC-grid with a nonlinear source characteristic (current source characteristic in the case of $I > I_n$, equal to a high source impedance) is a far greater threat than a PC event in a linear AC-grid. When neglecting transients, a PC DC++ to DC+ can be expressed by

$$V_{TOV} = \frac{V_{DC++}}{R_{i++} + R_{L++}||R_{L+}||R_{i+}} \cdot R_{L++}||R_{L+}||R_{i+} + \frac{V_{DC+}}{R_{i+} + R_{L++}||R_{L+}||R_{i++}}$$

$$\cdot R_{L++}||R_{L+}||R_{i++}$$

$$(8.10)$$

An AC to DC PC can be described by the following equation:

$$V_{TOV} = \frac{\underline{V}_{AC} \cdot \underline{Z}_{LAC}||R_{LDC}||(\underline{Z}_{LDC} + jX_{C_{DC}}||R_{iDC})}{\underline{Z}_{iAC} + \underline{Z}_{LAC}||R_{LDC}||(\underline{Z}_{LDC} + jX_{C_{DC}}||R_{iDC})}$$

$$+ \frac{V_{DC} \cdot R_{LDC}||Re(\underline{Z}_{LAC})||Re(\underline{Z}_{iAC})}{R_{iDC} + Re(\underline{Z}_{LDC}) + R_{LDC}||Re(\underline{Z}_{LAC})||Re(\underline{Z}_{iAC})}$$

$$(8.11)$$

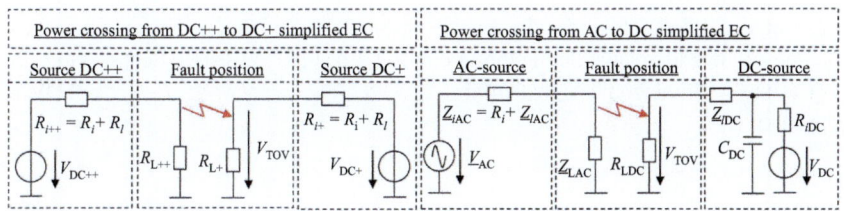

Figure 8.10   Possible power crossing events from AC to DC or DC++ to DC+ (Schork, 2019)

Table 8.3 Overview of the worst case threatening parameters in DC-grids (Schork, 2019)

Range of parameters			DC-grid			
Parameter category	**Parameter**	**Unit**	**Unipolar TN-C(-S) TT**	**Bipolar TN-C(-S) TT**	**Unipolar IT**	**Bipolar IT**
Transient overvoltages (surge voltage)						
Amplitude	$V_{1.2/50\,\mu s}$	kV	1.5 (@V_n=50 V) to 12 (@V_n=1 kV)			—
Rate-of-rise	$dv/dt_{1.2/50\,\mu s}$	kV/µs	1.2 (@V_n=50 V) to 10 (@V_n=1 kV)			—
Surge currents						
Amplitude	$I_{8/20\,\mu s}$	kA	0.1–10			
Rate-of-rise	$di/dt_{8/20\,\mu s}$	A/µs	12.5–1,250			
Commutation slope	$-di/dt_{8/20\,\mu s}$	A/µs	From −6.6 to −656			
Amplitude per pole	$I_{10/350\,\mu s}$	kA	25–50	17–33	—	—
Rate-of-rise	$di/dt_{10/350\,\mu s}$	kA/µs		up to 5	—	—
Grid follow currents						
Permanent	I_f	A	For DC/DC-converter type 1.1 to 1.5·I_n / For battery-buffered systems type kA-area		—	—
Transient	I_f	kA	Independence of C_{out} kA-area			
Temporary overvoltages (TOVs)						
Short circuit ($I_f/I_n<1.5$)	V_{TOV}	V	—	—	—	—
Short circuit ($I_f/I_n\gg1.5$)	V_{TOV}	V	—	1.5·V_n	—	1.5·V_n
TOV-duration during a short circuit	t	s	—	ms	—	ms
M-conductor-disruption	V_{TOV}	V	—	2·V_n	—	2·V_n
Duration during M-conductor-disruption	t	s		∞		∞
Earthing fault	V_{TOV}	V	—	1.5·V_n $I_f/I_n\gg1.5$	—	2·V_n
Duration during an earthing fault	t	s		ms		∞
Power crossing (PC) DC++/DC+	V_{TOV}	kV	0–1.5 depends on the source characteristic			
PC-AC/DC	V_{TOV}	kV	0 to ±1.0 depends on the source characteristic			
Duration at PC	t	s	Depends on the source characteristic			

The maximum possible TOV during PC depends on the source characteristic and voltage. If the considered DC-grid is under the umbrella of low-voltage systems (European Parliament and of the Council, 2014), the maximum system voltages of 1.5 kV DC and 1 kVA AC are the maximum expectable TOVs.

8.2 SPDs for DC-networks and specific applications

Table 8.3 summarizes the range of operation and stress parameters to be handled by SPDs. The actual application specifically the installation position, the grounding situation, the grid-form, etc. may require other parameters.

8.2.1 Effectiveness of SPD in DC-grids

In DC-grids the same parameters for surge voltages of EN 60664-1 apply as for AC-grid. The main differences with respect to the propagation of surge voltages in DC-grids are based in the greater grid capacitors in comparison to AC-grids. Capacitive energy storage components affect the surge voltage parameters and may cause oscillations. To examine this, the simulation model in Figure 8.11 is used.

In a simplified approach in Figure 8.11 high input impedance $Z_{DC/DC} \gg Z_L$ was assumed. For a typical half bridge converter, it is a realistic approach as long as the output capacitor of the converter is not included in $Z_{DC/DC}$. In addition, the surge voltage source in Figure 8.11 was simulated as an ideal voltage source as the lighting current could be considered constant current source.

The surge voltages at the converter input then depends only on the line impedance Z_L (R, L, C) as well as on the output capacitor C_{out}. Figure 8.12

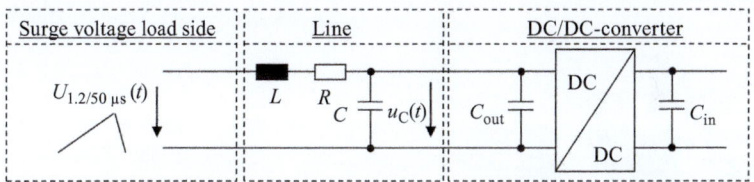

Figure 8.11   Equivalent circuit to estimate the surge voltage on DC/DC-converters

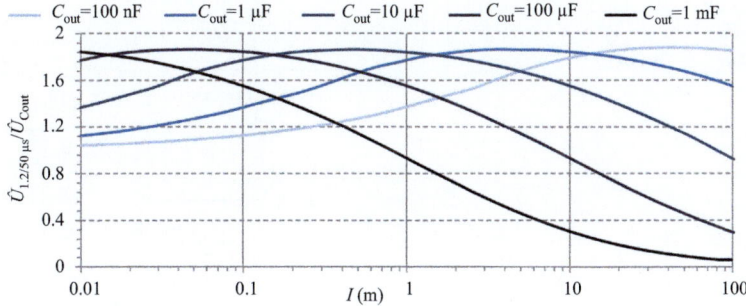

Figure 8.12   Voltage overshot factor K_U as a function of line length

illustrates the effect of huge capacitive components in an AC- or DC-grid. The voltage overshoot due to oscillations could go up to a factor of 2 at the end of the line and is well known from RLC oscillation circuits and does not depend on the grid voltage (AC or DC). Due to the high grid capacitors in DC-grids in contrast to AC-grids, oscillations can be expected at short line length. For example, DC networks with capacitors larger than 10 µF show a high risk of oscillations and a surge voltage load twice as high as in AC networks can be expected.

As a conclusion of the results in Figure 8.12, the protection range of SPDs in DC-grids decreases with increasing capacity values.

8.2.2   Selection of SPDs for controlled DC-grids

Figures 8.13–8.15 show examples of installations in controlled DC-grids equipped with SPDs at different points of installation to protect from overvoltages.

8.2.2.1   Example 1 – DC storage

In order to protect the equipment installed in a DC-storage plant as shown in Figures 8.13, the following specific requirements apply:

SPD installed in front of the battery rack:

- high volatile grid voltage depending on the state of charge (SOC) up to 1.5 kV;
- very low protection level $U_p \leq 3$ kV L+/L− to PE to protect the battery internal foil safely;
- asymmetric protection level needed depending on the converter technology;
- short-circuit current I_{SC} up to 10 kA per rack and $I_{SCtotal}$ up to 200 kA for all racks in a container.

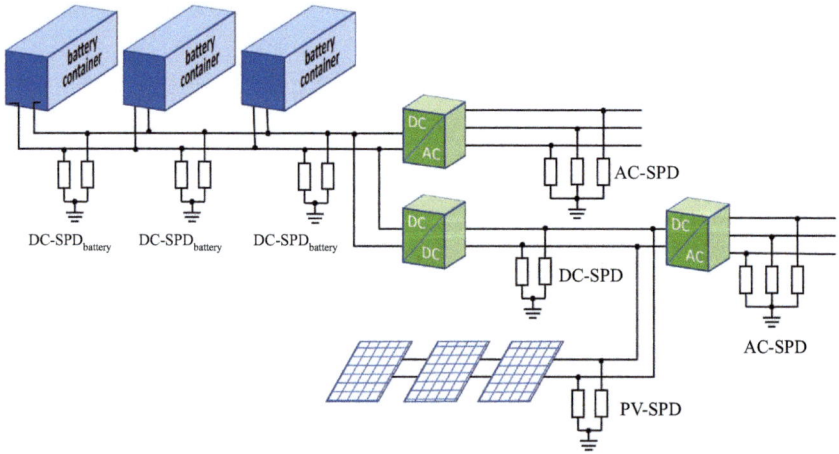

Figure 8.13   DC storage with battery containers, PV-power supply and AC-grid connection

SPDs installed on the photovoltaic (PV) side:

- See PV requirements.
  SPD installed on the AC side:
- See AC requirements.

8.2.2.2 Example 2 – EV DC charging

In order to protect the equipment installed in the EV charging station as shown in Figure 8.14, the following specific requirements apply:

SPD-installed downstream the active front end:

- grid voltage up to 1.5 kV;
- protection level requirement according to the relevant overvoltage category;
- asymmetric protection level among the L+, M, L− depending on the converter technology;
- high short-circuit current rating (SCCR) depending on the medium voltage transformer.

SPDs installed on the charging side:

- 350-kW charging power;
- 950-V charging voltage;

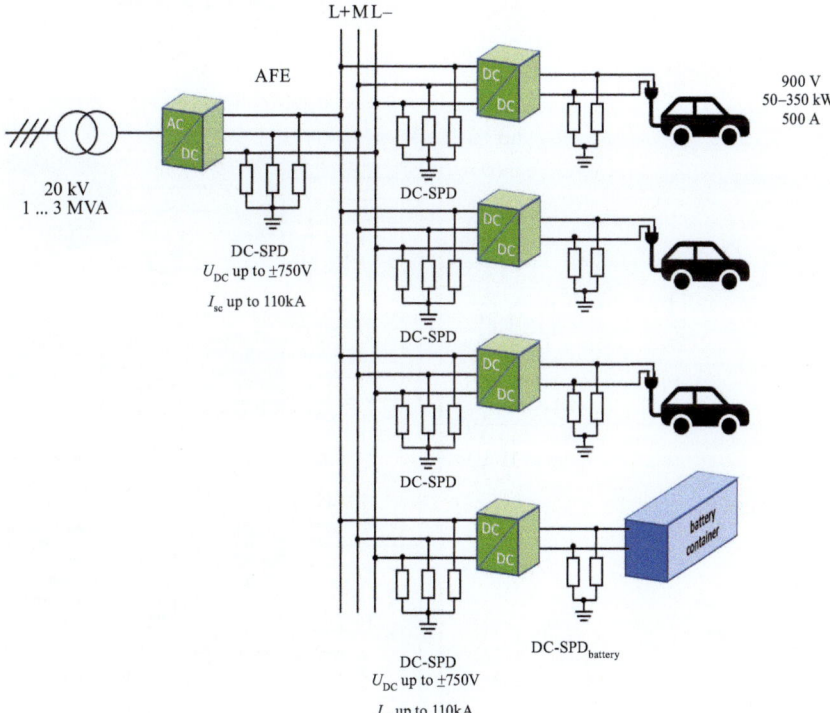

Figure 8.14 EV charging station with AC-grid connection via active front end (AFE) and DC storage to reduce peak loads

- up to 500-A charging current,;
- asymmetric protection level needed;
- SCCR 1.1–1.5×I_n.
    SPD installed on the battery side:
- See requirements of batteries.

8.2.2.3   Example 3 – DC industry

SPD-installed downstream the AC/DC converter as shown in Figure 8.15:

- grid voltage up to ±750 V;
- protection level requirement according to the relevant overvoltage category;
- asymmetric protection level among the L+, M, L− depending on the converter technology;
- high SCCR depending on the medium voltage transformer.
    SPDs installed on the charging side:
- grid voltage up to 1.5 kV;
- asymmetric protection level needed;
- SCCR 1.1–1.5×I_n.
    SPD installed on the battery side:
- See requirements of batteries.

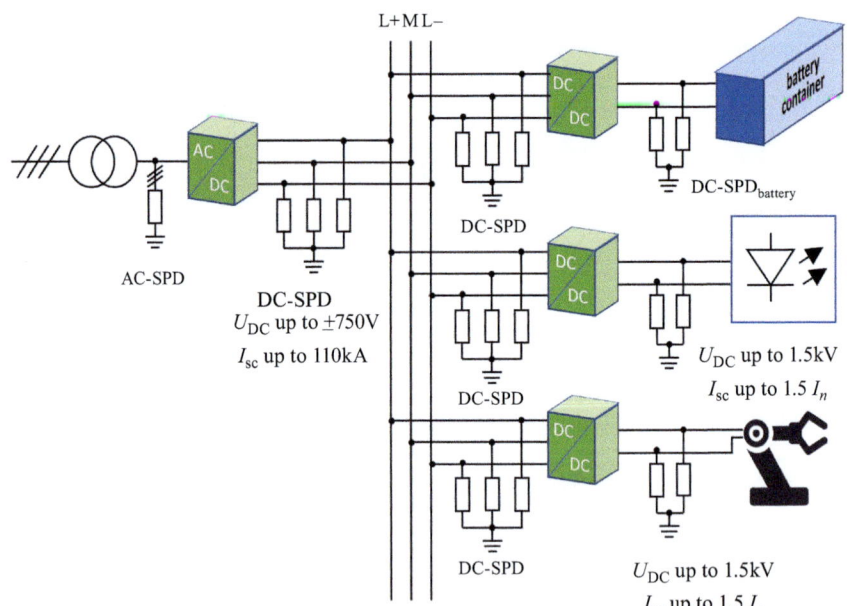

Figure 8.15   DC-power supply with battery container and AC-grid connection for industry application

8.3   Photovoltaic

8.3.1   Introduction

Over the past few years, PV applications become one of the leading developments in renewable energy. A PV plant due to the large area that it occupies and in conjunction with the sensitive electrical systems that it is composed of makes it vulnerable to lightning currents and surge overvoltages. Also a roof mount PV installation being on the roof is also potentially vulnerable either to direct lightning or to nearby lightning. The aim of this chapter clause is to describe the most common PV applications and to outline some specificities in the selection of SPDs in PV applications, which may vary compared to AC or DC applications.

8.3.2   Coupling of lighting overvoltages to AC and DC wiring on PV installations

In general, there are two types of PV installations – the roof mount and the open field PV power plant installations. Possible lightning overvoltage threats the roof mount and for the open PV field power plant installations are shown in Figures 8.16 and 8.17, respectively. Coupling between lightning and the PV installations can be ohmic due to direct lightning on the PV wiring or on the incoming grid power lines, inductive and capacitive due to the electromagnetic fields generated by the lightning flash.

8.3.3   Repetitive switching overvoltages in PV installations

An additional source of surge overvoltages in PV installations are switching type overvoltages mainly generated by the switching behaviour of high-power electronic devices used in DC to AC inverters. These switching overvoltages have a repetitive

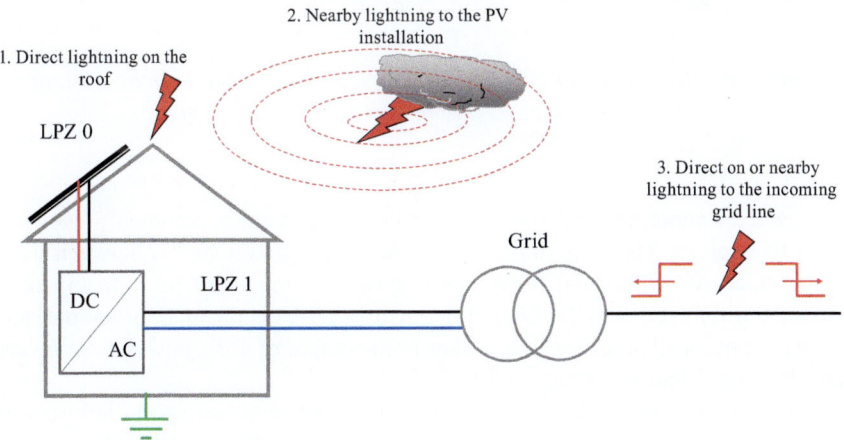

Figure 8.16   Coupling of lightning overvoltages to AC and DC wiring in roof mount PV installations

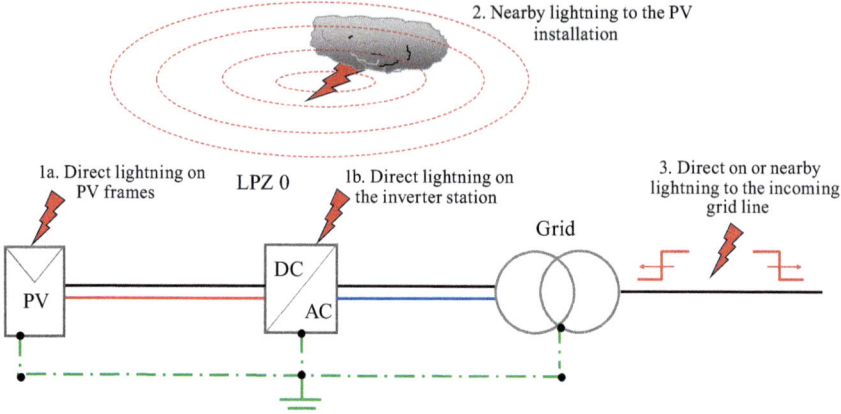

Figure 8.17   Coupling of lightning overvoltages to AC and DC wiring in open field PV power plant installations

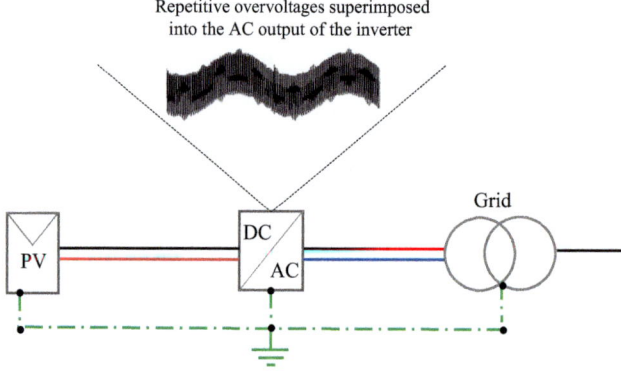

Figure 8.18   Repetitive switching overvoltages superimposed into the AC output of the inverter (the AC output may be single or 3-phase)

form and are superimposed to the AC output of the inverter, commonly known are repetitive spikes. The magnitude of the repetitive spikes is higher between live to earth, since AC filtering diverts high-frequency components to earth creating a potential difference with the filtered live parts (phase conductors). Some measured switching overvoltages/repetitive spikes at the output of a PV power plant inverter are shown in Figures 8.18 and 8.19.

The magnitude of the repetitive spikes may not cause insulation failure to the connected equipment and are much lower than the lightning overvoltages; however, they play a significant role to the selection of the SPD characteristics used at the AC output of the inverter, which will be explained later in this chapter.

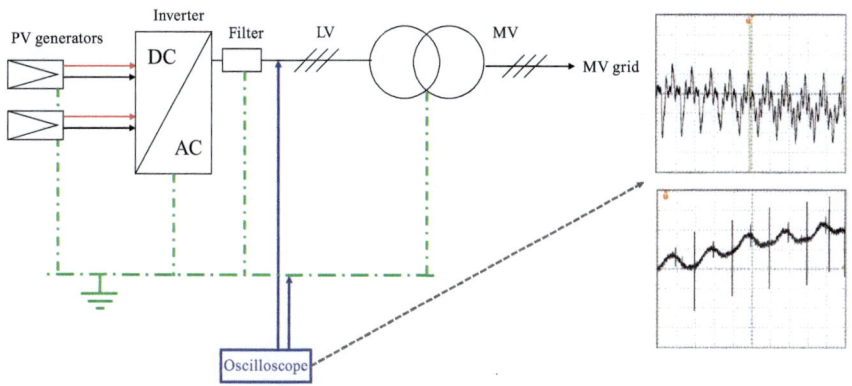

Figure 8.19 Switching overvoltages and repetitive voltage spikes measured at the output of a DC to AC inverter

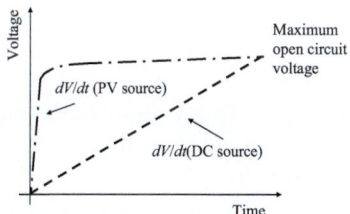

Figure 8.20 Main differences between PV and DC in voltage vs time characteristic

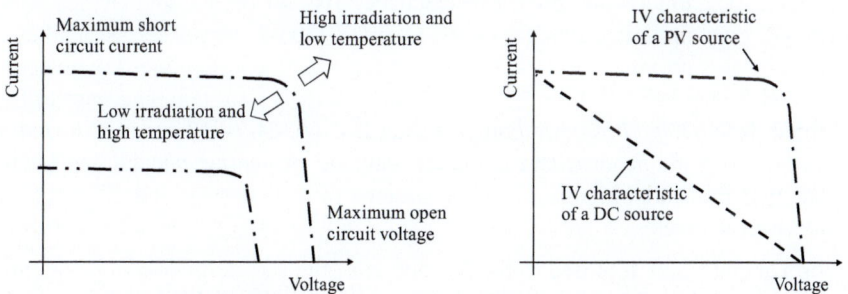

Figure 8.21 Main differences between PV and DC in current vs voltage (IV) characteristic

8.3.4 Main differences between PV and DC

The voltage vs time characteristic of a PV source compared to the DC source shown in Figures 8.20 and 8.21. A DC source is linear with respect to the rate of rise of the voltage; however, a PV source is non-linear having approximate five times faster rate of rise of voltage compared to a linear DC source. Also the current vs voltage characteristic has a similar non-linear behaviour, but it is also dependable on the irradiation and temperature.

These differences between PV and DC have created the need to produce different SPDs used in PV and DC systems. A PV-type SPD fulfils different testing criteria especially regarding the power source connected to the SPD during operating duty test and the end of life behaviour. For example, at the end of life, for an open-circuit failure mode SPD, its disconnector shall isolate the SPD from the PV system, considering the PV source characteristics rather the DC source.

8.3.5   Risk assessment for lightning overvoltage protection in PV systems

In order to evaluate the actual risk from overvoltages due to lightning, a risk assessment is needed. The purpose of the risk assessment is to compare the actual risk with the tolerable risk. The tolerable risk is a fixed value and is provided by the relevant standard.

- IEC 62305-2: Include all the required information to perform a compete risk assessment for all potential sources of coupling; however, it is more complex and the designer shall select many parameters to obtain the final risk result.
- IEC 60364-7-712 chapter 443 and IEC 60364-4-44 chapter 443: Include basic risk assessment calculations in order to calculate the actual risk from overvoltages due to lightning either from the DC (7-712) or the AC (4-44-443) incoming wiring only.

Practically IEC 60364-7-712 and IEC 60364-4-44 risk assessments provide a simplified calculation for the actual risk from overvoltages due to lightning from incoming AC and DC lines only. The calculated risk for the AC and DC incoming lines will give similar results compared to IEC 62305-2. Nevertheless, neither of these standards provides any risk assessment for other incoming lines (i.e. telecom, data) and for neither direct nor nearby lightning to the PV system, and refer to IEC 62305-2. IEC 62305-2 is more complex than IEC 60364-7-712 and IEC 60364-4-44 but is the most detailed and complete standard providing calculations for the actual risk from all possible coupling sources (direct, nearby and all incoming conductive services).

8.3.6   General PV applications and SPD location and main selection parameters

The following four case studies describe the general PV applications, which nowadays are available. More applications may become available or other hybrid applications (PV–wind-grid connected) are also currently available; however, the aim of this chapter is to provide some general principles and common issues that can be found in all PV applications. The PV circuit may be floating and isolated from earth (IT) or single polarity earthed (TN or TT), while the AC circuit may be TN, TT or IT (Figure 8.22).

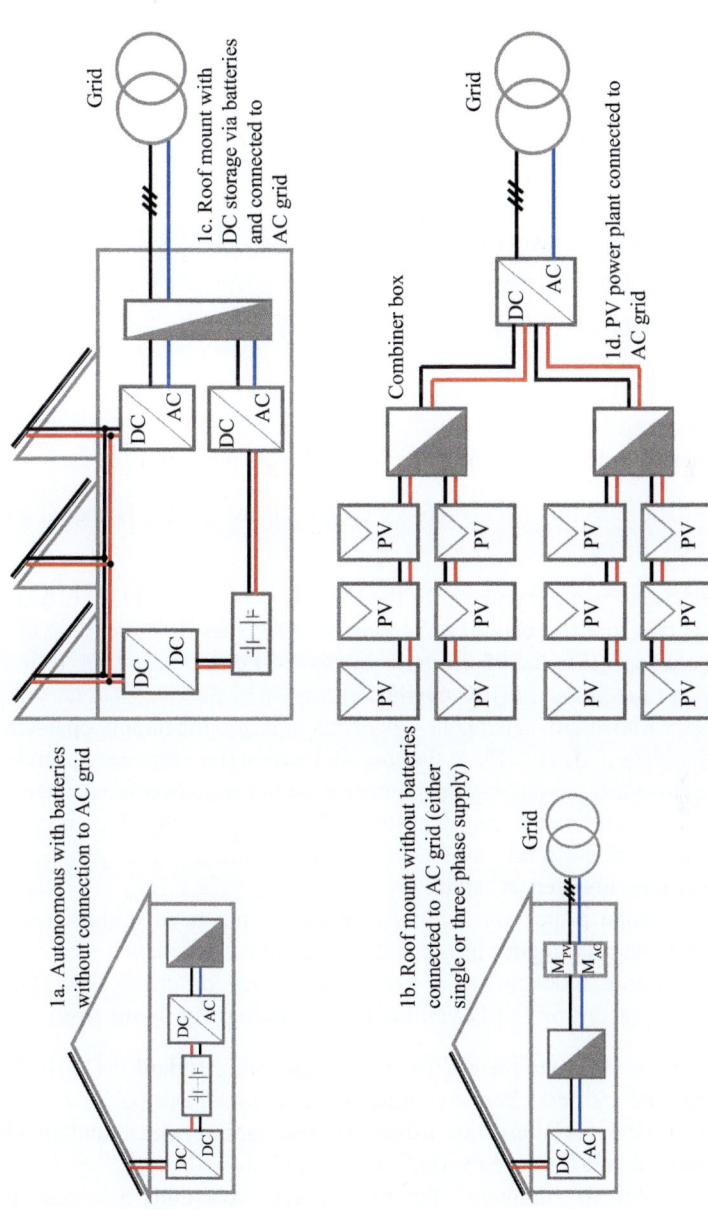

Figure 8.22   Common PV applications, autonomous, with AC grid connection, with energy storage and hybrid

Independent of the PV system (i.e. roof mount or PV power plant), there are some common locations and parameters for the selection of the SPDs, which are as follows:

(A) PV Side (IEC 61643-32)
   - PV panels
   - Combiner boxes (if present)
   - Inverter or/and converter PV input (DC side)
   - Batteries (if present)

(B) AC side (IEC 61634-12 and IEC 61643-32)
   - Inverter (s) output
   - Auxiliary circuit AC panels
   - Main connection point to grid (LV side)

(C) Signalling and communication (IEC 61643-22)
   - Inverter monitoring communication
   - Combiner boxes monitoring communication
   - Alarm and CCTV
   - Other measuring equipment (i.e. irradiation, temperature, etc.)

The SPD main selection parameters for each of the earlier (A, B and C) are as follows:

(A) SPD selection main parameters for the PV side (test standard IEC 61643-31)
   - SPD type and test class (i.e. T1/class I or T2/class II): According to risk assessment (IEC 62305-2 or IEC 60364-7-712) and according to installation rules standard (IEC 61643-32).
   - U_{CPV}: Maximum continuous operating voltage: Maximum open-circuit voltage ($U_{OC\ MAX}$) of PV at the specific location (i.e. string level, combiner box, inverter) considering the higher irradiation and lower temperature.
   - I_{SCPV}: Short circuit current rating: According to the maximum prospective short-circuit current at the specific location (i.e. panels, combiner box, inverter DC side).
   - U_p: Voltage protection level: According to the rated impulse voltage (overvoltage) category of the under protection equipment.
   - SPD failure mode (short-circuit failure mode or open-circuit failure mode): According to SPD failure mode requirement by the user.

(B) SPD selection main parameters for the AC side (test standard IEC 61643-11)
   - SPD type and test class: According to risk assessment (IEC 62305-2 or IEC 60364-4-44) and according to installation rules standard (IEC 61643-12 and IEC 61643-32).
   - U_C: Maximum continuous operating voltage: According to the maximum output voltage of the inverter and to the low voltage earthing system arrangement (TT, TN, IT).
   - I_{SCCR}: Short circuit current rating: According to the maximum prospective short-circuit current at the specific location.

- U_p: Voltage protection level: According to the rated impulse voltage (overvoltage) category of the under protection equipment.
- SPD failure mode (short-circuit failure mode or open-circuit failure mode): According to SPD failure mode requirement by the user.
- TOV rating (connection mode, voltage level, time duration of TOV, magnitude of repetitive spikes): According to the maximum output voltage of the inverter and to the low-voltage earthing system arrangement (TT, TN, IT).

(C) SPD selection parameters for signalling SPDs (test standard IEC 61643-21)
- For signalling SPDs, there is no specific requirement in a PV application compared to any other application. Detailed information for signalling SPDs can be found in Chapters 6 and 7.

8.3.7 Specific selection requirements for SPDs used in PV systems – U_{CPV}

The U_{CPV} of the selected SPDs on the PV side depends on the maximum open-circuit voltage $U_{OC\ MAX}$, which is the open-circuit voltage U_{OC} at standard test conditions ($U_{OC\ STC}$) of the PV circuit at a location (i.e. string level, array level, inverter input) multiplied by a safety/correction factor K_U. The K_U factor depends on the highest irradiation and lower temperature of the PV panels, which will affect the $U_{OC\ STC}$. If K_U factor is unknown a safe approximation of $U_{OC\ MAX} = 1.2 \times U_{OC\ STC}$.

In addition, the U_{CPV} depends on the selected protection mode (connection scheme assembly according to IEC 61643-32) of the SPDs compared with the DC side polarity earthing.

If one polarity (i.e. −ve) is earthed at the inverter under normal operating conditions, only the SPD +ve will be under the full U_{OC}, since for the SPD −ve, both poles will be at earth potential.

However, in the case of a +ve polarity solid earth fault, the Earth Fault Interrupter – EFI, will isolate the −ve polarity from earth. If the +ve polarity will remain earthed, the U_{OC} will remain between the poles but reversed. Therefore, the U_{CPV} of both polarity SPDs shall be rated at $U_{OC\ MAX}$ (Figure 8.23).

In a floating system, the full $U_{OC\ MAX}$ will be between the live polarities (−ve to +ve); however, between the live poles to earth, the voltage will be $U_{OC\ MAX}/2$ (i.e. 50% of $U_{OC\ MAX}$).

Under normal operating conditions, both SPDs (SPD +ve and SPD −ve), when connected between live poles to earth, will be under the 50% of $U_{OC\ MAX}$. Therefore, the U_{CPV} of both SPDs under normal operating conditions will be at 50% of $U_{OC\ MAX}$.

However, in the case of a +ve polarity solid earth fault, the system from floating will become earthed. Then the voltage between the −ve polarities to earth will become equal to $U_{OC\ MAX}$. The similar case will be for a −ve polarity solid earth fault but will be reversed polarity. Therefore, the U_{CPV} of both SPDs for a floating PV system shall be rated at full $U_{OC\ MAX}$ (Figure 8.24).

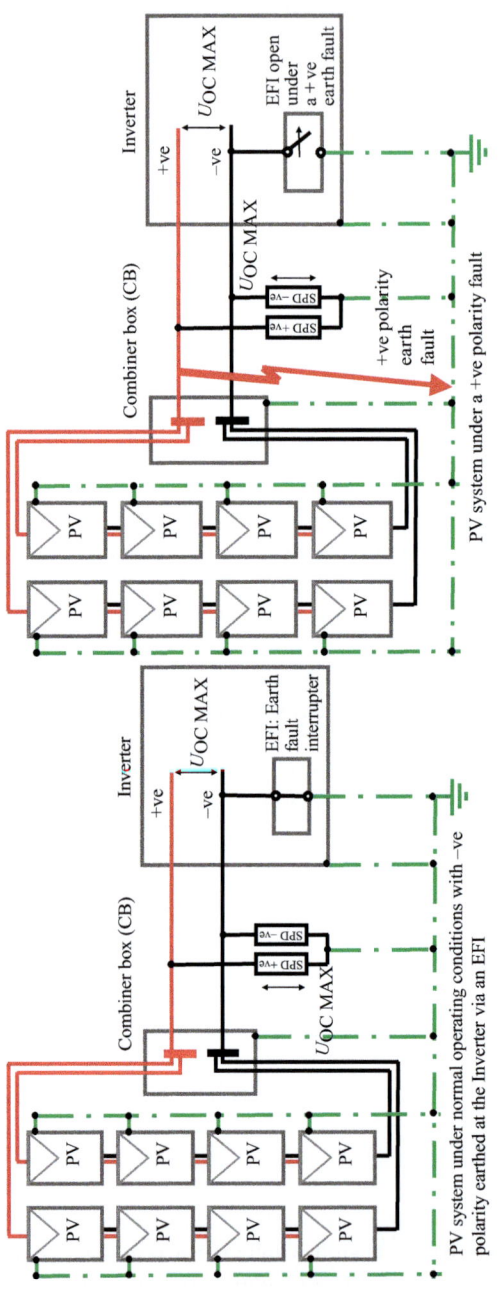

Figure 8.23   PV SPD U_{CPV} selection for the PV side of single polarity earthed photovoltaic applications

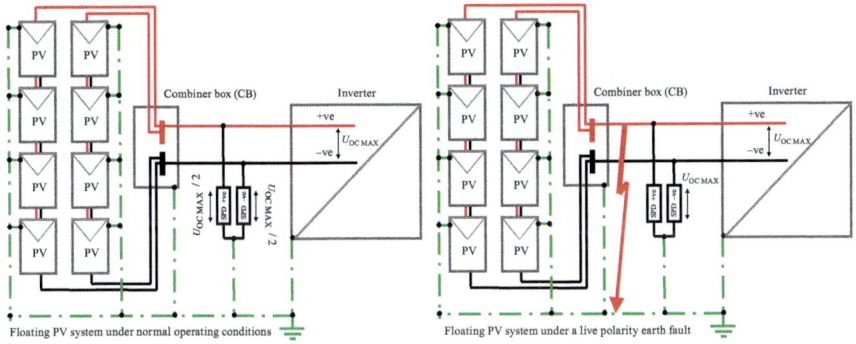

Figure 8.24   PV SPD U_{CPV} selection for the PV side of floating polarity photovoltaic applications

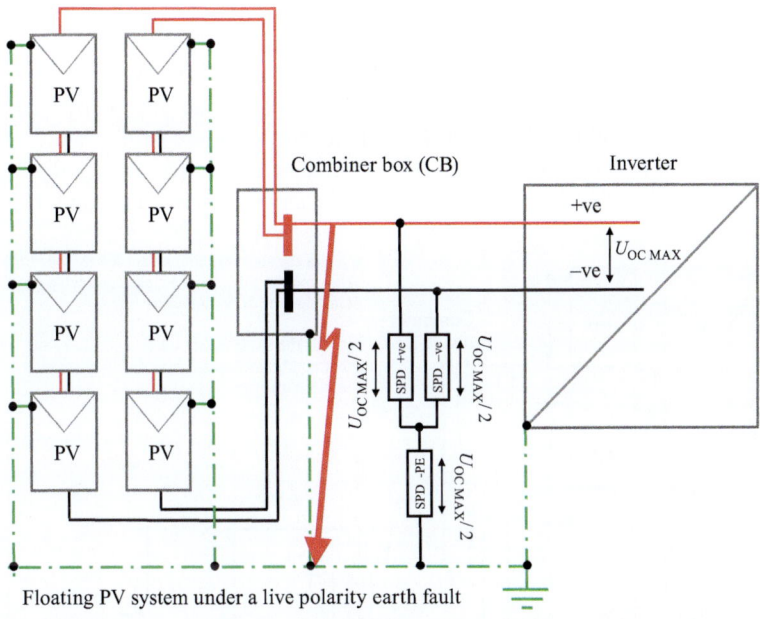

Figure 8.25   Poplar Y connection/assembly scheme of a PV SPD with $U_{CPV}/2$ selection for the each mode of the SPD assembly scheme

Another common approach is the Y (2+1) connection assembly scheme. This connection scheme contains three SPDs and the selected U_{CPV} for each one shall be rated at $U_{OC\ MAX}/2$ since, under either normal or fault condition, there will be always two SPDs in series between either live poles or live poles to earth (Figure 8.25).

8.3.8   Specific selection requirements for SPDs used in PV systems – I_{SCPV}

If a PV SPD fails in short circuit, the selection of the I_{SCPV} is a very important parameter. PV SPDs may run in short circuit at the end of life or under overstress leading the SPD to failure mode.

When the SPD runs in a failure mode, the selected I_{SCPV} of the PV SPD shall be higher than the maximum prospective short circuit at the SPD location.

In an earthed PV system, if the selected PV SPD has an open-circuit failure mode, the internal disconnector of the SPD shall open and isolate the faulty SPD. The I_{SCPV} rating of the PV SPD shall be higher that the prospective short-circuit current at the installation location, and there is no need to install any back-up fuse in-line with the PV SPD.

If the SPD has a short-circuit failure mode, an external disconnector shall clear the fault. The fault current will be cleared either by the back-up fuse of the SPD (F_3) installed in-line with the SPD, or by other overcurrent protection devices (i.e. F_2 for I_{SC_2} or by the EFI for I_{SC_1}) of the faulty system. In any case, the short-circuit failure mode SPD shall be able to withstand any prospective short-circuit current for the required time, until an overcurrent protection device clear the fault. Therefore, the I_{SCPV} of the PV SPD shall be higher than any of the maximum prospective short-circuit current at the SPD location (Figure 8.26).

In a floating PV system, a faulty SPD connected between live poles to earth (most common protection mode) will not create any short-circuit current. However,

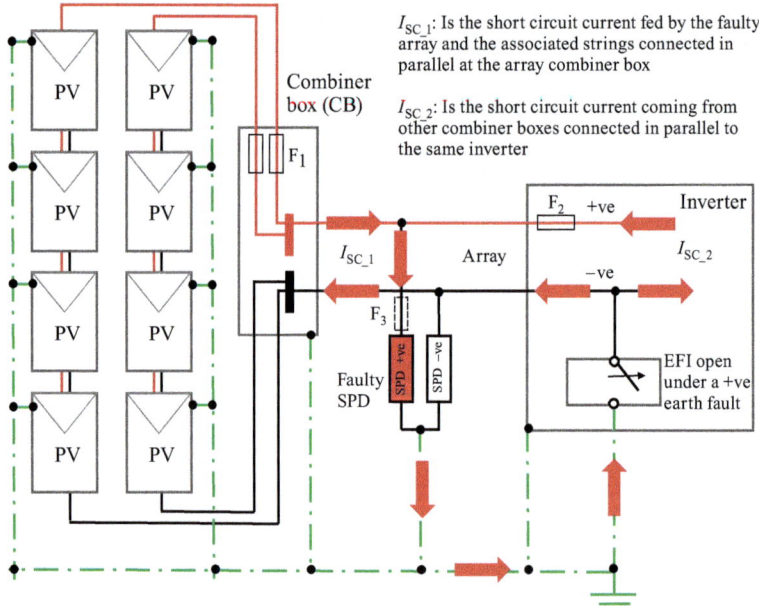

Figure 8.26   PV SPD I_{SCPV} selection for the PV side of single polarity earthed photovoltaic applications

it is vital to consider that when removing the faulty SPD, part of the $U_{OC\ MAX}$ will appear across its poles; therefore, the SPD shall be isolated from both sides (PV and inverter) before maintenance.

However, if there is a faulty SPD between live poles or if there is a simultaneous earth fault for both SPD polarities the short-circuit withstand capabilities of the SPDs are different. In this case, the earth fault (I_{SC_2}) will be interrupted by the input fuses (F$_2$) at the inverter.

However, the (I_{SC_1}) cannot be interrupted by the sting input fuses (F$_1$) at the combiner boxes. The only possible way to interrupt the fault is by using an open-circuit failure mode SPD, which will isolate the faulty SPD and will clear the fault current, provided that the SPD-declared I_{SCPV} is higher than the prospective short-circuit current.

Conversely, by using a short-circuit failure mode type SPD, it will allow the fault current to flow permanently through the PV SPD, provided that the permanent short-circuit current will not cause overheating to the PV SPD or the associated wiring. In this case, overcurrent coordination may be required so as to trip the external fuse (F$_3$) of the short-circuit type PV SPD (Figure 8.27).

8.3.9 U_C selection for the SPD on the AC side of the inverters considering the repetitive switching overvoltages

In the AC output of PV inverters, especially PV power plants (i.e. central inverters MW output or multi-string kW inverters connected in parallel), the voltage may be higher than the standard 230/400 V. Common voltage outputs are 480 up to 800 V$_{AC}$. It is also common that the LV side of the transformer, which is connected to the output of the inverter to use IT earthing system, plays a significant role in the selection of the U_C (see example shown in Figure 8.28).

A common degradation of the SPDs connected to the AC output of high-power inverters is due to repetitive switching overvoltages (spikes), superimposed into the AC output of the PV inverter, caused mainly by the fast switching of the power electronics.

To overcome the problem from the repetitive voltage spikes, the selected SPDs shall

(A)   have a U_C which will be higher than the magnitude of the repetitive spikes to avoid frequent conduction. This may be a mandatory condition for open-circuit failure mode SPDs containing limiting type components (i.e. varistors or diodes).

(B)   have the ability to conduct during the repetitive spikes but due to the frequent conduction to be able to dissipate the thermal energy without setting on its overcurrent protection and without degradation of its overvoltage protection. Usually a short-circuit type failure mode SPD has such behaviour.

(C)   use internal components that are not influenced by the repetitive spikes (i.e. selecting spark gaps with impulse sparkover voltage higher than the repetitive switching overvoltage to avoid conduction) (Figure 8.29).

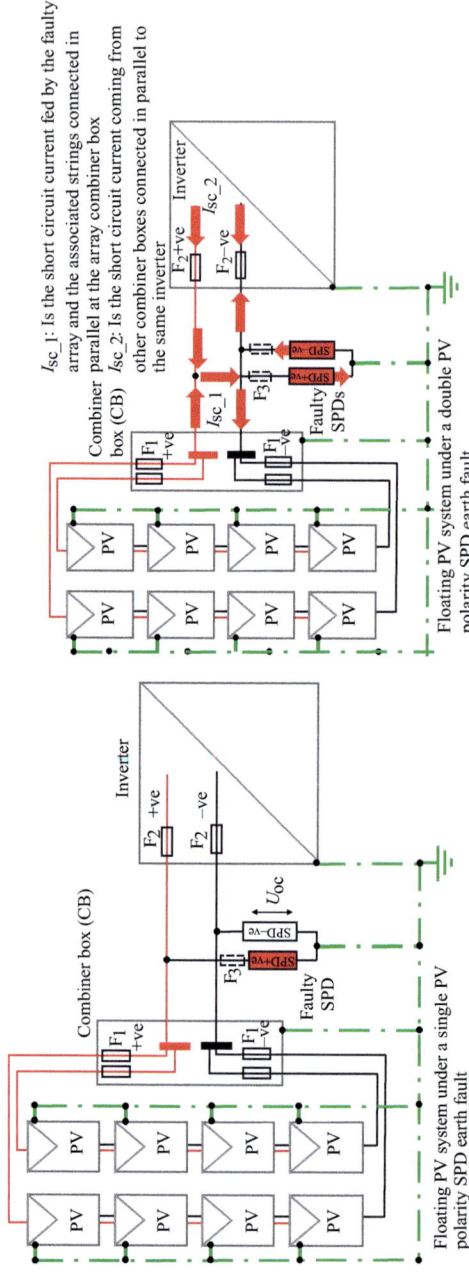

Figure 8.27 PV SPD I_{SCPV} selection for the PV side of floating polarity photovoltaic applications during a single or double polarity fault

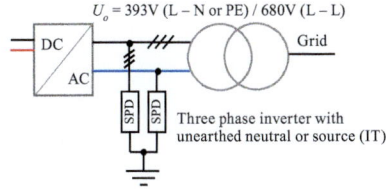

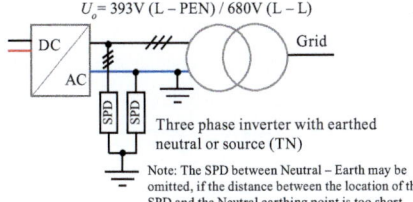

In IT systems or in systems without neutral, in case of a single phase to earth fault, the SPD will see the L – L voltage across its poles, therefore U_c shall be > 680 V × 10% and selected as defined in IEC 61643-12 and IEC 60364-5-53 (clause 534).

In TN systems, in case of a single phase to earth fault, the SPD will not see a variation to the voltage across its poles, therefore U_c shall be > 393 V × 10% and selected as defined in IEC 61643-12 and IEC 60364-5-53 (clause 534).

Figure 8.28 *Example of SPD U_C selection for the AC side of the inverters*

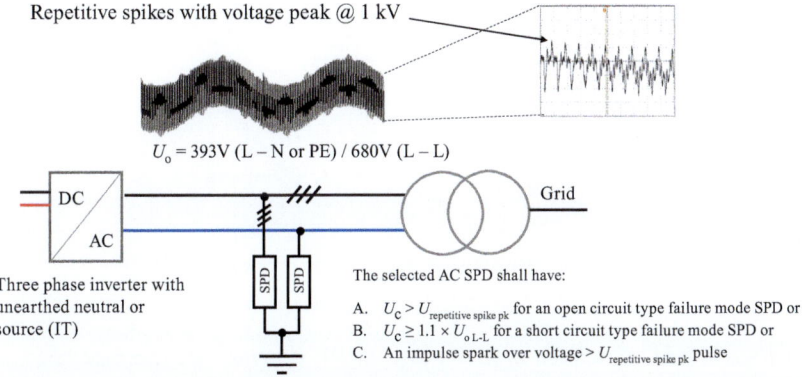

The selected AC SPD shall have:

A. $U_c > U_{\text{repetitive spike pk}}$ for an open circuit type failure mode SPD or

B. $U_c \geq 1.1 \times U_{\text{o L-L}}$ for a short circuit type failure mode SPD or

C. An impulse spark over voltage > $U_{\text{repetitive spike pk}}$ pulse

Example of repetitive spikes superimposed to the AC output of an inverter with an operating voltage of 680V (L-L) RMS with superimposed repetitive spikes having a peak of 1kV

Figure 8.29 *Example of repetitive spikes superimposed to the AC output of an inverter with an operating voltage of 680 V (L–L) rms with superimposed repetitive spikes having a peak of 1 kV*

8.3.10 Coordination of SPDs in PV applications with respect to overvoltage protection

SPD coordination rules in PV systems are similar to that in AC systems as defined in IEC 61643-12. Any upstream SPD shall absorb higher energy (i.e. current) than the downstream SPD. Also, the voltage protection level (U_p) of the selected SPD shall be selected so as to be at least 20% lower than the rated impulse voltage (U_w) of the protected equipment.

In the case of extended cable lengths (>10 m) between the SPD and the protected equipment, either a set of two SPDs shall be selected at both cable ends or an insulation coordination study shall calculate the need of an additional SPD. For more information about insulation, coordination rules refer to Chapter 7 (Figure 8.30).

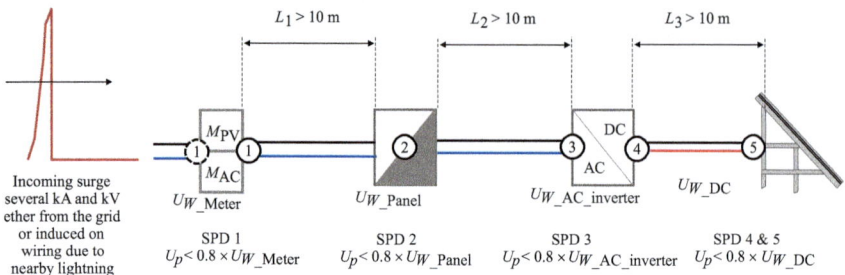

Figure 8.30   Selection of U_p for SPDs used in PV systems

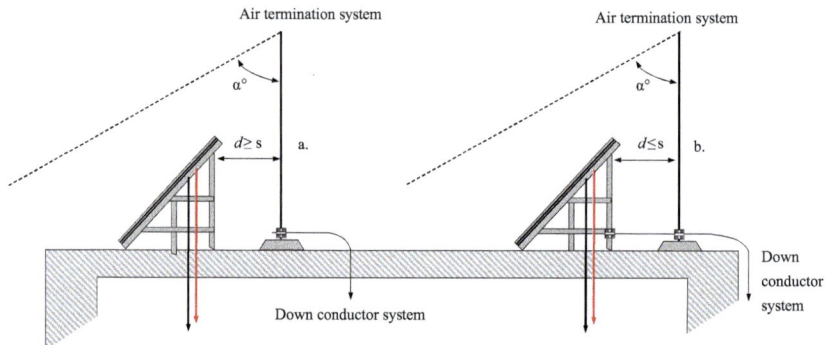

Figure 8.31   Selection of SPDs used in PV systems with external LPS

In the case of an external lightning protection system (LPS), used to protect the PV against a direct lightning flash, an important parameter to define the required type of the SPDs is if the LPS maintains the separation distance with the metallic parts of the PV system. If the LPS is attached to the metal parts of the PV system, then all SPDs shall be T1 in both PV and AC sides. However, if the LPS maintains the separation distance, then all PV side SPDs can be T2 and on the AC side only SPDs located at the grid connection point need to be T1 (i.e. location 1), all other AC SPDs can be T2. The required ratings (values) for I_n (T2) and I_{imp} (T1) shall flow IEC 61643-32 and IEC 61643-12 requirements. However, IEC 62305-2 risk assessment will determine the lightning protection Levels I, II, III or IV of the LPS, which is needed by IEC 61643-32 and IEC 61643-12 in order to select the required I_n and I_{imp} values (Figure 8.31).

8.3.11   Examples of SPD selection and location in common PV applications

The required type and location of SPDs for protecting a roof mount autonomous PV with batteries but without connection to AC grid primarily depend on if there is an external LPSs and if it is isolated or non-isolated from the PV installation. If it is a non-isolated LPS, then all SPDs shown in Figure 8.32 shall be T1. If it is no LPS or

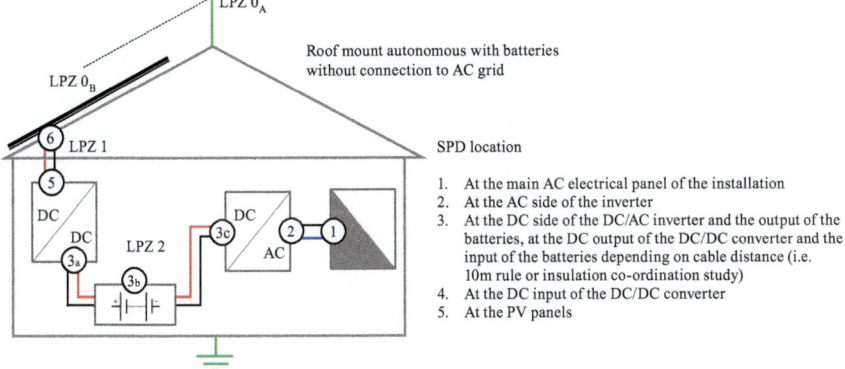

Figure 8.32 Example of SPD location in a roof mount autonomous PV with batteries without connection to AC grid

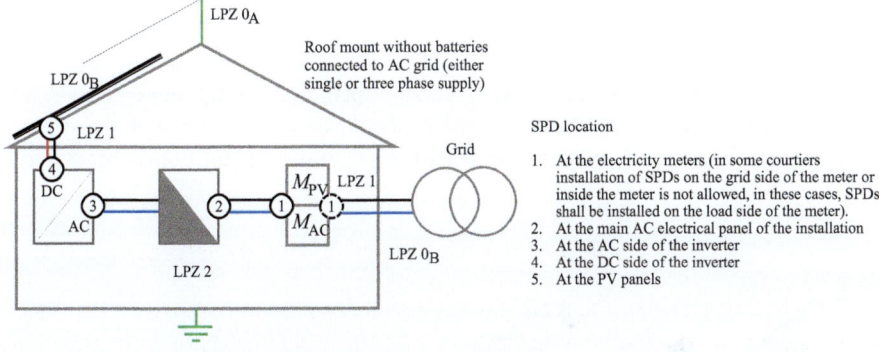

Figure 8.33 Example of SPD location in a roof mount autonomous PV with direct connection to AC grid

if there is an isolated LPS, for an autonomous PV without connection to AC grid, then all SPDs may be only T2.

An additional popular example especially for net metering applications is a roof mount PV installation with a direct connection to AC grid. In this case, the required type and location of SPDs depend also on if there are external LPSs and if it is isolated or non-isolated from the PV installation. If it is a non-isolated LPS, then all SPDs shown in Figure 8.33 shall be T1. However, if there is an isolated LPS, then all SPDs may be only T2 except the SPD at location (1), which shall be T1. If it is no LPS then all SPDs may be only T2 except the SPD at location (1), which may be T1 or T2, depending on electrical installation rules for the specific case (i.e. if there is a high risk of incoming partial lightning current through the AC Grid line, T1 SPD shall be used, see Figure 8.16 case 3). Typical examples are included in IEC 60364-4-44 chapter 443 (Figure 8.33).

A hybrid PV application may include both direct connection to the AC grid and energy storage via batteries. Such a case is shown in Figure 8.34. The selection and location of SPDs is a combination of the previous two examples.

In large-scale PV power plants (10–200 MW or larger), due to the extensive cabling and apparatus especially the number of PV panels, a risk assessment is essential in order to evaluate the real risk and to compare it to the tolerable risk. Following practical SPD installation rules may lead to extensive use of SPDs, which in many cases may not be a cost-effective solution compared to the real benefit of the offered protection. Therefore, any possible effort to reduce lightning overvoltages shall be considered during the construction of the PV plant by means of reducing cable loops, running cables through any possible metallic screen or facades, running buried cables preferably screened, etc. Additionally the technology of string and array inverters (central inverters) is constantly expanding introducing new techniques such as hybrid complexes, including DC/AC inverter, LV AC switchgear, LV/MV transformer, MV switchgear all in one system, which require integrated SPDs inside the equipment by their manufacturer. The following example shows a typical PV power plant using either central array inverters (i.e. 1 MW), including string combiner boxes. A similar application may be a PV application with string high-power inverters (i.e. 200 kW) without the combiner boxes. The types of SPDs depend on the risk assessment and on the presence of external LPS. If the PV panels use tracker systems, the electric part of the motor may require protection. A typical risk assessment of a PV power plant considering that the

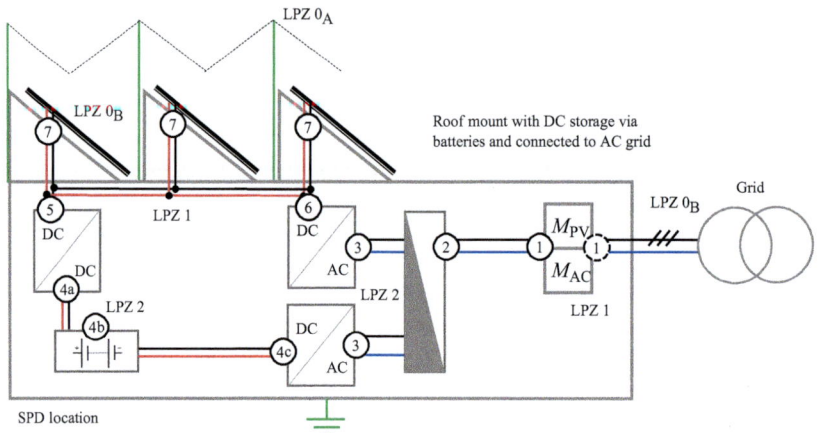

SPD location

1. At the electricity meters (in some courtiers installation of SPDs on the Grid side of the meter or inside the meter is not allowed, in these cases, SPDs shall be installed on the load side of the meter).
2. At the main AC electrical panel of the installation
3. At the AC side of the inverter (s)
4. At the DC side of the DC/AC inverter connected to the batteries, at the batteries and at the DC output of the DC/DC converter depending on cable distance (i.e. 10m rule or insulation co-ordination study)
5. At the DC input of the DC/DC converter
6. At the DC input of the DC/AC inverter with direct connection to AC grid
7. At the PV panels

Figure 8.34   Example of SPDs location in roof mount PV application with DC storage via batteries and connection to the AC grid

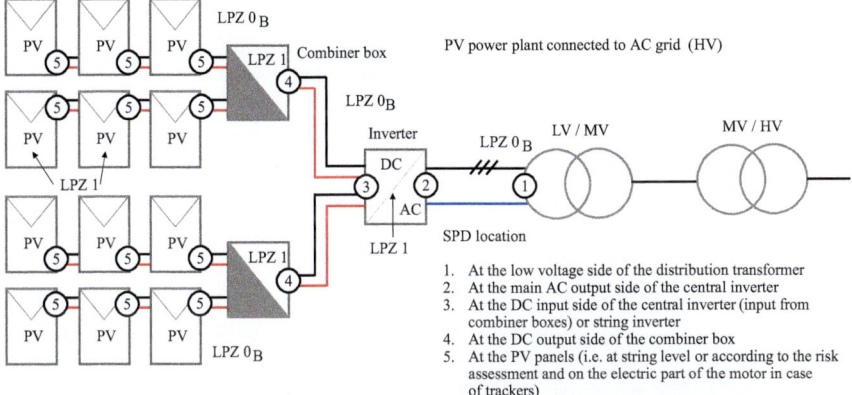

Figure 8.35 Example of SPD location in a PV power plant connected to AC grid

presence of people is restricted and low as well as the loss of service to the public is not essential may require a Level IV lightning protection. However, there are many cases that due to practical reasons external LPS is not possible to be installed (i.e. tracker movement, shading issues, etc.); therefore the use of overvoltage protection techniques (internal lightning protection) may be the only protection against lightning currents and overvoltages (Figure 8.35).

8.4 Wind turbines used in wind power plants

8.4.1 Introduction

This chapter clause outlines the basic risk threats and protection principles for wind turbines used in wind power plant against direct and nearby lightning flashes. Due to location (i.e. isolated on mountains or offshore) and due to the physical size and height of a high-power wind turbine, the risk of a direct lightning flash mainly on the blades is a rather common phenomenon. Additionally, a wind power plant has extensive overhead and buried cabling interconnecting the wind turbines together and feeding their produced power via a local MV/HV substation to the transmission grid (Figure 8.36).

8.4.2 Risk assessment for lightning overvoltage protection in wind turbine systems

A significant risk variation of a wind turbine compared to ordinary applications is that any damage or equipment replacement is very difficult due to practical constrains and has very high maintenance costs. Also the out-of-service downtime is very crucial when it comes to income losses due to the high-power rating of a wind turbine compared to an inverter of PV application.

In order to evaluate the actual risk from overvoltages due to lightning, a risk assessment is needed. The purpose of the risk assessment is to compare the actual

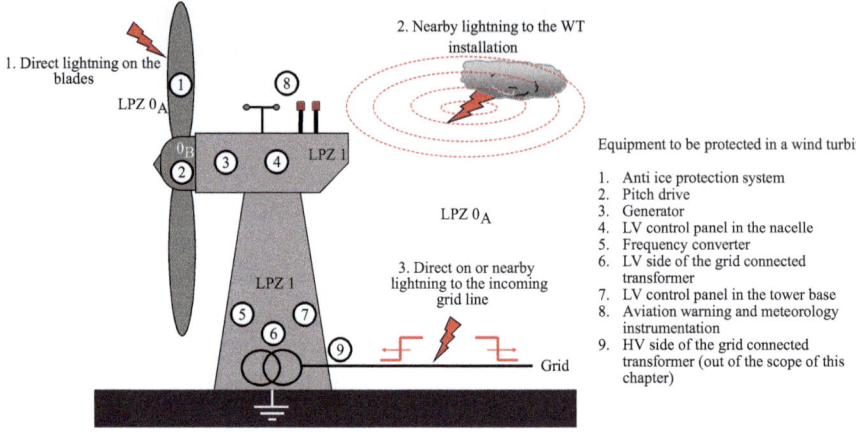

1. Direct lightning on the blades

LPZ 0A

2. Nearby lightning to the WT installation

LPZ 1

LPZ 0A

3. Direct on or nearby lightning to the incoming grid line

LPZ 1

Grid

Equipment to be protected in a wind turbine

1. Anti ice protection system
2. Pitch drive
3. Generator
4. LV control panel in the nacelle
5. Frequency converter
6. LV side of the grid connected transformer
7. LV control panel in the tower base
8. Aviation warning and meteorology instrumentation
9. HV side of the grid connected transformer (out of the scope of this chapter)

Figure 8.36   Coupling of lightning overvoltages to a wind turbine, power plant installation and critical equipment to be protected

risk with the tolerable risk. The tolerable risk is a fixed value and is provided by the relevant standard.

- IEC 62305-2: Include all the required information to perform a compete risk assessment for all potential sources of coupling; however, it is more complex and the designer shall select many parameters to obtain the final risk result.
- IEC 61400-24: Include a methodology to determine the lightning exposure assessment, very similar to risk assessment calculations. However, this methodology provides the required test levels and verification tests defined in IEC 61400-24, Annex D. These tests mainly involve impulse voltage testing and impulse current testing at very high levels (i.e. 200 kA–10/350 μs).

Practically, a wind turbine used in wind power plant being isolated (i.e. off-shore) and a high rise structure (>60 m) is highly probable to require a lightning protection Level II or I.

8.4.3   Main equipment to be protected inside a wind turbine

A wind turbine consists of three main parts – the blades, the nacelle and the tower. The nacelle and the tower are metallic installations, while the blades are made of insulated material. Inside as well as attached to them, there are electrical and electronic equipment that may be destroyed by a direct lightning flash. The most critical and vulnerable parts of a wind turbine are the blades. Having been made out of insulating materials, a direct lightning will cause serious damage, which may require even a blade replacement. Over the past few years advanced impulse current and impulse voltage testing techniques have been introduced by the relevant standards (i.e. IEC 61400-24) improving significantly the direct damage of them. However, due to the flow of the lightning current through the wind turbine (i.e. from the blades to the nacelle then to the tower and finally to the earthing system), there is ohmic and induced (i.e. inductive and capacitive) coupling with the internal wiring running in parallel. Using screened cables, metallic ducts and avoid inductive loops helps in reducing the induced effects; however, the ohmic coupling is not possible to be avoided in a wind turbine.

The most critical equipment that are influenced by a direct lightning flash to a wind turbine are shown in Figure 8.37.

8.4.4 Example of SPD technical characteristic and location in wind turbines

The main technical characteristics of the SPDs used in wind turbine applications are as follows:

- SPD type and test class: According to risk assessment (IEC 62305-2 or IEC 61400-24) and according to installation rules standard (IEC 61643-12 and IEC 61400-14), special care shall be taken for locations 1 and 2, since the SPD type as well as the I_{imp} is each time highly dependent on the specific application.
- U_C: Maximum continuous operating voltage: According to the maximum output voltage of the generator and the low voltage earthing system arrangement (TN or IT). In wind turbines, usually there are two low voltage systems. The generating systems (i.e. generator, frequency converter and grid connection transformer) commonly are operating at 400/690 V_{AC} (TN C or IT), while all the control systems are operating at 230/400 V_{AC} (TN C or TN S). Depending on the nominal operating voltage of the system, the appropriate characteristics of the SPD shall apply.
- I_{SCCR}: Short circuit current rating: According to the maximum prospective short-circuit current at the specific location considering also the contribution of the wind turbine and the grid fault level.
- U_p: Voltage protection level: According to the rated impulse voltage (overvoltage) category of the under protection equipment and to the nominal operating voltage of the system (i.e. 690 or 400 V_{AC}).
- SPD failure mode (short-circuit failure mode or open-circuit failure mode): According to SPD failure mode requirement by the user.
- TOV rating (connection mode, voltage level, time duration of TOV): According to the nominal operating voltage of the low-voltage systems (i.e. 690 or 400 V_{AC}) and to the earthing system arrangement at the transformers (most common systems may be TN C, TN S or IT).
- Withstand to extensive vibrations caused by the mechanical moving parts as well as during acceleration and braking of the rotating parts. This may influence the torque on the connected terminals as well as the unplugging of

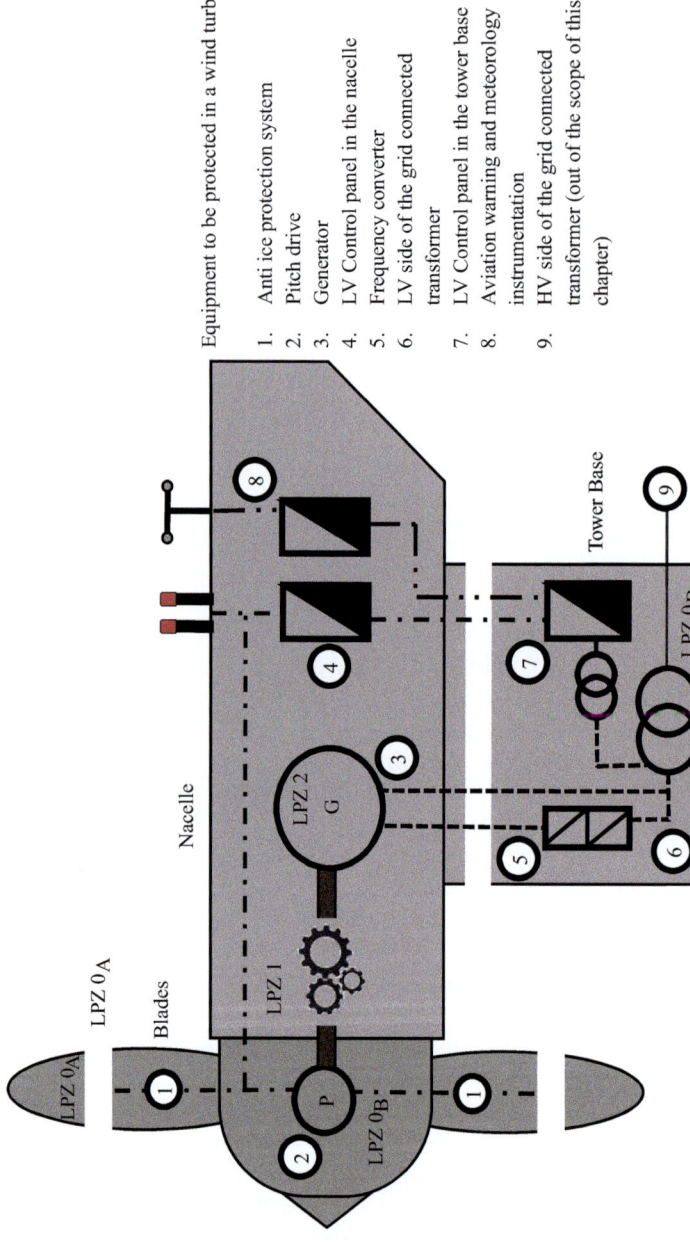

Equipment to be protected in a wind turbine

1. Anti ice protection system
2. Pitch drive
3. Generator
4. LV Control panel in the nacelle
5. Frequency converter
6. LV side of the grid connected transformer
7. LV Control panel in the tower base
8. Aviation warning and meteorology instrumentation
9. HV side of the grid connected transformer (out of the scope of this chapter)

Figure 8.37   Critical equipment to be protected inside a wind turbine against direct lightning effects

modular SPDs, which shall have some type of anti-vibration mechanism and high tightening torque terminals.

The SPD type (i.e. T1 or T2) in wind turbine applications is also commonly selected by applying the lightning protection zone concept, which require incoming active lines for both power and signalling between LPZ 0_A and LPZ 1 boundaries to use SPDs enabling it to discharge partial lightning current (i.e. Type 1). At downstream boundary zones (i.e. LPZ 1 to LPZ 2), the lightning current will be reduced; therefore, lower energy current discharge capability SPDs may be used (i.e. Type 2). However, it is important to consider that in a wind turbine, there are multiple parallel paths, which are bonded via SPDs to the tower, acting as a down conductor, offering parallel paths to lightning current flow towards the earthing system. Of course the impedance of the tower is significantly lower than the parallel cables, reducing the lightning current flow via the cables. Due to this, it is expected that high cross-sectional area cables (i.e. between generator and LV grid connected transformer) offer a low impedance path to earth and may carry higher energy level compared to other cable links between the tower base and nacelle (i.e. LV control panels) which will carry lower energy lightning currents. Table 8.4 summarizes the location, the lightning protection zone and the SPD type, considering also the parallel path effect.

For locations (1) anti-ice protection system and (2) pitch drive, the required SPD type depends on the separation distance between the down conductors, which is installed internally and attached to the blade and the metallic parts of the apparatus in the previous two locations. More information about the selection of the SPDs in these two locations is explained in Section 8.4.5.

8.4.5   SPD selection for wind turbines located inside and attached to the blade

An important parameter to the selected SPD type for equipment that are inside and attached to the blade such as the anti-ice protection is if the separation distance between the down conductor, which is fixed inside on the internal surface of the blade and the electrical supply of the above system, is maintained.

Also, if the pitch drive AC supply (if provided) is common with the anti-ice protection system, a similar approach may be needed for the AC SPD selection for the pitch drive supply.

If the separation distance (s) is not maintained, then the selected SPDs for the anti-ice protection (heating) and the electrical supply for the pitch drive shall be T1. However, if the separation distance is maintained, then all SPDs for these particular locations can be T2 (Figure 8.38).

As an example, considering that the separation distance (s) is not maintained, then the selected SPDs for the anti-ice protection (heating) shall be T1 (Figure 8.39).

Table 8.4 SPD location and selection parameters for wind turbines

Equipment to be protected inside the wind turbine	SPD category	LPZ	Specific notes for SPD type requirements
1 Anti-ice protection system inside the blades	Power and signalling	LPZ $0_A \rightarrow$ LPZ 1	The SPD type (T1 or T2) depends if the separation between the down conductor and the anti-ice protection is maintained or not, if it is maintained, the SPD may be T2, otherwise it shall be T1 and the I_{imp} shall be calculated based on the lightning current distribution rules
2 Pitch drive	Power and signalling	LPZ $0_B \rightarrow$ LPZ 1	The SPD type (T1 or T2) depends on if the separation between the down conductor and the anti-ice protection is maintained or not, if it is maintained, the SPD may be T2; otherwise, it shall be T1, and the I_{imp} shall be calculated based on the lightning current distribution rules
3 Generator	Power and signalling	LPZ 1 $\rightarrow$ LPZ 2	The SPDs along location (3) and (6) provide an additional low impedance parallel path to the metallic tower for the lightning current to flow to earth. Therefore, it is suggested to be T1 but with limited I_{imp} requirements, since the bulk of the lightning current will flow via the metallic tower
4 LV control panel in the nacelle	Power and signalling	LPZ $0_B \rightarrow$ LPZ 1	The SPD type shall be T2 or low-energy T1. The I_{imp} shall be calculated based on the lightning current distribution rules, since the bulk of the lightning current will flow via the metallic tower, which is a parallel path to earth of a non-isolated LPS
5 Frequency converter	Power and signalling	LPZ 1 $\rightarrow$ LPZ 2	The SPD type shall be T2 or low energy T1. The I_{imp} shall be calculated based on the lightning current distribution rules, since the bulk of the lightning current will flow via the metallic tower, which is a parallel path to earth of a non-isolated LPS
6 LV side of the grid connected transformer	Power	LPZ 1 $\rightarrow$ LPZ 2	The SPDs along location (3) and (6) provide an additional low impedance parallel path to the metallic tower for the lightning current to flow to earth. Therefore, it is suggested to be T1 but with limited I_{imp} requirements, since the bulk of the lightning current will flow via the metallic tower

(Continues)

7	LV control panel in the tower base	Power and signalling	LPZ 1→LPZ 2	The SPD type shall be T1. The I_{imp} shall be calculated based on the lightning current distribution rules since this point is a node, where the sum of multi-parallel active lines connect to earth
8	Aviation warning and meteorology instrumentation	Power and signalling	LPZ 0_B→LPZ 1	The SPD type shall be T2 or low-energy T1. The I_{imp} shall be calculated based on the lightning current distribution rules, since the bulk of the lightning current will flow via the metallic tower, which is a parallel path to earth of a non-isolated LPS
9	HV side of the grid connected transformer	HV distribution grid. Selection of HV and MV surge arresters shall be performed according to insulation co-ordination studies according to IEC 60071 series (insulation co-ordination for lines >1 kV$_{AC}$)		

For signalling SPD characteristics and selection categories refer to Chapters 6 and 7.

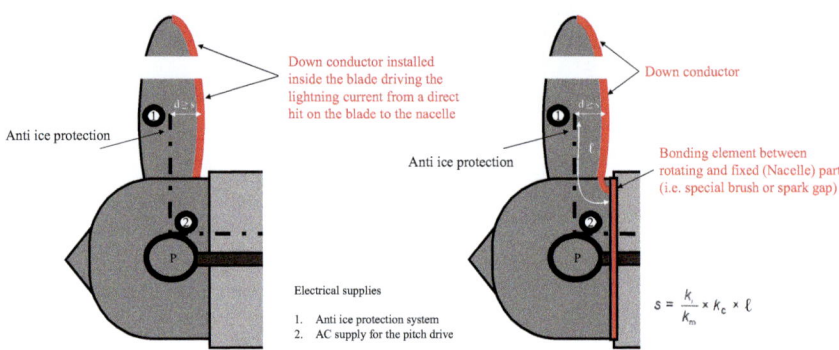

Where,

k_i Depends on the lightning protection level (LPSclass) according to risk assessment

k_m Depends on the electrical insulation of the material between the LPS(down conductor) and the metallic system (i.e. anti ice protection equipment)

k_c Depends on the number of down conductors

ℓ Is the length, in meters, along the down conductor from the point where the separation distance is to be considered, to the bonding point between the down conductor and the Nacelle

Figure 8.38   SPD selection for equipment attached to the wind turbine blade

8.5   Electric vehicle and boat charging stations

8.5.1   Introduction to electric vehicle (EV) charging stations

Electric mobility nowadays is expanding more and more, creating a need for charging stations not only either for residential or commercial buildings but also as a service in public parking buildings, in open parking lots and in street parking. Although an electric vehicle has the option to be charged directly from a 230-V_{AC} outlet by using either a portable DC charger or its internal charging system, the charging time is too long (i.e. hours depending on the EV batteries capacity). However, using fast and ultrafast DC charging stations and DC voltages up to 1,000 V_{DC} reduces the required charging time from hours just into few minutes for partial charge or a maximum of 15–30 min for full charge. On the other hand, the infrastructures for fast and ultrafast charging stations not only require high and expensive technologies such as inverters, monitoring and control systems, but also a dedicated power supply having in many cases a long cable network, which increases the need for overvoltage protection for both the charging station and the electric vehicle. In many countries, the national electrical requirements have as mandate the use of overvoltage protection in power lines feeding EV charging stations. The same applies in IEC 60364-7-722, which also indicate the overvoltage protection requirement.

8.5.2   Example of EV charging station in residential buildings

In a residential building, the protection of the EV charging station is part of the electrical installation. Therefore, an optimum protection achieved by providing

As mentioned in the risk assessment clause, the lightning protection level of a wind turbine may be LPLI, which will require protection for a lightning current of 200 kA. Considering that the blade has one down conductor the 200 kA will split by a factor of 2. The 100 kA will flow via the down conductor and the other 100 kA will flow via the SPD protecting the anti ice feeding power line. There fore I_{imp} shall be 100 kA (all poles). If it is a 3 pole (L, N, PE) feeding system, the I_{imp}per pole shall be 33 kA or 20 kA for a 5 pole (3L, N, PE) feeding system. Software engineering tools may use more detailed simulations for lightning current distribution, however a common practice defined by standards is an equal current split between conductive parallel conductors leading most of the times in safe results (i.e. overestimated SPDs values compared to the values calculated by software simulations).

Parameters are selected according to IEC62305 – 3 paragraph 6.3 (Table 10, 11 & 12)

k_i = 0,08 (LPS Class I)

k_m = 1 (Insulation material air)

k_c = 1 (Number of down conductors one)

ℓ = 40 (The length is 40 meters)

s = 3,2 m (the real distance, d < 3.2 m) therefore the required separating distance is not sufficient and T1 SPDs shall be used with I_{imp} calculated by the lightning current distribution rules.

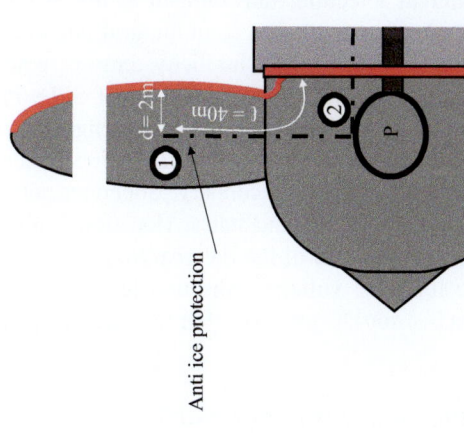

Anti ice protection

Electrical supplies

1. Anti ice protection system
2. AC supply for the pitch drive

Figure 8.39   Example of SPD selection requirements for the anti-ice system located inside a blade which is not isolated from the down conductor

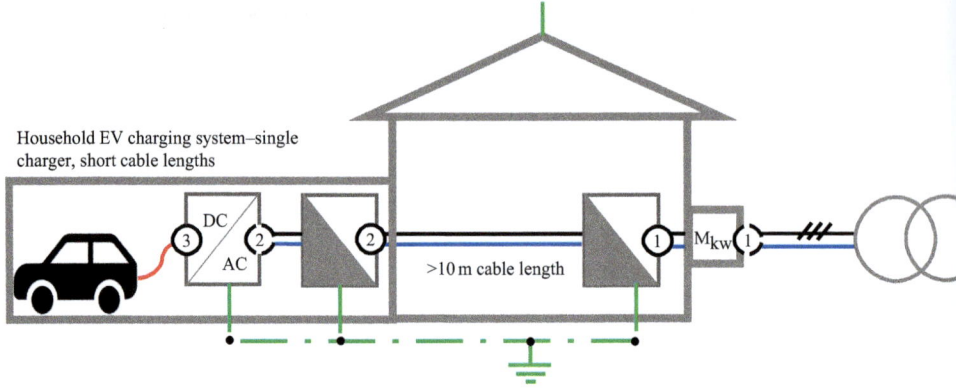

Household EV charging system–single
charger, short cable lengths

*Figure 8.40   Location of SPDs to protect EV charging systems – residential
application*

overvoltage protection in the entire electrical installation. Figure 8.40 shows a typical application. The required SPD locations are as follows:

1. At the electricity meters (in some courtiers, installation of SPDs on the grid side of the meter or inside the meter is not allowed; in these cases, SPDs shall be installed on the load side of the meter or inside the main electrical panel).
2. At the AC secondary electrical panel feeding the EV charger unit if any, otherwise it may also be installed internal to the charger.
3. At the DC side to protect the DC charging components (SPD usually is located internal to the charger).

The SPD types depend on common selection rules defined in IEC 61643-12 (i.e. the presence of external LPS, risk of overvoltages from the grid due to atmospheric origin, etc.). In a residential application, for simplicity reasons, practical rules defined by standards are mainly applicable in the selection of SPDs rather detailed insulation coordination studies. For example, if the cable length between the main electrical panel (location 1) and the secondary panel (location 2) is more than 10 m, SPDs shall be installed at both ends. If a secondary panel does not exist, the SPD shall be either next or inside the EV charging station (location 2 in dotted cycle line shown in Figure 8.40). At the DC side of the EV charging station, a DC SPD shall be installed in order to limit the voltage protection level (U_p) at the required insulation withstand level (U_w) for DC parts (i.e. the inverter components and the electric vehicle).

8.5.3   Example of EV charging stations public parking buildings

In public parking buildings, the EV charging stations may have a more extensive use. Also in the power and communication network feeding, the charging stations

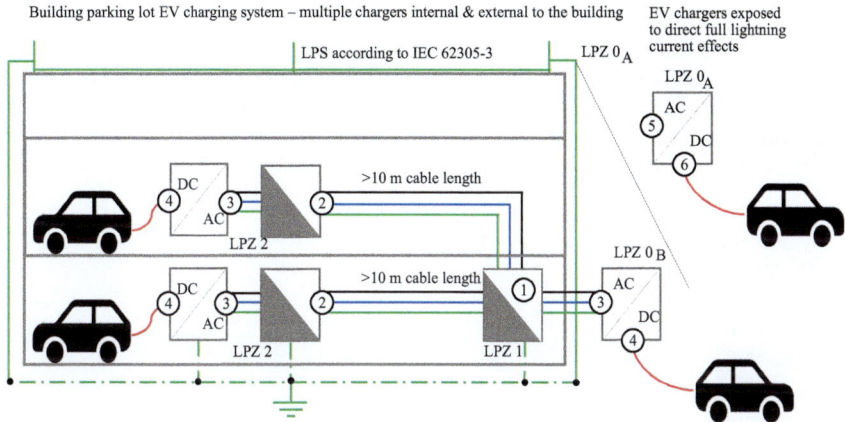

Figure 8.41   Location of SPDs to protect EV charging systems – public parking building application

have a wide range and may include multiple locations to be protected by different SPD types. The most common locations are shown in Figure 8.41.

1. At the main AC electrical panel of the building.
2. At the AC secondary electrical panel(s) feeding the EV charger units.
3. At the AC side of the EV charger unit (it may also installed internal to the charger).
4. At the DC side to protect the DC charging components (SPD is located internal to the charger).
5. At the AC side of an outdoor EV charger unit that is exposed to full lightning current effects (i.e. LPZ 0_A).
6. At the DC side of an outdoor EV charger unit that is exposed to full lightning current effects (i.e. LPZ 0_A).

The selection of SPDs in locations (1)–(4) may follow common selection rules defined in IEC 61643-12 same as in the residential building example. However due to the extended use of EV charging stations, an insulation coordination study may provide a cost-effective and comprehensive solution. Care shall be taken in locations situated outside the direct lightning protection zones (i.e. LPZ 0_A) which may be subjected to either to direct lightning or exposure to full lightning current effects (i.e. high electromagnetic field without reduction). In such location (5) and (6), the required SPDs should be T1.

8.5.4   Example of EV charging stations in open parking and street parking

In open space locations situated such as open parking lots and street parking, the EV charging stations are most likely to be installed in a direct lightning protection zone (i.e. LPZ 0_A) which may be subjected to either to direct lightning or exposure

Open space parking lot EV charging system – multiple chargers, exposed cables to full EMI by a lightning flash

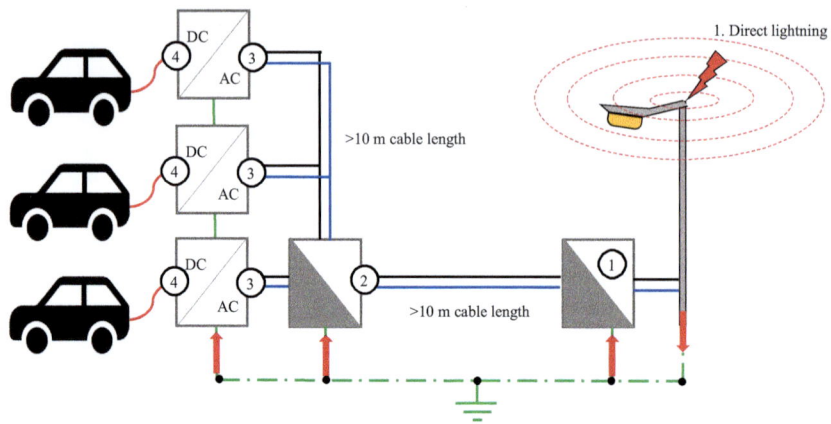

Figure 8.42   Location of SPDs to protect EV charging systems – open parking lots and street parking application

to full lightning current effects (i.e. high electromagnetic field without reduction). In such cases, all required SPDs are recommended to be T1.

1.   At the main AC electrical panel.
2.   At the AC secondary electrical panel(s) feeding the EV charger units.
3.   At the AC side of the EV charger unit (it may also installed internal to the charger).
4.   At the DC side to protect the DC charging components (SPD is located internal to the charger) (Figure 8.42).

8.5.5   Introduction to boat charging systems used in marinas

Modern marinas offer a variety of public services to the boat owners. Some necessary public services are water and electric supply. Similar to electric vehicles, boats contain internal chargers for their batteries; however, they are rather slow in charging compared to new electric boat charging stations installed at the dock offering a direct and fast DC link. In marinas the cable lengths are much longer than in EV parking applications. Also in large marinas, there are multiple distributions points (few hundred of points) starting from MV down to LV (AC and DC). Also, most of the distribution and final circuits are outdoor and therefore exposed to direct and full lightning effects. The required location and technical characteristic of the SPDs should be defined upon a risk assessment and a comprehensive insulation coordination study since following practical installation and selection rules may lead to high-cost solutions. However, a boat compared to an electric car contains elements that are more expensive, and all these parameters that include the presence of human inside the boat during charging (i.e. overnight) shall be considered during the risk assessment. Applicable standards to perform the risk assessment are IEC 62305-2 (Figure 8.43).

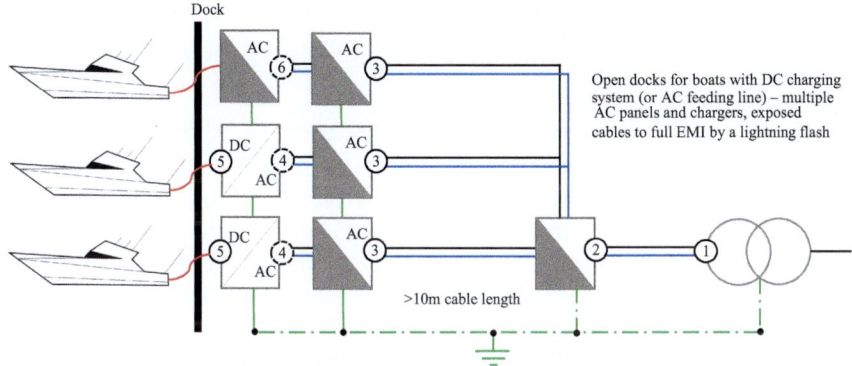

Open docks for boats with DC charging system (or AC feeding line) – multiple AC panels and chargers, exposed cables to full EMI by a lightning flash

Figure 8.43   Location of SPDs to protect boat charging systems in marinas

8.6   Active cathodic protected oil and gas pipelines

In past few years, the insulating components used as coatings on metallic oil and gas transmission pipelines have been improved with respect to their insulating resistance performance due to the high resistivity values (i.e. 10^9 Ωm) of recently developed insulating compounds. The improved insulating coatings have increased the corrosion resistance of the metallic pipelines but have decreased their immunity to electrical stresses generated either due to lightning flashes of due to electromagnetic interference (EMI) caused by power lines running near the pipelines. A poor insulating coating material will allow a voltage stress to be relieved along the length of the pipeline; however, a high resistivity insulation coating will allow a high voltage to build up along the pipeline leading to an electric coating stress and insulation damage.

Any insulation damage either due to mechanical or electrical reasons will create a weak spot with respect to corrosion since the metal surface will be directly exposed to environment. It is therefore necessary to apply a cathodic protection to the pipeline, which aims to protect the metallic parts against corrosion in the event of partial damages. Cathodic protection may be achieved either by connecting DC rectifiers of by connecting surficial anodes to the pipeline, which will polarize the pipeline. Efficient cathodic protection (either by rectifiers or sacrificial anodes) on a pipeline can only be achieved if the pipeline is isolated from earth. Therefore, a high resistance to earth is increasing the cathodic protection performance. On the other hand, any electrical stress on the insulated coating shall be reduced at a safe level and the only way to do that is by applying a comprehensive earthing system allowing a low resistance path for the electrical stress to be relieved along the pipeline. Now, this leads to a contradiction, since simultaneously, a pipeline need to be isolated but also connected to earth. A solution to this problem is the use of DC-decoupling devices, which block any DC current need by the cathodic protection, while it allows any surge or AC current to flow to earth.

Additionally the cathodic protection DC rectifier and the cathodic protection and corrosion monitoring system, since they are located along the pipeline length, commonly fed by an isolated power distribution line, are exposed and vulnerable to lightning overvoltages either from the power line or via the pipeline.

8.6.1 Introduction to EMI coupling on pipelines with active cathodic protection

EMI is a common problem when an insulated coated metallic pipe is running in parallel or crossing power lines, either overhead or underground. The induced voltages from the power line to the metallic piping can cause a series of problems, but the most severe and dangerous are when the induced voltage reach an unsafe level for people working along the pipeline, in faulty conditions of the power line or during a lightning strike, the effect becomes more hazardous.

Figure 8.44 illustrates the possible sources lightning, may cause damage to the coating of the pipeline, as well as on the cathodic protection system (i.e. DC supply, rectifier and cathodic protection and corrosion monitoring system).

Figure 8.45 demonstrates EMI sources mainly generated by steady-state current flow through overhead or underground (cables) power lines, but also due to line to earth faults. A line to earth fault may cause both ohmic and inductive/capacitive coupling to the pipeline.

The limitation of the lightning and EMI effects is succeeded by applying mitigation wires, which are earthing conductors either vertical or horizontal along the pipeline (see Figure 8.46). The mitigation wires are offering a low resistance path to earth. However, they are not directly connected to the pipeline, but via a DC decoupling device, which is a hybrid device incorporating an SPD to divert lightning currents and an AC filter allowing only AC current flow while blocking any DC component.

The DC decoupling device, during lightning current conduction, shall be able to discharge up to 100 kA 10/350 and protect the pipeline coating or the insulating joint against damage, while a decoupling circuit will provide the safe coordination with the rest of the device (see Figure 8.47). Also, the voltage protection level under lightning currents and overvoltages depend on the coating stress limit.

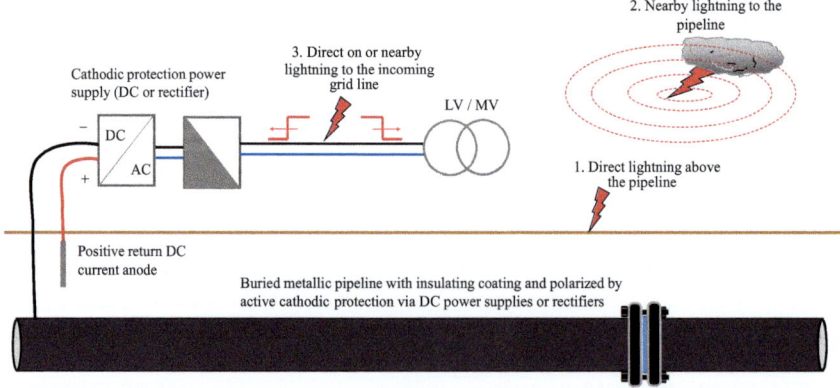

Figure 8.44 Lightning overvoltage coupling to cathodic protected pipelines

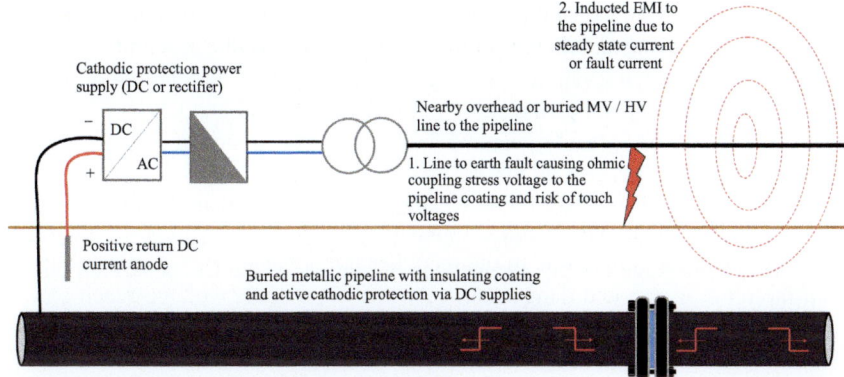

Figure 8.45   EMI overvoltage on cathodic protected pipelines due to faults and steady-state currents

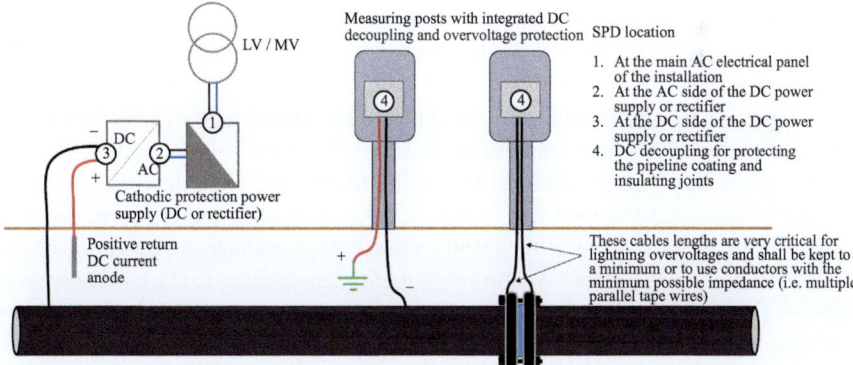

Figure 8.46   Location and selection of overvoltage and EMI protection in cathodic protected pipelines

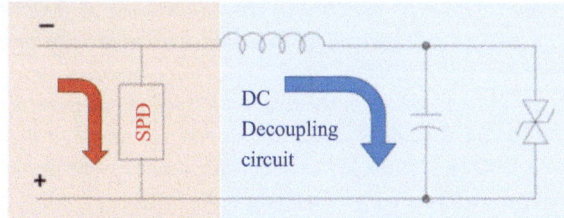

Figure 8.47   General concept DC decoupling devices function

During increased AC voltage along the pipeline, the DC decoupling circuit should act as a properly selected filter which will decrease the residual voltage into low AC voltage levels (i.e. <1.5 V$_{rms}$), protecting the insulating coating, but it should also conduct the appropriate current density up to a level which will not increase AC corrosion of metallic parts of the pipeline. Any AC current flow higher than 30 A/cm^2 may lead to accelerated AC corrosion damaging the metallic layer of the pipeline. Therefore, the selected locations for mitigation wires and DC decoupling devices shall be a subject of an AC mitigation study following EN 50443 and EN ISO 18086 standards. The selected AC steady state and fault current conduction of the DC decoupling device shall also be defined by the AC mitigation study.

Nevertheless, it is vital for the cathodic protection that the DC decoupling operates with limited DC leakage current to earth (i.e. <10 µA or lower). An increased DC leakage current reduces the efficiency of the DC power supplier and rectifiers.

8.7   Nuclear facility protection demonstration

Section 3.3 has defined the need for the protection of nuclear facilities. As indicated, it covers not only power plants but also all types of nuclear facilities such as laboratories and production plants. In many places in the world the nuclear activities are covered by specific agencies that define their own rules for lightning and surge protection.

A nuclear facility is first of all a facility like others, where lightning protection is provided by LPS components or by the use of natural components (component being part of the structure that has the ability to handle a lightning stress up to the required lightning protection level) or a mix. As such, all incoming lines should be protected by Type 1 SPDs at their entrance into the structure.

Note: It is often considered that using natural component LPS would prevent the need of Type 1 SPDs on incoming lines. It is a mistake because Type 1 SPDs are used for equipotential bonding between the LPS earthing system and the incoming lines. Even, if natural components will reduce the voltage generated by the lightning current inside the LPS (this is why separation distance is not calculated when natural components are used), the earthing system will still experience a high voltage rise (e.g. for a mild 10-kA surge and a typical resistance of 10 Ω), the voltage generated by the resistance not even taking into account its impedance that is generally much higher will be 100 kV). Equipment connected at the same time to this earthing system and to a low voltage power line will experience a voltage drop of around 100 kV between phase and earth. This is far greater than the usual LV systems insulation withstand (6 kV maximum for a 230/400-V system) and a sparkover will occur at the weakest point of insulation. Type 1 SPDs are installed to avoid this and provide a low-impedance path for the lightning current to escape. Earthing system has never a resistance equal to zero. To obtain 1 Ω is often a challenge for an earthing system and thus the lightning current will dissipate not only into the local earth but also in the other earthing systems connected to the incoming lines.

Additionally, the statistical risk analysis could demonstrate the need of coordinated sets of SPDs to reduce the stress inside the structure and provide protection for safety equipment (a coordinated SPD set generally includes Type 1 SPDs at the entrance of the line – generally at the main panel board – and then Type 2 SPDs in distributions panels and additionally Type 2 – or sometimes Type 3 – SPDs located near sensitive equipment). SPDs on a same line should be coordinated in energy and in voltage protection level. A list of equipment important for nuclear safety is also defined during the risk analysis and should be protected by specific Type 2 (or Type 3) SPDs.

It is often considered that for reliability reasons, the number of SPDs should be minimized. As a matter of fact, an SPD is by definition a weak point to protect the installation and as such its mean time between failures is a key parameter. There is a way to minimize the number of SPDs:

A Type 1 SPD on a connecting line could be skipped provided the line follows a few rules (see *Figure 8.48*):

• This line should be enclosed in a conduit (either metal tube or reinforced concrete duct) where both ends are equipotentially bonded with the structure and its earthing system at both extremities.

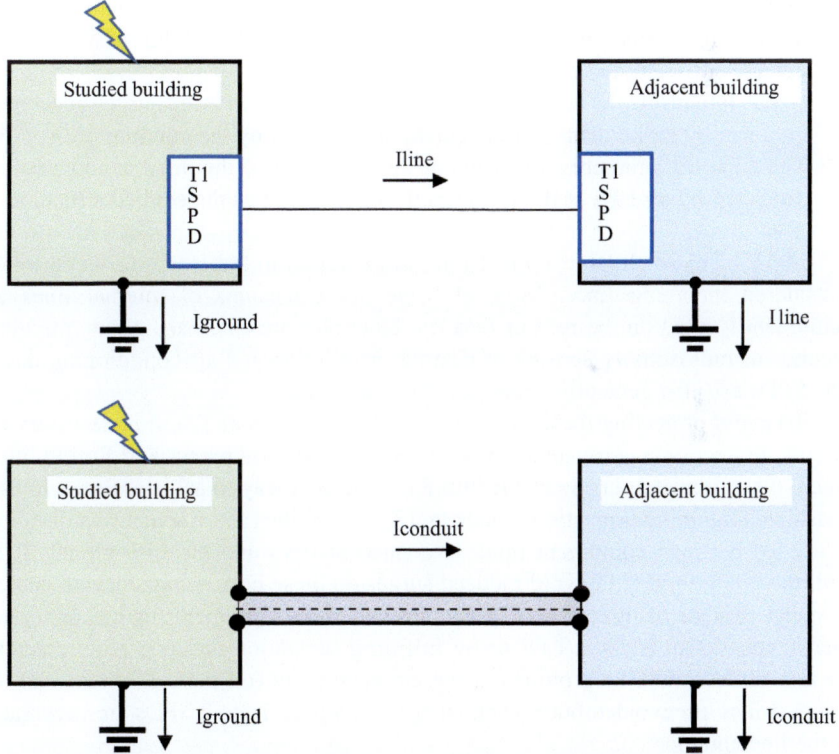

On top of the figure, T1 SPDs are used on the incoming line from adjacent building. Because the current will flow from the ground of struck building to the adjacent building thanks to the line, a second T1 SPD at entrance of adjacent building is recommended. On bottom part of figure, the current will flow from ground of struck building to the adjacent building thanks to the metal conduit, thus protecting the line.

Figure 8.48   Use of a metal conduit to avoid using T1 SPDs on an incoming line

- When a reinforced concrete duct is used, the reinforcing steel must be inter-connected and dense enough to provide shielding at susceptible frequencies.
- The conduit should not be impacted by a direct lightning strike: to avoid impact on the line best it to use underground conduit but it is not always enough.

    Note: on 7 November 2008 an underground gas pipe located on East of France was impacted by lightning to the point to create a leakage and a fire. The gas pipe was buried about 1 m deep under a tarmac path. Thus, being underground is important but not enough.

    The conduit can be protected from a direct strike by installing a 50-mm^2 copper or steel protection conductor in the ground above the conduit at some distance from it. Alternatively, it can be accepted that the conduit is impacted (a part of the lightning current will escape in the soil and only a reduced current will attach to the conduit), but it must be ensured the metal thickness is sufficient to resist puncture. If the conduit is punctured, the internal line will be impacted and T1 SPDs are needed. It should be ensured that the characteristics of the conduit will ensure that the voltage drop along the conduit will not reach a value sufficient to result in sparkover to the line.

    The line inside the conduit will not be disturbed by induced surges generated by a nearby strike thanks to the shielding provided by the conduit.
- The adjacent structure, connected at the other end of the line, should also be protected by an LPS at the same level of protection as the studied structure.

Many of these structures will have lines external to the structure that are not considered incoming lines. Most of these are extensions of internal lines to equipment located on the roof or facades. Examples are external lighting circuits, security or radioactivity sensors or heating, ventilation and air-conditioning devices. SPDs are also generally necessary on these lines.

To avoid protecting these lines by an SPD (either T1 or T2), it is necessary to provide protection against direct strike for the outside equipment and for the lines where they circulate outside of the building. The best way to achieve this is to use an isolated lightning protection system and if it is not the case, a separation distance is needed between equipment (and its connected line) and the LPS circuit. It is additionally necessary to avoid induced surges on these lines using shielded cables or metal conduit. If direct strikes are not avoided on equipment (or its connected line) thanks to an isolated LPS or by fulfilling separation distance requirements, Type 1 SPDs should be provided at the entrance point (on the roof or façade). If direct strikes are avoided but no shielding is provided, Type 2 SPDs are necessary at the line entrance.

Very often, it is not possible to install SPDs at the entrance due to lack of space so the tendency is then to install the SPDs in the nearest panel. However, the first panel could be a significant distance from the entry point, allowing coupling of surges on the incoming line onto other internal lines, creating additional damages. It is critical that these lines are protected before being routed through metal cable trays or metal conduit, especially those containing cabling to supply important safety equipment (ISE). A specific analysis (not an easy one) is needed to check that all these internal disturbances can be accepted by the system.

The main difference between nuclear activities and other type of structures is that surge protection needs to be demonstrated by calculation when generally for most structures, the protection efficiency is inferred from the application of known protection schemes. Nuclear regulations very often impose that the safety of installations be demonstrated. This includes the surge protection of important devices that play a key role for maintaining the barriers and of sensors installed to check the barrier integrity. Two demonstrations are basically needed:

- Protection of equipment by a set of coordinated SPDs:

  First, coordination between SPDs on the same line needs to be demonstrated (this may be a challenge when SPDs are not of the same brand or when SPDs are of the same brand but SPD manufacturer rules to achieve coordination cannot be 100% fulfilled. For example, the wiring of the T1 SPD is longer than requested or the distance between two SPDs is shorter than requested).

  Coordination must also be demonstrated between the last SPD and safety equipment to be protected. The last SPD is not a black hole, it allows a surge (current and voltage) towards downstream safety equipment. This surge is decreased due to many SPDs in cascade in the upstream circuit, but it is not zero. Even though low, this surge can create disturbances on the input circuit of the equipment. The effect of this surge to equipment is mainly covered by electromagnetic compatibility rules. Disturbances due to this surge may mean damages but it may also mean false order (opening a valve for example). It may also damage an internal small varistor that has been purposely added in equipment by its designer to pass the EMC regulations.

  Coordination between SPDs, especially between the last SPD and the equipment to be protected may be covered by a calculation technique called let through energy.

  The let-through energy calculation allows to transform the stress generated by an SPD into an equivalent Combination Wave Generator and if the with-stand of an SPD or of a piece of equipment, when tested by a CWG, is known, it is possible to confirm that coordination is achieved. This process is strongly dependent on the SPD manufacturer because the key parameters to feed that method, as well as the tolerances on the various parameters, can generally only be obtained from the manufacturer.

  Another popular way to demonstrate coordination is to perform tests. In that case it is difficult to be sure that the tested case covers the tolerances of the various SPDs and especially when the front SPD is at the lowest part of the manufacturing range and the second SPD at the highest part of the manu-facturing range in terms of residual voltage, leading to underestimated results. In addition, tests between SPDs and equipment to be protected are always a challenge to avoid damaging the equipment during the test.

  A third way to demonstrate coordinated of a set of SPDs is to perform simulations using dedicated software. In that case, it is easier to take care of the tolerances especially on the SPD connecting lead lengths. It may be necessary to perform a few simple tests at low level to check simulation or obtained a few missing data, especially regarding equipment to be protected.

The coordination rules will help determining the real probability of damage P_{SPD} associated with specific equipment at a specific location, when for less sophisticated studies, default values of P_{SPD} are used for calculating the risk.

● Demonstration that the SPDs will not degrade the reliability of the installation:

It may seem contradictory because SPDs are used to protect equipment, but to achieve this protection, these are used as voluntary weak points. It should be the first element to be damaged, if any, before equipment to be protected. And a weak point, by definition, degrades the reliability. This is why very often, SPDs are limited in nuclear facilities. The more you add SPDs, the more you degrade the reliability. Reliability studies then include the following:

− Identify dangerous situations for functions which contribute to nuclear safety objectives.
− Prioritize the criticality of dangerous scenarios (probability and severity).
− Demonstrate the acceptability of risks and compliance with nuclear safety requirements.

A failure mode and effect analysis is performed by investigating the effect of lightning and surges on the various components that are needed to perform a function. Then annual frequencies of occurrence are defined for various causes that may degrade the function and lightning and surge is one of these causes. As many other causes may degrade the function, it is necessary to check that the lightning and surge influence remains under control to a certain percentage of the tolerable frequency of damage. The influence of the various SPDs will have first a beneficial effect (hopefully using SPDs remains important to ensure the protection of equipment) by using the associated P_{SPD} probability of damage but other parameters will degrade this protection efficiency:

* Low magnitude surges: such surges that may be too low to allow Type 1 SPDs to react. For example, if the front T1 SPD is a spark gap, a surge below the sparkover voltage of this spark gap will propagate freely downstream. This is one of the worst cases for SPD coordination known as blind spot (Figure 8.49). A surge below the sparkover voltage of the front gapped SPD and thus below the maximum surge that this front SPD can handle will damage the second level SPD that is not designed for that surge. But a higher surge that will activate the front spark gap SPD will lead to a much lower stress on the second level SPD that will survive.

* Influence of the SPD failure rate: this is linked to the probability of failure of the SPD and therefore to their mean time before failure (MTBF). This is well known for electronic manufacturers but generally only considers ageing due to external influence such a temperature. Influence of surges is almost never considered in reliability calculations but a cumulated series of low amplitude surges can degrade not only active components of SPDs such as varistors but

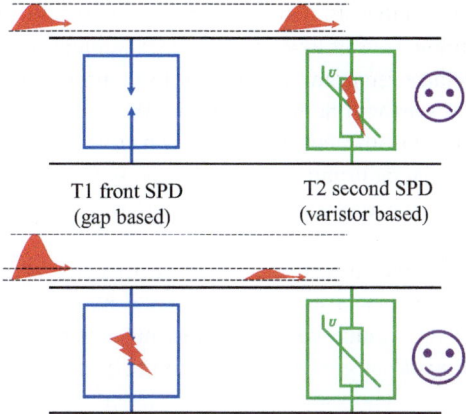

T1 front SPD
(gap based)

T2 second SPD
(varistor based)

Figure 8.49   Blind spot example where a mild 10/350 surge may destroy the downstream T2 SPD when a bigger surge will lead to proper coordination of SPDs on an incoming line

may also degrade the electronic circuit used more and more by SPD manufacturers to improve the SPD efficiency and monitor the SPD state. Typical values for SPD MTBF range between a few hundred of thousand hours and 10 billion hours.

* Influence of the SPD lead length that will degrade the protected level U_p ($U_{p/f}$ concept defined in standard, see Chapter 7).
* State of the SPDs: are the SPDs well connected? Are the SPDs periodically inspected? Is the inspection made by a skilled person? Is the skilled person qualified or even certified to perform the inspection? Is the disconnector open or close?

As it can be observed, the failure mode of the SPDs is a critical point. An SPD that fails open does not protect anymore until it is changed. Eighty per cent of surges consist of multiple pulses and thus, if an SPD fails, there may be no opportunity to replace it before the second surge occurs (a few 100 of ms later). An SPD that fails short will lead to opening of the upstream circuit breaker and thus will not continue to supply the equipment that was supposed to be working all the time to ensure its function. An SPD disconnector is designed to disconnect the SPD from the line as soon as the SPD fails then plays an important role in the strategy to maintain protection. Specific disconnectors can help to some extent to alleviate the problem (refer to Section 9.2 for more details) by disconnecting a failed SPD before the upstream circuit breaker. Monitoring of SPDs, early replacement and inspection policy are then critical for a few SPDs in front of sensitive and important equipment. A preventive action may be needed (refer to Section 9.1 for more details) to remove the SPD from the circuit and install a new one before this SPD fails (ageing, too many low surges, etc.).

Hopefully, all equipment that are important for safety will not need to be protected against surges.

An ISE is an element (physical or procedural) of which the absence or lack of control can lead to a major accident. A risk control measure (RCM) is a set of activities and procedures intended to guarantee the presence and effectiveness of one or more ISE.

Such equipment must be protected in a deterministic manner, independently of the risk analysis which is based on a statistical approach. It is therefore important to limit this list to the minimum necessary and therefore to list only ISE that may be damaged by lightning (either direct or induced) called below LISE. To determine the list of LISE, it is necessary to start from the list of RCM provided by the nuclear facility operator and to exclude those that are not subject to a direct strike or influenced by lightning or surges due to lightning. It is then possible to exclude from the list of RCM:

- Procedures (by nature).
- Those associated with scenarios for which lightning cannot be a triggering or aggravating factor (e.g. linked to a leak of a non-flammable liquid).
- Those that rely on mechanical equipment (therefore, neither powered nor connected to a data or communication network) insofar as lightning could not impact them directly (a mechanical brake or a valve, e.g. protected from direct impacts by the structure).
- Those that are powered or connected to a data or communication network, when their failure does not create a dangerous situation during a storm (e.g. when portable alternative means can be used or when the measures are redundant so that at least one of the measures is not affected by lightning or overvoltages [such as an emergency siren as long as there are more than one that can be heard from anywhere on the site]).
- Those which are provided with positive safety measures and whose failure will be detected immediately (e.g. a fire detector sensor: a sensor damaged by surges will be interpreted as a fire and will launch alarms by the fire management system).

All these rules will allow to determine a surge protection plan and a maintenance policy.

8.8   Data centres

For data centres, the protection plan is not as complex as it is for nuclear facilities because there is no or limited risk for the environment and problems are mainly related to economics. However, it is of prime importance to maintain the service provided and it is fundamental not to lose any data. Protection is therefore mainly related to protecting the various sources of energy (primarily the power utility feed and secondary sources such as power generators and PV panels) as well as the cooling system (computers and mainframes do not like high temperatures but generate a lot of heat). Of course, protection systems against intrusion and fire should also be protected. Most of the UPS should also be protected from their supply side (including the by-pass if any) and also from their distribution panels (damage to equipment connected to a UPS should not damage the UPS or inhibit its ability to provide energy to other equipment).

It is more and more common to provide protection against lightning using an isolated LPS. There are generally many equipment on the roof for communication or for the cooling and air conditioning. Even if most of the data flow is transferred via copper or fibre optics underground links, there are generally back-up communication tools on the roof (satellite dishes or ground antennas) to communicate with specific agencies or customer premises. In addition, there are also GPS antennas for synchronism as well as many other antennas. If the distance between one piece of equipment on the roof or its connected cables and pipes and the LPS circuit is lower than the separation distance, then there is a need to provide protection of this circuit with Type 1 SPDs. As mentioned in Section 8.7, these SPDs should be located at the line entrance because if not, an in-depth analysis of the routing of this circuit should be made to check which circuits may be disturbed by this disturbance emitter. T1 SPDs at entrance will protect the structure against disturbances but:

- Will require a coordination study be performed between this T1 SPD and downstream Type 2 SPDs on the same circuit.
- Will not protect the piece of equipment on the roof itself. To protect it, other SPDs are needed near the equipment or shielded methods be used as explained in Section 8.7.
- Additionally, the number of SPDs will be large, requiring a more complex maintenance scheme, especially when located on the roof.

By using an isolated LPS, T1 SPDs are not needed and depending on the type of circuit on the roof (shielded or not), Type 2 SPDs or even no SPD at all on the roof will do the job.

The protection plan will then start by defining the T1 SPDs on all incoming lines at ground level.

Note: A line is considered an entering line even if it supplies a neighbouring building. The flow of energy is not important, rather it may be towards the studied building when it comes from the power utility or towards another building or equipment. Even if the line is no longer used, it will be a path for lightning current until it is removed or grounded.

It includes the HV lines from power utility, the LV lines connected to neighbouring buildings, the data and telecom copper lines and the line connected to the external power sources such as a PV plant or a back-up power generator (a power generator located inside the structure will also need to be protected by a Type 1 SPD if the gas exhaust chimney may be struck by lightning).

As indicated previously, when the studied building is struck by lightning, its local earth will rise in potential. For example, with a 100-kA lightning current, even if the earth resistance is 1 Ω, the voltage at the earthing terminal will be 100 kV; well in excess of the LV voltage or sometimes even of the HV voltage. Thus, there is a risk of flashover to an incoming line. A partial lightning current, assumed to be as a first approximation, 50% of the lightning current that struck the building will then flow to the adjacent structures connected by the incoming lines. It may also be considered first approximation that the current shares equally between all these paths (including the metal pipes). Inside each path, the current will also share equally between all the conductors (e.g. the current of the path will be divided by 5 in a 3-phase+neutral+PE cable). In general, a typical value found in the market for SPDs is I_{imp}=12.5 kA. One can then ask why it is necessary to calculate the I_{imp} current and why not to use the default value of 12.5 kA? Reason is that the disconnector (when it is not included in the SPD assembly) used in series with the SPD is designed by this calculated current. The disconnector should have the same withstand as the SPD itself. If the calculated I_{imp} for the T1 SPD is 2.8 kA, a gG fuse rated 40 A will do the job. When using the default value of 12.5 kA, the fuse should be rated 160 A! The task is even more complex for circuit breakers because the surge withstand for 10/350 waveshape is not generally available in data sheets, so it is necessary to refer to the CB manufacturer for the information.

HV lines are often forgotten in risk analysis. Their high insulation withstand makes it appear they are immune from lightning stress and frequently one relies on the protection means provided by the utility. However, as discussed before a 100-kA on a 1-Ω earthing system will reach 100 kV that is just above regular insulation withstand of 20 kV HV lines (95 kV). When the earthing is 10 Ω (often recommended as a maximum value when an LPS is used), a simple 10 kA will reach this level and considering the highest lightning current of 200 kA considered in standards for LPL I, this value will raise up to 2 MV that exceeds a lot of HV systems insulation.

In the same way, a surge coming from the HV system will propagate to the data centre in the absence of HV surge arresters, and then, the transformer may be damaged. It is not acceptable for a data centre to lose its primary supply source. Furthermore, a part of the incoming surge (assumed to be 20%) will propagate through the windings of the transformer to stress lower system voltage inside the data centre. Considering installation HV surge arresters in such circumstances is a good idea.

To avoid losing power supply from the main source, T1 SPDs should be installed as close as possible to the entrance point of the electrical installation. SPD should be installed with their appropriate disconnector but often they will also be protected by the overcurrent protective devices of the installation (back-up protection), especially regarding indirect contact protection. Other SPDs may be used in front of circuit breakers supplying part of installation: in that position SPDs are able to protect these CB even if the CB surge withstand is not known. Such SPDs should be installed with their appropriate disconnector (if not included in the SPD itself). For TT system, an Residual Current Device (RCD) should be present on parts of the circuit, but the surge withstand of RCD is very low (generally 3–5 kA 8/ 20) and they will not handle a 10/350 surge. Installing the T1 SPD upstream of the RCD will then protect the RCD but the SPD is not protected by the RCD in the case of a fault to ground (indirect contact risk). For TT systems, SPDs not protected by an RCD should be of connection type CT2 to avoid creating a direct path between phase and earth in case there is a failure of the SPD.

Installing an SPD in front of a CB is particularly interesting for lines going outside of the structure, such as light poles in parking lots. Light poles can be tall and numerous and as such prone to get struck by lightning. A surge current will then propagate towards the lighting electrical panel inside the building. Circuit breakers used in location usually have a low rating and resulting low surge withstand. The T1 SPD located in front of the CB will protect it against this damaging surge.

Note: 'Upstream' or 'in front' when addressing surges means that the SPD is located between a possible surge source and a sensitive device, which is not related to the flow of power current.

Coordinated Type 2 SPDs should typically be provided for the following circuits:

- UPS as well as their by-pass circuit and distribution panels.
- Cooling system (cold water for mainframes as well as venting system for all rooms).
- Firefighting means (pumps) as well as the fire detection system.
- Pumps to avoid the risk of flood.
- Copper data and telecom lines.

As discussed in Section 8.7, there is no need to protect all sensors related to intrusion or fire detection if they remain inside the structure. If they are located outside of the protected zones, Type 1 SPDs may be necessary on the line, not to protect the sensors but because it should be considered an entrance point for surges.

Case study

A data centre is located at the ground floor of a higher building. Due to the flash ground density, it was decided to install a T1 SPD in front of the HV/LV utility transformer. The building earth dedicated to the data centre is very good (2 Ω) and a copper bar surrounds the sensitive rooms to allow direct connection of equipment to earth. The transformer earth is different from the data centre earth. A Type 2 SPD is installed in front of the UPS supplying the most sensitive equipment. Another circuit supplies normal equipment (lights, power sockets and alike). A neighbouring building (10 m from the data centre building) is protected by an LPS but has a much bigger earth (15 Ω). When a lightning occurred on the lightning rod of that building, lighting circuit located in the ceiling was producing a lot of sparks and the T2 SPD was found destroyed. In fact, the coupling between the earthing system of the two buildings led to an injection of a partial lightning current inside the data centre earthing system. If partial lightning current of 10 kA was assumed (T1 SPDs were rated I_{imp} 12.5 kA and were not destroyed), the voltage generated at the earthing system was 20 kV. This was too low for sparking over the transformer insulation. Only a small part of the current flows to the utility circuit, mainly by capacitive coupling between transformer windings. The T2 was destroyed by bad coordination between the T2 and the T1 SPDs (but UPS did not fail). The voltage was high enough to create sparks between the lighting circuit in the roof and other circuits such as fire sprinkler that were bonded to another earthing. The T1 SPD avoided a sparkover at the transformer that would have been able to damage strongly the transformer and need to be maintained. A coordinated T2 SPD should replace the present T2. Equipotential bonding of all metal circuits entering the data centre would help avoiding sparks and offer an alternate route to the partial lightning current (Figure 8.50).

If there is a risk of induced surges directly on subsidiary circuits, T2 SPDs may be used in front of sensitive equipment. If this fear is negligible, the T2 SPDs on the secondary side of the UPS panel will be enough.

Note: It is assumed that important sensitive equipment are supplied by UPS.

Very frequently, isolation transformers (transformer with two windings and a grounded screen between them) are used to change the type of earthing system inside installation. In that case, a T2 SPD is needed in front of the transformer.

If the induced surges can be negligible inside the structure (e.g. a tight mesh of rebars in a concrete structure, a metallic shelter, cables routed in metal trays or other solutions avoiding induced surges in the structure from nearby strikes) and no circuit being disturbed due to coupling with another circuit where surges circulate (e.g. the PE conductor of Type 1 SPDs), no other T2 SPDs are needed.

On the reverse a sensitive equipment connected directly to an external line should be protected by a coordinated T2 SPD (as detailed earlier a T1 SPD is necessary at line entrance in that case) located as close as possible to the equipment but never at more than 10 m from it.

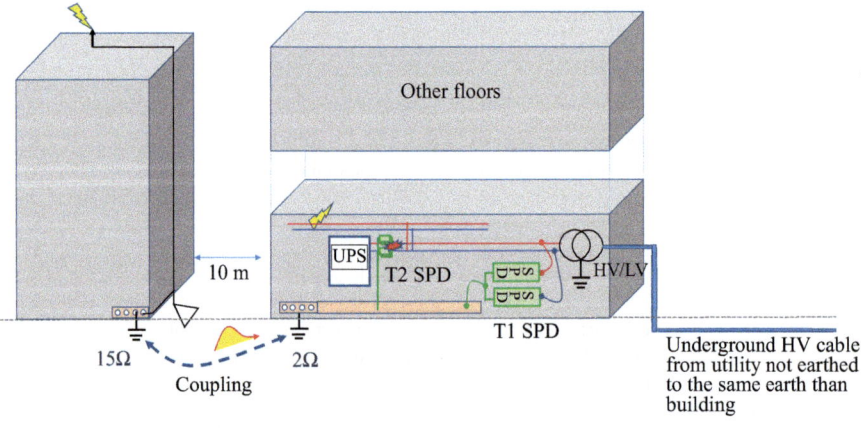

Figure 8.50   Data centre case

8.9 Country specificities

With the increase in world globalization, manufacturers of SPDs are increasing looking to markets beyond their domestic borders to promote their products. The assumption is often made that this is a relatively simple matter; with little more needed than to ensure that the SPD's maximum continuous operating voltage, U_C, is suitable for connection to the nominal system voltage, U_n, of the country in question.

While important, this is only one of many electrical and legislative require-ments which must be considered! For example, within the USA it is mandatory that the NEC* be followed with few exceptions.[†] The NEC requires that an SPD intended for installation on the electrical system be 'listed for its application', meaning that it must be tested in accordance with the relevant ANSI[‡] standard covering SPDs, which is UL[§] 1449.

This code requirement makes compliance with UL 1449 mandatory for SPDs installed on the US electrical network. The situation is similar in European coun-tries, which at the CENELEC level require compliance with relevant low-voltage directives, such as EN 61643-11 for SPDs installed on the power network.

8.9.1 Review of the requirements governing the installation of surge-protective devices on the US Electrical Distribution Network

In many countries around the world the low-voltage power distribution system is nominally 230 V_{AC} derived from the secondary of a 3Ph Star (WYE) wound transformer. In the United States, Latin America and certain Asian countries, the situation is more complicated with at least seven different power systems being in common usage – refer to Figure 8.51.

---

*The National Electrical Code (NEC) is a regionally adoptable standard covering the installation of electrical wiring and equipment in the United States and published by the National Fire Protection Association as NFPA 70. It is not a federal law; however, states typically adopted the code to enforce safe electrical practices. The state's authority having jurisdiction (AHJ) inspects for compliance with the standard.

[†]Certain bodies, such as those under the federal government, railways and power utilities, are exempt from following the requirements of the NEC. In addition, certain installations are not covered by the scope of the code, for example: ships or watercraft, railway rolling stock, aircraft, vehicles other than mobile homes and recreational vehicles, underground mines and equipment under the control of com-munications utilities, to name a few.

[‡]The American National Standard Institute (ANSI) is a private non-profit organization, headquartered in Washington, DC, overseeing the development of voluntary consensus standards for products in the United States. It works to coordinate these national standards with international standards to ensure that US products can be used worldwide. These standards ensure that the characteristics and performance of products are consistent that people use the same definitions and terms and that products are tested the same way. ANSI also accredits organizations that carry out product or personnel certification in accor-dance with requirements defined in international standards.

[§]Underwriters Laboratories (UL) was established in 1894 as a bureau of the National Board of Fire Underwriters. It is now an independent company headquartered in Northbrook, IL, participating in the safety analysis and certification of many products and technologies.

Power network description	Source configuration (secondary)	Secondary side low voltages
Single phase 1Ph, 2W+G	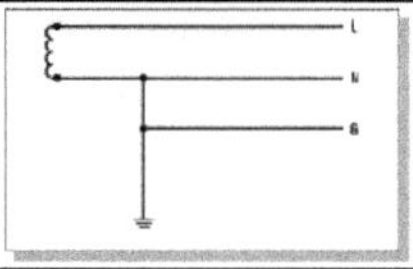	Residential 110V[a], 120V[a], 127V[b], 220V[a], 240V[a,c] (L–N)
Single phase Split phase or Edison 1Ph, 3W+G	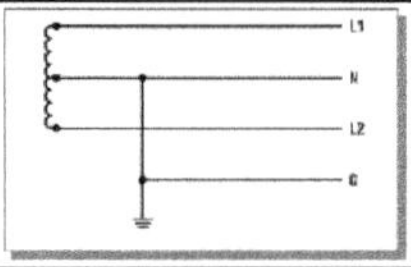	Residential 120/240V[a,b] (L–N / L–L)
Three phase WYE No neutral 3Ph Y, 3W+G	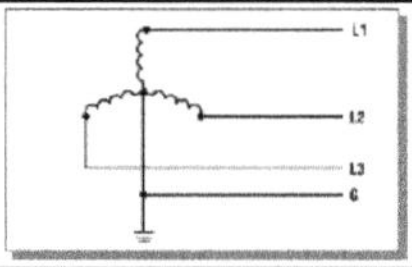	Heavy industrial 480V [a,e] (L–L)
Three phase WYE With neutral 3Ph Y, 4W+G	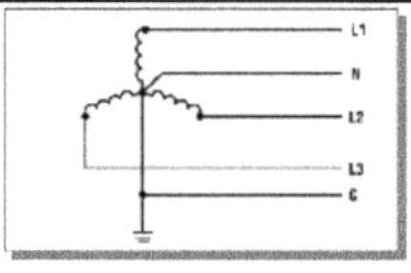	Industrial 120/208V[a], 127/220V[c], 220/380V[f], 230/400V[d], 240/415V[e], 277/480V[a], 347/600V[c] (L–N / L–L)
Delta High-Leg 3Ph Δ, 4W+G	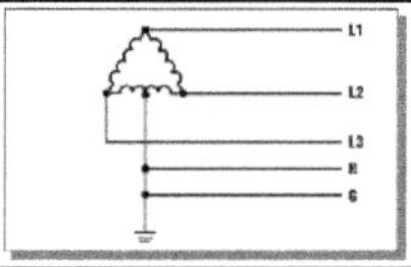	Commercial and industrial 120/240V[a] (L–N / L–L)
Delta Ungrounded 3Ph Δ, 3W+G	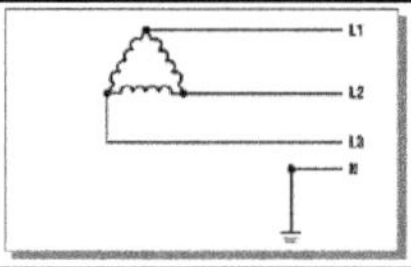	Heavy Industrial 240V[a], 480V[a] (L–L)
Delta Corner grounded 3Ph Δ, 3W+G	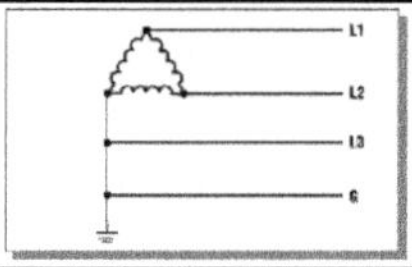	Commercial and Industrial 240V[a], 480V[a] (L–L)

Some countries where adopted:

[a] United States, American Samoa, Bahamas, Belize, Brazil, Caribbean, Colombia, Japan, Philippines, Taiwan, (60Hz)

[b] Mexico, Venezuela, Aruba, (60Hz)

[c] Canada, (60Hz)

[d] Europe, CENELEC, South Africa, United Kingdom, (50Hz)

[e] Australia, Malaysia, ASEAN, India, New Zealand, (50Hz)

[f] Argentina, China, Hong Kong, Singapore, Thailand, UAE, Vietnam (50Hz)

Figure 8.51   Various power distribution networks commonly encountered in North America and other countries

Most residential and light commercial installations in the United States are supplied with a 120/240 V_{AC}, 1Ph, 3W+G system derived from the secondary winding of a centre-tap transformer. This results in two 120 V_{AC} line voltages which are out of phase by 180 degrees with each other. The system neutral conductor is connected to the transformer centre-tap and grounded. A voltage of 240 V_{AC} can be obtained by connecting the load between the two 120 V_{AC} lines. These lines are often designated L1/L2 and described as 'hots'. This system is sometimes described as a split-phase or Edison single-phase system.

A common problem faced by this supply occurs when the neutral connection becomes corroded or disconnected and is known as a 'loose neutral'. Under such conditions if the loads connected between L1–N and L2–N are not balanced, the zero point of the system shifts causing the voltage across one 'half' of the load to increase while that across the other to decrease. Such an event is often referred to as a TOV, although this is probably inappropriate as the condition can be sustained for many hours. If an SPD is installed on such a system, this increase in voltage may exceed its maximum continuous operating voltage U_C causing the internal metal oxide varistors to rapidly accumulate heat and become a potential fire hazard. It is for this reason that so much effort is dedicated to testing an SPD's ability to withstand, or safely disconnect, from system TOV events.

The ungrounded delta or WYE is a different form of distribution network that has been used for many years in the USA. It has its history in the days of cotton mills of the south, where a complete factory might operate on a single large motor with overhead axel running the length of the warehouse with periodical pulleys and belts feeding machinery below. Under such conditions it was vital that a fault in this main motor would not result in all production stopping. The ungrounded system was introduced to address this concern by allowing one winding (phase) of the motor to fault to ground and keep the plant running until maintenance could be scheduled.

Unfortunately, such ungrounded systems are inherently unstable when subject to an intermittent phase fault (often called an arcing fault). Under such circumstances, the magnetic energy stored in the inductive component of the system, and the capacitive energy stored in the windings of the supply transformer, can initiated a resonance under transient conditions. This resonant swing has been known to result in the phase-to-ground (pseudo ground) voltage rising as much as 5 p.u.

As with the loose neutral TOV on the split phase system, these resonant TOVs on the ungrounded system can stress any connected SPDs and cause them to fail. It is for this reason that SPD standards such as UL 1449 and IEC 61643-11 place special emphasis on the testing of SPDs under conditions of abnormal over-voltages (TOVs).

8.9.2   National Electrical Code®

The NEC is the primary authority in the USA which regulates equipment connected to the utility network. It is administered by the National Fire Protection Association under NFPA 70 and follows a 3-year code revision cycle. Section 90 of the NEC states that the primary purpose of the Code is the safeguarding of persons and property from hazards arising from the use of electricity. It explains that the code is not intended as an instruction manual for untrained persons but does provide a sound basis for the study of electrical installation procedures.

The NEC is adopted in some form or another by every state in the United States, but it is not mandated that all adopt the latest edition[||] and certain organizations are exempt all together from having to follow the Code – two of the main ones being power utilities and railways. The AHJ is the organization or individual responsible for certifying equipment and signing off on a new installation. Electrical inspectors enforce the Code's implementation. In general, articles of the Code are divided between those which are mandatory (contain the words, *shall* and *shall not*) and those which are permissive but not necessarily required.

Outside the United States, the adoption of the NEC has gained importance with Mexico and Venezuela incorporating it into their national laws. Other countries adopt similar codes to promote electrical safety. Canada has the Canadian Electrical Code, and the International Electro-technical Commission (IEC) administers electrical standards under technical committee TC 64. Likewise, Australia and New Zealand adopt the AS/NZS 3000 Standard for Wiring Rules.

The selection and installation of SPDs is required to comply with the 'Listing' requirements of the NEC. Their use is regulated under a new Article 242, added to the 2020 Edition of the Code, titled *Overvoltage Protection*. This article combines the requirements contained under former Article 280 which covered *surge arresters*, and Article 285 which covered *SPDs*.

Article 242 is arranged under three logical parts:

- Part I covers general requirements, installation requirements and connection requirements for overvoltage protective devices.
- Part II covers SPDs permanently installed on premises wiring not exceeding 1,000 V_{AC}.
- Part III covers surge arresters permanently installed on premises wiring over 1,000 V_{AC}.

The NEC states that an SPD must be 'Listed' for its application, meaning that before it can be connected to the US electrical network, it must carry the mark of a Nationally Recognized Test Laboratory (NRTL) signifying that it has been tested to the appropriate Listing requirements for this class of equipment. UL is one such NRTL which is permitted to conduct SPD listings. Others include ETL (Intertek Testing Services) and CSA (Canadian Standards Association), which test to the UL 1449 standard but can only place their own mark on the product. CSA and UL have an agreement allowing a dual logo arrangement (cULus). To address counterfeiting of the UL mark, UL's Listing of SPDs requires that a special holographic label, strictly controlled by UL, be applied to the product regardless of the country of manufacture.

UL uses a system of Category Control Numbers (CCNs) to identify categories of products that have been evaluated. A CCN typically consists of four or more

[||]As of 1 June 2020, the 2020 edition of the NEC was in effect in 1 state (with 21 states under update), the 2017 edition in 34 states, the 2014 edition in 9 states and the 2008 edition in 3 states.

alpha-numeric digits, followed by a title to identify the product group. Each category has a published 'guide card' which provides information on the applicable standard and criteria for use of the UL Mark. CCNs under UL 1449 to which SPDs are assessed include the following:

- VZCA – Type 1 and Type 2 SPDs intended for hardwiring to service entrances panels rated 1,000 V_{AC}, or 1,500 V_{DC}, or for use with PV and LPSs.
- DIMV – moulded case circuit breakers combining SPD functions rated 600 V_{AC} or less.
- OWIW – moulded case SPD.
- VZCC – SPD certified to IEC publications.

8.9.3 NEC articles regulating the installation of SPDs

Article 242.10 SCCR:

One of the more significant requirements governing the installation of SPDs is covered under Article 242.10 of the NEC. It requires that the SPD's SCCR be coordinated with the prospective fault current available at its point of installation on the electrical network. It states, 'The SPD shall be marked with short-circuit current rating and shall not be installed at a point on the system where the available fault current is in excess of that rating'.

This requirement was introduced some years ago due to concerns that SPDs, which had not been evaluated for safe end-of-life failure, were being installed on service panels where the prospective current could be as high as 200 kA, 60 Hz. The introduction of this requirement forced UL 1449 to modify the test standard and begin evaluating an SPD's ability to safely disconnect itself during abnormal overvoltage conditions, and to be labelled with the SCCR rating to which it was tested. This rating now regulates the maximum prospective fault current to which the SPD can be connected. For example, an SPD marked with an SCCR of 65 kA (per UL 1449 testing) can only be installed at a point on the network where the symmetrical rms fault current is equal to, or less than, this value.

If this coordination cannot be achieved, one option is to add a series-rated, current-limiting circuit breaker, between the SPD and the source. It is important to understand that UL requires the SPD to be tested in conjunction with any intended overcurrent protection device, and that any Listing subsequently obtained applies only if this precise combination is used. Installing it using a different overcurrent device violates the Listing's conditions of acceptability (COA).

To verify the SCCR the SPD manufacturer may wish to obtain for his product, UL requires that the product be energized at an abnormal overvoltage[¶] while connected to a supply capable of delivering this amount of symmetrical RMS current (into a short circuit). The product must demonstrate a safe-failure behaviour, either through operation of an external overcurrent protector (such as fuse or circuit breaker); or through operation of an internal thermal disconnector; or

---

[¶]Ref. Table 44.1 Test voltage selection table, UL 1449 Ed 4, Fourth Edition, Dated 20 August 2014. Revised 1 August 2018.

through containment of any explosion and subsequent fire hazard (usually via means of a metallic enclosure).**

The need for SPDs to safely disconnect during abnormal overvoltage conditions of the network has forced manufacturers to design into their product's various disconnection technologies. These generally fall into two categories:

- **Thermal disconnection** – which ensures safe behaviour under conditions of abnormal over-voltages where the current is limited to a few tens of amperes (such as in the case of a loose neutral), and the temperature of the internal varistors gradually increases until disconnection results.
- **Over-current disconnection** – which ensures safe behaviour under conditions where an internal failure of the SPD may suddenly draw a significant amount of current from the network requiring very rapid disconnection. This is usually achieved by using fuses or circuit breakers having a fast enough response to limit the i^2t energy to a point where disconnection occurs before the internal varistors explode.

The need to include fast acting overcurrent protection to allow an SPD to safely and quickly disconnect itself from an abnormal condition of the network must often be traded off against the designer's aim of also achieving a high discharge current rating I_n. A well-designed SPD will have both a high SCCR, and a high I_n rating.

8.9.4   *Underwriters Laboratories Inc.*

The primary charter of ULs is safety. For this reason, the UL 1449 standard 'SPDs', and the test procedures described therein, are constantly under review and update. The standard is currently in its fourth edition and was last revised in 1 August 2018.

UL 1449 classifies SPDs into 'SPD Types'[††]:

- Type 1 SPDs are permanently connected devices certified for installation at any location between the secondary of the utility transformer and the line-side of the facility's primary overcurrent disconnect (service entrance breaker). They may also be installed on the load-side of this primary disconnect without requiring a dedicated fuse or breaker. The nominal discharge current rating I_n of Type 1 SPDs is either 10 or 20 kA.
- Type 2 SPDs are permanently connected devices certified for installation on the load-side of the service entrance primary overcurrent disconnect. Type 2 SPDs may, or may not, require the use of a dedicated fuse or breaker. Such conditions are usually spelt out under UL's COA for that device.

**Those familiar with the plastic form-factor, typical of DIN SPDs used in IEC compliant countries, are often surprised by the metal enclosures with flying leads and conduit fitting, typical of SPDs installed in the USA. One reason for this is that a metallic enclosure can assist a manufacturer in passing the abnormal overvoltage testing requirements of UL 1449.

[††]Note: It is important not to confuse the usage of SPD Type under UL 1449 with the same terminology used in CENELEC markets where SPDs are also classified as Types 1, 2, 3 in accordance with EN 61643-11! In CENELEC markets SPD Type is synonymous with the SPD's test classes I, II, III in accordance with IEC 61643-11; while under UL it is a classification related to the location where the SPD may be installed on the network.

The manufacturer is generally required to provide these same details in his installation prints, but this is not always followed. An installer is advised to check such details before installation. The nominal discharge current rating I_n of Type 2 SPDs is: 3, 5, 10, or 20 kA.

- Type 3 SPDs are point-of-use devices, intended to be installed at a conductor length 10 m or greater from the panel board. These devices are typically cord connected or receptacle outlet SPDs.
- Type 4 SPDs are SPD component assemblies that are incomplete in some aspect which prevents them from obtaining a UL Listing. Generally, such SPDs are components without an outer enclosure and used to assemble a complete Listed SPD. Very often Type 4 SPD assemblies have not undergone short–circuit[‡‡] and/or intermediate current[§§] testing, since they are deemed only to be a part of the final SPD.
- Type 5 SPDs are SPD components in their raw state, such as metal oxide varistors. They are intended to be integrated with disconnectors and terminals to form a complete listed SPD.
- Types 1, 2, 3 component assembly SPDs (Recognized components) are intended to be factory installed into end-use equipment. They are evaluated for use in Type 1, 2 or 3 SPD applications and generally tested in the same way as listed Type 1, 2 or 3 SPDs, apart from not being evaluated under the intermediate current and short-circuit current test regimens. While being compliant with safe failure, these Type 1, 2 and 3 component assembly SPDs have COA, for example exposed terminals, that require them to be housed in a listed assembly to provide protection from exposure to live parts.

This brief review of UL 1449 would not be complete without some words of explanation on the difference between 'Recognized' and 'Listed'. UL recognition applies to a 'component', while listing applies to the final 'product'. A recognized component is designated with a symbol *RU* while a listed product is designated with the symbol *UL*. This distinction can become blurred, particularly in the case of SPDs. Put simply, a *recognized component* is not intended to be used as a stand-alone device but rather installed as a component in a final assembly which itself will be *Listed*. Very often a recognized component will have aspects – for example exposed live terminals – which require that it be installed in an enclosure to obtain safety. It is for this reason that the typical European-style SPDs intended for installation on DIN rails are generally Recognized and not Listed under UL 1449.[‖‖]

[‡‡]UL 1449 Section 44.2 – short-circuit current testing.

[§§]UL 1449 Section 44.3 – intermediate current testing.

[‖‖]Only being able to have a product *Recognized*, and not *Listed*, can be a substantial barrier-to-entry into the US market as it places a burden on the end-customer to have the final assembly Listed. For example, if an overseas manufacturer sells his SPD to a panel board manufacturer for inclusion in say a motor control cabinet, and this SPD only has UR, the panel board manufacturer will have to have to pay to have the final assembly Listed. If on the other hand, the SPD is listed as a product (UL mark) then it can be connected direct to the motor control centre without any need for further testing. The situation has recently changed somewhat as UL has introduced a moulded-case SPD classification which can be listed.

8.9.5   SPDs used in US lightning protection installations (LPS)

The US code governing lightning protection installations is NFPA 780. It details all aspects of structural protection from the effects of lightning, as well as the installation of SPDs. A counterpart to this code is the UL Master Label program under UL 96A Installation Requirements for LPSs. Lightning protection installation can be certified under the voluntary *UL Master Label* program. A UL field inspector will assess all aspects of the installation (per UL 96A[¶¶]) and the components used (per UL 96) and sign off on this.

The use of SPDs on such an installation is also governed by UL 96A, which under clause 13.2 states 'A surge arrester, protector, or antenna-discharge unit shall be installed on each electric and telephone service entrance and each radio and television antenna lead-in'.

It should be noted that only SPDs Listed under UL category VZCA, as Type 1 or Type 2, and having a nominal discharge rating I_n of 20 kA, may qualify as an approved service entrance SPD for use on a UL 96A lightning protection installation.

8.9.6   US position RE IEC test class I

In dealing with country specificities, it is important to note that differences will always exist, often with historical reason. For example, the US IEEE*** introduced the 8/20 waveshape used in the evaluation of SPDs under test class II, while the IEC introduced the 10/350 waveshape used in the evaluation of SPDs under test class I when intended to carry partial, or direct, lightning currents.

US surge protection standards, such as those developed under the IEEE C62 low-voltage series, have chosen to base their requirements and evaluation purely on the 8/20 waveshape. In addition, the United States, as a P member of the IEC and Secretariat of IEC/SC37, has included an 'in some countries clause' into the IEC 61643-xx series stating that the class I test waveshape is not adopted in the United States.

While harmonization should always be one of the goals of standardization committees around the world, we often must accept differences while seeking common ground where possible. This remains one such area to those of us involved in the work of surge protection.

8.9.7   Asian countries adopting US systems

The power distribution system encountered in most Asian countries is based on the IEC/European 230/400 V, 50 Hz, 3-phase, WYE low-voltage distribution. As such, SPD tested in accordance with IEC 61643-11 can generally be connected to these power networks, provided they are also compliant with relevant local and national regulations.

¶¶UL 96A is largely derived from NFPA 780: standard for the installation of lightning protection systems.
***Institute of Electrical and Electronic Engineer (IEEE) was formed in 1963 and is the world's largest association of technical professionals with more than 423,000 members in over 160 countries around the world. The Surge Protective Devices Committee falls under the IEEE Power & Energy Society and is responsible for the creation and maintenance of SPD Standards, Guides and Practice documents.

This said, there are a few countries within Asia – primarily Japan, Taiwan and the Philippines – that have instead adopted the 120/240 V, 60 Hz, split phase, residential power distribution network used in North America – refer to Figure 8.51.

The Japanese national electrical scheme is a unique system and differs between east and west Japan. The low-voltage residential system is usually derived from the secondary of a 3-phase, WYE transformer. In the east of the country the secondary 3-phase L–L voltage is 200 V, 50 Hz, resulting in a single phase L–N voltage of 115 V, 50 Hz, while in the west the secondary 3-phase L–L voltage is either 200 V, or 210 V, 60 Hz, resulting in a single phase L–N voltage of 121 V, 60 Hz. A split phase 100/200 system is also used with L1/L2 at 180 degrees phase angle. In general, SPD intended for the voltage systems used in the US market can also be used on the Japanese network.

IEC 61643-11 Annex B details expected TOV conditions to which SPDs should be tested. Table B1 lists typical values for 230/400 V-based networks, Table B2 lists values for US networks and Table B3 lists values for the Japanese network.

8.9.8 Summary

This section has sought to provide an overview of some of the more important aspects to be considered when installing SPDs on the US power distribution system, as well as providing some insight into the legal and legislative framework which govern such installations under the National Electric Code and ULs. In addition, it has briefly addressed some countries adopting similar voltages and frequency to North America, where SPDs designed for the US market are often installed.

Bibliography

R. Weiss, L. Ott and U. Boeke, "Energy efficient low-voltage DC-grids for commercial buildings," in ICDCM, Atlanta, USA, 2015.

J. Kaiser, C. Strobel, H. Mann, *et al.*, "Joint project "DC-Schutzorgane" – Development of a comprehensive protection concept and protective devices for future low voltage direct current grids," in Albert-Keil-Kontaktseminar, Karlsruhe, 2019.

F. Schork, Ph.D. Thesis: Stoßstromanwendungen von Leistungshalbleitern im Überspannungsschutz, Ilmenau: TU-Ilmenau, 2019.

K. Dinesh, Z. Friuz and G. Arindam, "DC Microgrid Technology: System Architectures, AC Grid Interfaces, Grounding Schemes, Power Quality, Communication Networks, Applications and Standardizations Aspects," IEEE Access, 2017.

IEC, IEC TR 63282 ED1 – Assessment of standard voltages and power quality requirements for LVDC distribution (Draft), Geneva: IEC, 2020.

F. Schork, R. Brocke, T. Böhm and M. Rock, "Requirements on surge protective devices in modern DC-grids," in 34th International Conference on Lightning Protection, ICLP, Rzeszów, Poland, 2018.

IEC, IEC 61660-1: Short-circuit currents in d.c. auxiliary installations in power plants and substations – Part 1: Calculation of short-circuit currents, Geneva, Switzerland: IEC, 1997.

IEC, IEC 60947-1:2020 – Low-voltage switchgear and controlgear – Part 1: General rules, Geneva: IEC, 2020.

IEC, IEC 62305-1:2010-12, Protection against lightning – Part 1: General principles, Geneva: IEC, 2010.

F. Schork and R. Brocke, "Asymmetric overvoltage protection levels in DC-grids," in Conference on Sustainable Energy Supply and Energy Storage Systems, NEIS2020, Hamburg, 2020.

J. Kaiser, F. Schork, K. Gosses, L. Ott, K. Bühler and T. Böhm, "Safety considerations for the operation of bipolar DC-grids," in INTELEC, Broadbeach, Queensland, Australia, 2017.

J. Kaiser, K. Gosses and L. Ott, "Grid behavior under fault situations in ±380 VDC distribution systems," in 2nd Int. Conf. on Direct Current Microgrids, ICDCM 2017, Nuremberg, Germany, 2017.

Y. Han, J. Kaiser, L. Ott, *et al.*, "Non-isolated three-port DC/DC converter for +-380 V DC microgrids," in PCIM-Europe, Nürnberg, 2016.

A. Makkieh, A. Emhemed, D. Wang, A. Junyent-Ferre and B. Graeme, "Investigation of different system earthing schemes for protection of low-voltage DC microgrids," in The 7th International Conference on Renewable Power Generation (RPG 2018), Lyngby, Denmark, 2018.

European Parliament and of the Council, Directive 2014/35/EU of the European Parliament and of the Council, Brussels: Official Journal of the European Union, 2014.

IEC 61643-31, Low-voltage surge protective devices – Part 31: Requirements and test methods for SPDs for photovoltaic installations, 2018.

IEC 61400-24, Wind Energy Generation Systems – Part 24: Lightning Protection, 2019.

IEC 62305-3, Protection against lightning – Part 3: Physical damage to structures and life hazard, 2010.

IEC 60364-7-722, Low-voltage electrical installations – Part 7-722: Requirements for special installations or locations – Supplies for electric vehicles, 2018.

EN 50443, Effects of electromagnetic interference on pipelines caused by high voltage A.C. electric traction systems and/or high voltage A.C. power supply systems, 2011.

EN ISO 18086, Corrosion of metals and alloys – Determination of AC corrosion – Protection criteria, 2019.

NFPA 70, 2020 Edition, National Electric Code.

NFPA 780, 2020 Edition, Standard for the Installation of Lightning Protection Systems.

Underwriters Laboratories Inc., UL 1449: Standard for safety – Surge protective devices, Fourth Edition, 2014.

Underwriters Laboratories Inc., UL 96A: Standard for installation requirements for lightning protection systems, 2016.

Underwriters Laboratories Inc., UL 96: Standard for lightning protection components, 2016.

IEC 61643-11, Low-voltage surge protective devices – Part 11: Surge protective devices connected to low-voltage power systems – Requirements and test methods, 2011.

A.J. Surtees, M. Caie and V. Murko, "A review of requirements governing the installation of surge protective devices on the electrical distribution network within the USA," in ICLP, 2006.

Chapter 9

New trends

Qibin Zhou[1], Ralph Brocke[2] and Alain Rousseau[3]

9.1 Smart SPDs

9.1.1 Background

Surge-protective device (SPD) is the core device to protect the equipment against over-voltage and surge current in the electrical and electronic systems. The safe and stable operation of SPD is very important for the lightning protection of the electrical and electronic systems. Therefore, the real-time monitoring of SPD working status is necessary to ensure the reliable operation of SPD. However, manual periodical inspection requires a professional inspection team and a large number of inspection equipment, which is time-consuming and costed. The safe operation of SPD cannot be fully guaranteed during the inspection period due to incorrect installation or the self-degradation failure during the period.

With the rapid development of sensors and Internet of Things (IoT) in recent years, the periodical inspection of SPD is gradually upgraded to real-time online monitoring. Some kind of SPD with monitoring, communication and analysis functions appears in the market. It can not only monitor the surge information, the leakage current, the working temperature of SPD, the disconnection status of disconnectors and so on, but also to give warning or life prediction message when SPD comes to degradation or failure status through IoT. This new type SPD is usually called Smart SPD by the manufacturers and users.

9.1.2 Configuration of smart SPD

Smart SPD usually includes surge protective module, monitoring module and communication module.

9.1.2.1 Integrated configuration

When these modules are integrated together, it becomes a complete product. Figure 9.1 shows three typical examples of integrated configuration.

[1]Electrical Engineering Department, Shanghai University, Shanghai, China
[2]DEHN SE + Co KG, Neumarkt, Germany
[3]SEFTIM, Vincennes, France

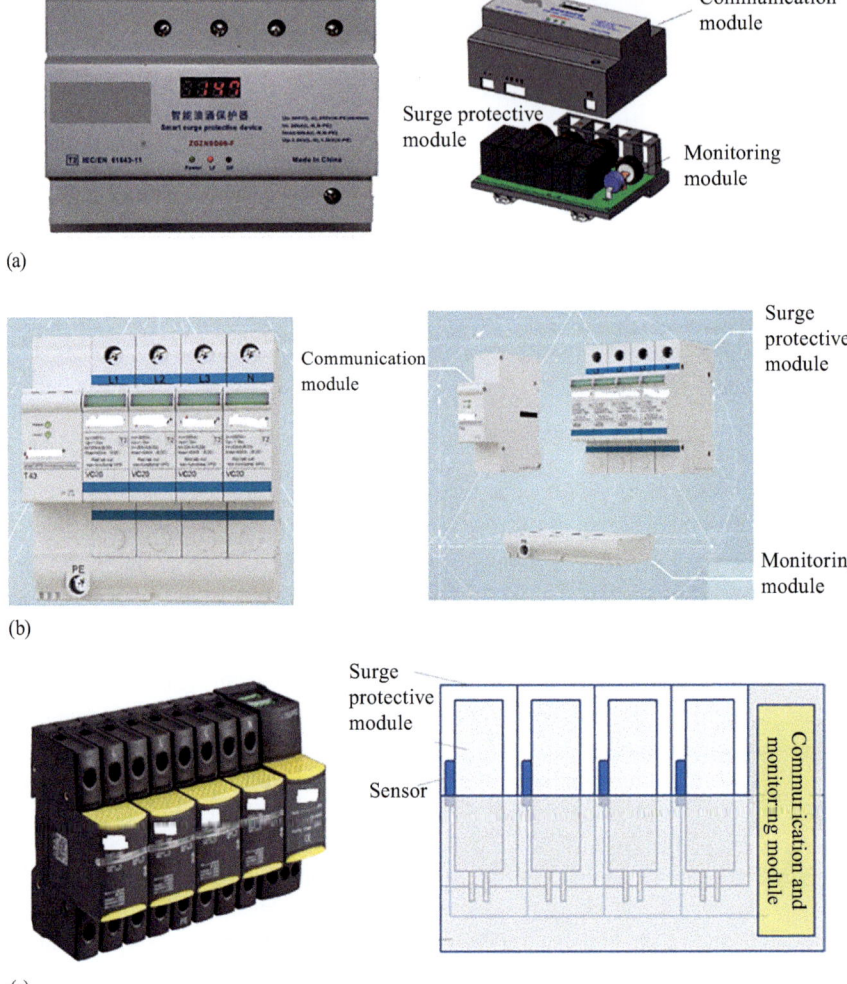

Figure 9.1   Three examples of integrated configuration: (a) Example 1;
(b) Example 2; (c) Example 3

9.1.2.2   Separated configuration

When the surge protection module (SPD itself) and the other two modules (the monitoring module and the communication module) of smart SPD are separately installed, they can also achieve the same purpose as the integrated configuration. At this time, the monitoring and the communication modules are combined into a product named SPD monitoring device. Figure 9.2 shows an example of separated configuration of smart SPD.

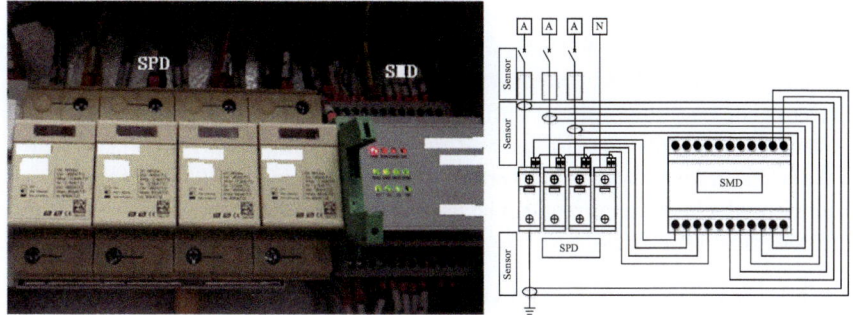

Figure 9.2   Examples of separated configuration

9.1.3   Functions of smart SPD

9.1.3.1   Basic function

Smart SPD shall have the normal function of a basic SPD, meet the performance requirement of IEC 61643-01/-11/-21/-31/-41 and pass the corresponding test.

The additional modules and accessories shall not impact the normal function as a basic SPD.

9.1.3.2   Additional functions

Smart SPD may include the following functions:

- fault monitoring and alarming;
- disconnector status;
- SPD working status monitoring (triggering number, disconnector status, working leakage current, SPD temperature and system voltage);
- surge monitoring (surge number, peak value, waveform, energy, etc.);
- other information monitoring (environment temperature and humidity, power quality, etc.);
- smart SPD shall have communication interface and can be connected to a network through field bus, Ethernet and so on. The target is to integrate smart SPD into the Web of Things. The communication interface would remain the responsibility of the SPD manufacturer.

Some smart SPDs declare degradation analysis and life prediction functions.

9.1.4   Degradation analysis of smart SPD

The degradation analysis function is declared by some smart SPD manufacturers. At present, they are two basic methods to analyse the degradation of SPD based on the monitored leakage current and surge current.

9.1.4.1   Leakage current analysis method

Leakage current of an SPD refers to the small current passing through the component metal oxide varistor (MOV) inside SPD under its working voltage. When SPD operates continuously or suffers from some kind of overvoltage stress, the leakage current will increase as well as the temperature on the SPD. When the leakage current is greater than a certain value, the heating of SPD increases rapidly. Therefore, the leakage current can be used as an important parameter to monitor the SPD degradation process.

9.1.4.2   Surge current analysis method

Surge current flowing through an SPD can be monitored by surge current sensor. The energy inside the surge current can be calculated by the time domain integration of the voltage and current. When the total energy flowing through the SPD exceeds the threshold, a warning message is issued. The energy threshold comes from the surge withstand capability tests on SPD. From the tests, the relationship among the withstand number N, the duration t_r and the magnitude I_{max} of square-wave impulse current will be built. Figure 9.3 shows an example of the relationship among the duration t_r and the magnitude I_{max} of the square-wave impulse current for a certain number N withstand capability of a certain MOV.

9.1.4.3   Other methods

Some other methods for degradation analysis based on temperature monitoring or the combination of earlier methods are also provided by some manufacturers. Since SPD degradation is a complicated procedure, there is no a perfect method at present

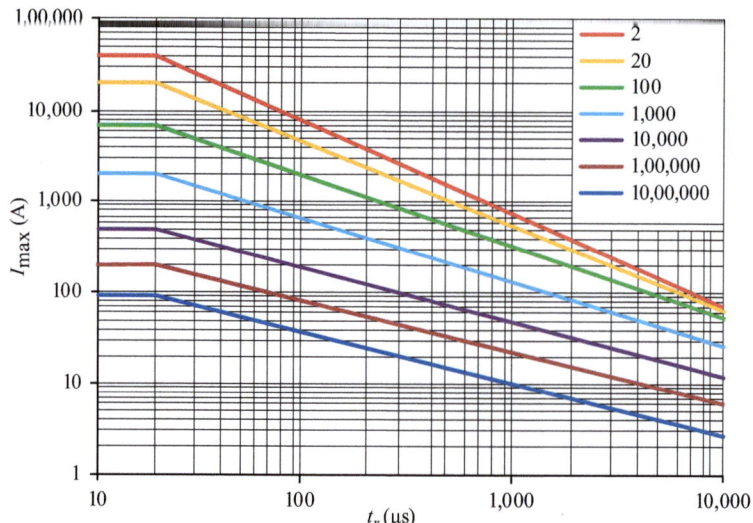

Figure 9.3   *Relationship among the duration t_r, the magntude I_{max} under the withstand number N*

to predict the SPD failure. All methods provide a qualitative prediction as a reference for the users.

9.1.5   Application of smart SPD

The application of smart SPD is an epitome of the IoT. The whole system includes the sensing, network, application and the platform layers as Figure 9.4 shows. The four-layer structure enables SPD to be online monitored remotely to ensure the safe operation of SPD in substations, hospitals, banks and other important locations.

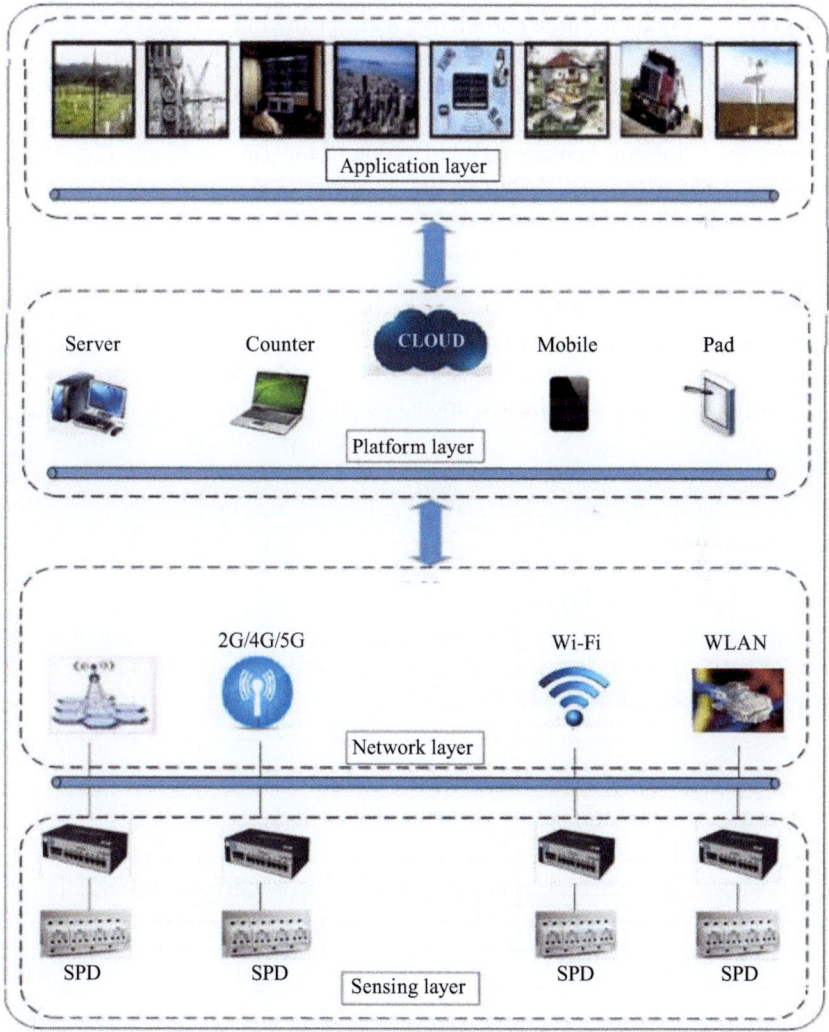

Figure 9.4   Four-layer structure of smart SPD application system

9.1.5.1   Sensing layer

The sensing layer consists of SPD and sensors. The sensors measure voltage, current, temperature, environment humidity and other information. Through field bus, Bluetooth, infrared and other short-distance transmission technology, the information will be transmitted to the upper information integration terminal; USB interface can be installed to download or upload data to realize data transmission between monitoring station and SPD monitoring terminal. Meanwhile, administrator can manage SPD terminal through gateway. The local screen displays surge times, overvoltage amplitude and life warning indication.

9.1.5.2   Network layer

The network layer solves the problem of data transmission obtained by the sensing layer. Network technologies such as long-distance wired and wireless communication technologies are applied to carry out data transmission through high-speed networks such as 4G/5G, Wi-Fi, Ethernet and NB-IoT to realize the interconnection between the sensing and the platform layers. The data such as surge current waveform, leakage current and so on are uploaded to the platform layer through above-mentioned high-speed networks.

9.1.5.3   Platform layer

The platform layer is responsible for data integration and data processing. It is composed of database server and information management platform. The data from the monitoring layer is analysed and processed. The analysed results are transmitted to PC or mobile equipment from the cloud computing platform for real-time monitoring and warning. For example, the engineers can obtain real-time information of SPDs at anytime and anywhere and immediately take countermeasures such as disconnecting circuit breakers or reminding nearby colleagues to change the degraded ones.

9.1.5.4   Application layer

The application layer means the possible application fields of the previous three layers. These fields include the buildings, communication, transportation, agriculture, power substations and so on.

9.2   Specific SPD disconnector

9.2.1   Introduction

SPDs are suitable for discharging and limiting overvoltages due to switching operations or inductive coupling. The main protective components of Type 2 SPDs are usually high-performance MOVs which are characterized by a relatively high discharge capacity and do not allow any follow currents.

These technologies have been around for years and are employed by many different manufacturers. The relevant product standard IEC 61643-11 (IEC, 2011) describes the minimum requirements for SPDs to ensure that the devices fulfil the specified parameters in the relevant application and moreover that they behave in a defined and safe manner in the case of failure due to overload or at the end of their service life.

However, the changing requirements in the installation environment, such as the increased application of high-performance electronic components, fluctuating short-circuit capacity and instable network conditions, mean that, in future, different and higher demands will be put on SPDs which could lead to system overloading or failure. For this reason, there are additional requirements with regard to the behaviour of SPDs in the case of failure (Brocke and Zahlmann, 2015; Nakata *et al.*, 2015).

9.2.2   Failure scenarios for SPDs

9.2.2.1   Degradation by multiple discharge impulses

Generally, one can differentiate between failure of SPDs as a result of overloading or simply because they have come to the end of their service life. While failure at the end of life is usually the outcome of a creeping process, e.g., brought about by numerous discharge processes, overloads often involve individual and short-lived events. Overload can, for example, be caused by the occurrence of high impulse currents which exceed the designated discharge capacity of a device (Rousseau et al., 2014).

Fundamentally, the failure scenarios for SPDs depend upon the technology employed. Spark-gap-based SPDs do not age, for example, as a result of the given network voltage and are not as sensitive to superimposed, high-frequency interference such as can be invoked by switching power semiconductors (Chicco *et al.*, 2005) (Figure 9.5).

MOV-based SPDs are also directly connected to the mains voltage, but they are much more sensitive to fluctuations or superimposed interference. The curve of

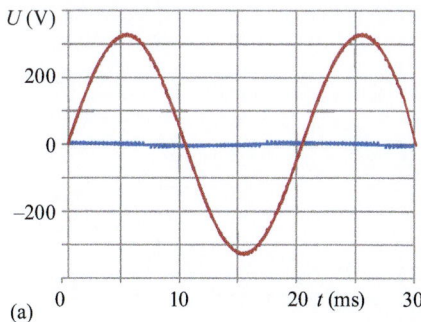 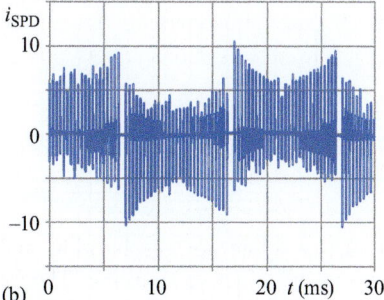

Figure 9.5   *(a) Voltage and current on an MOV-based SPD with $U_c = 275$ V, installed at the output of a pulse width modulation (PWM)-controlled inverter and (b) current through the MOV*

an MOV can be influenced by, e.g. high energy impulses (amplitude or duration), by a number of impulses with a lower energy content (Figure 9.3) or by permanently increased leakage currents (Rousseau *et al.*, 2015).

A slow fall of the *U/I* curve of an MOV generally leads to a continuous increase of the leakage current and thus to a rise of the temperature of the MOV. This effect is well known as thermal runaway. Before the temperature can reach an intolerable level, however, the SPD is safely and reliably disconnected by the integrated thermal disconnector (TD).

The sudden, complete breakdown of an MOV as a result of an impulse with high energy content, however, leads to the irreversible destruction of the grain boundaries of the MOV ceramics and, in turn, to a low impedance fault pattern. The resulting short-circuit current must be dealt with and disconnected by an integrated or external overcurrent protective device (OCPD). Figure 9.6 presents a schematic summary of the changes to the *U/I* characteristics due to an ongoing degradation and an MOV breakdown.

The applicable product standard (IEC, 2011) already contains tests, which check the failure behaviour. MOV-based SPDs are usually not only protected by integrated TDs but also by internal or external OCPDs such as, for instance, fuses or circuit breakers.

In order to ensure that the upstream installed OCPD (fuse or MCB) will not be tripped or even overloaded by the impulse current itself, the OCPD must have a high current rating. If the impulse current exceeds a given limit, the OCPD will be destroyed (Figure 9.7).

However, if one examines the disconnection characteristics of a fuse or circuit breaker, it becomes apparent that the disconnection time depends upon the occurring fault current. For example, seconds pass before fault currents in the order of a few hundred amperes are disconnected, whereas large short-circuit currents are interrupted in a matter of milliseconds.

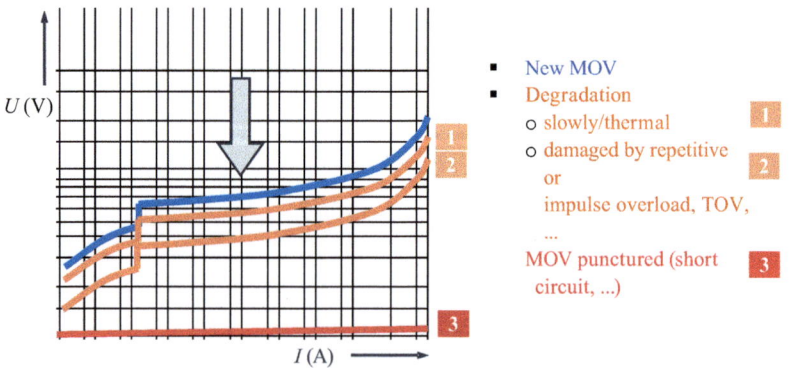

Figure 9.6   Change of V/I characteristic

Fuses overloaded by impulse currents

Power supply with MCBs overloaded by impulse currents

Figure 9.7   Fuses and MCBs overloaded by impulse currents

9.2.2.2   Disadvantages of existing solutions

In modern SPDs with MOVs as protective elements, disconnectors are usually thermally coupled with the varistor ceramics for simple and effective interruption of leakage currents of up to several amperes.

However, if the leakage current increases too quickly, the TD may not be activated quickly enough. This is, for example, the case when the characteristic change of the MOV results from a long-lasting temporary overvoltage (TOV). Such temporary, power frequency overvoltages are caused by faults in the low-voltage network, like the loss of the neutral conductor or short circuits between the phase conductor and the neutral conductor. If the resulting leakage current exceeds a certain amount, a very rapid change in the characteristics of the varistor will occur – the varistor breaks down and becomes conductive. In this case, the fault current rises rapidly, and the TD fails to disconnect the SPD from the low-voltage power supply. Furthermore, the respective OCPD (fuse, circuit breaker) may also be incapable of providing reliable protection for the MOV, hence the SPD because the OCPD may not be triggered due to too high impedance fault currents.

Figure 9.8 schematically shows the ranges of slow ⬛ and faster changes of MOV V/I-characteristics ⬛ up to short circuit ⬛ as well as the relevant protection ranges.

A further fault scenario, which could lead to the inadmissible overload of an MOV-based SPD, is the undefined failure behaviour of the MOV itself. If an MOV breaks down due to a very large impulse current, a significant residual resistance, depending upon the impulse energy introduced and the homogeneity of the MOV ceramics, may remain, which limits the rising short-circuit current (Figure 9.9). In this case, too, the upstream fuse or circuit breaker cannot protect the SPD because there is either no or very belated tripping.

9.2.3   SPD disconnectors

The standard IEC 61643-12 (IEC, 2020) introduces guidelines how to select and install external SPD disconnectors. However, the selection of the appropriate SPD

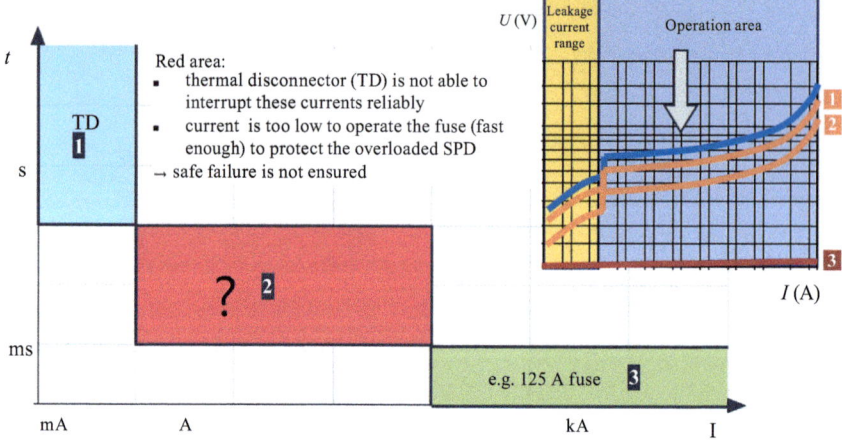

Red area:
- thermal disconnector (TD) is not able to interrupt these currents reliably
- current is too low to operate the fuse (fast enough) to protect the overloaded SPD
→ safe failure is not ensured

Figure 9.8   Standard application using a fuse

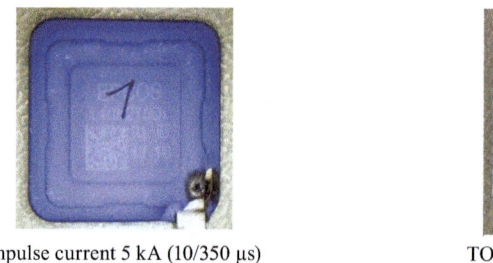

Impulse current 5 kA (10/350 μs)   TOV fault current 50 A (4 s)

Figure 9.9   Effects of failure of an overloaded MOV

disconnector may be difficult. In general, SPD disconnectors may be located either in the branch of the SPD or upstream as a part of the electrical installation. To make use of the full surge carrying capability of the SPD, the disconnector should have the same surge withstand as the SPD. Especially for SPDs having a high surge current rating, this may lead to too high ratings and in such a case, a coordination with the upstream OCPD, which is part of the installation, may not be achieved and thus the disconnector is useless (Kaito and Suzuki, 2015). This frequently leads to a 'gap in protection' which depends upon the nominal current rating of the fuse and the actual short-circuit current (Table 9.1).

In many cases already for Type 2 SPDs, tested with 8/20 impulses, the selection of an appropriate SPD disconnector is quite a challenge. Due to the much higher energy content of Type 1 SPDs when tested with 10/350 impulses, the conflicting requirements are

- a low rating to disconnect a failed SPD before the SPDs are destroyed (and especially before tripping any upstream OCPD),
- a high rating to withstand 10/350 impulse currents.

Table 9.1 I_{imp} capacity of NH fuses

Nominal values for NH fuses	Impulses (8/20 µs), which cause the fuse to trip (kA)	
	I_n (A)	I^2t_{min} (A^2 s)
35	3,030	14.7
63	9,000	25.4
100	21,200	38.9
125	36,000	50.7
160	64,000	67.6
200	104,000	86.2
250	185,000	115

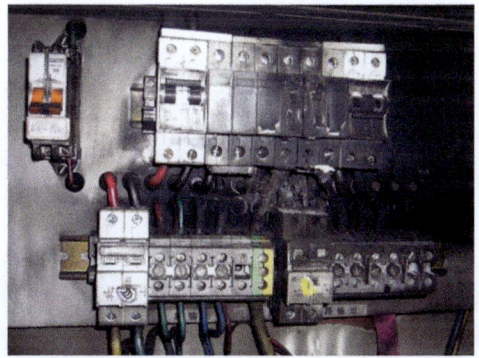

Figure 9.10   Overloaded SPDs and distribution board destroyed by short-circuit current

Internal fuses, which are better adjustable to the possible failure behaviour of the SPD, are often used in modern SPDs; however, a certain gap in protection may still remain.

If the actual short circuit is limited either by a limited short-circuit current capability of the mains or by a relevant fault impedance of the SPD itself, relatively long disconnection times t_{off} occur generating a very high amount of energy W_{total} in the SPD, although the maximum value of the existing fault current I_{fault} is low. In the case of failure, this energy in the SPD turns into pressure and heat and can overload the SPD (Figure 9.10) before it is safely disconnected from the low-voltage power supply (Nakata *et al.*, 2015; Kaito and Suzuki, 2015; Xin *et al.*, 2015).

During the last years, several new technologies show up on the market that try to close the gap between the safety-related fast disconnection of an overloaded SPD and a sufficient surge current carrying capability.

Table 9.2   *Electrical ratings and t–I characteristic of SFDs (Nakata et al., 2015; Sato et al., 2015)*

Electrical ratings

Rated current (A)	Surge withstand capability	Maximum voltage drop due to surge	Surge wave form	Breaking capacity
30	20 kA: 15 times	300 V @ 20 kA		
28	15 kA: 15 times	260 V @ 15 kA	8/20	AC 250 V 100 kA
23	10 kA: 15 times	200 V @ 10 kA		

Time–current characteristics

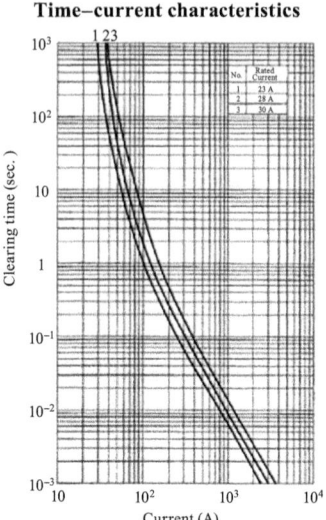

9.2.4   New concepts for SPD disconnectors

9.2.4.1   SPD fusing disconnector

Specific fuses can be found on the market that have a low rating and a high surge withstand, but there is generally no indication on their time to disconnect making the coordination with other OCPDs impossible. Requirements and tests as well as selection and application principles for SPD fusing disconnectors (SFDs) are described in Japanese Standards (JEITA, 2013a; JEITA, 2013b). Table 9.2 shows electrical ratings and time–current characteristics of different SFDs that comply with the Japanese Standard (JEITA, 2013a).

From an electrically point of view, SFD should have

- breaking capacity larger than prospective short-circuit current of a power supply where SPD is installed,
- large surge withstand capability to prevent unwanted tripping,
- low-rated current with time–current characteristics for overcurrent coordination with upstream OCPD and safe disconnection in the case of SPD failure,
- rated current coordinated with a conductor connected to the SPD,
- lower voltage drop due to surge current (Figure 9.11).

9.2.4.2   Specific protective devices for low-voltage SPDs

Specific protective devices for low-voltage SPDs (SSDs) are defined as disconnecting devices connected in series with an SPD having a surge withstand coordinated with the discharge current of SPD to be protected.

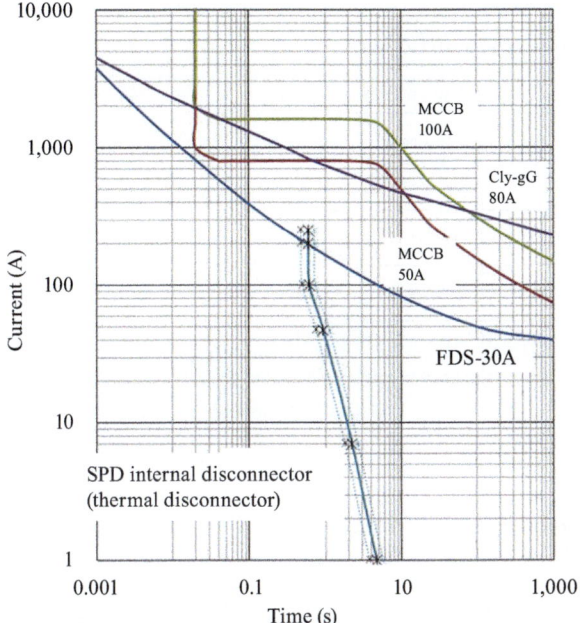

Figure 9.11 Example of the time–current characteristics of a fuse, MCCBs and an SFD (Nakata et al., 2015)

There is a contradiction between the need for OCPDs having a high surge current rating to ensure that an incoming surge does not operate the OCPD and a fast acting OCPD that protects an SPD also at relatively low fault currents. The requirements for SSDs should be as follows (Figure 9.12):

- a sufficient surge withstand capacity (at least the same as the SPD to be protected);
- a low-voltage drop, when tested with surge currents to provide a good overall protection (effective protection);
- a low tripping current, to protect overloaded SPDs;
- a time–current characteristic which is coordinated with the upstream installed OCPD to ensure the SSD operates first.

The technical concept of a fast-acting and surge resistant breaker has been published in Xin *et al.*, 2015. There is a Chinese industrial standard that defines the requirements and tests for SSDs (NB/T 42150 2018).

9.2.4.3 Short-circuit disconnection components

The disadvantages described when protecting overloaded SPDs can also be overcome by integrating a high-performance switching element into the SPD (Brocke and Schoepf, 2019). Such a switching element must, on the one hand, be capable of rapidly interrupting or even preventing currents in the milliampere or ampere range

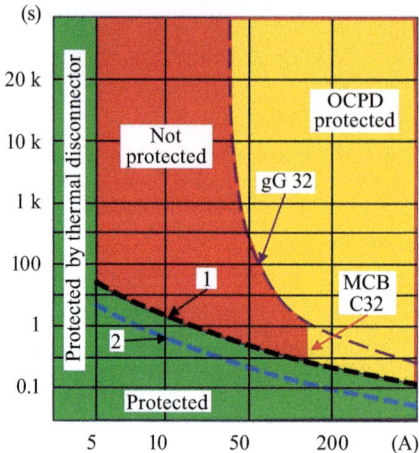

Figure 9.12   Time–current characteristics of a fuse, an MCB and an SSD (2) in comparison with overload limits (1) of an SPD

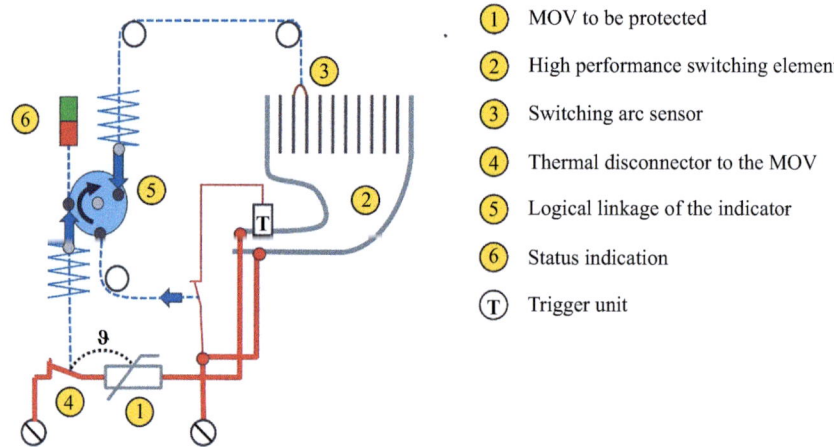

Figure 9.13   Basic components of an SPD with integrated high-performance switching element

and, on the other, quickly and reliably interrupt fault currents in the order of prospective short-circuit currents.

While mechanically activated switching devices, such as circuit breakers, always have a time lapse from detecting fault current, triggering to finally moving the switching contacts, the switching element introduced here is activated by the surge itself. At the same time, it shall only minimally or even not at all affect the protective behaviour of the SPD, i.e., neither negatively affect the discharge capacity nor the protection level of the SPD. Figure 9.13 schematically shows the

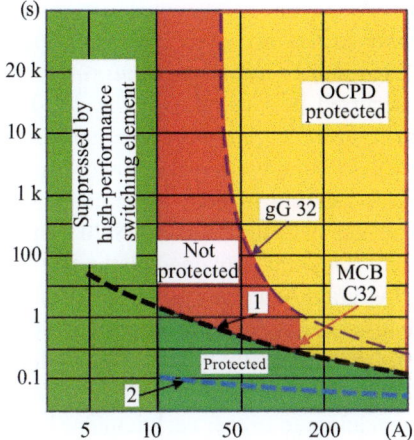

Figure 9.14 *Time–current characteristic of an SCDC (2) in comparison to a gG 32 A fuse, 32 A MCB and the overload limits (1) of an SPD*

basic components of such an MOV-based SPD with an integrated, high-performance switching device and deactivation circuit.

Figure 9.14 shows the time–current characteristic of a short-circuit disconnection component (SCDC) in comparison to a gG 32-A fuse and a 32-A MCB. The fault current range up to appr. 10 A is suppressed by the switching element.

In addition, the integrated switching unit prevents leakage currents because it provides reliable galvanic separation under normal conditions. This further provides a high TOV withstand, far higher than the values required by the product standard (IEC, 2011).

Also, for this kind of devices, a Chinese Standard that defines the requirements and tests exists (CEEIA, 2019).

9.3 Multi pulses

Multiple pulses (in short multi pulses) are known as an important cause of disturbance and especially for digital systems, due to the cumulative stress within a short time and also due to the steeper rise time for subsequent strokes. It is known since 2013, thanks to CIGRE that defines lightning parameters for engineering applications, that more than 80% of negative lightning stroke are composed of three to five surges, with a geometric mean time interval between surges about 60 ms. However, there are limited studies on the effect of multiple pulses on SPDs. A cumulative stress could be more difficult to handle by an SPD, if a new surge occurs before the temperature of the SPD decreases below a certain level. The operating duty test includes a series of 15 impulses in 3 groups of 5 impulses but with 8/20 μs impulse and only for nominal discharge current (or peak value of I_{imp} for Type 1 SPD) and for Type 1 SPDs a group of 10/350 μs impulse but with cooling down between each impulse.

Note: Only a small number of observed positive flashes have more than one stroke. Normally, positive flashes are associated with the 10/350-μs waveshape when negative flashes have shorter duration impulses.

The first studies occurred in Australia in the 1990s for HV systems and addressed the performance of the high-voltage arresters (mainly distribution type) under this particular stress. It is known thanks to work made by two Australians, from one side by Prof. Matt Darveniza and from another side by Rick Gumley, that an MOV-based SPD that can withstand 20 kA 8/20 can be destroyed by multi impulses at 5 kA 8/20 only. This was explained by the fact that the time being short between two impulses, the grain boundary in the MOV block has no time to cool down before a new impulse occurs. They had both developed a generator able to produce these multi impulses very easily by using the various capacitors of a usual 8/20 μs generator and triggering each of them one after the other. One of the generators was using a pendulum to trigger each impulse by passing in front of each capacitor. This was efficient and very simple but at the same time limited to low energy cumulated stress (a few kA with an 8/20-μs waveform). Progresses in generator design and manufacturing allow by now to perform these multi pulse tests with various waveshapes, including 10/350 μs. New lightning current test generators allowing multiple pulses are by now available and allow new studies. These generators can generate a maximum of ten surges, and the interval time is from 1 to 1,000 ms. Generally, first and last impulses have the same magnitude when intermediate impulses have a half peak value (Figures 9.15 and 9.16).

The studies developed using these new generators, first concentrated on MOV (how they behave under such a stress, what about the evolution of the residual voltage as well as the reference voltage at 1 mA and so on).

Tests have shown that MOV blocks that are part of SPD can catch fire or be destroyed under multi-pulses conditions. This can explain failure modes encountered sometimes in field but other causes can also explain these failure modes such as TOVs. Further studies are then needed that should include measurement in field and probability of occurrence of this type of stress. It is clear that lightning discharge can be multi impulse but it needs to be demonstrated that these multiple

Figure 9.15   Multi-pulse generator (photo GrandTop)

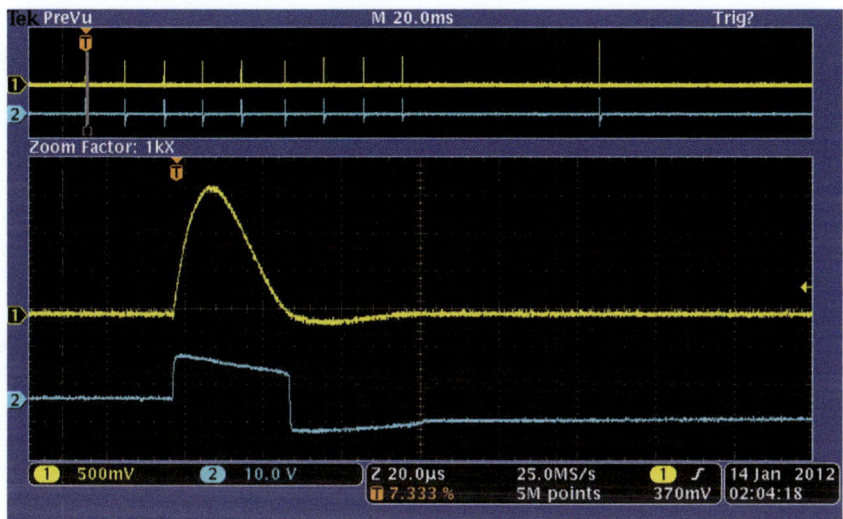

Figure 9.16 *Example of ten impulses injected on a varistor (yellow: current, blue: voltage)*

pulses can propagate to the LV system from the power lines and be a significant source of stress for SPDs. HV arresters installed in line on HV pylons and Type 1 SPDs are close to the lightning current source and are probably more affected by multi pulses than others. However, in line HV arresters are often used in series with spark gap and that may prevent negative effect of multiple pulses on the long-term behaviour of these SA. Type 1 SPDs are also often of the gap type. Study should then determine if the gap-type SPDs are also influenced by this type of stress that is so far concentrating on the MOV heating failure process.

Such studies should also be performed to check the behaviour of all types of SPDs, including trigger spark gaps and combination-type SPD where the effect of such multiple pulses could not only degrade the SPDs but also other components. These components may not be directly connected to the main protection function of the SPD but to a subsidiary function that may affect the long-term behaviour of the SPD.

Bibliography

IEC, 61643-11 Ed. 1, Low-voltage surge protective devices – Part 11: Surge protective devices connected to low-voltage power systems; Requirements and tests, Geneva: IEC, 2011.

R. Brocke and P. Zahlmann, "Use of SPDs with short circuit failure mode behavior in mains with volatile short circuit power," in *9th Asia-Pacific International Conference on Lightning (APL)*, Nagoya, Japan, 2015.

N. Nakata, N. Amano, S. Araki and A. Sato, "The short circuit current coordination between internal and external disconnector of SPD," in *9th Asia-Pacific International Conference on Lightning (APL)*, Nagoya, Japan, 2015.

A. Rousseau, X. Zang and M. Tao, "Multiple shots on SPDs – Additional tests," in *32nd International Conference on Lightning Protection (ICLP)*, Shanghai, China, 2014.

G. Chicco, J. Schlabbach and F. Spertino, "Characterisation and assessment of the harmonic emission of grid-connected photovoltaic systems," in *Proc. IEEE Power Tech*, St. Petersburg, Russia, 2005.

A. Rousseau, F. Cruz, X. Zang and L. Dongbo, "Degradation analysis of MOV type SPDs with surge supurposing AC operating voltage," in *International Symposium on Lightning Protection (XIII SIPDA)*, Balneário Camboriú, Brazil, 2015.

IEC, *IEC 61643-12 Ed. 2.0: Low-voltage surge protective devices – Part 12: Surge protective devices connected to low-voltage power distribution systems – Selection and application principles*, Geneva: IEC, 2008.

K. Kaito and J. Suzuki, "Operating current coordination between SPD and SPD disconnector," in *9th Asia-Pacific International Conference on Lightning (APL)*, Nagoya, Japan, 2015.

L. Xin, R. Chen and B. Li, "Failures of SPD external overcurrent protectors and solutions," in *9th Asia-Pacific International Conference on Lightning (APL)*, Nagoya, Japan, 2015.

JEITA, *RC4501: SPD disconnectors for low-voltage surge protective devices performance requirements and test methods for SPD fusing disconnector (SFD)*, Tokyo, Japan: JEITA, 2013a.

JEITA, *RC4502: SPD disconnectors for low-voltage surge protective devices selection and application principles for SPD fusing disconnector (SFD)*, Tokyo, Japan: JEITA, 2013b.

A. Sato, N. Morii and H. Sato, "Development of a fuse-type SPD disconnector and guidelines for selecting it in coordination with an over current protection device," in *9th Asia-Pacific International Conference on Lightning (APL)*, Nagoya, Japan, 2015.

China Electrical Equipment Industry Association, *NB/T 42150-2018 (Energy Industry Standard) – Specific protective devices for low-voltage surge protective devices*, Beijing, China: China Electrical Equipment Industry Association, 2018.

R. Brocke and T. J. Schoepf, "Advanced Circuit Interruption Technology (ACI) – an innovative design concept for SPDs," in *11th Asia-Pacific International Conference on Lightning (APL)*, Hong Kong, China, 2019.

CEEIA, *T/CEEIA 390-2019: SPDs with integrated overcurrent protection function based on spark gap*, China Electrical Industry Association Standard, 2019.

M. Darveniza, D. Roby and L. R. Tumma, "Laboratory and analytical studies of the effects of multipulse lightning current on metal oxide arresters," *IEEE Transactions on Power Delivery*, Vol. 9, No. 2, 1994.

M. Darveniza, L. R. Tumma, B. Richter and D. A. Roby, "Multipulse lightning currents and metal-oxide arresters," *IEEE Transactions on Power Delivery*, Vol. 12, No. 3, 1997.

H. Norinder, *Duration of lightning strokes and occurrence of multiple strokes*, Institute of High-Tension Research, University of Uppsala, 10 December 1949.

Y. He, Z. Fu and J. Chen, *Investigation of the effects of multi-waveform multipulse impulse currents on MOV for class I SPD through operating duty tests*, 33rd ICLP, 2016.

A. Rousseau, L. Dunxun, F. Cruz and X. Zang, *Optimized analysis of the heating transfer of MOV based on natural lightning tests*, CIGRE, 2016.

A. Rousseau, X. Zang, L. Zhang and M. Tao, *Multiple shots on SPDs-test methods and results*, 31st ICLP, 2012.

Chapter 10

Ongoing issues and possible solutions

Alain Rousseau[1]

As indicated in Chapter 5, a great deal of surge-protective device (SPD) protection characteristics is depending on the surge protection components that are used inside and, in that direction, main evolutions are coming from a combination of components to achieve better protection characteristics rather than from break-through technologies. It is often expected that new surge protection components will approach better the ideal protection characteristics (impedance equals zero when there is a surge and open circuit for the remaining time) but basically, based, for example, on varistors, there are little improvement to notice at the present time. It is well known that it is possible to improve one characteristic of a varistor to meet a specific goal or to better suit an application, but it is generally associated with a degradation of another characteristic. Main progress in terms of varistors is presently coming from their shape that will better fit to the available space inside an SPD to maximize efficiency and minimize the size of SPDs. Size of SPDs is always a challenging issue and we will come back to this later when discussing installation rules. Regarding gaps, progress is mainly coming from their size that also decreases, from their surge withstand capability increase and the follow current (AC current that follows the flow of a surge when a gap operates) reduction.

Until a new surge protection technology emerges, SPDs consist of proved surge protection components and probably more and more on combination of components to improve protection while reducing size and mitigating drawbacks associated with each component.

What is expected is mainly a lower voltage protection level to improve efficiency especially when SPDs are far from equipment to be protected, when equipment surge voltage withstand is uncertain or when the cabling of SPDs may degrade their efficiency.

In that direction, technology progress goes in the direction of a much lower voltage protection level for Type 1 SPDs and it may happen that in future Type 1 SPDs are replaced by Type 1+2 SPDs (Type 1 SPD able to divert a partial direct lightning current while protecting at a low level of voltage as it is expected from a Type 2 SPD). When Type 1+2 SPDs will become smaller and cheaper, they may as

[1]SEFTIM, Vincennes, France

well replace Type 2 SPDs at installation entrance. In fact, partial lightning current is injected in the power installation when, for example, a light pole is struck by lightning. A light pole may act as a lightning rod due to its height and also due to the fact that there may be many light poles on a parking lot. If the power supply for the light pole is coming from the building, a partial lightning current may flow towards the installation and this may require a Type 1 SPD on that line entrance. In the same way, a lightning that strikes another building protected by a Lightning Protection System and connected to the same utility power supply will inject a partial lightning current in the buildings in the vicinity. It is the same thing, when a lightning strikes a nearby tree and there is coupling between the tree roots and the building earthing system. In all these cases, a Type 1+2 SPD would be helpful even if not required by standards.

There are two cases that deserve a better attention on voltage protection levels:

- For IT systems (LV power system where the transformer impedance is either not grounded or grounded through a high impedance), there is a high voltage generated on phases when a ground fault occurs on one phase. Due to this high voltage that is long enough to be considered permanent, the voltage protection level is high. However, there is no indication that equipment used on these systems have a higher surge withstand voltage than equipment to be used on other system such as TN or TT systems. To obtain the required low level of protection expected for sensitive equipment, it is generally necessary to use a cascade of SPDs. SPDs for IT systems with a lower voltage protection level would be helpful for the market.
- For applications where high-frequency impulses are superimposed to the AC power voltage, the best way at the present time is to increase the voltage that the SPD can support permanently to such a high value that these impulses do not degrade the long-term behaviour of the SPD. However, as a consequence, it means that the voltage protection level of the SPD is also very high and protection efficiency deserves a special attention (installation rules, cascade of SPDs to meet the expected low level of protection, surge withstand analysis of the equipment to determine whether it can withstand a higher voltage level, etc.).

The development of SPDs with better voltage protection characteristics is important but this benefit can easily be degraded by installation practice. First of all, SPDs are not always incorporated at the planning stage. When it is necessary to incorporate an SPD in a panel that is already designed, it is likely that the well-known rules (shortest connection conductor length) will not be applied in full. Thus, the protection efficiency will be degraded. When the size of an SPD is too large, it may not fit inside the panel and either another SPD with smaller characteristics will be used instead (e.g. Type 2 instead of Type 1) or the SPD will be installed in another cabinet that will also degrade the protection. Even, if SPDs are integrated at the panel design stage, the smallest SPD will be used because a too large SPD may imply using a larger panel that will increase the cost.

The development of SPDs with mall size without any degradation of the announced characteristics will avoid partly these installation problems. Another improvement of the SPD installation rule is related to the absence of external SPD disconnector (either because it is integrated in the SPD or because it is not necessary). This will facilitate the SPD cabling by reducing the lead length and will also avoid the space dedicated to the SPD disconnector that may be bulky, for example, for Type 1 SPDs.

The question of SPD disconnector remains an important topic for SPD applications. The requirement for the SPD disconnector may appear sometimes as a limitation to the SPD use, and especially for Type 1 SPD, that requires quite high rating for the SPD disconnector. A simple picture of a Type 1 SPD associated with its SPD disconnector generally shows that the size of the disconnector is twice as large as the SPD size or even more. It may seem strange that a safety device such as an SPD needs another bigger device to provide its safe end of life. To avoid this drawback, many SPDs integrate their own SPD disconnector. But this built-in disconnector still needs to be coordinated with the upstream overcurrent protective device, and this mean that detailed characteristics of the incorporated SPD disconnector need to be known by the installer. The publication of these characteristics or coordination rules with upstream overcurrent protective device would help avoiding the misuse of these SPDs. More and more overcurrent-protective device announces their surge withstand even if not requested by standards and that helps the installer to decide of the best strategy in the case of SPD end of life. The development of Specific SPD Disconnectors as discussed in Chapter 9 may also help to select an SPD disconnector that provides a safe end of life of the SPD, whereas not degrading its surge protection characteristics and disconnect a failed SPD quicker than the upstream overcurrent-protective device.

To help installer to install SPD with short lead lengths, many possibilities are presented in standards and they have been summarized in Chapter 7. However, one of these possibilities is not yet developed as a commodity product: low impedance conductor. This would help installers because it is not rare to meet installations where SPD efficiency is doubtful either because rules were not known or more frequently because they were known but too difficult to apply. SPDs are not acting as no-way signs: SPD will mitigate the effect of surges but will not force the surge to go back. Surges will flow whatever is done to reduce their disturbance, and a badly installed SPD will simply badly protect. The development of SPDs designed in such way that it helps installers to succeed in installing SPD with full respect of the rules would lead to a rather significant protection improvement. Built-in SPD disconnector is one way to achieve this, but progress in the direction of facilitating installation is still expected.

What happens when a badly installed SPD fails to protect? Very often, the SPD itself will appear as being responsible, and another SPD of another type will be used instead. But, if the problem is related to installation rules, this will not solve the problem. It is surprising to see that in spite of enormous progress made by SPD manufacturers and standards during the last decades to improve the products, test them with stringent requirements and achieve one of the highest reliability levels, SPDs are often considered strange animals to say the least. Of course, SPDs' installation rules are very specific but are by now quite well known and SPD field experience is good. It is then important for SPDs to develop in such a way that a plug and play installation is possible, at least for the more common application. Monitoring of SPDs such as developed in the smart SPD concept, presented in Chapter 9, may also help to use SPDs more efficiently. We will discuss that later.

SPDs developed specifically for particular applications will probably develop even if the largest market will be dedicated to commodity products. It is not rare that surge protective components are integrated inside equipment to help pass a few tests as, for example, EMC tests. However, these are so far only components and they do not provide the safety level associated with SPDs. For example, varistors are barely associated with thermal disconnector when incorporated inside an equipment or on printed circuit board. Such components may then have a bad end of life in addition to complicating the coordination with the upstream SPD. To avoid these drawbacks, a new trend appears that consists in including an SPD tested according to SPD standards inside an equipment. This could solve a lot of problems and especially lead length connection or SPDs too far from equipment.

New applications will lead to new SPDs and associated new standards. For example, DC SPDs need to be tested in a specific way (e.g. DC permanent voltage may age the varistor) and especially when a powerful battery is present, it should be included in the end-of-life tests.

Monitoring of SPDs may also be a new challenge and opportunity. It is mainly the purpose of the smart SPD concept. A basic smart SPD will be an SPD with additional function such as counting the surges. But it may be much more than that. SPDs are located at installation entrance and they are well located to monitor the power quality in addition to protect downstream equipment and electrical installation. They can also react to an SPD state change and inform the user that something happened and that inspection is needed. It can also record the surge stresses that the installation has been exposed to and try to indicate a cumulative energy or charge to predict maintenance. For safety, it is always better to replace a weak SPD before it fails. This will just apply to SPDs that are degrading with time and is not able to avoid a too large surge to occur and degrade an SPD. So, SPD disconnector and end-of-life tests will remain necessary. However, it may help, under certain circumstances, removing SPDs from the installation before it fails for critical installations. In the 2000s, some tramway operators have been using that method to avoid MOV-based surge arrester to become faulty and potentially dangerous due to high DC short-circuit current when surge arresters were near people in tramway stations.

Ideally, the smart SPD could also help the installer to validate that its installation is compliant with the standard (lead length, type of system ...) and the specific rules imposed by the SPD manufacturer. Instead of simply informing that an SPD is not protecting anymore (e.g. because disconnected), it may inform the user that the SPD protection is degraded or the installer that the installation conditions justify another level of protection in another circuit.

Basically, SPDs have greatly improved their characteristics over the last decades, and it is likely that this trend will continue for better efficiency and to take care of the worst conditions. When SPDs fail, it is often because SPDs are not compliant with standards or because they have been misused (wrong system, voltage variation greater than normal, Type 2 SPDs when a Type 1 would be needed, etc.). It happens from time to time tricky situations that can cause SPD damages (e.g. unexpected temporary overvoltages) but, in that case, standards are quick to react to cover this case or a specific application. In general, field experience is good. It does not mean that research is not going one to cover uncovered cases or provide new solutions. For example, the effect of multiple pulses (see Chapter 9) on SPD's failure mode is still under study.

Index